World Sustainability Series

Series editor

Walter Leal Filho, Hamburg, Germany

More information about this series at http://www.springer.com/series/13384

Walter Leal Filho · Ulisses M. Azeiteiro
Sandra Caeiro · Fátima Alves
Editors

Integrating Sustainability Thinking in Science and Engineering Curricula

Innovative Approaches, Methods and Tools

Springer

Editors
Walter Leal Filho
HAW Hamburg
Hamburg
Germany

and

Manchester Metropolitan University
Manchester
UK

Ulisses M. Azeiteiro
Department of Sciences
and Technology
Universidade Aberta and CFE
University of Coimbra
Porto
Portugal

Sandra Caeiro
FCT/UNL
Universidade Aberta
and CENSE
Lisbon
Portugal

Fátima Alves
Department of Social Sciences
and Management
Universidade Aberta
Porto
Portugal

ISSN 2199-7373 ISSN 2199-7381 (electronic)
ISBN 978-3-319-09473-1 ISBN 978-3-319-09474-8 (eBook)
DOI 10.1007/978-3-319-09474-8

Library of Congress Control Number: 2014949354

Springer Cham Heidelberg New York Dordrecht London

Printed on acid-free paper

Springer is part of Springer Science+Business Media (www.springer.com)

Preface

The book "Building Sustainability Thinking into Science and Engineering Curricula—Innovative Approaches, Methods and Tools" is a further volume of the new "World Sustainable Development Book Series" and contains some of the papers presented at the "World Sysmposium on Sustainability in Higher Education" (WSSD-U-2014) held in Manchester, UK, during 3–5 September, 2014.

If we want to build sustainability thinking into the science and engineering curriculum, and hence encourage students in science and engineering to think more about green issues and become more sensitive towards matters related to sustainability, we need new approaches. The ones used to date have not fully worked, due to a number of reasons. These include:

(a) Lack of due emphasis on sustainability in the curriculum of science and engineering courses
(b) Lack of appropriate materials to support curriculum delivery
(c) Lack of training of the staff who should deliver sustainability issues in science and engineering courses
(d) Limited availability of materials showcasing how it can be done

This book is meant to address the above needs, in an integrated way. Consisting of case studies, descriptions of practical experiences and empirical analyses, the book is structured around two main parts:

- Part I looks at matters related to curriculum innovation and the effectiveness of current training and education programmes, showcasing some innovative methods and tools currently being used by universities all over the world in order to include sustainability in the curriculum as a whole, but also of science, engineering and other courses in particular.
- Part II handles examples of good practice, where the handling of sustainablity issues goes beyond the trivial, and engages on matters of strong social, political and economic relevance. In particular, Part II refers to cross-cutting approaches,

outlining examples of successful programmes, including replicable case studies and the examples of good practice which can convincingly show that the integration of sustainability issues in science, engineering and other courses is a feasible goal.

Students will benefit from a more sustainability-focused approach in the science and engineering curriculum in a number of ways. For instance, they will learn how to use natural, physical and economic resources more efficiently and more effectively. In addition, they will be sensitised about the need to seek multiple views when handling sustainability challenges. Moreover, they will be in a better position to recognise and manage risks, so as to minimise adverse impacts of projects on people or environment. Finally, they will be encouraged to contribute towards building a sustainable society, where they—as students now and professionals later—will be able to play a key role.

We want to thank the authors for sharing their work, their know-how and their expertise to a world audience, and hope the many experiences amassed in this book will help to support the work of sustainability lecturers, researchers and students working in this very important field.

Autumn 2014

Walter Leal Filho
Ulisses M. Azeiteiro
Sandra Caeiro
Fátima Alves

Contents

Part I
Curriculum Innovation

Preparing Teachers for Sustainable Development in Higher Education

Iolanda Bueno de Camargo Cortelazzo

Abstract

Education goes through many challenges in this technological society. The teacher development paradigm being followed remains the one based on the industrial society; and teachers in all levels, including higher education, do not understand what sustainable development in higher education is. Some of them are "*green pacers*"; others have never heard about the subject. This paper presents some results of a project on teacher a development program which deals with the adequate use of digital technologies in technical, vocational and higher education at public school in the state of Paraná and describes some of the changes that have been occurring both at elementary and secondary schools. The emphasis is not only on the use of digital artifacts but on the changes that digital technology provokes in knowledge management and in the learning and teaching process. It discusses the necessity of teachers being able to understand sustainability as a survival issue, which cannot be forgotten in education; on the contrary, it needs to be incorporated by teachers and also by students with autonomy, creativity, respect and collaborative work. The project aims to work with teachers in an interdisciplinary dimension incorporating transmedia and digital curation. Transmedia permits different narratives of the same subject to be represented in different media and digital curation permits to gather scientific knowledge on a defined issue according to students and teachers levels in an ocean of garbage available on the web. This paper will be useful to people looking for a new trend on preparing teachers for higher education from sustainable development viewpoint.

I. Bueno de Camargo Cortelazzo (✉)
Universidade Tecnológica Federal do Paraná - UTFPR, CEP: Av. Sete de Setembro 3165, Rebouças, Curitiba 80230-901, Paraná, Brazil
e-mail: icortelazzo@utfpr.edu.br

W. Leal Filho et al. (eds.), *Integrating Sustainability Thinking in Science and Engineering Curricula*, World Sustainability Series, DOI 10.1007/978-3-319-09474-8_1

Keywords
Higher education · Teacher development · Sustainability · Digital curation · Autonomy

1 Introduction

Taking into consideration the statement renewed in the Report of the United Nations Conference on Sustainable Development, resulting from the conference held in Rio de Janeiro Brazil in June 2012 of "commitment to sustainable development and to ensuring the promotion of an economically, socially and environmentally sustainable future for our planet and for present and future generations" (United Nations 2012, p. 1), higher education has the great challenge of preparing professionals and researchers to work in their fields, not only to reach their professional goals, but also to work with the view on sustainability.

Although most of students at the higher education may have autonomy, objectiveness and pro-activity, teachers limit their role focusing only on professional knowledge and skills. The third axe of competence, which is attitude, needs to be part of their development, since it can help professionals to work in and maintain a sustainable world.

In the same Report, in item 230, there is the recognition of the need to prepare teachers and schools to promote education for sustainable development.

> We recognize that the younger generations are the custodians of the future and the need for better quality and access to education beyond the primary level. We therefore resolve to improve the capacity of our education systems to prepare people to pursue sustainable development, including through enhanced teacher training, the development of sustainability curricula, the development of training programmes that prepare students for careers in fields related to sustainability, and more effective use of information and communications technologies to enhance learning outcomes. We call for enhanced cooperation among schools, communities and authorities in efforts to promote access to quality education at all levels (United Nations 2012).

During the first 10 years of 2000, mankind faced challenges due to rapid technological transformations that made the world smaller and put different cultures face to face. Religion, ethics, moral values and education have struggled to survive and as a species, we have struggled to maintain respect and collaborative work. Collaboration has become a new value, necessary to lead people to solve problems, to organize their present and future, to guarantee better conditions for their children and to make a sustainable world possible.

Although research and studies proliferate in the Research Centers and life has become easier and more comfortable; the education expansion increased in numbers but their outcomes have been less and less effective. In many parts of the world, education does not take people to better life conditions in terms of health, social rights, jobs, and justice. In fact, in those areas, teachers do not participate in professional development programs, but in "continuing formation" programs which just repeat what they had learnt in their undergraduate or graduate programs. They go on teaching in an anachronistic paradigm, throwing lots of contents and demanding their students to digest them without meaning, and interrelation with the real world or even any connection to the practical life. Individualized or group answers just added one to another, without any debate, argumentation or reflections are common in many of the teacher education programs in public higher education in Brazil.

In many institutions, Teacher Colleges emphasize theory from the ideological not scientific point of view. It is necessary to rethink education, to evaluate the quality of teacher professional development, to integrate theory so as to practice in the real world and to understand what sustainable development means.

Taking again the United Nation Report in Rio 2012 into consideration, it is important for the teacher colleges and other educational institutions to work with teacher professional development stimulating them to integrate sustainability across disciplines:

> We strongly encourage educational institutions to consider adopting good practices in sustainability management on their campuses and in their communities with the active participation of, inter alia, students, teachers and local partners, and teaching sustainable development as an integrated component across disciplines. 235. We underscore the importance of supporting educational institutions, especially higher educational institutions in developing countries, to carry out research and innovation for sustainable development, including in the field of education, to develop quality and innovative programmes, including entrepreneurship and business skills training, professional, technical and vocational training and lifelong learning, geared to bridging skills gaps for advancing national sustainable development objectives (United Nations 2012, p. 45).

This paper also refers to the movement around the world of promoting Education for Sustainable Development ESD. This recognizes the people's rights to the kind of education which empowers them as constructors of a sustainable society and contributors to new scenarios where human beings can live and work collaboratively, realizing their necessity to be "social".

> ESD entails including key sustainable development issues into teaching and learning and requires innovative, participatory teaching and learning methods that empower and motivate learners to take action for sustainable development. ESD promotes skills like critical thinking, understanding complex systems, imagining future scenarios, and making decisions in a participatory and collaborative way (UNESCO 2013, Annex 1, p. 2).

Sustainable development does not refer only to ecological and economical issues but mainly to social development.

A presentation to bring to discussion is the need to rethink higher education teacher development programs, in order to make teachers aware of their responsibility to prepare professionals for a sustainable world both in secondary school and in higher education. It presents the first findings and reflections of a collaborative research on a teacher development program in Curitiba, Paraná in the South of Brazil.

The research project is aimed to instigate teachers to change their way of approaching knowledge in order to have their students develop new approaches too. During the twentieth century, the integration of different thinking theories permitted the development of the systemic thinking. "From a systems perspective, the goal is to understand the whole and many levels of interrelationship that characterize the system (Remington-Doucette et al. 2013, p. 405). Considering this kind of thinking is a basic competence, teachers need to learn about and develop this and other different scientific approaches not presented at teacher colleges.

This work starts presenting some considerations on educational and cultural approaches towards to a "pseudo inclusion" and "pseudo-knowledge' construction as well as the context in public higher education institutions in South of Brazil. Some questions and information on teacher's use of technology, followed by a theoretical information on new learning approaches and new teacher roles that served as basis for the case studied and described here, are presented. The main part of this work describes some results of a teacher development program in Curitiba. At the end, final considerations indicate some possibilities for higher education sustainable development.

2 Sustainable Educational Development: Cultural and Technological Issues

Examining public higher education institutions, one of the great problems refers to the cultural, educational, and technological approaches.

2.1 Cultural Issues and Definitions

Diversity, minorities and multiculturalism must be understood in the sense of interculturalism, since informal and formal kinds of education are marked for the dictatorship of minorities (gender, race, age, etc.) and influence the behavior and even the creation of laws. As Cantle says, "this phenomenon (multiculturalism) is difficult, though manageable, when the number of minorities was limited, but has become extremely problematic in an era of super diversity" (Cantle 2013). Explaining both phenomena, this author states:

> Interculturalism is more than a simple set of policies and programmes. It is about changing mindsets by creating new opportunities across cultures to support intercultural activity and it's about thinking, planning and acting interculturally. Perhaps, more importantly still, it is about envisioning the world as we want it to be, rather than be determined by our and separate past histories (Cantle 2012).

The public educational policies have led institutions to drive their actions to the Program "education for all"—everyone "has to" access higher education, transforming the right of attending higher education into a duty, considering those who did not attend higher education as second class people. Based on this assumption, decisions were made and actions taken. This paper presents an analysis based on the need for changing some rigid cultural approaches in basic education (elementary and secondary levels), by developing new teacher programs like "Education for All".

Hierarchy and knowledge have been displaced by the cult of diversity in the sense of multiculturalism. Laws and movements take out the parents' right to correct and to conduct their children. New waves of pedagogy put the learner in the center of the education process and teachers have lost more and more of the respect and recognition without reaching the aims of meaningful learning and better performance. Knowledge has been compacted; reading has become a secondary issue at school; and counting skills have been supported by calculators ignoring the high order cognitive functions development. Google is the "Master". Inclusion is the word and exclusion is the "non word".

However, statistics show an increasing number of people included in schools, even at the higher education level and certified as professionals in various levels, mainly in developing countries. A considerable number of professionals have been excluded in the job market because of their lack of competence (knowledge, skills and adequate attitudes) for the jobs they apply. There is a shortage of drivers, hairdressers, dressers, technicians, bakers, carpenters and bricklayers, besides other occupations; lots of lawyers, engineers, administrators and teachers are without jobs or even employed in other occupations because they do not have the preparation to practice their profession. These people earn low wages and are not satisfied because they do not work in the profession they chose to be prepared for.

Taking all these issues into consideration, the definition of sustainability chosen to base this work is that presented by Moore:

> Sustainability is a concept, a goal, and a strategy. The concept speaks to the reconciliation of social justice, ecological integrity and the well being of all living systems on the planet. The goal is to create an ecologically and socially just world within the means of nature without compromising future generations. Sustainability also refers to the process or strategy of moving towards a sustainable future (Moore 2005, 327 apud Baughan 2013, p. 3).

There are different views of sustainable development in higher education. As Baughan (2013) explains when he refers to the participants of his research, most of the teachers in many public higher education institutions in Brazil "often tend to consider sustainability in terms of pro-environmental behaviour, more so than other areas and issues." Those who already work with sustainable issues in their course understand the relevance of sustainability—as shown above—to higher education.

In this work, sustainable development includes sustainable education. The comprehension of the school social function in preservation of tradition and identity in parallel with the transformation and development must reinforce teacher professional development to incorporate these issues to their teaching: keep tradition and change what is necessary in the new trends according to a rigorous analysis of the context.

Referring to sustainability and sustainable development, Table 1 is borrowed from Leal Filho (2011) and shows the influences on attitudes toward sustainability. Those attitudes must be a main goal to be attained in a teacher professional development program to prepare for education for sustainable development.

Table 1 Some factors which influence attitudes towards sustainability

Knowledge	Information on the meaning of sustainability and its implications
Background	Of receptivity to sustainability issues
Experience	Previous experience with environmental and social issues facilitates understanding of the role of sustainability
Perception	The integrated view of environmental, political and economic elements enables a broader perception of sustainability
Values	Due to its high degree of complexity, an individual's values often determine whether his/her attitudes are favourable to sustainability
Context	Sustainability is not only related to ecological components per se, but also entails items such as economics, politics, and social matters. However, links with the latter are often ignored by universities
Continuation	Sustainability is not about short term, isolated efforts, but continuous, long term ones

Source Leal Filho (2011)

2.2 Communication and Education Issues

Considering the technological development faced by society in the last 30 years and the historical perspective, mankind faces the end of a period and the dawn of a new era where beliefs and perspectives of different generations are in conflict. No generation can throw away the culture constructed by previous generations, but they are doing so due to a lack of understanding about that culture. They have not been taught their history. They are permitted to be themselves with unlimited freedom and superficial information distributed by media anywhere at any moment.

On one hand, technological transformations permit a social communication much more intense, more constant and multidirectional. Multi diversity has become a characteristic of today's social communication. Different people, no matter their age, ethnicity or sex can participate in social networks and communicate to each other instantly, no matter how far they are from each other. Many of the social networks have a flow of communication on frivolities; people waste lots of time without purpose. Children and youths are "multitask" individuals and dispersion destroys concentration. People can post their images or other people's images and videos without permission or censorship. Many social networks influence the increasing consumption of irrelevant products. People involve themselves in competing in possessing more updated items which make them less and less satisfied.

There are more and more people who do not talk or even pay attention to each other around a table in restaurants, in restrooms, movie theaters, at the bus stop, in the tube, in a living room, or in student's and teacher's lounges because they are talking to other people by mobile and surfing in their tablets and ipods. They surf the surface. They do not go deeply in their relationship or into their search on the web.

Most of teachers who are against technologies ignore the changes and the context they are inserted in; many students are not motivated, they are dispersed and not interested; most of them have not realized that there is a new paradigm. There are new student profiles, since the demands they receive from media are quite different of those in the nineties.

Although students are exposed to pressures of failure, deprivation and dropout, many educational institutions do not understand the changes face by society in the last decades. Sobrinho Filho states:

> In consequence, their academic pathways have constantly been threatened by risks of failure, interruption, and dropout. This new higher education setting imposes that the institutions understand the impacts new amounts of students produce in university daily routine. Those institutions must rethink their new roles to answer society demands to them. From student point of view many are their difficulties. To reach any upward social mobility through education, students with economic and social deprivations–many of them are the first in their families to access higher education–need to push to overcome the quality and quantity of schooling knowledge their parents had (Dias Sobrinho 2013, p. 117–118).[1]

This author also recalls the problem with teacher professional development, indicating their poor schooling and teaching preparation: "An important difficulty comes from the lack of institution accomplishment mainly in schooling and teaching in the teacher professional development. The forced teacher, curriculum and physical infrastructure improvisation have been forfeited the meaning of public higher education…" (Idem, 118).

Research shows that working on teacher professional development with the focus on their empowerment on different approaches and uses of digital technologies as a medium to ubiquitous learning can change this context, because they may realize they can improve themselves as teachers and professionals wherever they are and whenever they have some free time.

Teachers, who take new technologies in their hands, create new possibilities, collaborate to solve daily or scientific problems, use digital mobile and ubiquitous technologies, design joint projects, work in the dissemination of good ideas, break walls down and facilitate collaboration. Students can learn more deeply about new discoveries and research outcomes and immediately bring their application to real world.

To verbal language, new technology adds audio, visual, audiovisual digitalized and through hypermedia provides new comprehensions of the world and different ways to express feelings and knowledge. Since the end of the 1990s decade, digital technologies have permitted a convergence of different media and recently, the creation of virtual worlds. These new technologies have a direct impact on education which is neither considered in formal schooling nor in informal education.

2.2.1 Teacher's Roles

At this critical moment, in this changing and instable world, what is then, the role of teacher? What is the identity of these professionals responsible for adult, teenagers

---

[1] This citation was translated from Portuguese by Cortelazzo.

and children education? How do they carry out their professional development? What kind of communication they have socially with their families, friends and scholarly with their peers and students? Which physical and symbolic technological artifacts do they use? What does educate mean to them? Do they still believe that school is the place where people are educated? How do they deal with the social means of communication? How do they face the issue of digital technology, instantaneous communication and sustainability?

All this questioning around teacher professional development, social and scholar communication has to deal with meaningful and social learning and sustainable development. New learning approaches, methodologies and technologies need to be integrated and educational professionals need to study and learn about them to change their practice and attract students to learn, know and aggregate new values to their personal, social and professional performance (Downes 2012).

2.2.2 New Approaches: Connectivism, Transmedia, Digital Curation for Social Learning

From this view, connectivism is presented as one possible learning approach which requires the teacher, among other roles, to work as a digital curator, as a director of transmedia production as well as promoter of student autonomy and pro-activity. Transmedia is a new approach to integrate media in communication. Recently, many teachers found out transmedia may serve as one of the possibilities for students to express their knowledge. Social learning is the approach used to base the project to be described in the next pages. This kind of learning is meaningful and implies collaboration and knowledge production. These issucs imply deep learning as presented by Warburton: "Deep learning is a key strategy by which students extract meaning and understanding from course materials and experiences. Because of the range and interconnectedness of environmental, social and economic issues, and the importance of interdisciplinary thinking and holistic insight, deep learning is particularly relevant in the context of education for sustainability" (Warburton 2003, p. 44).

The conceptions of connectivism and connective knowledge on which this work is based were taken from Downes: "Connectivism is the thesis that knowledge is distributed across a network of connections, and therefore that learning consists of the ability to construct and traverse those networks" (Downes 2012, p. 9).

The epistemological foundation of this approach is connective knowledge: "a property of one entity must lead to or become a property of another entity in order for them to be considered connected; the knowledge that results from such connections is connective knowledge (Downes 2012, p. 29) which implies diversity, autonomy, interactivity and openness.

As individuals set connections to the group or to the community they belong, they try to share multiple visions of phenomena from different perspectives and understand the different faces of a situation. When collaborative learning leads to transform learning in ideas, the connectivism permits that learning ecology is extended with new possibilities of new knowledge creation, local problem solution appropriateness, new learning setting organization, and new narratives.

2.2.3 Technological Issues: Mediating Education

Transmedia is known as content communication in different media treated in different narratives expressed by different languages. Ibrus and Scolari (2012) indicate that this new form of communication embeds the social, pedagogical and cultural dimensions; it is an emergent form. A particular phenomena may be expressed in different narratives using written verbal language (book), audiovisual language (film), oral verbal language and gesture (theatre), RPG (game), digital (virtual world, simulation, clip, etc.). They do not exclude or replace one another, on the contrary, they complement, since each of them brings new knowledge about the phenomena.

This way, transmedia is a powerful tool to teachers involved with connectivism, connected knowledge and deep learning. The conception of knowledge as a dynamic process carried out by an individual, a social group or an organization is shown by Siemens (Siemens 2006) as a flow cycle of creation, co-creation, and dissemination, communication of ideas, personalization and implementation.

Digital Curation is another concept Web 3.0 brings to higher education and most of the teachers have never heard about.

There are lots of definitions for content curation, but the one chosen to base the project is:

> … an active process whereby content/artifacts are purposely selected to be preserved for future access. In the digital environment, additional elements can be leveraged, such as the inclusion of social media to disseminate collected content, the ability for other users to suggest content or leave comments and the critical evaluation and selection of aggregated content. This latter part especially is important in defining this as an active process (Antonio et al. 2012, p. 1).

Connectivism, transmedia, connective knowledge, content curation and deep learning elements incorporated to teacher development may be significant to draw higher education towards the transformation of professionals profile for a sustainable future.

3 Professional Teacher Development: Government, University and Teachers Partnership

The explanations in the former narrative were necessary to understand the research in a special program in Curitiba—Education Development Program (PDE)[2] Paraná held by the Secretary of Education of the State of Paraná in partnership with some private and public universities in that city.

[2] "Programa de Desenvolvimento Educacional (PDE) is a state public policy ruled by the Complementary Law n. 103 of March 15, 2004, setting the dialogue between higher education and basic education teachers through theoretical and practical activities resulting knowledge production and qualitative change in teaching practice in local schools" (SEED 2012).

Table 2 Number of participants in PDE

Year	Number of participants
2007–2012	9.096 Teachers (finished)
2013	1,963 Teachers (second year)
2014	1,956 Teachers (first year)
(2007–2013)	19,337 Thematic notebooks and articles

The goal of such program is to offer public school teachers methodological and theoretical foundations to develop educational organized actions, to result in a new construction of their practice. Every year since 2007, principals, pedagogical coordinators and teachers have been invited to participate in a 2 year project as part of their professional development. In Table 2, numbers show the amount of teachers involved (13.025) and their production which is composed of thematic notebooks and articles elaborated during the Program.

The Program requires the participants to be out of their school during the first year to study, to prepare an educational project and to elaborate the pedagogical material to be used during the project implementation; they are 100 % of their work time in the program; the second year, teachers return to school part time, remain in the Program for 25 % of their work time and carry out their projects for 6 months; then, in the last 6 months of the Program, they have to write an article to publish and disseminate the outcomes of their projects.

As an extension program, the university puts their teachers, researchers—updated with new methodologies, technologies and processes—into contact with teachers (practitioners) who are in elementary and secondary public schools, far from research and scientific studies in courses to update their knowledge, to present new trends and possibilities. They make school teachers reflect on their practical experience, present their difficulties, recognize their problems and find out possible solutions to specific contexts.

The studies and orientations presented to public school teachers in one of the phases of the Program at UTFPR by some of the university teachers intends to focus on the three competences presented by Remington Doucette et al., "the ability to integrate multiple perspectives into one's viewpoints"; "the ability to holistically analyse complex systems across the different sustainability domains of environment, society, and economy that span a variety of temporal and spatial scales" and "the capacity to recognize the values underlying the actions of individuals and the structure of social systems" (Remington-Doucette et al. 2013, p. 409). These abilities are basic for teacher work.

Since the orientations follow what is set in these authors' studies, there is a strong agreement that the integrated approaches and the competences "encompass a diversity of perspectives borne from different value and belief systems, which creates conflict in reaching solutions to sustainability challenges" (idem, 409). University teachers orientations also aim to have teachers and later their students, to "be able to discern and critically reflect on the values that punctuate these perspectives" (idem).

This work question is: besides the fact of leaving the school to prepare themselves and their project, why is this program worthwhile to be presented as a movement of higher education for sustainable development?

Remembering the United Nations Report of 2012 statement "*We strongly encourage educational institutions to consider adopting good practices in sustainability management on their campuses and in their communities with the active participation of, inter alia, students, teachers and local partners, and teaching sustainable development as an integrated component across disciplines*" (United Nations 2012), when the university adopts public school teachers to work with them in their particularities, making them research their own practice under the lights of theory already developed from other similar difficulties, the university intends to lead them to change their practice, and bring back the outcomes of this program to debate at the higher education level. Those outcomes certainly impact on new guidelines and action for the sustainable development of that society.

In the last 3 years at UTFPR, this program worked with about Two hundred teachers out of 3,600 and their projects on indiscipline, class management, school management, inclusion, minority's education and gender, mass media communication influencing education, environmental preservation, ecology and sustainable environment alongside lots of other subjects. Many of these subjects were related with sustainability in the sense of "reconciliation of social justice, ecological integrity and the well being of all living systems on the planet" (Moore and Kearsley 2005, 327 apud Baughan).

3.1 Methods

This study was carried from the point of view of collaborative work under the lights of phenomenology which permits important issues emerge from the participants' narratives. The sample for this study was composed of 94 teachers in functions at school as school principals, class teachers and pedagogues during 6 months in the first year, and just 9 teachers during 2 years (2012/2013). Treating data collected from 94 teacher narratives; some categories were selected and information interpreted, bringing some conclusions which may be generalized to be applied in other programs.

During the first period, participants should attend some face to face classes and to work at distance using Moodle, a Virtual Learning Environment. This VLE is where participants let their narratives in activities using tools such as "online tasks", forums of discussion, and wikis. School teachers, pedagogical coordinators and principals did their activities after reading some resources in different languages—audiovisual, digital or printed. They should read, think, discuss, and produce texts collaboratively. The university teacher coordinated the face to face classes, prepared and accompanied the work on the VLE Moodle returning them their feedback. This researcher taught about fundamentals, evaluation and planning to 94 participants and advised 9 teachers in a interval of 3 years—two men (a system analyst and a

Math teacher) and 7 women (two school principals, two pedagogues and three teachers); and worked with Moodle as instrument to collect data for this case The communication between advisor and advisee during the Program also provided narratives used to understand the development of the advisees during the program.

3.2 Preparation, Interlocution and First Findings

The analysis of those narratives indicated that many of the participants started complaining about bad work conditions, salary and did not recognize their responsibility on that. They blamed this on the "others"—government, students, their families, from an ideological point of view, some were angry, others bored, and others silent. In class and on Moodle, they were instigated to share their school problems and search collaboratively solutions to each other school problems. Better, they started to give voice to those who had success in overcoming problems to generate and disseminate solutions and possibilities. Participants were demanded to be active, collaborative and learnt to be pro-active.

This researcher and other university teachers as *content curators* prepared different environments, including virtual, to have their school teachers participate and develop their project. Guiding and advising the realization of the project, the advisor with three or four advisees created a community of practice, forcing them to dialogue to each other, making them recognize the other and his/her problems and difficulties. They used blogs, VLE, email, discussion lists, social networks to discuss, to clarify their duties, to produce knowledge and to demonstrate their learning.

The advisor moderated the dialogue when necessary. These dialogues happened both face to face and on virtual environments in forums of discussion. Instead of lectures of up-to-down ideological indoctrination, school teachers were guided to debate on real practical problems and to identify their origins, causes and specificities (forums of discussion—virtual or in class). During these debates, individually or collaboratively, they found out their limits, their lack of knowledge, or skill or attitude to act and change their context. They were suggested to review their beliefs under the light of new approaches. They were submitted to new kinds of learning and connections.

Networks were created, communities were founded and they were motivated to visit, integrate a community and transit in others. Different learning ecologies have been developed to attend particular needs. School teacher acted in their school in different roles (principal, pedagogical coordinator, class teacher, regional supervisor); sharing their anxieties and concerns in the theoretical courses they had taken during the first 6 months of the Program with all the participants, with the advisor and the small team each advisor guides. They reflected, planned, debated, and designed their own project, sharing their findings, their doubts and their hopes.

School teachers were introduced to new digital technologies and started to understand the role of technology as a support to them and to their students to research, to communicate, to create, to produce and to express their new knowledge in different languages in various media. They learnt that besides verbal language they can express themselves and share their knowledge using social networks and transmedia which permit understanding different perspectives of a particular object, event or person.

During the special lectures, participants were not passive receivers or passive witnesses; they were instigated to be autonomous, proactive, participant as well as critical. They shared experiences they had lived and they contributed to each other project. They were taught to escape from indoctrination, enlarging their views from research on the Web, reading the classics and the authorized authors in their knowledge fields.

Taking part in a community of practice, they developed connected knowledge not only on their special subject but also on the other subjects creating a new learning ecology facilitated by digital technology. In a learning community, they discovered their strengths and their fragilities and were advised how to overcome both by the advisor and their peers. They experienced to be learners in new environments, dealing with new technologies and work in new contexts. They learnt the difference between multiculturalism and interculturalism. They learnt new attitudes: respect, tolerance solidarity, compassion, collaboration, constructive criticism, sharing and commitment.

As other advisors, this researcher instructed their advisees to design a collaborative action research project to make their intervention in their local schools.

3.3 Discussion: Education for Sustainable Development Can It Be Effective?

The project can impact and bring concrete results if the school takes part since the beginning in the planning phase, in the decision takings, in defining the research and intervention tools. Belonging feeling motivates and permits commitment and success.

As university and school teachers went through this collaborative path, they constructed a comprehensive meaning of sustainability in the way that it reflected in the preparation of the teaching or implementing materials and, later, in their own intervention at school.

Other relevant point of the Program is that school teachers should register in two Networked Work Groups (NWG): in a group as coordinator, in another as user, according to their interest. During the design of their material, they have to prepare a plan and material to work with their peers on the objects of their project. When they returned to school to implement their project, they had to coordinate a NWG and to participate in another, constructing their connected knowledge on specific subjects.

During the design and production of the material to be used they know other possible kinds of communication resources beside the printed ones. They might construct maps, methodologies, games, digital environments, audiovisual clips. They learnt to manage, design and produce; they learnt to distinguish what they need and who can help. They learnt to work in collaborative teams and use their talents.

After a year studying, researching and working far from their local schools, they returned to their workplace, presented their projects to their peers and got their approval to start the implementation of the project.

During the period of implementation, advisors and advisees met to share the outcomes and discuss the barriers and difficulties. During this period, new learning was developed: how to deal with the establishment, how to convince people who remained in the daily routine to adopt new behaviors, new goals, new technologies, and new methodologies? How demand the principal of the school rethink school management as part of education for sustainable development? How to contribute to overcome bad policies?

From the constant dialogue with school teachers, university teachers concreted deep learning on their practice and realized in what aspects and dimensions they also have to change to be part of a successful education for sustainable development. A critical view of higher education was constructed during the process of the programme development. When university teachers met and shared the results of their work with school teachers, they constructed connected knowledge not only about their subjects and the program, but mainly on the impact of teaching on the sustainability of the society. They could understand the contributions higher education may bring to sustainable development in the future.

In the last 6 months, university and school teachers have been working together to conclude the project; they collect data, reflect on it and produce their findings to share with the society through the articles to be published on the Web. New learning, new realization and new concepts were concreted during this journey.

Both higher education and basic education gained experience and connected knowledge and meaningful social learning emerged, provoking changes in educators and school manager roles. Many teachers understood that higher institutions and basic schools must go on being partners. Universities and schools became places of development of solutions and recommendations, to change curricula for a sustainable education. The programme brought a meaningful contribution to rethink under-graduation curricula and teacher initial education as well as teacher professional development.

Collected data in forum discussions, in the virtual learning environment and essays preparing the articles show that most basic school teachers were prepared by traditional Pedagogy and continuing education programs based on ideological indoctrination with low practice on teaching practice and pre-service training. Considered as passive receivers during their studies in higher education institutions or passive witnesses in their internship, most of teachers had neither developed autonomy, pro-activity nor innovative initiatives; they just repeated models and reproduced traditional behaviors. They did not have any idea of what sustainability is; they did not understand the relevant role they have as educators for the sustainable development in the future.

The assessment of reflections and activities carried out both with the first 94 teachers and later with the 9 out of the whole group, demonstrated that changing people attitudes takes more time than developing skills or producing new knowledge. Although those teachers had been selected according to some characteristics such as activeness, having an open mind and interest, their attitudes at the beginning of the project were of passivity, dissatisfaction and frustration, A the end of this study, the 94 teacher narratives permitted to understand at least three types of attitudes: academic (theoretical—2 teachers), collaborative and pro-active (90 teachers); and passive (2 teachers). Academic attitude refers to those teachers maintaining a theoretical attitude in their teaching practice, prioritizing scientific knowledge. Passive attitude refers to those who did the minimum to complete the project plan they have to follow. Pro-active and collaborative attitudes refer to the fact that most teachers changed their attitudes, acting, collaborating, studying and participating in the discussions, in the search for solutions, and presenting intervention projects.

4 Conclusions

Programme participants brought their findings to be discussed in meetings about changes in curricula to respond actual demands of a society struggling with problems of pollution, water preservation, food shortage, lack of social justice; and to contribute to create new answers, new possibilities, and new guidelines to face those challenges. The use of web tools and different languages, the collaborative tasks used to stimulate teachers to work as teams and the debates they were demanded to do, instigated them to practice opened new possibilities; plan their classes and make students collaborators and even protagonists of the teaching and learning process. They were demanded to engage in deep learning when the used approach to work with them was connectivism. As Warburton wrote emphasizing the attitudes people should have to learn:

> Although it has general pedagogical significance, deep learning is particularly crucial in the case of sustainability education, where holistic insight and an ability to organize and structure disparate types of information into a coherent whole are central to the whole exercise. Deep learning involves paying attention to underlying meaning. It is associated with the use of analytic skills, cross-referencing, imaginative reconstruction and independent thinking. (Warburton 2003, p. 45)

At local school, new practices and new actions were replicated, transforming their teaching and producing meaningful learning that may aggregate values to student personal and social performance and to teacher professional performance. They indeed learnt from each other.

At the university, the teachers of this Program introduced new technologies, new approaches and attitudes in their classroom practice as well as in their inquiry objects. Those teachers who researched their own practice enriched their production with innovative contributions.

A new concept of teacher continuing education was installed at university: teacher development as a professional responsible for the sustainable development of the future of society. This professional impacts other professionals and students during their education in the elementary and secondary schools; and directly the teachers preparation in the higher education mainly in what refers to social justice.

Contributions of higher education to sustainable development through teacher professional development have just begun in Brazil. There are much more to be constructed and digital technologies, content curation, social learning and social networks are tools to make this happen effectively.

To conclude, teacher professional development focusing education for sustainable development is fundamental to empower and motivate not only learners as UNESCO states,[3] but mainly teachers, "to take action for sustainable development" since "ESD promotes skills like critical thinking, understanding complex systems, imagining future scenarios, and making decisions in a participatory and collaborative way".

References

Antonio A, Martin N, Stagg A (2012) Engaging higher education students via digital curation. In: Brown M, Hartnett M, Stewart T (eds) Future challenges, sustainable futures. Proceedings ASCILITE Wellington, pp 55–59. Available at http://www.ascilite2012.org/images/custom/antonio,_amy_-_engaging_higher_education.pdf. Last accessed on 31 Mar 2014

Baughan P (2013) Exploring the relationship between sustainability and university curricula: a small-scale study. Learn City J 3(1):54–64. Available at http://openaccess.city.ac.uk/2625/1/Exploring_the_relationship_between_sustainability_and_university_curricula_%2D_Patrick_Baughan.pdf. Last accessed on 31 Mar 2014

Cantle T (2012) About Interculturalism Community Cohesion and Relations. Available at http://tedcantle.co.uk/resources-and-publications/about-interculturalism/. Last accessed on 28 Mar 2014

Cantle T (2013) Interculturalism as a new narrative for the era of globalisation and super-diversity. In: Barrett M (ed) Interculturalism and multiculturalism: similarities and differences Strasbourg: council of Europe. Available at http://tedcantle.co.uk/wp-content/uploads/2013/03/082-Interculturalism-as-a-new-narrative-for-the-era-of-globalisation-and-super-diversity-2014.pdf. Last accessed on 28 Mar 2014

Dias Sobrinho J (2013) Educação superior: bem público, equidade e democratização. Avaliação (Campinas) (online) 18(1):107–126. Available at http://www.scielo.br/pdf/aval/v18n1/07.pdf. Last accessed on 28 Mar 2014

Downes S (2012) Connectivism and connective knowledge essays on meaning and learning networks. National Research Council Canada. e-book. Licensed by creative commons CC BY-NC-SA, 2012. Available at http://www.downes.ca/files/books/Connective_Knowledge-19May2012.pdf. Last accessed on 28 Mar 2014

Ibrus I, Scolari CA (eds) (2012) Crossmedia innovations: texts, markets, institutions. Peter Lang, Frankfurt am Main, Berlin, Bern, Bruxelles, New York, Oxford, Wien. Available at http://www.peterlang.es/download/extract/65167/extract_262228.pdf. Last accessed on 20 Dec 2012

Leal Filho W (2011) About the Role of universities and their contribution to sustainable development. Higher Education Policy 24:427–438. Available at http://www.haw-hamburg.de/

[3] UNESCO 2013, Annex 1, p. 2.

fileadmin/user_upload/FakLS/6Forschung/FTZ-ALS/PDF/HEP_hep201116.pdf. Last accessed on 20 Feb 2014

Moore J (2005) Seven recommendations for creating sustainability education at the university level: a guide for change agents. Int J Sustain High Educ 6(4):326–339

Moore M, Kearsley G (2005) Distance education: a system view. Wadsworth, Belmont

Remington-Doucette S, Connell MKYH, Armstrong CM, Musgrove SL (2013) Assessing sustainability education in a transdisciplinary undergraduate course focused on real-world problem solving: a case for disciplinary grounding. Int J Sustain High Educ 14(4):404–433. Available at http://www.emeraldinsight.com/journals.htm?issn=1467-6370&volume=14&issue=4&articleid=17096010&show=abstract&PHPSESSID=46n7la5osj2d5l73s6lvaqsts6. Last accessed on 15 Apr 2014

SEED. Documento Síntese PDE (2012) Secretaria de Estado da Educação do Paraná. SEED, Curitiba. Available at http://www.gestaoescolar.diaadia.pr.gov.br/arquivos/File/pde_roteiros/2013/documento_sintese_2014_incorporando_avaliacao.pdf. Last accessed on 20 Mar 2014

Siemens G (2006) Knowing knowledge, e-book. Licensed by creative commons. Available at http://www.elearnspace.org/KnowingKnowledge_LowRes.pdf. Last accessed on 15 Apr 2014

UNESCO (2013) Proposal for a global action programme on education for sustainable development as follow-up to the United Nations decade of education for sustainable education (DESD) after 2014. UNESCO general conference. 37C/57, 4 November 2013. Available at http://unesdoc.unesco.org/images/0022/002243/224368e.pdf. Last accessed on 20 Feb 2014

United Nations (2012) Report of the United Nations conference on sustainable development. Conference held in Rio de Janeiro Brazil from 20 to 22 June 2012, New York, United Nations. Available at http://www.uncsd2012.org/content/documents/814UNCSD%20REPORT%20final%20revs.pdf. Last accessed on 26 Mar 2014

Warburton K (2003) Deep learning and education for sustainability. Int J Sustain High Educ 4(1):44–56. Available at http://www.emeraldinsight.com/journals.htm?issn=1467-6370&volume=4&issue=1&articleid=839800&show=pdf. Last accessed on 14 Apr 2014

Author Biography

Professor Iolanda Bueno de Camargo Cortelazzo has been working with teacher development for 40 years and researching on Technology in Education for 20 years. She received her doctorate in Education from the Universidade de São Paulo, Brazil, in 2000 where she researched collaboration, team work and new technologies, then coordinated programs of Distance Education both in private and public higher education, publishing articles and books on the subject. She joined the Department of Education and the Coordination of Technology in Education at UTFPR in 2010 and started her participation in the Education Development Program in a partnership of UTFPR and Secretary of Education in the Paraná State in the south of Brazil. She also manages the Program for Technicians in Teaching Multimedia and she develops her course on Technology and Education at Universidade Tecnológica Federal do Paraná. She is one of the founders of RCE in Curitiba—Paraná, Brazil.

A Minor Programme on Sustainability for the Engineering Curriculum at the University of Chile

Luis S. Vargas and Claudia Mac Lean

Abstract

The present paper presents a Minor Programme proposal on sustainability in the current engineering degree programmes at the University of Chile. Students in the Civil Engineering curriculum at the University of Chile can take any minor, as long as they comply with the course requirements. The Minor on sustainability is conceived as part of the undergraduate electives included in the curriculum of the School of Engineering, and it is expected to enter into the curricula on spring 2014. The paper discusses different alternatives to introduce sustainability-related topics in the curricula, the design process of the Minor and its organisation. The Minor has the following structure: one introductory course, three elective courses which can be selected from a given list and a final workshop. The paper describes the design process, support, and dissemination of the Minor and the main challenges found along the road.

Keywords

Education · Engineering · Sustainable development · Curriculum

L.S. Vargas (✉)
Department of Electrical Engineering, University of Chile, P.O. Box 8370451, Santiago, Chile
e-mail: lvargasd@ing.uchile.cl

C.M. Lean
Office of Engineering for Sustainable Development, Faculty of Physical and Mathematical Sciences, University of Chile, P.O. Box 8370451, Santiago, Chile
e-mail: cmaclean@ing.uchile.cl

W. Leal Filho et al. (eds.), *Integrating Sustainability Thinking in Science and Engineering Curricula*, World Sustainability Series, DOI 10.1007/978-3-319-09474-8_2

1 Introduction

In December 2002, the United Nations General Assembly proclaimed the UN Decade of Education for Sustainable Development, emphasizing that education is an indispensable element for achieving sustainable development (DESD-United Nations 2005–2014). Universities play a significant role in promoting the concept of sustainability in different avenues, such as new educational programs, research, cooperation with other entities and the use of demonstration projects. In fact, education for sustainable development has been on the agenda of many engineering faculties since 1990s (Segalas et al. 2010).

Much progress has been made over the last decade in developing ways to introduce the concepts and issues surrounding sustainable development to engineering students and to challenge them to think about sustainability-related problems in different ways (Fenner 2013). However, while education for sustainability literature shows there have been calls for embedding sustainable development content throughout engineering curricula (Allen et al. 2010), there has been little of strategic and systemic integration (Byrne et al. 2013).

One approach to introducing sustainability concepts to engineering education is incorporating sustainability-related topics directly in the curricula. Such mechanism has been implemented by many universities; see for example the case of the University of Nottingham Engineering Faculty. In another approach, the engineering undergraduate curriculum can incorporate a professional specialisation, for example the Energy, Sustainability and the Environment specialisation at the University of Cambridge. A third case worth-mentioning is to create a whole new academic unit dedicated to sustainability-related topics, for instance, the Department of Civil and Environmental Engineering at the Massachusetts Institute of Technology.

In regards specifically to Minors in Engineering Schools, a wide range of such programmes can be found worldwide. To mention a couple: Minor in Sustainability Engineering at the Binghamton University—the State University of New York—and also a Minor in Sustainable Engineering at the Columbia University in the City of New York.

The issue of sustainability in higher education from the learning and institutionalisation processes point of view has been revised in (Wals 2013). A major reflection in this work is the classification of "bolt-on" and "built-in" approaches. The first one consists of adding new courses and modules that have elements of Education for Sustainable Development (ESD), whereas the second one corresponds to integrating sustainability in existing studies and research programmes as well as in staff development. As the "built-on" responses seem to be becoming extinct, in the proposed Minor presented in this paper the "built-in" approach is selected. Therefore, this Minor programme is formed by using the current Minor structure of the Engineering Programme and its courses are mostly already available in the curricula.

International evidence has shown (Wals 2013) that most of the universities that engage in sustainability are universities that have a focus on education, rather than on research. In the case of the University of Chile, its main focus is on research, and experience has shown that it has been an important challenge to open space for the introduction of sustainability at the Engineering School.

The main purpose of the paper is to present a field experience in assembling a Minor on sustainability in a research-oriented Engineering School, which is discussed in light of the international analysis and scholar evidence. Regarding the participatory approach, a wide consultation and collaborative practice have been viewed as important to achieve consensual agreement to harmonise collective views.

The remainder of this paper is organized as follows. In Sect. 2 a description of the engineering curricula at University of Chile's Engineering School is presented. In Sect. 3 the main aspects of the design process are described. Section 4 focuses on the Minor's structure proposal. Finally, Sect. 5 summarizes the main conclusions of this work.

2 Engineering Curricula Description

The engineering studies at University of Chile last 6 years. The main specialisations comprise mechanical, industrial, computer science, civil construction, electrical, chemical, mining, and biotechnology engineering. The first 2 years are set as a core programme in mathematics, physics and computer science. In Chile the secondary school education varies greatly in terms of contents and quality, so these 2 years serve as leverage for the students as well.

This two-year programme is denominated Common Plan, and the courses are described in Fig. 1.

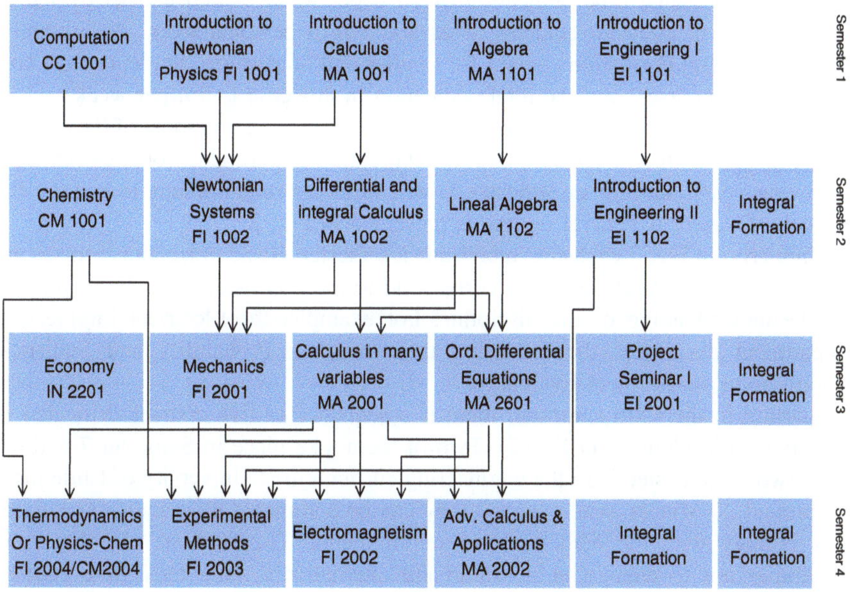

Fig. 1 Common plan

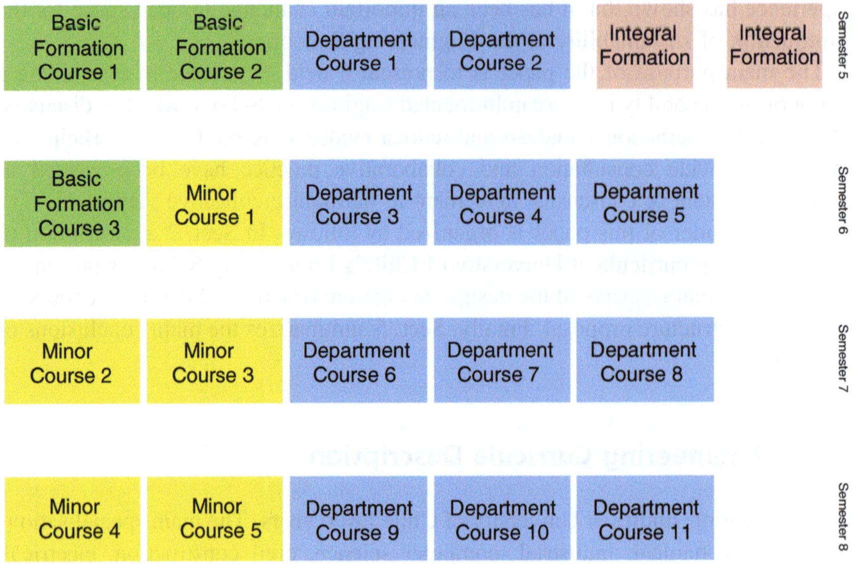

Fig. 2 The third and fourth year of the bachelor programme

In Fig. 1, each row represents a semester, and the academic year has two semesters. Lines with arrows indicate courses requirements. For instance, the course FI 2002 Electromagnetism has three requirements: FI 2001 Mechanics, MA 2001 Calculus in Many Variables and MA 2601 Ordinary Differential Equations. Every course has the same time dedication of 10 h per week, which consist of 4 h of teaching, 3 h of assistant-guided classes and 3 h of personal work. Additionally, there are Integral Formation courses for English and humanities. Students must take four of these courses, and the total dedication in this case is 5 h per week.

In third year students choose their field of specialisation and most of the courses they take are given by a specific Department (called Department Courses). The two following years, starting in semester 5, will fulfil the requirements to acquire the Bachelor Degree. A general outline is presented in Fig. 2.

Similarly, in Fig. 2 each row represents a semester. Basic Formation courses are defined by each Department, and correspond to those considered to be significant for the understanding of the discipline. For example, the Electrical Engineering Department designates the following three courses: Probability and Statistics, Optimisation, and Project Evaluation.

Starting in the sixth semester, there is a sequence of five courses dedicated to Minors (Minor Courses in Fig. 2). Two of them take place in Semester 7 and the other two in Semester 8. In the engineering curriculum at University of Chile there are around 20 Minors available, and they can be completed by students as long as they comply with the Minor course requirements. The students also have the possibility not to engage in a Minor and alternatively, take those five courses according to their own election freely.

3 The Design Process Description

In the present chapter the design process of the Minor in Engineering for Sustainable Development is described.

The origin of the idea of a Minor on sustainability is a directive given by the Dean of the Engineering School to the head of the Office of Engineering for Sustainable Development. The Minor would be implemented by putting together a set of courses, which could be offered as a Minor in Engineering for Sustainable Development. This indication responds to a very sensitive need expressed by engineering students to incorporate sustainability concepts and contents into the curriculum of the Engineering School.

It consists of the first step towards introducing sustainability in the curricula, and it was seen as an opportunity to introduce the sustainability concepts in a way which posed less resistance and barriers to implementation. The incorporation of sustainability topics in the Common Plan of the Engineering School is being currently developed, through working sessions with the coordinators and professors of the following courses: Computation, Introduction to Engineering, Chemistry and Electromagnetism.

4 Stage 1: Setting the Task Force

A group of professors were invited by the Head of the Office of Engineering for Sustainable Development to an initial brainstorming session to commence designing the Minor. The participating professors teach sustainability-related courses in their own fields at the Faculty. The departments involved in this initiative are Industrial Engineering, Electrical Engineering, Civil Engineering, Chemical Engineering, Mathematical Engineering, Mining Engineering, and Geophysics.

The School of Engineering has a Teaching and Pedagogical Area. The main task of this office is to assist professors to enhance their pedagogical skills. In addition, it organises and manages the Minors at the Engineering School.

The Teaching and Pedagogical Area promotes a competence-based approach in the training of engineers, therefore they recommended the Minor in Engineering for Sustainable Development should integrate a similar methodology. The Area also suggested integrating students in the working sessions, in order to be able to embrace from the beginning of the design process the students' thoughts and recommendations. Accordingly, students involved in environmental student groups were invited to participate in the design sessions of the Minor.

As a result, the task force in charge of designing the Minor is led by the Head of the Office of Engineering for Sustainable Development and the whole team is composed of 10 Professors, 3 Professionals and 4 students.

5 Stage 2: The Design Process

Five monthly working sessions were organised by the Head of the Office of Engineering for Sustainable Development. These sessions would typically last 2 h and have a couple of objectives to be discussed and agreed.

The team went through a typical evolution of forming, norming and storming, and it eventually reached a state of settled relationships and expectations. It is worth mentioning that vertical authoritarian leadership styles where not commonly observed throughout the sessions, which contributed greatly to the depth of the discussions and to achieving consensus. This process has followed a social learning approach (Barth and Rieckmann 2012), which is a learning system in which people learn from and with each other and, as a result, become individually and collectively more competent.

As an organisational change is needed, the involvement of staff as an important stakeholder group in this stage is seen as crucial, and empowering these stakeholders is considered as a critical step to attain the organisational change (Barth and Rieckmann 2012).

The key outcomes of the working sessions are a list of competences students should have by the end of the Minor, the structure and courses of the Minor, and the responsible teams of the various tasks.

6 The Sustainability Minor Proposal

The competences students should have by the end of the Minor are the following:

- The student links sustainability to complex systems analyses, in its social, environmental and economic dimensions.
- The student assumes sustainability as a professional attribute related to critical thinking and responsibility.
- The student integrates complexity and sustainability educational tools in the practice of its own field of specialisation.
- The student discusses, promotes, and proposes sustainable solutions to its local environmental.

The previously described competences have determined the definition of the structure of the Minor—which has 45 credits. The structure of the Minor aims not only to introduce basic sustainability contents and topics, but also to permit students to specialise in their own fields of interest, and to experience the complexities linked to sustainability through a practical workshop.

In terms of methodological approaches (Littledyke et al. 2013) argue that wide consultation, consensual agreement and collaborative practice have been viewed as important to achieve collective views and coordinated action for education for sustainability. The overall structure of the Minor is shown in Fig. 3.

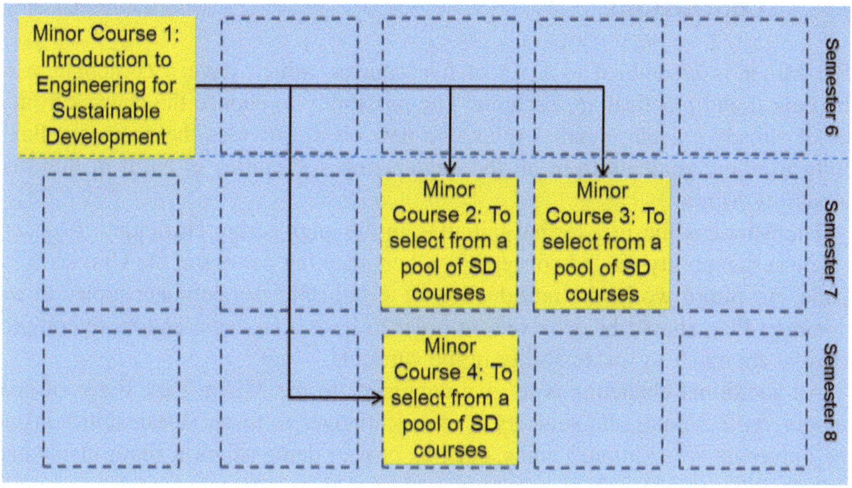

Fig. 3 Structure of the minor in engineering for sustainable development

Firstly, there is a mandatory introductory course to Engineering for Sustainable Development, which purpose is to introduce a common background in terms of engineering and sustainability to the students. There is a wide range of new forms of learning in ESD programmes (Wals 2013), including trans- and interdisciplinary learning, social learning, project-based learning, gaming, computer simulations, distance learning, backcasting, case-studies, policy-laboratoires, problem-based learning, bootstrapping, values education, ecological footprint analysis, experiential approaches, reflective journal writing. In our case, we have adopted interdisciplinary learning, project-based learning, values education, and the ecological footprint analysis.

Secondly, three elective courses must be chosen from a pool of approximately 15 courses offered by the various Engineering Departments within the Faculty. These elective courses comprise subjects such as climate systems, environmental engineering, environmental economics, sustainability in construction, innovation for sustainability, sustainability in mining, social project evaluation, and renewable energies. This initiative also led to the creation of two new elective courses in the undergraduate programme.

The Minor ends with a practical project where students propose a sustainability intervention to their surroundings (home, university, neighbourhood, etc.). The design of this workshop followed the ideas of (Segalas et al. 2010) in the sense that sustainability courses at technological universities should focus their content on the social and institutional aspects of sustainable development, and apply a constructive and community-oriented pedagogical approach.

The Minor in Engineering for Sustainable Development at the Engineering School of the University of Chile is scheduled to be launched in August 2014.

7 Conclusions

The Minor is assembled as a set of five courses, which have a combination of theoretical and practical applications. The possibility to choose the courses from a set of already existing courses (plus two new ones) provides the necessary flexibility to allow students from all the engineering specialities to enrol in this sustainability Minor.

The Minor is the first formal initiative in the engineering curriculum to incorporate sustainability-related concepts in all engineering careers at the University of Chile. As future work, the introduction of sustainable development topics in the Common Plan should be achieved, in order to have a strategic and more comprehensive approach to sustainability in the curricula.

An additional challenge is to give continuity to the Minor Task Force of professors, professionals and students in other initiatives to foster sustainability inside the university in additional fields, such as project demonstration (both inside and outside the campus), in incorporating sustainability contents and new courses in every Department, and in promoting multidisciplinary research within the Engineering School.

References

Allen D, Allenby B, Bridges M, Crittenden J, Davidson C, Hendrickson C, Matthews S, Murphy C, Pijawka D (2010) Benchmarking sustainable engineering education: final report, center for sustainable engineering, Syracuse, NY. Available at www.csengin.org/csengine/sub.html?skuvar=147. Accessed 18 Aug 2011

Barth M, Rieckmann M (2012) Academic staff development as a catalyst for curriculum change towards education for sustainable development: an output perspective. J Clean Prod 26:28–36

Byrne EP, Desha CJ, Fitzpatrick JJ, Hargroves K (2013) Exploring sustainability themes in engineering accreditation and curricula. Int J Sustain High Educ 14(4):384–403

DESD-United Nations (2014) Decade of education for sustainable development-DESD. http://www.desd.org/. Last accessed on 24 Apr 2014

Fenner D (2013) Letter of introduction. Engineering education for sustainable development-EESD13 book of abstracts, 22–25 Sept 2013, Cambridge UK

Littledyke M, Manolas E, Littledyke RA (2013) A systems approach to education for sustainability in higher education. Int J Sustain High Educ 14(4):367–383

Segalas J, Ferrer-Balas D, Mulder KF (2010) What do engineering students learn in sustainability courses? The effect of the pedagogical approach. J Clean Prod 18:275–284

Wals AEJ (2013) Sustainability in higher education in the context of the UN DESD: a review of learning and institutionalization processes. J Clea Prod. http://dx.doi.org/10.1016/j.jclepro.2013.06.007. (Article in press)

Authors Biography

Professor Luis S. Vargas received the Electrical Engineer diploma (1985) from the Universidad de Chile, Santiago, Chile. He obtained his M.Sc. degree from Universidad de Chile in 1987 and his Ph.D. degree in Electrical Engineering from the University of Waterloo, Canada. From 1994 he has worked at the Universidad de Chile where currently he is an Associate Professor.

Claudia A. Mac-Lean obtained the Industrial Engineer diploma (2010) from the Universidad de Chile, Santiago, Chile. She completed the M.Phil in Engineering for Sustainable Development degree from the University of Cambridge in 2012. From 2013 she works at the Engineering Faculty of the Universidad de Chile, implementing an Office of Engineering for Sustainable Development.

Food for Thought: A University-Wide Approach to Stimulate Curricular and Extracurricular ESD Activity

Helen Puntha, Petra Molthan-Hill, Aldilla Dharmasasmita and Eunice Simmons

Abstract

Sustainability and Higher Education have been the focus of much recent academic and professional research as there has been a growing expectation that Higher Education institutions will produce 'sustainability-literate graduates' (Lacy et al. in A new era of sustainability. U.N. Global Compact-Accenture CEO Study, 2010; Sky in The sustainable generation: the sky future leaders study, 2011; Scott et al. in Turnaround leadership for sustainability in higher education, 2012) and a growing demand from students for future-proof skills (Drayson et al. in Student attitudes towards and skills for sustainable development. NUS/HEA, 2012). The process of embedding Education for Sustainable Development into curriculum is however challenging, and for some disciplines more than others. This paper examines how Nottingham Trent

H. Puntha (✉)
NTU Green Academy Project 'Food for Thought', Centre for Academic Development and Quality, 202 Dryden Centre, NTU, Nottingham NG1 4BU, UK
e-mail: helen.puntha@ntu.ac.uk

P. Molthan-Hill
Management Division, NTU Green Academy Project 'Food for Thought', Room 703 Newton, NTU, Nottingham NG1 4BU, UK
e-mail: petra.molthan-hill@ntu.ac.uk

P. Molthan-Hill · A. Dharmasasmita
Nottingham Business School, Room 703 Newton, NTU, Nottingham NG1 4BU, UK
e-mail: aldilla.dharmasasmita@ntu.ac.uk

E. Simmons
Department of Animal, Rural, and Environmental Science, Nottingham Business School, Room 703 Newton, NTU, Nottingham NG1 4BU, UK
e-mail: eunice.simmons@ntu.ac.uk

W. Leal Filho et al. (eds.), *Integrating Sustainability Thinking in Science and Engineering Curricula*, World Sustainability Series, DOI 10.1007/978-3-319-09474-8_3

University has adopted a unique approach to centre the development of Education for Sustainable Development around the specific topic of food. The paper will share the model for engaging students and staff members across an institution with sustainability using a unifying theme which constitutes a critical global challenge of relevance to all disciplines. Details will be given of the process and challenges of the approach which has sought to facilitate personal, disciplinary and inter-disciplinary sustainability literacy. The approach has been largely successful in its aim of developing new processes and content to lead to the embedding of Education for Sustainable Development across the formal and informal curriculum as well as the institutional culture.

Keywords
Education for sustainable development (ESD) · Sustainability literacy · Curriculum · Online learning · Virtual learning environment (VLE) · Video · Food

1 Introduction

In this paper we will share how we at Nottingham Trent University (NTU) addressed the challenge of embedding sustainability across an entire institution through our use of the over-arching theme of food during phase one of our 'Food for Thought' project. The project included two strands: 'Appetite for Change' (formal curriculum) and 'Sustain Yourself' (informal curriculum/extra-curriculum). Within the 'Appetite for Change' (formal curriculum) strand we used food as the focus topic for an online 'Sustainability in Practice' certificate which was open to all students and staff as well as a starting point to create co-curricular activity to enhance the student experience. The paper will focus on an explanation of the certificate as this has been the most heavily-resourced and successful aspect of phase one of the project. We will explain the curriculum design model which could be adapted for use at other institutions both in online and offline settings to support sustainability literacy. The strength of the curriculum model lies in the use of a theme which immediately establishes the project as 'real-world' rather than theory-driven and is flexible to encourage personal, disciplinary and interdisciplinary understanding of sustainability as well as providing stimulus for extra-curricular activity such as research and community-based action. The design of the certificate enabled the facilitation of sustainability literacy as well as other skills and attributes and provided ground-breaking opportunities for students to co-create curriculum, providing impact beyond the immediate certificate participants through the 'recycling' of participant work into teaching material.

For a long time, Higher Education (HE) has been a catalyst for change; creating exemplars for governments and business organisations to follow and having an influence in policy decision-making. There is one area however which recent research has identified as a challenge with regard to its integration into HE strategy and especially into its curricula; sustainability (Cortese 2003; Stubbs and Schapper 2011). Lozano et al. (2013: 10) argue that, 'In spite of a number of sustainable development (SD) initiatives and an increasing number of universities becoming engaged with SD, most higher education institutions (HEIs) continue to be traditional, and rely upon Newtonian and Cartesian reductionist and mechanistic paradigms'.

Several academics have argued that HEIs are segregated into highly specialised yet specific 'areas of knowledge' and that this has resulted in disjointed learning as departments focus solely on incentives such as tenure and research, and are often deterred from trans-disciplinary collaboration. Consequently HEIs may be producing graduates who know only about their specific area with a focus on—'individual learning and competition… professionals who are ill prepared for cooperative efforts'. (Cortese 2003: 16; Winter and Cotton 2012; Djordjevic and Cotton 2011).

Orr (1994: 5) contends that 'The kind of education we need begins with the recognition that the crisis of global ecology is first and foremost a crisis of values, ideas, perspectives, and knowledge, which makes it a crisis of education, not one in education'. This is echoed in Cortese (2003: 16) who further argues that 'Higher education institutions bear a profound, moral responsibility to increase the awareness, knowledge, skills, and values needed to create a just and sustainable future. Higher education plays a critical but often overlooked role in making this vision a reality'.

Many HEIs have achieved success during the last decade in addressing sustainability through their estates. Whilst there have also been notable successes in the area of curriculum and co-curriculum development it would seem that the embedding of sustainability into the whole curriculum and not only into individual modules and degree courses presents a challenge to HEIs.

2 Background

NTU has been rated as one of the greenest universities in the UK, based on Green League Table from 2008–2013 (Green League 2013). It has an EcoCampus Platinum award and participates in Green Impact and the LiFE Index. Our institutional mission to provide 'education and research which shapes lives and society' reflects our existing commitment to sustainability. NTU has made major achievements in the area of sustainability in recent years with clear related strategy and activity in the areas of estates, procurement, waste, volunteering and catering. Our current aim is to embed education for sustainable development (ESD) into the curriculum in line with our Graduate Attributes, one of which is on the subject of Global Citizenship and calls for inclusion of ESD in the curriculum. In 2010 NTU began to integrate ESD with the introduction of a steering committee, the Sustainability Action Forum (SAF).

In September 2012 NTU invited renowned expert Geoff Chase to run a primer workshop for embedding ESD. The workshop achieved some success in motivating academics to engage with sustainability; however some common obstacles were identified in the workshop feedback. Participants expressed concern that the diversity and complexity of sustainability issues meant that integrating sustainability into the curriculum in any meaningful way posed great challenges for example in terms of where to begin, which of the many urgent sustainability challenges to tackle first and the amount of time required to undertake the changes within their courses and modules. Related to this, some participants or their colleagues were of the view that sustainability was an 'add-on' to core curriculum and should not be prioritised. Through our own research undertaken in late 2010 we had identified a need to provide structure for embedding ESD in line with our related strategy and this need seemed to be reflected in the workshop feedback.

In November 2012 we applied to take part in the second round of the Higher Education Academy (HEA) Green Academy Change Programme with the intention that we would use the programme to undertake work to support the embedding of ESD into curriculum whilst addressing the existing challenges identified through the research and workshop feedback. We envisaged that the project should therefore provide the following:

- ESD curriculum which would facilitate not only the learning of sustainability content but wider skills and/or knowledge with the potential to support students' wider learning and employment
- teaching resources which are either already contextualised or could be easily contextualised into the various disciplines to reduce the time required of academics
- wider impact beyond the immediate project participants particularly in terms of mainstream curriculum development
- flexibility of access in terms of who when and how staff and students engage
- appeal to wide pool of staff and students by the nature of the project—we wanted to do something unusual that would therefore add value to the usual work/study experience

In addition, we viewed the programme as a potential opportunity to join up some of the existing good practice within and outside the curriculum for example estates and catering. Our application was happily selected and we commenced our project. At the time of writing we have completed the first phase of the project and the second phase is due to end in October 2014.

3 Sustainability and Food

As stated, we decided to use the over-arching theme of food. It was felt that a theme would constitute an engaging and flexible way of embedding ESD. In contrast to previous theory-driven approaches the adoption of a theme would:

- offer a 'way in' to sustainability for students and staff

- enliven the subject for those with no previous knowledge or interest in sustainability
- facilitate student disciplinary and interdisciplinary knowledge and skills
- develop a collective understanding of sustainability which would support cooperative and collaborative work
- provide opportunities within research and extra-curricular activities as well as curriculum
- reduce the effort required of academics to contextualise sustainability theory into practical issues and/or application
- offer a unique opportunity to be part of an institution-wide project addressing a critical local and global sustainability challenge

Food production and consumption is an urgent global sustainability challenge with far-reaching implications for everyone (Foresight 2011). The theme is so broad as to encompass all three aspects of sustainability: economic, social and environmental; and can be made relevant to all academic disciplines as it covers such vast topics as agriculture, food production methods, nutrition, transport, waste, energy, climate change, bio fuels, biotechnology, commodity prices, biodiversity etc.

Food initially provided an authentic topic with which staff and students could engage to develop their understanding of sustainability as a concept. The intention is that this initial grounding will lead to further activity beyond the theme of food.

4 Appetite for Change

The first phase of the 'Appetite for Change' (formal curriculum) strand consisted of an optional online 'Sustainability in Practice' certificate. The second phase which is being embarked upon at the time of writing involves the embedding of ESD into the mainstream curriculum through the work of four interns together with academics (this is discussed in more detail in the 'student-as-co-creator' section). The certificate is explained below. The certificate was open to staff as well as students; where the paper uses the term 'participant' this refers to the experience of both staff and student participants within the certificate; where the paper uses the term 'student' with regard to the certificate this refers to aspects of the certificate experience relating solely to student participants.

4.1 Certificate Design and Promotion

Given that the certificate was not credit-bearing it was felt that it should provide a unique experience which would add value beyond the participants' usual work or course of study. In the case of the certificate this included the opportunity to develop digital literacy skills, enhance their CV, produce a video which could be shared with potential employers, contribute to NTU Global Week and win prizes! These benefits were emphasised in promotional work which took place during NTU

Welcome Week in September 2013 and NTU Green Week (when the certificate was officially launched) in November 2013. We anticipated that we might gain 90 participants for the certificate however we had over 2,000 students access session one and 70 complete the certificate. The reasons for the drop-off will be explored in future research.

4.2 Online Format

The certificate was open to all NTU students and staff and appeared automatically in the students' Virtual Learning Environment (VLE) homepage, staff needed to be enrolled manually. It was a fully online course comprised of 4 sessions and a video assessment piece (discussed in more detail in a later section). The estimated time commitment required to complete the certificate was 20 hours including 30–60 minutes per session and up to 16 hours for the video. Participants could complete the course in their own time alongside their usual course or work over a 3 month period between November 2013 and February 2014. Each new session was released to the participant once they had completed the preceding session. Completion of the Fourth and final session unlocked the Dropbox for participants to submit their video assessment piece.

It was decided that an online format would be the most appropriate for the certificate. This format offered an effective way to reach all students and staff without the logistics of timetabling, room booking etc. It also offered participants the experience of online learning which for many was their first experience of a fully online course.

Our initial intention was to hire an external company to build the certificate to our requirements. It proved impossible to secure the expertise and as a project team we employed an individual to build the certificate learning room within our institutional VLE. Having never created an online course before this was a massive challenge and we gratefully received a lot of support for example from the NTU Continuing Professional Development Department who provided training as well as some hands-on support for the VLE. There were many advantages to housing the learning room within our institutional VLE rather than an external platform; it allowed us to monitor more accurately which students were accessing or not and the demographics of those students. It allowed us to automatically enrol all students making it more convenient for them to commence the certificate and it encouraged people to view the certificate as part of the 'everyday' business of the institution since the learning room was fully integrated within the VLE.

As the certificate was based online it was necessary to provide a great deal of on-screen information that otherwise might have just been spoken to in a face-to-face teaching session. We had to pre-empt what the participants might want to know and might need and this required some consideration in terms of the wording and layout appropriate to an online environment. The learning room itself included extensive sections on support and course information as well as assessment guidelines. We also set up a twitter feed to keep the learning room more 'live'; we tweeted twice or more each week day with links to food and sustainability news items, websites, news updates about the certificate itself and some more entertaining tweets such as Instagram photos of cooking experiments which had failed, pictures of pandas etc. We gained several Twitter followers including Network NCN (a business networking event aiming to bring local entrepreneurs together with students, Dairy Farmers of America (DFA) and the Nottingham Evening Post newspaper.

In addition to the Twitter feed, we employed some further techniques to keep the learning room 'live' to encourage engagement and retention. The first was that we held prize draws for participants wherein participants could win credit for their NTU smartcards to spend on printing, catering etc. on campus. The aim of this was both to motivate participation and to encourage participants to stay on 'track' with the certificate since entry into the prize draws was dependent on them completing activities within certain time periods. In session one we awarded credit of higher monetary value to participant prize winners who attempted more challenging tasks (this session had three alternative activities to complete which ranged from basic, intermediate and advanced). Second, we used videos in several different ways to make the learning room live and personal:

- each session began with a welcome video which outlined the aims of the session and reminded participants of the prize draw
- each session ended with a closing video which summarised the session and incorporated participant comments and ideas from the discussion forum
- the videos for each session featured different team members every time to enable the participants to 'meet' different members of the team.
- the Session One opening video was placed on the homepage
- we filmed a 'Christmas-themed' reminder video which we placed on the homepage

We tried to make the videos interesting to watch e.g. use of props, one team member talking whilst standing on their head and we produced a video montage of the 'bloopers' which we added to the homepage of the learning room at the close of the certificate.

4.3 Pedagogical Approach

Constructivists such as Novak (1998) believe that for learning to be meaningful it should encompass thinking (cognition), feeling (affect) and acting (motor or psychomotor). The certificate was designed to encompass all three aspects with the aim of providing a meaningful and transformative learning experience which would broaden the participants' perspective of their self and subject as well as teach them knowledge. Related to this, the certificate was designed to facilitate graduate attributes. As indicated earlier in this paper, like many HEIs, NTU has a set of 'Graduate Attributes' i.e. a list of qualities, skills and competencies which the institution seeks to facilitate within its students. One of the Graduate Attributes relates to global citizenship and encompasses sustainability, international awareness and leadership capacity.

In recognition that sustainability can be used as a tool to support students' learning and skills in all sorts of areas and also that some staff do not view sustainability as a curriculum priority, the certificate aimed to support not only the global citizenship Graduate Attribute but others such as communication skills, critical thinking and international awareness needed for students to thrive in a world of 'supercomplexity' (Barnett 2000).

We designed activities to support the development of participants' sustainability literacy as outlined by Stibbe (2009: 10f):

> … **the skills, attitudes, competencies, dispositions and values** that are necessary for surviving and thriving in the declining conditions of the world in ways which slow down that decline as far as possible. Gaining practical skills requires a form of learning which goes beyond memorising and repeating facts. It requires **active learning**, a broad term used to refer to **self-reflection, self-directed enquiry, learning by doing, engagement with real life issues,** and **learning within communities of practice**. (emphasis added)

Table 1 indicates the strategic design of the certificate content, and how the certificate aimed to support participants in their exploration of sustainability challenges through the lens of their personal, disciplinary and interdisciplinary perspectives; in this way encouraging a real-world view and a thorough exploration of complex issues.

Table 1 Overview of certificate content

	Topic	Aim	Example activity
Session one	Student experience of sustainability and food	Engage students on a personal level	Explore the most 'sustainable' options for a chicken stir fry
Session two	Sustainability and food in the disciplines	Facilitate disciplinary understanding of sustainability	Source an online video which relates to food, sustainability and the student's discipline
Session three	Connections between disciplines; identifying challenges	Facilitate interdisciplinary understanding of sustainability	Source an online video which relates to food, sustainability and two or more disciplines with reference to the food life cycle
Session four	Global and local solutions	Identify disciplinary/ interdisciplinary solutions to sustainability challenges	Source local or global food sustainability solutions

4.4 Content Design

As the certificate was open to all staff and students it was important that it was accessible to all as well as being an appropriate level of difficulty and relevant to all disciplines and this presented many interesting challenges during the content design stage. The content was developed collaboratively with staff from across the university. We had a core team who did the initial planning during a one-day intensive session which included reps from the Student Union, Environment team, Business School, Centre for Academic Development and Quality. We then held a planning session for each session to develop content and this attracted other staff in addition to the core team. It was important that participants could 'see themselves' and their disciplines within the certificate, that they could perceive the relevance of the activities. We were fortunate to be given support from many colleagues across the university who provided over ninety online examples (links to news stories, journal articles, websites etc.) of sustainability challenges relating to food within the various disciplines. Two examples of such resources include a web link to a news story about the effects of light pollution on crop production which was provided by a lecturer in Astrophysics and a web link to a news story about the rise of food bank use in the UK and the suggested causes for this, provided by a lecturer from Social Sciences.

4.5 Use of Digital Pedagogies

We did not want to have a learning room full of text documents but wanted the certificate to be as interactive as possible. The certificate was designed in the style of a Massive Open Online Course (MOOC). As stated previously the certificate was based entirely online. In addition we employed various digital pedagogies to support participant learning. Digital artefacts such as videos, prezis, quizzes and discussion forums were the basis of the certificate learning activities and participants has to conduct their own online research to complete a task for each session. Such artefacts and accompanying tasks were designed to keep the learning environment active and to facilitate skills such as critical thinking and communication skills for example:

(i) the use of a prezi in session three on the subject of the life cycle of a strawberry yoghurt demonstrated the interconnections between the different stages of the food life cycle as well as the connections between the various disciplines and how they relate to different stages of the food life cycle and to each other in terms of different sustainability challenges relating to food. This level of complexity would have been difficult to communicate effectively through a textual document; the motion of prezi allowed a visual representation of the connections to be drawn and considered.

(ii) a quiz was employed in session three following the prezi on food life cycles. The quiz offered a new but relatively easy task to provide a quick check on what participants had learned from the prezi as well as providing some new information not contained within the prezi.
(iii) a zee map of the world was created in session four. When participants completed their tasks and posted their solutions to the discussion forum we then added these to the world map which we displayed during NTU Global Week and other events.
(iv) discussion forums were established for each session as well as within the support section of the learning room. The discussion forums performed several functions: they provided a location for participants to submit their work at the end of each session, they provided a venue for students to meet and discuss ideas (the activities sometimes stipulated this for example the session two activity involved participants posting a video to the discussion forum and commenting on someone else's video), they allowed us to monitor student engagement and satisfaction and they enabled us to add a release function to the session material i.e. participants would have the next session released to them once they posted to the discussion forum.

4.6 Video Assessment Piece

We opted to use the innovative assessment medium of video for the certificate. Participants were asked to submit a video of 3 min or less as their final piece of work. This was the only piece of work which was assessed. There were several reasons for the choice of video as medium:

- since the certificate was fully online the video medium was in keeping with the participant experience of the certificate
- as the certificate was non-credit-bearing we wanted it to provide an interesting experience beyond the participants' usual work or course of study
- it was felt that a short video would provide an innovative addition to participants' CVs
- it was considered that the end videos would provide an unusual and interesting addition to teaching material

From session two onwards participants were asked to source videos online as part of the session activities to introduce them to video as a medium of presentation in preparation for their end assessment. The videos were judged as pass or fail by school panels. Multidisciplinary and staff videos were judged by the project team. The videos were judged according to achievement of the following learning outcomes:

- demonstrate engagement with sustainability as a concept
- interpret the theme of Food for thought and contextualise this through the subject of their degree
- devise an interesting, innovative, creative or perceptive means of displaying this concept to a wide audience

All submitted videos were entered into a competition to win restaurant vouchers for a local restaurant with strong sustainability credentials. During NTU Global Week in March 2014 the videos were displayed on the NTU Student Union website and were voted for by students. According to the number of votes received a student winner was selected for each of the nine academic schools as well as an overall student winner and an overall staff winner. On the last day of NTU Global Week we held a Video Awards Night where we screened the 10 winning videos, presented school winners with some small tokens and presented the restaurant vouchers to the overall staff and student winners. The event also provided an opportunity to stimulate interest for further work in the area and to thank all the colleagues who had contributed to the certificate and present them with tokens of appreciation—selections of locally-produced products which had been donated by 'The Great British Food Group' a local support group for independent caterers and food/drink producers as initiated by our Guest of Honour for the evening local café owner Wendy Baird.

We received some anecdotal feedback that some potential participants were put off starting or completing the certificate as they were reticent about producing a video. We are exploring options for future rounds of the certificate e.g. providing some hands-on support for the making of the video or offering an alternative assessment medium.

4.7 Sustainability as Interdisciplinary Working

Inter-disciplinary working was an important and unique feature of the certificate designed to provide participants with new perspectives and opportunities which might not be open to them on their usual course. It was considered that this feature would support preparation for inter-disciplinary employment environments as well as supporting solutions for food sustainability challenges which require expertise from many different fields.

Some participants took the interdisciplinary experience to a further level and created and submitted a multidisciplinary video for example a student from Arts and Humanities teamed up with a Nottingham Business School student to co-produce a video on the theme of waste which included potential solutions based on their disciplinary expertise.

4.8 Students as Co-creators of the Curriculum

In the spirit of both the circular economy of a sustainable food life cycle we designed the certificate to have minimal waste and reusing/recycling wherever possible i.e.:

- the work which participants undertook throughout the certificate repeatedly returned to their course of study so that they could tie it in with other work on their course whether knowledge or skills.
- where course activity did not correspond directly to a students' course it brought them into contact with staff and students from other disciplines so that even where they were learning content that they may not come back to they were discussing it in the context of inter-disciplinary thinking.
- the next phase of the project will see participant work from the certificate being recycled into teaching materials. We have employed four interns all assigned to different academic schools to undertake this work which it is intended will include ready-made seminars, tutorial activities, case studies and other resources based on the student videos and discussion forum ideas.

As stated previously, staff could participate in the certificate and through their participation could gain a greater understanding of sustainability in order to then feed this into their curriculum. Given previous feedback from workshops and research as described earlier we understood that many staff are not interested in and/or do not prioritise sustainability within the curriculum. For those teaching staff who did not participate in the certificate, the student-as-co-creator model offers the opportunity for a role reversal where students can teach staff about sustainability following their participation in the certificate. Through the teaching materials to be developed from the certificate work students can directly or indirectly (depending on their involvement at this stage) support staff in embedding it into the curriculum. This model reduces the time and effort required of academic staff to do this themselves as well as giving students a unique opportunity to be involved in curriculum development. Through the teaching materials staff will have access to real-life examples of sustainability challenges and solutions from within and outside their disciplines to support the embedding of the graduate attribute of global citizenship within the curriculum as well as facilitating other graduate attributes.

The interns will also support individual academics and course teams to develop other new curriculum outside of the certificate through bespoke training, guidance for course development approval processes etc.

5 Sustain Yourself

From project conception it was planned to support the certificate with various activities relating to the informal curriculum i.e. to facilitate sustainability literacy through extra-curricular activities. One project known as 'Sustain Yourself' aimed to engage students in ESD through the themes of food and health with a series of optional cookery classes. It was intended that the sessions would support students to adopt healthy lifestyles, develop important life skills, socialise with other students, feel more ‘at home’ at the university and place their activities and consumption patterns in the context of wider local and global networks e.g. in terms of food supply and security and food miles. The project which built on an existing cookery course led by Beverley Lawe from the School of Education complemented existing initiatives such as work by NTU catering on sustainable food and student cookery books which were developed by NTU Student Support Services.

The cooking classes achieved some success on the city campus in terms of the ‘train the trainer’ model i.e. training students to train other students, though take-up of the classes was not as high as hoped. Future activity is planned, some of it using a different approach to encourage wider take-up. At the time of writing, preparation is underway for some ‘brain food’ events wherein students will be given ideas for nutritious food to eat during the exam period in summer 2014. We are also looking at the possibility of community-based cooking activities perhaps in tandem with the mobile Citizens Advice Bureau as suggested by NTU law students at the NTU Global Week world café; the world café was a further extra-curricular activity led by the project interns as part of NTU Global Week in March 2014 wherein attendees were invited to discuss ideas relating to food, sustainability and the future of the university and to write their comments on paper tablecloths as they moved around different tables which were hosting various discussion topics.

6 Conclusion

Undoubtedly the use of the over-arching theme of food within the project has been a key element of the success of the project in terms of encouraging good participation from across the institution from staff and students and in ensuring coherence between different aspects of the project—curriculum, co-curriculum, extra-curriculum, events etc. It is difficult to imagine that a more theory-based approach would have achieved the same level of success as this approach rooted in the critical challenge of food sustainability; an issue which relates to and implicates all of us both personally and professionally. The flexibility and perceived significance of the theme then has ensured participation and coherence. In the future we may look to include a different theme constituting another sustainability challenge either alongside or instead of food.

Whilst the use of the theme of food established the project as an action rather than theory-driven endeavour phase one of the project which we have described in this paper remained largely theory-driven with the exception of the cooking courses. The certificate itself facilitated sustainability literacy. It encouraged and in some cases will have resulted in action as it may have impacted on the perspective and behaviour of the participants. A fuller evaluation will illustrate the extent of this impact. The content and tasks of the certificate though remained largely theoretical and abstract. Participants researched or formulated solutions regarding food and sustainability and in some cases may have enacted these solutions but the model of the certificate in phase one did not formally integrate action with theory. During phase two we are looking to potentially integrate parts of the certificate with volunteering activity as well as disseminating the ideas of participants both as teaching material and to interested parties e.g. researchers, industry, charities etc. who may be able to enact the solutions with or without the input of participants.

The interdisciplinary nature of the activity content and the fact that the certificate was open to all staff and students provided an unusual opportunity for participants to interact with students and staff from other disciplines in an academic setting and to learn from the starting point of a problem rather than from a discipline perspective i.e. the starting point for the certificate was 'how do we feed the world in a sustainable way?' and participants, through the various activities and assessment piece, were invited to interrogate this question from personal, disciplinary and interdisciplinary perspectives some of which changed over the duration of the certificate as more levels of complexity were added.

Anecdotal feedback from participants who completed the certificate suggests that the curriculum design model described above—personal, disciplinary and interdisciplinary—was appreciated by participants who found that this added depth to their knowledge and experience. We have not yet gained feedback from participants who did not complete however and this is a topic we would like to know more about. There was a large drop-off between session one which focussed on personal perspectives on sustainability and session two which introduced disciplinary perspectives of sustainability. There are many possible reasons for this for example, despite our encouragement to participants to stay with the certificate; that they would begin to see the relevance of the certificate to their discipline as the certificate unfolded some may have found session one which focussed on the sustainability of a chicken stir fry to be not academic enough to hold their interest. Alternatively participants may have enjoyed session one but have found session two 'too academic' to hold their interest for an optional non-credit bearing certificate. Once we have conducted a more thorough evaluation we may then amend the design for the next phase of the certificate.

The certificate was based completely online; overall the online format provided an interesting and flexible format with a lot of potential for active learning styles such as problem-based learning. A great deal of technical expertise was needed however to build and maintain the learning room. In addition a certain pedagogical approach different to in-person learning was needed to ensure effective engagement and learning. We utilised various means described in more detail in earlier sections to encourage engagement, a feeling of academic community and a 'personal touch' within the online setting such as discussion forums, videos, a support forum, twitter feed etc. For most students however, online learning particularly a fully online course constitutes a new way of learning and in the next phase of the certificate we may look to provide regular 'live' tutorial sessions based online within the learning room and also some possible face-to-face sessions both of which may support further engagement and a greater sense of academic community. In addition we are looking to set up a Facebook page for participants so that they can continue to network and explore collaborative research and practice opportunities with other certificate participants.

The inclusion of the innovative assessment method of video stimulated debate with the project team in terms of the advantages and disadvantages of the method. Overall it was felt that such a method was appropriate as it was in keeping with the MOOC-style online format, encouraged parity of submission and assessment process particularly with the interdisciplinary nature of the certificate and would provide the participants with a unique addition to their CV. Anecdotal feedback suggests that the video format may have discouraged some participants from completing the certificate. With the understanding that this constitutes a new method of assessment for most if not all participants we are looking to introduce more support in addition to the current learning room guidance for future phases of the certificate e.g. online or in-person workshops on video production. We will also provide videos from previous participants as examples and may offer an alternative assessment method.

The student-as-co-creator model adopted within the certificate was a unique aspect of the curriculum design and one which will ensure that the certificate has a wider impact beyond the immediate participants and we look forward to the work of certificate participants being fed forward into future curriculum. The challenges we will encounter with this aspect of the project will likely include the transference of knowledge between disciplines in ways which are relevant to different subject areas and in keeping with the subjects being taught on the various courses. For each video or discussion forum idea which will be developed into teaching material, careful consideration will need to be given as to which course/module learning outcomes the video can support in terms of knowledge and skills and how to utilise the video most effectively to support these learning outcomes.

One unanticipated outcome of the certificate was that a few academic colleagues have asked if they can use material from the certificate for their future research and it might be that we can strengthen the links between the certificate and research in the future as this was not a feature by design in phase one of the project. A second emergent piece of work is that the cooking classes will continue, some in different formats and perhaps involving some community-based work again which constitutes a new direction for the project.

As alluded to, a formal evaluation will follow with staff and student participants and contributors and this will no doubt provide further ideas of how to improve and expand the project for future phases. It is considered that aspects of the project might be useful for adaptation and adoption at other institutions and we would welcome both discussion and collaboration with other institutions on this project.

Acknowledgments The authors would like to thank everyone who has contributed to the Sustainability in Practice Certificate in particular Professor Chris Pole, Grant Anderson, Laura Scott, Katie Wright, Charley Greening, James Lindsay, Seraphina Brown, Kelly Osborne, Amy Scoins, Kayleigh Smith, Trevor Welsh, Graham Thomas and the eLearning team from the NTU Centre for Academic Development and Quality and all the academics across the institution who researched and provided web links, supported the development of content and promoted the certificate to their students. We would also like to thank everyone involved in promoting and running the cooking classes in particular Beverley Lawe, Christine Walker, Fiona Dick and the Welcome Week reps.

References

Barnett R (2000) Realizing the university in an age of supercomplexity. Open University Press, Buckingham

Cortese AD (2003) The critical role of higher education in creating a sustainable future. Plan High Educ 31(3):15–22

Djordjevic A, Cotton DRE (2011) Communicating the sustainability message in higher education institutions. Int J Sustain High Educ 12(4):381–394

Drayson R et al. (2012) Student attitudes towards and skills for sustainable development. NUS/HEA. Available at http://www.heacademy.ac.uk/assets/documents/esd/Student_attitudes_towards_and_skills_for_sustainable_development.pdf. Accessed 8 Apr 2014

Foresight (2011) The future of food and farming: final project report. The Government Office for Science, London. Available at http://www.bis.gov.uk/assets/foresight/docs/food-and-farming/11-546-future-of-food-and-farming-report.pdf. Accessed 24 July 2013

Green League (2013) Full people and planet green league 2013 tables. Available at http://peopleandplanet.org/green-league-2013/tables. Accessed 8 Apr 2014

Lacy P et al (2010) A new era of sustainability. U.N. Global Compact-Accenture CEO Study 2010. Available at http://www.accenture.com/SiteCollectionDocuments/PDF/Accenture_A_New_Era_of_Sustainability_CEO_Study.pdf. Accessed 24 July 2013

Lozano R et al (2013) Declarations for sustainability in higher education: becoming better leaders, through addressing the university system. J Clean Prod 48:10–19

McNamara KH (2010) Fostering sustainability in higher education: a mixed-methods study of transformative leadership and change strategies. Environ Pract 12(01):48–58

Novak J (1998) Learning, creating and using knowledge: concept maps as tools to understand and facilitate the process in schools and corporations. Lawrence Erlbaum Associates, New Jersey

Orr D (1994) Earth in mind: on education, environment, and the human prospect. Island Press, Washington, D.C

Scott G et al (2012) Turnaround leadership for sustainability in higher education. Available at http://www.iau-hesd.net/sites/default/files/documents/le11_1978_scott_report_2012.pdf. Accessed 14 Jan 2014

Sky (2011) The sustainable generation: the sky future leaders study, Dec 2011. Available at http://corporate.sky.com/documents/pdf/publications/2011/sky_future_leaders_study. Accessed 8 Apr 2014

Stibbe A (2009) The handbook of sustainability literacy: skills for a changing world. Green Books, Totnes, UK

Stubbs W, Schapper J (2011) Two approaches to curriculum development for educating for sustainability and CSR. Int J Sustain High Educ 12(3):259–268

Winter J, Cotton D (2012) Making the hidden curriculum visible: sustainability literacy in higher education. Environ Educ Res 18(6):783–796

Authors Biography

Helen Puntha is a Research Officer in the Centre for Academic Development and Quality at Nottingham Trent University. Her current work is in the areas of research informed teaching, assessment and feedback and sustainability in the curriculum. She is Deputy Lead of the University's Food for Thought Green Academy (an HEA change programme) and is a Steering Group member for the British Conference of Undergraduate Research (BCUR).

Dr. Petra Molthan-Hill is Principal Lecturer in Business Sustainability at Nottingham Business School and as NBS Sustainability Coordinator she is responsible for embedding sustainability and business ethics into the curriculum of the business school. Currently she is leading the HEA Green Academy 'Food for Thought' Project for Nottingham Trent University developing teaching methods on how to embed sustainability into various disciplines from Business Studies through Medieval History to Astrophysics. Her main research interest has been experiential learning methods and the integration of sustainability into the curriculum but also the behaviour and sense-making of managers in cross-cultural comparisons.

Aldilla Dharmasasmita has been teaching Undergraduates and Postgraduates in various modules, where sustainability has been fully embedded. She is also supervising students in Greenhouse Gas Management Consultancy Projects. Her research interests are applying Institutional Theories for comparative studies in Corporate Social Responsibility (CSR) and Sustainability. Currently, she also plays a key role in designing and embedding innovative online pedagogies for the University's curriculum across all faculties (as part of an HEA initiative), the Sustainability in Practice Certificate.

Professor Eunice Simmons is the Dean of the School of Animals, Rural, and Environmental Science (ARES) at NTU. She has qualifications in Biology, Soil Science and Forestry plus a PGCE and two years' teaching experience in a rural comprehensive. Posts in five very different Universities have given Eunice expertise in maximising educational opportunities for students on every level and type of course.

Sustainability and Employability: Alliances at the University of Bedfordshire

Diana Pritchard and Mark Atlay

Abstract

Reviews from across the globe note that following the UNESCO-decade of Education for Sustainable Development, advances have been limited within higher education. Emergent research seeking to understand and overcome prevailing barriers, includes attention to the constraints within institutions related to crowded curricula and competing higher education agendas. Since 'graduate employability' is a key policy concern of many universities in the UK, understanding how it links to, or impacts on, the promotion of education for sustainability is crucial. This paper offers insights from experiences developing at the University of Bedfordshire which has a 'widening participation' mission to provide for students from families which are under-represented in higher education. In these circumstances, the relative priority which university managers, academics and students assign to the delivery of commitments to education for sustainability and to enhanced graduate employability are thrown into relief. Conducting preliminary stakeholder analysis of the policies, meetings, workshops and group discussions, we examine the different perspectives and approaches of managers, academics and students regarding education for sustainability and its connections with employability; comparing the scenarios in 2013 and 2014 which represent pre- and post-coordination of efforts. Our findings document the institutional drivers, and the initiatives that these distinct stakeholder groups adopt, which effect shifts that now more closely

D. Pritchard (✉) · M. Atlay
Centre for Learning Excellence, University of Bedfordshire, Park Square, Luton, LU1 3JU, UK
e-mail: diana.pritchard@beds.ac.uk

M. Atlay
e-mail: mark.atlay@beds.ac.uk

W. Leal Filho et al. (eds.), *Integrating Sustainability Thinking in Science and Engineering Curricula*, World Sustainability Series, DOI 10.1007/978-3-319-09474-8_4

link education for sustainability to the employability agenda. Key factors were support for academic-led proposals, student participation in co-learning activities and inter-stakeholder collaborations. This preliminary study contributes to sector wide discussions about approaches to the implementation of education for sustainability.

Keywords
Employability · Widening participation · Stakeholder analysis · University of Bedfordshire

1 Introduction

Consistent with international frameworks stimulated by the 2005–2014 UNESCO Decade of Education for Sustainable Development, higher education authorities in the UK provide evolving types of support for institutions to enhance "graduates' capabilities to contribute to sustainable and just societies" (HEA 2005). Education for sustainability has become part of mainstream higher education policy in the UK (HEFCE 2013) although, as in other countries around the globe, advances have been limited. Whereas education for sustainability is widely understood to encompass various aspects (including knowledge of environmental, economic and other social issues of sustainability; nurturing ethos and values; and the adoption of participatory teaching and learning methods), in the UK progress has been largely confined to curriculum content changes (Cotton and Winter 2010; Tilbury 2011) relating to cognition and knowledge. While these elements may indeed deliver education *about* sustainability (Sterling 2004) they have failed to deliver on other ideas, understandings, values, behaviours and skills which are widely recognised to be important to prepare graduates to respond to future challenges and to be effective in the face of both predicted and unknown changes (Orr 1994). This means that more profound learning experiences are required if universities are to deliver deeper and transformative types of education *for* sustainability that can equip graduates to make "informed decisions in their home, community and working lives" (Fien and Tilbury 2002), with capacities required as much to respond to changing worlds but also to become agents of change (Sterling 2004). In this context, attention within higher education and research has turned to understand and overcome barriers, including the constraints imposed by perceptions of crowded curricula and competing agendas (Tilbury 2011).

Since 2006, graduate employability has been a central agenda in higher education institutions in the UK (Pedagogy for Employability Group 2006). To generate interest and find time within a packed curriculum, education for sustainability has to build on institutional priorities such as employability and the research and professional interests of staff that provide the drivers for their work. UK universities have traditionally considered graduate employment as an aspect of their relationship with the labour market and wider economy, enjoying a significant degree of independence in defining what this association is. More recently, employability is viewed as a primary definition of their function (Boden and Nedeva 2010). In the context of high tuition fees, student expectations that their degree and university experience will deliver enhanced employability has increased, and a plethora of publications, guidelines and case studies has emerged to illuminate different institutional responses and promote perceived best practices (Pegg et al. 2012). Data on the employability of graduates, publically available in the Destination of Leavers of Higher Education survey reports, comprise a resource which potential students and their parents can access to inform decisions and compare across institutions via the statistical data (comprised in the Key Information Set). University branding, especially for those not nationally acknowledged as primarily research institutions, often includes statements about how the institution delivers on this criteria (Thomas and Jones 2007).

This is the case at the University of Bedfordshire (UoB) which is amongst many in the UK dedicated to 'widening participation'. This means it provides access for students who come from families which have no, or limited prior tradition in higher education, with economically marginal backgrounds. That it is amongst the top five in the country for financial support provision to its students (Office of Fair Access 2012) attests to the high proportion of widening participation students amongst its student population. Its 24,000 students include an ethnic diversity of both local and overseas students, representing over 100 countries, while 34 % students are over 25 years old (University of Bedfordshire 2014) many of whom work at least part time and already have rich experiences on which to draw. Coming to study with limited experience of professional workplaces and appreciation of the relevance of the skills they must acquire for employment or alternative career paths, they invest resources in a degree with the expectation that it will deliver social mobility through enhanced employment prospects. Responding to these student demographics, the strategic priorities of the university emphasise enhancing the student experience and employability. Potential and current students at UoB are assured through its marketing that graduate employability is the institution's "highest concern" and they are encouraged to "influence and shape their education" (University of Bedfordshire 2014). To these ends, the University had developed, as a result of prior work of its Centre for Excellence in Teaching and Learning, a curriculum framework called CRe8 with five interwoven strands that identified effective practice and institutional expectations in the areas of: learner development, curriculum enhancement, teaching practices, employability, and assessment (Atlay et al. 2008). The University's Learner Experience Strategy (2013–2018), amongst its clauses, establishes the pedagogic need to "apply creative approaches" for curriculum enhancement and teaching practices to support its "diverse learner populations", to develop employability and other skills and a learning and teaching culture where co-creation of knowledge amongst students and academics is valued. From another perspective, an internal task force, the Employability Action Group (EAG), headed by an executive dean, developed an "employability framework" as a tool to promote and audit departmental offerings regarding a series of activities in the domains of practitioner and transferable skills, professional and employer engagement and personal development, with a strong emphasis on the importance of subject-related work experience. As perhaps for other universities concerned with widening participation, the employability agenda is prioritised and focuses on graduate skills to provide for employer needs. A further aspect, in light of the University's approach to involve students and their representatives in the decision-making processes, representatives of the student union (BedsSU) are on the EAG, amongst other major committees.

In this context, how are sustainability commitments connected to the employability agenda at UoB? This paper sets out to examine the linkages and who and how they are developed. Understanding from the outset that universities are complex institutions, we identify the respective perspectives, tools and processes adopted by composite stakeholder groups, namely management, academics and students.

2 Data Collection and Analysis

This paper draws on qualitative content obtained from a variety of University documents, including strategic and policy statements and minutes from meetings. It also uses information gathered during participation of the authors in a series of events which have involved a total 72 people from the University, including 4 university managers, 47 academics (including 1 dean and 6 professors), and 21 students. Amongst the events were two full day workshops, two half day workshops, three student focus groups, and ten departmental meetings taking place between June 2013 and March 2014. These activities provide information that allows a comparison of the scenarios prevailing at these moments and—reflecting a preliminary stakeholder analysis—documents the particular tools and approaches that different groups adopt to engage with education for sustainability. As with other organisations (Freeman 2010), universities are complex institutions comprising distinct communities, or groups, which have both shared and distinct interests and perspectives. Understandings these are key to understand how change takes place. In turn, and consistent with an action research approach, summaries of findings as they emerged were subsequently circulated, presented and discussed in meetings with a view to enrich discussions and build consensus about proposed future measures to embed sustainability within the institution. The following provides a characterisation of these which are also represented in Table 1.

(1) The scenario pre-coordination

The University's Strategic Plan (2012–2017) included "education for sustainability" and "its integration into the curricula and professional development". This formed part of its broader commitment (for which it has received national recognition in the 'Green League') to deliver on sustainability goals through carbon emission reduction. While this comprised a vital institutional commitment, it could

Table 1 Stakeholder positions, instruments and outcomes

	Before 2012	Change processes 2013–2014	Outcomes and plans for 2014
Management	Strategic commitments to education for sustainability but with limited synergy, coordination and shared actions	Coordination with interested individuals to enhance the University's strategic approach	Strategy revised to emphasise synergies with future skills that address both education for sustainability and employability
Academics	Reflecting own academic research and other interests, isolated efforts incorporating sustainability–related material in existing units	Organised a sustainability forum establishing principles of horizontal and co-learning and multidisciplinarity; proposing ways forward to continue this bottom-up initiative and to embed sustainability in teaching and learning and curriculum development	Validation of importance of respect for academic autonomy and professionalism; efforts from other stakeholders to take its proposals forward; identification of potential transdisciplinary collaboration and courses
Students	BedsSU focus on own agenda encouraging green and social behaviours and responsibilities	Student collaboration with academics and CLE through the sustainability forum and student focus groups	Support for new teaching collaborations and assessments with students for co-learners and production of materials
Departments	No courses dedicated to sustainability Limited number of units dealing explicitly with sustainability No mechanism to support inter-departmental working and course development	Sustainability forum activity helps clarify thinking around course and unit content and teaching activities Ideas for new courses emerge through dialogue amongst interested academics	New units and courses being developed based around multidisciplinarity. All courses to consider education for sustainability issues

(continued)

Table 1 (continued)

	Before 2012	Change processes 2013–2014	Outcomes and plans for 2014
CLE	No prior engagement with sustainability agenda	Reconceptualisation of the curriculum framework to embrace education for sustainability Finding connections to EAG focus on employability Identifying ways to recognise the sustainability forum	Amended framework to provide the basis for institution-wide activities which link employability and sustainability Use of 'world of work' and 'work of the world' distinctions to build alliances and in staff induction Recognition of sustainability forum as integral to professional development Support pilot initiatives for academic-student collaboration to develop teaching materials that emphasis future skills for both sustainability and employability

not, of itself, implement teaching and learning or curriculum changes or initiatives. Indeed in a context where senior management expressed concerns that *any* course—even those in subject areas dealing with core knowledge aspects about the material, ecosystem or carbon reduction aspects of sustainability—"would not appeal to our potential students" there was reticence about curriculum offerings which explicitly announce delivery of education either *about* or *for* sustainability. This position can best be understood perhaps in relation to the close attention being paid to institutional recruitment targets, the development of courses specifically related to sustainability was seen to be high risk. Moreover, at this point, within the university's strategic and policy domains no connections were made linking the sustainability and employability agendas.

So across the University in 2013 there were no courses dedicated directly to any aspects of sustainability, and although all but one department recognised the significance of incorporating sustainability knowledge and understandings for future curriculum offerings, only a handful of units existed (including in business studies, life sciences, tourism, and construction) which explicitly delivered core knowledge aspects.

Within other departments, individual academics—as in universities elsewhere—engaged to various extents with the sustainability agenda, motivated by social, moral and professional concerns and responsibilities relating to the need to change the status quo, prepare students for the challenges ahead and to keep their subject relevant in light of the fact that "this is the single most important matter that we face". Typically they worked in isolation. The proposals of at least two academics in two departments (English and Sociology) to establish a course or even a unit dealing with aspects of climate and environmental changes were rejected by respective heads of department because it was believed there was an inadequate student demand. Nonetheless these academics, and indeed others interested to address core knowledge elements and some of the values and attributes they considered to be relevant to sustainability, incorporated relevant materials and stimulated critical thinking and reflections on climate and environmental change in their existing units because "students need to understand and engage with these issues". Notably, in one department, a small group of academics had already planned to work together to agree the social and environmental values that they considered were important to prepare students with sustainability skills for business management with a view to map these across their course offerings during a forthcoming course review processes.

For its part, the BedsSU had conducted activities that promoted behaviours which were compatible with environmental and social responsibilities, specifically involving ethical sourcing and fair-trade campaigning. In line with its organisation of a 'Go Green Week' on campus, it was preparing what turned out to be a successful submission to the National Union of Students (NUS) for funding which had an overt "green" focus, aiming to "promote pro-environmental behaviours amongst students, staff and the wider community" and on contributions to "green the curriculum" (BedsSU 2013). This manifested a prevailing concern reflected in the National Union of Students for behaviours and learning relating to the natural environment (Drayson et al. 2013).

Thus the position in 2013 was characterised by a desire on behalf of the University to address sustainability issues within and beyond the curriculum but where there was no clear definition of what was meant within the University by education for sustainability and no mechanism by which this could be accomplished.

(2) Growing sustainability from the grass roots

In June 2013, an academic based within the humanities took the initiative to organise the first university-wide workshop on sustainability (henceforth referred to as the sustainability forum). This brought together academics from all the faculties who were sufficiently motivated by research or other interests to dedicate a full day to identifying commonalities and synergies between them and their disciplines for research and teaching approaches on this topic. This single event, significant for being generated as an organic "bottom-up" initiative, effectively set in motion a series of other activities conducted by other stakeholders that have contributed to advance sustainability with teaching and learning and have made connections with the prevailing employability agenda through an array of instruments.

Key principles and understanding were established and developed at the sustainability forum which took place at workshops in June 2013 and January 2014, and, in the case of the latter involved home and overseas undergraduate and postgraduate students. Participants agreed that no discipline or academic level held privileged knowledge, and that everyone had relevant insights to offer from their respective backgrounds and experiences. This established multi-disciplinarity, "horizontal learning" (Freire 2005) and co-learning and collaboration as the fertile basis from which to explore the skills, critical facilities and modes of awareness that can equip students to respond and adapt to our changing world scenarios, make the curricula relevant for future challenges and propose ways to take the agenda forward in the institution. Given this, the sustainability forum had, effectively, defined itself as a non-hierarchical space of "democratic… processes of change"; a characteristic has been observed elsewhere in communities committed to education *for* sustainability (Tilbury 2011).

The sustainability forum also identified that beyond discipline specific core knowledge there were also generic elements. This included recognition that teaching needs to make explicit ethics and values, nurture critical and creative thinking, stimulate self-reflection and understanding of the changes and action taking place at local and global levels, the implications for resource distribution and justice and of diversity of cultures and knowledge systems. These have been more widely recognised as integral components of education for sustainability (Stirling 2004; Tilbury and Wortman 2004). As such, the forum had generated ideas which mirrored the intentions in the University's Learner Experience Strategy (2013/2014), which includes amongst its clauses, the need to "apply creative approaches" for curriculum enhancement and teaching practices to support its "diverse learner populations". As significantly, through collective reflection the forum also identified the overlaps with attributes that participants understood would also serve to prepare graduates for their working lives, such as imagination and creativity, adaptability and flexibility, collaboration and team work, and the assumption of individual responsibility: in effective capturing all the priority attributes that employers expect graduates to exhibit (Thomas and Jones 2007; Pedagogy for Employability 2006; Institute of Directors 2007). The forum also identified the challenges participants needed to address to stimulate futures thinking and of embracing uncertainty and complexity, given the importance of taking long term perspectives of inevitable and unpredictable changes in environmental and social systems but also of labour markets and economies.

Practices and ideas about appropriate teaching and learning approaches and materials were also shared. This included the adoption of active lessons which "get us out there"; the students emphasising their interest to learn "in the practice", beyond the confines of the classroom. In effect, the education for sustainability agenda this forum was forging reiterated the University's central institutional emphasis for enhancing graduate employability through work experience. Encouraged by this process, two further shorter workshops were organised by academics from the Business School to exchange experiences of teaching activities relating to work on-campus, in collaboration with the Estates department, and with community businesses and organisations to implement carbon emission reduction measures.

The forum also generated ideas about moving the education for sustainability agenda forward. It established that, in light of student demographics and the University mission, it should be linked with employability. In addition, it included proposals to develop new multi-disciplinary course offerings, professional recognition of the sustainability forum and ways to recognise student participation to motivate their attendance.

As significantly, the forum provided an important space for scholars. They expressed both excitement and fulfilment at participating in it because it provided a channel through which they were able to assert their academic autonomy and professional identity which scholars typically perceive as having been undermined over recent decades in UK universities (Henkel 2005; Slaughter and Leslie 1997).

From the inception, the University's professional teaching and learning services, the Centre for Learning Excellence (CLE) provided the sustainability forum with organisational, logistical and financial support. It also attended all its events, valuing the potential to deliver innovative teaching and learning approaches and material where initial initiatives had stalled (Atlay et al. 2008). Given its remit for teaching and learning, curriculum and staff development, the CLE began a series of meetings to take forward the proposals generated from the forum and build a common understanding of the issues. For this it adopted the "world of work" and the "work of the world" (Porritt 1992) as a conceptual distinction around which to structure discussions about common graduate and pedagogical approaches. Recognising the overlaps which deliver on sustainability and employability it began to make links with, and modify, existing pedagogies and frameworks to assimilate both dimensions. It also set out to support the forum by taking its recommendations forward.

In the meantime, the BedsSU had been awarded funds from the NUS for its initiative, the 'Bedfordshire Green Hub', enabling student officials to provide conceptual inputs and participation in the sustainability forum and assist academics and the CLE organise a series of student focus groups. These aimed to generate qualitative information about student understandings of, and interest in, sustainability, and ways to engage students via appropriate teaching and learning approaches. Students typically expressed wishes for more activities to develop their understanding and skills through "practical ways of understanding what sustainability is about". They provided feedback validating the use of poems and images and the production of timelines (where they plotted on the wall those events and changes they understood as relevant to global changes) as activities that can effectively stimulate reflection on geographical and times scales, the relationships between causes and effects of phenomena and the value of group work. These processes have produced a series of realised and ongoing outcomes.

(3) Nurturing sustainability as the future agenda

Senior management, in coordination with interested individuals, is reviewing its strategies and policies, as well as defining which actors should be involved in further shaping them. These documents will emphasise the linkages between education for sustainability and employability. Responding to the higher profile interest of academics across subject areas, it has invited departments to develop new courses offerings which, in the first instance, build on existing expertise in disciplines that deal with material themes *about* sustainability where an existing employer demand can be demonstrated for graduates who can provide "for the sustainability challenges of organisations" (University of Gloucestershire 2011).

With a view to embedding deeper education for sustainability provision, the Centre for Learning Excellence is linking sustainability and employability throughout its activities and frameworks, including into a revised and refreshed version of the curriculum framework (CRe8) and staff induction. It has recognised the sustainability forum as a 'special interest group' acknowledging its value as an ongoing space for continuing professional development purposes through the University's Professional Teaching Scheme. It is also piloting four associateships enabling students to work with academics to co-produce curricula material that documents and reflects on the acquisition of workplace relevant skills and understandings as relevant to discipline specific themes relating to sustainability.

3 Findings: Preliminary Stakeholder Analysis Reveals Potential Alliances

A number of issues emerge from the work at UoB on employability and sustainability relating to the distinct perspectives, instruments, processes and outcomes that stakeholder groups have developed to engage with these issues. Yet, the institutional approach to each is different and whereas employability is being centrally driven to deliver the University's 'employability commitment' to its students, the momentum and approach to education for sustainability has been defined more organically, through engagement by interested staff and students. Where these two agendas meet, interesting developments are beginning to emerge that have the potential to significantly enhance each area.

The sustainability forum has generated a vital impetus to advance the education for sustainability agenda at this institution. It has contributed to broadening understandings beyond the participant members, and developed incipient institutional approaches and local teaching practices that both advance education for (deeper) sustainability and contribute to employability in accordance with the institution's teaching and learning approach. Early indications would suggest that the forum represents a powerful "force for change" (Tilbury 2011) within the institution.

Employability is more than gaining employment. It is about having the skills necessary to maintain your career through multiple changes and of, potentially, managing a portfolio of contemporaneous roles (HEA 2012). Thus there is a dimension of employability that is about the 'sustainable self'. Working with students to understand this is an important part of the CRe8 curriculum (Atlay et al. 2008) and surfacing the issues in this way may help to further engage students in their responsibilities for their own learning and development.

Some specific teaching and learning approaches have emerged as important to connect links between sustainability and employability particularly the "world of work" and the "work of the world" dichotomy. This serves to build and deepen awareness that distinct sets of knowledge and understanding, skills and qualities, values and behaviours and capacities are required both to prepare graduates for an unknown and changing future world as well as preparing them to be able to help shape the world of the future. The piloting of co-produced curricula material developed in response to sustainability in real-life challenges looks set to deliver an approach and associated teaching and learning materials which take into account the expectations and priorities of key perspectives, namely those of students as well as academics and employers. This has been demonstrated elsewhere as being effective (University of Gloucestershire 2012). Interestingly, while debates have identified the significance of collaboration and teamwork as central skills to deliver on both agendas, there remains a vacuum about how to develop and assess such skills effectively and efficiently.

Additional issues are beginning to emerge from this work. There is a significant opportunity for students and academics to be involved in the co-generation of topics and issues around the education for sustainability agenda. Such activities change the relationships between the student, tutor and curriculum, requiring adaptation from all parties. Whilst education for sustainability remains a special interest of a few academics engaged in the forum it is unlikely to have the significant curriculum change desired. The development of the necessary attributes, skills and outlooks requires significant course team cohesion to see sustainability as a whole curriculum responsibility rather than one that is left to the interests of a few—a requirement it shares with the employability agenda.

The extent to which Employability and Sustainability issues are subject specific or are a generic outcome of the educational process is also emerging as an issue. As staff start working in an interdisciplinary manner, subject differences begin to matter less but the different perspectives that each brings can help to deepen understandings of the issues concerned, broadening the horizons of both students and academics.

4 Conclusions: Productive Synergies and Alliances

It is now well established that education for sustainability requires a major shift in the way students are taught and learn, requiring a more flexible approach combined with practical and applied learning opportunities. These not only complement other drivers in pedagogy in this direction, but, as this study has indicated and is validated elsewhere, much of this is "in line with what graduates will need in an increasingly complex work environment" (Cade 2008; HEA 2005) to meet the employability agenda.

In demonstrating the impact of the sustainability forum, this paper has highlighted the potential of groups of self-motivated academics to advance the sustainability and employability agendas in their institutions. Facilitating spaces where educators and researchers can explore these issues effectively empowers them to contribute to transformational education. As effective agents of social change, as was shown elsewhere, their respective institutions must support them to do so (UN and Economic Commission for Europe Strategy for Sustainable Development 2012: 11).

Yet, the issue of who, if anyone within the University (educational development, estates and facilities, student unions, etc.) 'owns' or is best suited to develop the sustainability agenda has been raised by this paper. Here there are further parallels with the employability agenda—how employability is framed depends on where it is owned (e.g. Careers, Educational Development, Registry). Our work shows the importance and effectiveness of collaborative working, defining and sharing agendas together which comprise valuable lessons about how universities organise themselves to help facilitate change. It remains to be seen whether such an approach can extend beyond the interested few to effect systematic and institution-wide change.

References

Atlay M, Gaitan A, Kumar A (2008) Stimulating learning—creating CRe8. In: Nygaard C, Holtham C (eds) Understanding learning-centred higher education (chapter 13). Copenhagen Business School Press, Copenhagen

BedsSU (2013) The Bedfordshire green hub—bringing students, community and society together. Project document. Available from BedsSU

Boden R, Nedeva M (2010) Employing discourse: universities and graduate 'employability'. J Educ Policy 25(1):37–54

Cade A (2008) Employable graduates for responsible employers. Research on the links between sustainability and employability in the graduate job market in relation to higher education teaching and learning. Report to the Higher Education Academy. http://www.heacademy.ac.uk/assets/documents/sustainability/EmployableGraduates2008.pdf. Last accessed 28 Feb 2014

Cotton D, Winter J (2010) 'Its not just bits of paper and light bulbs': a review of sustainability pedagogies and their potential use in higher education. In: Jones P, Selby D, Sterling S (eds) Sustainable education: perspectives and practice across higher education. Earthscan, London

Drayson R, Bone E, Agombar J, Kemp S (2013) Student attitudes towards and skills for sustainable development. National Union of Students/Higher Education Academy, London

Fien J, Tilbury D (2002) The global challenge of sustainability. In: Tilbury D, Stevenson R, Fien R, Schreuder D (eds) Education and sustainability: responding to the global challenge., Commission on Education and Communication. IUCN, Gland, Switzerland

Freeman R (2010) Strategic management: a stakeholder approach. Cambridge University Press, Cambridge, UK

Freire P (2005) Pedagogy of the oppressed. Continuum, New York

Henkel M (2005) Academic identity and autonomy in a changing policy environment. High Educ 49:155–176

HEA (2005) Sustainable development in higher education: current practice and future development. Higher Education Academy, York, UK

HEA (2012) Pedagogy for employability http://www.heacademy.ac.uk/assets/documents/employability/pedagogy_for_employability_update_2012.pdf. Accessed 20 Apr 2014

HEFCE (2013) Sustainable development in higher education: consultation on a framework for HEFCE. Higher Education Funding Council for England, England

Institute of Directors (2007) Graduate employability skills. Institute of Directors, London

Office of Fair Access (2012) Access agreements 2012–2013: final data including initial teaching training. OFFA. Available on http://www.offa.org.uk/press-releases/offa-announces-decisions-on-2013-14-access-agreements/. Accessed 30 Mar 2014
Orr D (1994) Earth in mind: on education, environment, and the human prospect. Island Press, Washington, DC
Pedagogy for Employability Group (2006) Pedagogy for employability learning and employability series one. ESECT and HEA, York
Pegg A, Waldock J, Hendy-Isaac S, Lawton R (2012) Pedagogy for employability. Higher Education Academy, York, UK
Porritt J (1992) Capitalism as if the world matters. Earthscan, Abingdon, UK
Slaughter S, Leslie L (1997) Expanding and elaborating the concept of academic capitalism. Organization 8(2):154–161
Stirling S (2004) Higher education, sustainability and the role of systemic learning. In: Corcoran P, Wals A (eds) Higher education and the challenge of sustainability: contestation, critique, practice and promise. Kluwer Academic, Dordrecht
Thomas L, Jones R (2007) Embedding employability in the context of widening participation, vol 2., Learning and employability. Higher Education Academy, York, UK
Tilbury D (2011) A global overview of commitment and progress. http://insight.glos.ac.uk/sustainability/Education/Documents/. Accessed 1 Apr 2014
Tilbury D, Wortman D (2004) Engaging people in sustainability., Commission on education and communication. IUCN, Gland, Switzerland
UN, Economic Commission for Europe Strategy for Sustainable Development (2012) Learning for the future: competences in education for sustainable development, Geneva, Switzerland
University of Bedfordshire (2014) http://www.beds.ac.uk/aboutus/facts
University of Gloucestershire (2011) Education for sustainability: a guide for educators on teaching and learning approaches. http://insight.glos.ac.uk/sustainability/education/documents/efs%20educators%20guide%20final%207july11.pdf. Last accessed on 28 Feb 2014
University of Gloucestershire (2012) Graduate employability and sustainability: policy trends and student perspectives. Project briefing. University of Gloucestershire Graduate Challenge Project, February

Authors Biography

Diana Pritchard has worked at the University of Bedfordshire in curriculum enhancement since 2013, focusing on employability and sustainability. Since 2009 she has also been a research associate at the University of Sussex. Her thirty year career has involved work in social development and environmental research, teaching and training, and in policy and project management, conducted primarily in the UK, USA, Spain, Mexico and Nicaragua. Between 1997 and 2003 she was based in Central America, collaborating with rural communities, local and national authorities, academics and international agencies in biodiversity conservation activities. Since 2000 she conducted various consultancies including with the Global Environmental Fund, United Nations Development Programme, the UN Economic Commission for Latin America, Centro de Investigacion para el Mejoramiento de Maiz, various international NGOs and the British Academy (UK). She publishes on international conservation policy and local community conservation and now turns research attention to the field of education for sustainability.

Mark Atlay is Director of Teaching and Learning. He spent seven years lecturing in chemistry at the University of Glamorgan before moving on to work on the development of distance learning materials at the Open University. At the University of Bedfordshire he has worked in a number of areas including Quality Assurance, Staff Development, Quality Enhancement and Educational Development. His interests are in effective curriculum design and the development and recording of students skills and attributes. His current role involves providing overall direction for the University of Bedfordshire's curriculum and supporting its implementation. Mark was Director of the University's CETL (see www.beds.ac.uk/bridgescetl) has led a HEFCE Good Management Practice project (GMP201) on effecting change in Higher Education, was a chair of Southern England Consortium and is a trustee of the UK's Centre for Recording Achievement. Since 1999 he has worked for the Quality Assurance Agency in a range of areas including institutional review, audit, IQER and Access.

The Environmental Sustainability of Brazilian Universities: Barriers and Pre-conditions

Luciana Londero Brandli, Walter Leal Filho,
Marcos Antonio Leite Frandoloso, Eduardo Pavan Korf
and Denise Daris

Abstract

The aim of this paper is to discuss the application of sustainability in Brazilian universities, emphasizing the preconditions for implementation and the barriers and difficulties for its application. The methodology approach is based in a theoretical background and on the gathering of empirical evidence. Due to the scope of the study and the nature of the methodological approach, the survey, by no means, can be regarded as representative. However, since it is combined with the review of the literature, it builds a rough profile of the status of the sustainability in Brazilian universities. Larger studies can be undertaken at a later stage, provided that the necessary staff time and funding are available. The results indicate positive experiences and opportunities; risks, threats and obstacles;

L.L. Brandli (✉) · D. Daris
University of Passo Fundo, PPGeng Campus I, Passo Fundo, RS 99052-900, Brazil
e-mail: brandli@upf.br

D. Daris
e-mail: dedaris@hotmail.com

W. Leal Filho
HAW Hamburg, Hamburg, Germany

W. Leal Filho
Manchester Metropolitan University, Manchester, UK
e-mail: walter.leal@haw-hamburg.de

M.A.L. Frandoloso · E.P. Korf
Faculty of Engineering and Architecture Campus I, University of Passo Fundo, Passo Fundo, RS 99052-900, Brazil
e-mail: frandoloso@upf.br

E.P. Korf
e-mail: eduardokorf@upf.br

W. Leal Filho et al. (eds.), *Integrating Sustainability Thinking in Science and Engineering Curricula*, World Sustainability Series, DOI 10.1007/978-3-319-09474-8_5

creativity and new solutions for Brazilian universities. This paper will be useful to academics and researchers interested in the topic of sustainability in higher education institutions in Brazil and other developing countries.

Keywords
Brazilian universities · Sustainability · Implementation

1 Introduction

The pressures on higher education today are intense (Barlett and Chase 2013), from legislators, society, work market and others, a fundamental transformation of higher education ensures that more students graduate with the skills and abilities to meet the challenges of the twenty-first century.

Implementing sustainability is not an easy task (Lozano 2006; Leal Filho 2011) it is confusing, time intensive, and with many uncertainties and various stakeholders that should be involved (Tilbury 2012 apud Waas et al. 2012). There is often resistance to change that is evident through the large number of barriers to change, that should be identified, addressed and overcome. Brazil, is not different, and maybe as a developing country, there are specific barriers that need to be explored.

At the end of the Decade for Education for Sustainable Development, UNESCO (2005–2014) many researches and analyzes are being conducted (Gonçalves et al. 2012) to show what have been developed in terms of objectives proposed 10 years ago. The focus of this paper is surrounding this topic in the moment that discusses the situation of Brazilian universities and brings some insights to promote sustainability.

Sustainability should not be understood to be a special subject but should be incorporated into the practices of each academic course (curriculum). This does not mean to insert into the curriculum a subject about environmental issues or sustainability but it means introducing an attitude and behavior pattern into each student as a future member of society and a future professional (Brandli et al. 2014).

According Leal Filho (2009), the levels of implementation of sustainability in university systems may be found at roughly three different stages of evolution. According the author these are as follows:

Stage 1: the principles of SD are not universally understood, there are not significant efforts towards promoting sustainability in university operations and no systematic projects that try to promote sustainability either holistically or in the context of specific subjects.

Stage 2: the principles of SD are widely understood and there are significant efforts towards promoting sustainability at university operations. There are projects which try to promote sustainability in the university as a whole or in the context of specific subjects, as well as in research and extension.

Stage 3: in this stage are the universities which fulfill requirements at level 2 and committed to sustainability on a long-term basis doing so by means of sustainability policies (ISO 14001 or EMAS for European universities), the existence of various senior members of staff who oversee its sustainability efforts and the existence of centrally-funded sustainability projects.

The author commented that it is not possible to measure the actual numbers of universities at the different stages, but to present an estimate based on the available literature and conferences on SD held over the past 10 years. The result is that around 20 % of the universities are found in stage 1, especially in developing countries (Brazil's case); around 70 % of the universities are found in stage 2, with an uneven distribution towards industrialized nations and towards North American and western European countries; up to 10 % of the universities are found in stage 3, almost exclusively in North America, Western Europe and Australia/Oceania.

Many universities have not succeeded in implementing the principles of sustainable development, according Leal Filho (2011) the reasons are lack of institutional interest, limited resources and staff involvement. The problems in implementing sustainability in higher education are influenced by social, political and economic issues. They can be categorized in perception, attitude, perspective, focus and cost-effectiveness (Leal Filho and Manolas 2012).

Waas et al. (2012) synthesize the barriers that higher education encounters in working toward sustainability. They show that these barriers could be related to understanding of sustainable development (SD), paradigm shift, resources and leadership (Fig. 1).

Disciplinary organizational structure hindering integrative thinking and interdisciplinary cooperation and learning
SD is perceived as an "add-on", not a built-in aspect of higher education
Lack of vision and prioritization/leadership of SD among higher education leaders
Lack of awareness, common understanding and knowledge of sustainability in higher education and its consequences
Perceived lack of scientific basis of sustainability
Confusion about SD
Broadness of SD
Lack of coordination and vision to change sustainability policies and education at government level
Little or no motivation or realism
Sustainability is considered to be radical
Changes into curricula are translated into budget claims
Overcrowded curricula
Sustainability is considered to have little or no relevance to the discipline, its courses and research
Lack of (financial) resources and uncertainty about the required efforts/resources to engage and implement sustainability
Threat to academic credibility of scholars and teachers

Fig. 1 Barriers that higher education encounters in working toward sustainability. *Source* Waas et al. (2012)

The aim of this paper is to discuss the application of the sustainability in the Brazilian universities emphasizing the barriers and difficulties of the implementation and the preconditions for this implementation.

2 The Brazilian Universities

The level of achievement of Brazilian universities does not compare with European universities (Leal Filho 2009), neither with North American universities (Barlett and Chase 2013).

The Brazilian higher education consists of 2,377 higher education institutions, according to data from the Ministry of Education (MEC). Of this total, 85 % are colleges, 8 % are universities, 5.3 % are technology centers, and 1.6 % are technological institutes. This means that there are around 190 universities in Brazil (Ranking Universitário Folha 2012).

In Brazil few universities have implemented Environmental Management Systems in their campuses. Although there are some examples that need to be mentioned as the UNISINOS, the first university in Latin America with ISO 14001 Certification (Venzke et al. 2012). This scenario is quite different from that showed in Disterheft et al. (2012b) for European universities, where current state of Environment Management System (EMS) implementation processes and practices is an advanced level with many universities using ISO14001, EMAS and EMS.

Generally, as commented by Tauchen and Brandli (2006) and Brandli et al. (2011), most of the environmental actions that have been adopted in Brazilian universities are isolated and do not consider a systemic vision.

As the largest Brazilian university and the third in Latin America with an academic community of over 106,000 people, since 1997 the University of São Paulo has the Permanent Program and Energy Efficiency—PUREUSP (PUREUSP 2012), evaluating the potential for 20 % energy savings for the University. In addition to prescribing some measures to control energy consumption, it has a monitoring system for consumption (SISGEN) that identifies indicators for the seven campuses. The cost impact of natural resources constitute a strong impulse also to environmental and social measures, dealing with paper consumption and with water management (PURA program), and to establish parameters for the calculation of the ecological footprint and CO_2 emissions (USP 2009).

Although cooperation exist with other international universities such as the Universidad Autónoma de Madrid (UAM), the USP does not yet have a formal EMS or ISO14001 certifications. This is mostly due to political and economic reasons.

3 Methodology

The research method used in this study consisted of a background analysis (with a review of the literature), complemented with the collection of empirical evidence. A questionnaire with four questions below was sent by email to a sample of 10 experts

Table 1 Surveyed institutions in Brazil

University	Federal state in Brazil	Month and year of foundation	Number of students
Federal university of Bahia	Bahia	April 1946	27,600 students
University of Brasilia	Brasilia	April 1962	30,727 undergraduates and 8,913 graduate students
Estadual university Paulista (Unesp)	São Paulo	January 1976	35,000 graduate students 12,000 mil postgraduate students
University of Southern Santa Catarina Unesul	Santa Catarina	November 1964	30,000 students in graduate, posgradutate and distance education
Estadual University of Roraima	Roraima	November 2005	3,000 graduate students
Federal Institute of Education, science and technology from Rio Grande do Sul	Rio Grande do Sul	December 2008	6,200 students

working in some Brazilian universities, of which 6 were returned. For the selection of the surveyed institutions was made a random sampling, considering the spatial distribution in country.

Due to the scope of the study and the nature of the methodological approach, the survey, by no means, can be regarded as representative. However, since it is combined with the review of the literature, it builds a rough profile of the status of the sustainability in Brazilian universities. Larger studies can be undertaken at a later state, provided that the necessary provisions for staff time and funding are available. The surveyed institutions can be seen in Table 1.

The questions posed to the experts were as follows:

What is the status of sustainability in your university?

How do you see the implementation of sustainability in Brazilian universities nowadays?

What are the main barriers to the implementation of sustainability in your university?

What must be done to promote sustainability in Brazilian universities?

The next section presents the results gathered, the analysis and its implications. The transcriptions in the text were freely translated to English.

4 Results and Analysis

4.1 Status of Sustainability in the Brazilian Universities

The respondents have shown that most universities do not have institutionalized sustainability programs within the framework of the management of the whole university. Sustainability actions are isolated, sometimes guided by ideologies.

Those universities which reported an institutional approach to sustainability have stated that their efforts are still in the early stages. About the green actions, the waste management is the action more cited among the universities.

This scenario means that there is not a formal policy for sustainability and structured framework for managing the environmental impact of the university activities like the a Environmental Management System (EMS).

As a respondent commented:

> …We have not yet … institutionalized programs focusing on the actions and decisions of sustainable orientation. However, the institution is taking several actions that aim to reduce direct and indirect impacts on the environment. But there are isolated actions that are not systemic or actually institutionalized through program to sustainable practices.…

> …In the Brazilian universities sustainability is seen from a really early stage…

Furthermore, most of the sustainability actions found, focus on green campus, missing the approach to sustainable education.

The sustainability is reported in the social balance sheet.

With regard to federal higher institutions, two comments are to be mentioned:

> The disorganization in public universities puts the theme of sustainability away.

> Federal universities are still discussing the creation of centers for these purposes.

4.2 Barriers for the Sustainability in Brazilian Universities

The respondents identified several commons barriers: lack of mandatory strategies from the institutional program that motivate staff, professors and students to engage in sustainability; the lack of interest in sustainability; lack of knowledge about sustainability; and the activities including teaching, research, and university management that leave no space for the implementation of sustainability measures.

Other problems may also be added the list:

1. Cultural change
2. Importance attributed to sustainability
3. Lack of cooperation networks between universities
4. Government policies to encourage the implementation of education for sustainability and sustainable practices on campus, especially from Ministry of Education.
5. Lack of resources or available funding for sustainability projects.
6. Lack of staff and a senior member of staff who can implement and oversee sustainability efforts. This could be in terms of size and team's background.
7. Lack of projects between companies and universities, and R&D projects.

Many of these barriers can be explained as are organized societal levels of participation presented by Disterheft et al. (2012a). According the authors, the educational sector at the macro level involves the international and national framework of educational policymaking; at the meso level, it involves the governance of the institutions and their organizational structure and at the micro level, it refers to the concrete learning settings and spaces for participation in institutions and their communities.

Beyond these, there are more contextual barriers that can be observed in development countries (macro level): there is a lack of sustainable-related practice in the day-to-day of the population; there is a lack of sustainability practices in the city life including sustainable transport, formal mechanisms for selective collection of municipal waste, investments in water and efficient energy consumption; and there are intrinsic difficulties for the population to (attend) comply with the legislation/and/or not sufficient environmental legislation. All of these issues are affected by the role of the university.

One of the challenges of the educators is developing in themselves and in their students, mindsets and habits that enable people to live sustainably. In this point, maybe the role of the university is more evident in developed countries.

The Fig. 2 presents the link between the participation processes in higher education for promoting sustainability practices and for fostering citizenship and democratic values, and the barriers in developing countries like Brazil.

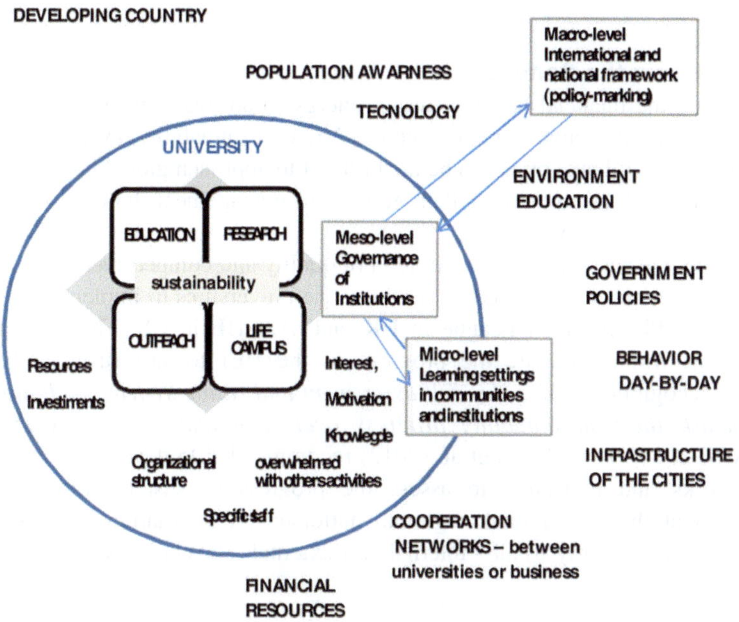

Fig. 2 Barriers that university encounter in implementing sustainability in developing countries. *Source* the authors

4.3 Pre-conditions for the Sustainability in Brazilian Universities

Brazilian universities must realize the importance of their role in society with respect to education for SD by educating leaders and advancing knowledge, and by developing an institutional understanding, culture, and planning for the integration of sustainability.

To improve the efforts for the implementation of sustainability in Brazilian universities, some points should be considered:

4.3.1 Investing in Support

In order to have a structure for the sustainability, top-down policies and full-time equivalent Environmental Management Staff must be considered.

Universities should o raise more support, in terms of staff capacity, financial support and investments.

Political support is closely related to the financial support for implementation of sustainable projects and as such, It is important that rectors and other officials of the HEI support sustainable initiatives and to add them to the institutional strategic planning.

One of respondents said:

> *Personally, I believe the campus is very large. So, such initiatives should be enforced because there is hardly any awareness by so many people.* This is an important point of view because successful experiences in sustainability show their need to influence people without authority (Newman 2013).

4.3.2 Know the Benefits

Understand the financial benefits (water, energy, food, waste management, etc.).

Understand the environment benefits; View sustainability as a commitment universities would take on not to be the best but to approach global issues including clime change, disasters and conflict, ecosystem management, harmful substances, and resource efficiency.

To view the sustainability as a criteria of quality and competitiveness, a national ranking about sustainability could motivate the universities to implement it, as the People and Planet Green League in UK, and AASHE in US. This standardized instrument would measure the progress of the HEI toward sustainability and achieve recognition. *Using this self-assessment and rating system, institutions can benchmark their sustainability progress over time and compare with others* (AASHE 2014). Also, Waas et al. (2012) recommended to develop an assessment frameworks and indicators to assess the progress of sustainability in higher education at the institutional, regional, national and international levels and to communicate regularly to all—internal and external- stakeholders.

4.3.3 Educate and Train

Develop an institutional understanding about the topic of sustainability, and include it in the educational process. According Porritt (2012) the starting point is staff training and preparation. *Only then can we begin to embed sustainability more widely into university programmes across the curriculum. As educators, we have a huge responsibility to embed sustainability into what we do across the board, not just for those students who choose to take subjects that directly relate to issues around sustainability, but for the whole student body. In essence, we should be preparing students for the work of the world, not just the world of work.*

As a respondent commented.

> *Change and learning. Aware teachers, managers and other staff are not only about the subject but also related to learning and education in general (greater understanding regarding these topics and also their relationship to sustainability). Go beyond the mainstream. Change into a place of knowledge generation and not Knowledge reproducing (as seems to be today).*

4.3.4 Involve People/Stakeholders

Sustainability must be funny and bring some news every time that to engage the students and staff in campaigns with a sustainability focus.

Engage the community and external stakeholders (as for example business-university partnerships). Universities could be very important actors in any specific sector, in terms of research as well as the education and training of employees. In exchange, they could get more clean technology investments in environmental management in their campus activities or even project funding (R&D).

4.3.5 Politics/Cooperation Networking

In terms of the country, and based on the barriers mentioned, there is an eminent necessity for the government policies to encourage and value the implementation of education for sustainability and sustainable practices on campus, especially from Ministry of Education. The National System of Higher Education Assessment (Sinaes) analyzes the institutions, courses and student performance. The evaluation process considers aspects such as teaching, research, outreach, social responsibility, management of the institution and faculty. As the information is used for institutional orientation of higher education and to support public policies, it could include some sustainability criteria.

In this way, it is important to promote the cooperation between networks in universities in Brazil and with other countries. There are good examples as Ecocampus in UK which provides and support a flexible approach to EMS implementation in HEI (http://www.ecocampus.co.uk).

Waas et al. (2012) suggests that government subsidy programes should stimulate sustainability in HEI initiatives and collaboration and networking at all levels (regional, national, international).

The universities should engage with the Regional Centers of Expertise (supported by the United Nations University). In Brazil, there are three regional centers, one in Curitiba/SC; Rio de Janeiro/RJ and São Paulo/SP.

5 Conclusion

Incorporating sustainability is a social issue that should obviously be born in the university as a resource for teaching, research and outreach, by doing so, disseminating a sustainable posture, skills, development, knowledge and technologies.

In Brazil there are many barriers be overcome, some inherent in the process of incorporating sustainability and cultural change widely discussed in the literature and other related to the current context of the country and state of involvement of these universities with the topic.

However, several successful sustainability efforts by HEI may show how the university is capable and dynamic, and how it could be applied to the Brazilian reality. With regard to the stage of sustainability in Brazilian universities, future research will be necessary to develop a better understanding of the current situation and ways to organize and promote sustainability in Brazilian universities.

Acknowledgments Acknowledgements the CAPES for the financial support for the research postdoctoral of Prof. Luciana L. Brandli under the supervision of Prof. Walter Leal Filho at Hamburg University of Applied Sciences (Process 9285-13-0).

References

AASHE (2014) Sustainability tracking, assessment and rating system. Association for the Advancement of Sustainability in Higher Education. https://stars.aashe.org/ (Last Accessed 27 Mar 2014)

Barlett PF, Chase GW (2013) Sustainability in higher education: stories and strategies for transformation. The MIT Press, Cambridge, p 316

Brandli LL, Frandoloso MAL, Roorda N, Fraga KT, Vieira LC (2014) Evaluation of sustainability using the AISHE instrument: case study in a Brazilian University. Brazilian J Sci Technol 1:4

Brandli LL, Frandoloso MAL, Tauchen J (2011) Improving the environmental work at University of Passo Fundo, Brazil—towards an environmental management system. Brazilian J Oper Prod Manage 8(1):31–54

Disterheft A, Caeiro S, Azeiteiro U, Leal Filho W (2012a) Implementing sustainability at the campus—towards a better understanding of participation processes within sustainability initiatives. In: Filho W (Ed) Sustainable Development at Universities: New Horizons. In the series Umweltbildung, Umweltkommunikation und Nachhaltigkeit—Environmental Education, Communication and Sustainability. vol 34. Peter Lang, Frankfurt. pp 345–361

Disterheft A, Caeiro S, Ramos MR, Azeiteiro U (2012b) Environmental management systems (EMS) implementation processes and practices in European Higher Education Institutions e top-down versus participatory approaches. J Clean Prod 31:80–90

Ecocampus The leading environmental management system and award scheme for the higher and further education sectors. http://www.ecocampus.co.uk (Last Accessed 25 Mar 2014)

Gonçalves F, Pereira R, Leal Filho W, Azeiteiro UM (eds) (2012) Contributions to the UM decade of education for sustainable development. Peter Lang Scientific Publishers, Frankfurt, p 430

Leal Filho W (2009) Sustainability at Universities: opportunities, challenges and trends. Peter Lang Scientific Publishers, Frankfurt, p 340

Leal Filho W (2011) Role of universities and their contributions to sustainable development. High Educ Policy 24:427–438

Leal Filho W, Manolas E (2012) Implementing sustainable development in higher education. In: Azeiteiro UM (ed) Gonçalves F, Pereira R, Leal Filho W. Peter Lang Scientific Publishers, Frankfurt, Contributions to the UM decade of education for sustainable development, p 430

Lozano R (2006) Incorporation and institutionalization of SD into universities: breaking through barriers to change. J Clean Prod 14(9–11):787–796

Newman J (2013) Sustainability strategic planning: establishing accountability in a word of distractions. In: Barlett PF, Chase GW (eds) Sustainability in higher education: stories and strategies for transformation. The MIT Press, Cambridge, Massachusetts, p 316

Porritt J (2012) Universities must lead the way on the sustainability agenda. The Guardian

PUREUSP(2012) Informe mensal. gestão de energia. indicadores de uso. mês: março/2012. En: PUREUSP. Programa para Uso Eficiente de Energia na USP. São Paulo: PUREUSP. http://www.usp.br/pure/scc/upload/informe_mar%E7o_2012.pdf . (Last Accessed 23 June 2012)

Ranking universitário Folha (2012) http://ruf.folha.uol.com.br/2012/ensinosuperiornobrasil/ (Last Accessed 27 Mar 2014)

Tauchen J, Brandli LL (2006) A Gestão ambiental em instituições de ensino superior: modelo para implantação em campus universitário. Gestão & Produção 13(3):503–515

UNESCO (2005–2014) United nations decade of education for sustainable development (2004–2015): draft international implementation scheme. 2005, UNESCO: Paris

USP (2009) Universidade de São Paulo. USP sustentabilidade: impacto ambiental na Universidade de São Paulo. São Paulo: USP Inovação

Venzke CS, Nascimento LFM, Gomes LP, Campani DB (2012) Environmental Management on University Campuses in Southern Brazil. In: Walter Leal Filho (ed) (Org). Sustainable Development at Universities: New Horizons 73. Frankfurt am Main: Peter Lang Scientific Publishers, vol 34, pp 885–898

Waas T, Hugé J, Ceulemans K, Lambrechts W, Vandenabeele J, Lozano R, Wright T (2012) Sustainable higher education—understanding and moving forward. Flemish Government—Environment, Nature and Energy Department, Brussels

Authors Biography

Professor Luciana Londero Brandli is graduated in Civil Engineering (1995), master's degree in Civil Engineering (1998) and Ph.D. in Production Engineering (2004). Pos Doctorial Research at Hamburg University of Applied Sciences (2014). She is currently Associate Professor in the University of Passo Fundo, south of Brazil, working in the Master Program in Engineering, Infrastructure and Environment. Her current research interests include sustainability in high education and green campus, environment management, management of urban infrastructure, sustainable cities and green buildings.

Professor Walter Leal Filho has a first class degree in Biology and a doctorate in environmental science (PhD), having also completed a post-doctorate programme on environmental communication. He also has a higher doctorate (Dr. rer. nat habil.) in environmental information (DSc), a DPhil in sustainable development and holds the titles of Doctor of Letters (DL), Doctor of Literature (DLitt) and Doctor of Education (DEd) commensurate with his scientific performance and outputs translated by over 300 publications among books, book chapters and scientific papers.

Professor Marcos Antonio Leite Frandoloso has a degree in Architecture and Urbanism at Federal University of Pelotas (1986), Master in Architecture at Federal University of Rio Grande do Sul (2001), and a course in Urban Ecology at the Universitat Oberta de Catalunya (2008). Currently he is PhD cadindate at Universitat Politècnica de Catalunya, Barcelona, Spain, focusing on eco-efficiency and environmental management at universities. Besides his Architectural office, he is Professor and researcher at the University of Passo Fundo—UPF—since 1995, nowadays he is Coordinator of Product Design course at the UPF. He has experience in Architecture and Urbanism with emphasis on building and environmental planning, acting on the following topics: energy efficiency, energy and environment, sustainable construction, bioclimatic architecture, architectural heritage, ecodesign and urban ecology.

Professor Eduardo Pavan Korf received his Master degree in Engineering from the Federal University of Rio Grande do Sul in 2011 and graduate in Environmental Engineering from the University of Passo Fundo in 2010. Since 2010, he has been a Professor in Environmental Engineering Course at the Passo Fundo University and has been researching the topics: environmental monitoring such as air pollution, environmental modelling and environmental geotechnics. He is currently a PhD student in the Engineering from the Federal University of Rio Grande do Sul, expected defense to March 2015.

Professor Denise Daris holds a degree in Food Engineering from Regional Integrated University of High Uruguay and Missions (2000). Specialization in Food Technology (2004) and Word Safety (2011). She is currently an engineer of Work Safety, Professor at the Federal Institute of Education, Science and Technology of Rio Grande do Sul, IFRS, Brazil. Student at Master in Environmental and Civil Engineering at Passo Fundo University since 2013 researching atmospheric pollution.

Implementing a Sustainability Strategy: A Case Study from the University of Leeds

Louise Ellis and Jessica Martin

Abstract

This paper outlines the extensive engagement processes undertaken with staff, students and external stakeholders in order to develop an integrated sustainability strategy at the University of Leeds. The engagement process inspired five themes which underpin commitment, indicators and methods of reporting, and will ensure sustainability is embedded throughout teaching, research and operations at the institution. The paper will discuss the holistic nature of the strategy and outline how sustainability is being embedded at the University of Leeds. In addition, the engagement process and the role this has in raising the profile of sustainability will be examined. Reflecting on organisational change and stakeholder engagement, this paper will discuss how the higher education sector can make a transformational contribution to sustainable development. The strategy is an engagement driven living document, and will provide a framework which will inspire collaboration for a positive future.

Keywords

Sustainable development · Universities · Strategy · Leeds

L. Ellis (✉) · J. Martin
University of Leeds, 31 Cromer Terrace, Leeds LS2 9JT, UK
e-mail: L.A.Ellis@leeds.ac.uk

J. Martin
e-mail: J.Martin@leeds.ac.uk

W. Leal Filho et al. (eds.), *Integrating Sustainability Thinking in Science and Engineering Curricula*, World Sustainability Series, DOI 10.1007/978-3-319-09474-8_6

1 Introduction

The University of Leeds is a large, complex institution, which has significant social, environmental, economic and cultural impacts both nationally and globally, and therefore has a responsibility to ensure the impacts of the organisation are positive. In order to achieve this, in 2013 we began a rigorous engagement process to underpin the development of an integrated sustainability strategy, which would function as a catalyst to embed sustainability into the heart of the institution, and ensure we have an overall positive impact on society. This paper will outline the theoretical justifications for such a participatory process, and will provide a case study of developing and implementing a strategy at such a complex institution with a huge physical presence and the demands required of a research intensive organisation. The paper will provide an insight into developing a strategy which integrates teaching, research and innovation as well as operations; an innovative and holistic approach to strategy development within the higher education sector.

Before discussing the University of Leeds and outlining the process of participatory engagement undertaken to underpin the development of a new integrated sustainability strategy, this paper will give a brief background overview to sustainability in higher education, the principles of stakeholder engagement and organisational sustainability.

2 Background

Sustainability in Higher Education Institutes (HEIs) has been in play in numerous forms for over a decade however, this has been in different forms ranging from environmental management, community engagement, to education for sustainable development. Until more recently, there have been limited approaches which take a more holistic perspective on both the true breadth of sustainability or full institutional integration. For a number of HEIs the focus began as one which addressed environmental management and the greening of campuses, this is demonstrated in Wals' (2014) outline of how this has been reflected in the themes covered in the first nine years of the International Journal of Sustainability in Higher Education (IJSHE), see Table 1 (Wals 2014: p. 10).

Campus greening has seen substantial focus on energy efficiency, waste management and to a lesser extent biodiversity. There are two fundamental challenges that arise from this type of approach; the first is that the true scope of sustainability is lost and the emphasis remains on environmental problems and the second is that it often fosters a culture of 'it's someone else's problem' with estates or central administration teams being seen as responsible for addressing the issue, rather than sustainability becoming truly embedded in the of culture of the HEI.

Like environmental management, incorporating sustainability into the curriculum has also been a key area of work within HEIs, Lozano and Young (2013) have identified four main approaches employed to achieve this:

Table 1 Thematic focus of articles published in the first nine volumes of IJSHE (Wals and Blewitt 2010; in Wals 2014: p. 10)

Area	Number of articles	Percentage (%)
Environmental management/ecological footprint/campus greening	44	25
Integrating sustainability in existing disciplines	31	17
Pedagogy, learning and instruction	31	17
Philosophy/principles/concepts	19	11
Community outreach/partnerships	15	8
Policy/organisational learning/institutional commitment	15	8
Course development/curriculum	7	4
Auditing, assessment, quality assurance	10	6
Research	3	2
Competencies, professional development	3	2
Total	178	100

1. "Some coverage of particular environmental and/or social issues and material in an existing course.
2. A specific sustainable development course added to the curriculum.
3. Sustainable development intertwined as a concept within pre-existing disciplinary-orientated courses, with the relevant sustainable development component issues matched to the nature of each specific course.
4. Sustainable development offered as a specialisation within the framework of particular faculties or schools within an institution." (ibid: p. 134)

Although these trends have been identified, incorporation of sustainability into the curriculum has been slow (Lozano and Young 2013), with some suggesting that embedding sustainability into the curriculum could take another two decades (Desha and Hargroves 2012). Traditionally, addressing the challenge of embedding sustainability into the curriculum has been the remit of select individuals in academic departments, and there has been limited linkage with research and innovation activity, and even less with institutional operations.

In order to truly address the question of sustainability in the higher education sector, HEIs must take an integrated view of sustainability, seeing it a mechanism by which to connect operations, student education and research and innovation, and therefore embedding sustainability into the heart of an institution. Key to embedding sustainability is ensuring that all HEI stakeholders are involved throughout the development and implementation of changes.

3 Stakeholder Theory

It is suggested that since Freeman's seminal work introducing the concept of stakeholder theory, "Strategic Management: A Stakeholder Approach" (1984), stakeholder theory has evolved from the original spoke-and-wheel design as developed by Freeman (1984), to the study of mutually engaged relationships, creating the groundwork for transparency and accountability (Andriof et al. 2002). The concept of mutuality has been introduced within stakeholder engagement theory, and it is suggested that if organisations are to be equipped to deal with creating transformational changes in sustainability, they "need better understanding of the dynamics and expectations fundamental to living, acting and working in a network of collaborative relationships" (Andriof and Waddock 2002: p. 21).

In order to gain an understanding of these collaborative networks, and consequently an understanding of how to achieve change within an organisation, thorough engagement with a wide range of stakeholders is necessary, and interaction with stakeholders needs to be integrated into decision making at every level in the organisation (Carroll and Buchholtz 2008). This ensures stakeholder engagement is embedded into the way in which the organisation functions, rather than a tokenistic exercise.

Whilst justifications and analysis of the benefits of stakeholder theory can vary greatly, they can largely be defined as descriptive, normative or instrumental, and are often a combination of the three (Donaldson and Preston 1995). Applying Donaldson and Preston's categories in the context of the University of Leeds, stakeholder engagement can understood to have descriptive, normative and instrumental value, which helps to support the justification and necessity of applying stakeholder theory and pursuing engagement in the context of developing an integrated sustainability strategy. Stakeholder theory can be understood as performing a descriptive function in the context of the University of Leeds as it has initiated a stakeholder analysis, encouraging us to develop an understanding of how we function as an organisation, and determine who our stakeholders are. There has also been an instrumental function, as engaging with our stakeholders has encouraged buy-in, increased awareness of the sustainability agenda throughout the University, and developed links within the organisation to increase collaboration and efficiency. Finally, applying stakeholder theory to our strategy development has performed a normative function, as it has allowed us as an organisation to exercise our belief that the interests of all stakeholders are of intrinsic value (Donaldson and Preston 1995), and has encouraged us to give a platform for stakeholders to express their views.

4 Organisational Sustainability and Strategy

Like any large organisation, embedding sustainability into HEIs requires organisational, including cultural, change and therefore a clear vision and strategy is needed to guide this change. However, this is not an area of sustainability in higher education which is well understood (Hoover and Harder 2014 and Stephens and Graham 2010).

Although, as Hoover and Harder (2014) and Denman (2005) point out HEIs are often complex organisations and whilst there is no one size fits all approach for processes of organisational change, lessons can be drawn from organisational sustainability literature and practice, identifying key elements of a strategic approach for implementing sustainability. Key reference points are engagement, gaining buy-in, understanding the context, identifying what matters, creating a vision, translating the vision into an action plan and developing a process of accountability. These reference points have been used as guides in the development of the University of Leeds integrated sustainability strategy. For more discussion on implementing organisational sustainability see Blackburn (2007), Grayson and Hodges (2004), Hitchcock and Willard (2008), Pohl and Tolhurst (2010).

4.1 Institutional Context

The University of Leeds is a large multi-disciplinary British university, with 30,761 students and 7,144[1] staff. The University is home to a number of leading science and social science centres of research, and is part of the Russell Group of research intensive universities. Institutionally it is structured across nine academic faculties which are supported by central administrative services including facilities, communications and IT. The University of Leeds is located in an urban centre, with its main campus area being situated in the city centre of Leeds. The estate is made up of a variety of building types ranging from over 200 years old to modern new buildings and covering uses as broad as a swimming pool and greenhouse to lecture theatres and office space.

In 2013 the University of Leeds began the process of developing a new integrated sustainability strategy, which will cut across the whole organisation and ensure that student education, research and innovation and operations are integrated in order to ensure that the university has a positive impact on society, and that the fundamental business of a university, to create knowledge and education is supported. To the University of Leeds sustainability means taking responsibility for it's economic, social and environmental impacts, whilst having an overall positive impact on society. Fundamental to developing a strategy to underpin this commitment was ensuring that an inclusive participatory process was undertaken, which will be outlined in the following section.

5 The Engagement Process

From the outset, participation and stakeholder engagement has been absolutely crucial to the development of an integrated sustainability strategy at the University of Leeds. Participation has been defined by those implementing sustainability

[1] Correct at time of publication in 2014.

initiatives in this context as referring to "student and faculty involvement, giving the institutional community the opportunity to shape an institutional transformational process towards a more sustainable campus" (Disterheft et al. 2012: p. 345). In order to achieve our aim of linking together operations, teaching and research and innovation we need to encourage collaboration and communication throughout the institution. Stakeholder buy- in and concepts of ownership have been considered imperative to developing a strategy which is relevant, achievable and transformational for the institutional community at the University of Leeds. For this reason, we embarked upon various methods of engagement to encourage meaningful participation from as many staff, students and members of the local community as possible.

5.1 Organisational Review

We began the engagement process with an organisational review, allowing us to take stock of sustainability initiatives and processes which were already in place, identify potential gaps in collaboration throughout the organisation, develop an understanding of our stakeholder demographics and compare our sustainability initiatives with similar institutions (Table 2).

5.2 Stakeholder Engagement

Having developed an understanding of the organisation in terms of both performance and stakeholders, we were able to determine which feedback mechanisms would provide us with the most meaningful and useful engagement from the widest range of participants. Stakeholder engagement can be said to be driven by a sense of political and civic morality, implicitly suggesting that stakeholders have the right to influence actions which will impact on them, and by presenting the maximum number of stakeholders with the opportunity to contribute, the process becomes as democratic as possible. Phillips is one proponent of this understanding of stakeholder engagement, suggesting "Stakeholder status as here conceived indicates the presence of an additional obligation over and above that due others simply by virtue of being human" (Phillips 1997, p. 53).

This moral obligation to present stakeholders with the opportunity to influence the strategy, combined with the sense of ownership stakeholders develop through contributing, drove the engagement process, and ensured we pursued as many avenues as possible to obtain thorough feedback.

As colleagues across the higher education sector, both nationally and internationally, have noted, "nothing less than a paradigm shift in organizational thinking is needed for colleges and universities to promote cultural transformation" (Marans and Shriberg 2012: p. 557). In order to achieve this paradigm shift, we wanted to involve as many stakeholders as possible in the earliest stages of strategy

Table 2 Organisational review

Organisational review	
Performance mapping	A thorough review of our current activities and performance to date was undertaken, to identify the strengths and weaknesses of the institution in relation to sustainability. The mapping also allowed us to determine opportunities to create links between operational activity, and research that is taking place in various parts of the organisation. This element of the process also help us to begin to understand the wide range of stakeholders who were either already involved in implementing sustainability at the University and those who weren't involved but hold key positions in ensuring success moving forward
Benchmarking	We benchmarked the University against other Russell Group[a] universities based on a wide range of sustainability activity and performance indicators
	This allowed us to ascertain our comparative performance in relation to similar institutions, and provided a national assessment of the sustainability progress within the sector
	We also collated examples of best practice from similar institutions, and reviewed our performance in external ranking schemes (including the people and plant green league, and the business in the community index) and national accreditation bodies
Internal governance review	We conducted an internal review of our policies, targets and reporting of issues that related to sustainability, taking into account their scope, efficiency, levels of transparency and accountability. This activity allowed us to map our internal policy landscaping identifying any areas of cross-over and any gaps which existed
Stakeholder analysis	Before initiating the engagement process, we analysed our stakeholders to develop an understanding of how to maximise participation, and provide inclusive, wide-ranging engagement mechanisms for people to influence the direction of the strategy
	Stakeholders were identified and then analysed in terms of the combination of their interest in and ability to implement sustainability initiatives, based on classic stakeholder mapping techniques (Manktelow 2005). This provided us with a comprehensive list of stakeholders which also differentiated between primary and secondary stakeholders, determining the extent to which they were able to achieve and interested in transformational change

[a]The Russell group "represents 24 leading UK universities which are committed to maintaining the very best research, an outstanding teaching and learning experience and unrivalled links with business and the public sector" (Russell Group 2014)

development, this was key in ensuring that we developed a collective sense of feel responsible for shaping the vision for cultural transformation and ensuring that we captured the insights and ideas of our stakeholders. For real transformational change to be realised it is vital that innovation and vision from across the institution (and externally) is capture as no one individual group can create transformation alone (Table 3).

Table 3 Stakeholder engagement

Stakeholder engagement	
Workshops	We organised initial workshops with groups of staff and students, at the very beginning of our engagement process
	The workshops comprised of a materiality assessment, allowing participants to determine the most significant impacts of the organisation from their varying perspectives. Understanding what is material to our organisation, like others, is key as this ensure that we focus on the key environmental, social and economic impacts (Pohl and Tolhurst 2010)
	We then completed a visioning task, asking what participants thought a sustainable university would look like in 2020 and 2050
	Through these initial workshops we engaged with approximately 70 people. As a result of the workshops we were able to produce an overall materiality index which worked out the most cited and most important impacts, as well as a visioning index which displayed the most cited ideas for a sustainable university, and organised them thematically
	Both indexes were populated continuously throughout the engagement process
Postcards	We designed a postcard asking "What would a Sustainable University look like?" with a section on the back to allow feedback. These were distributed amongst staff through our network of Environmental Co-ordinators, and were sent back to the Sustainability office through internal mail
	Static displays with the postcards were placed in all catering outlets, gyms and libraries and in 23 schools and other buildings. These displays had wooden trees made from recycled university furniture to which people tied their postcards to make a visual impact
	Stalls were located in The refectory (the largest and most central catering outlet on campus) to specifically engage students during meal times
	Postcards were available during open days for prospective students and their families to provide feedback
	A stall was also included at Skyride; a huge city-wide cycling event that came through campus, which allowed us to engage with non-university community and local residents
	In total 266 responses were received and the responses where collated and split into themes
Online engagement	Internet responses were collated via our own website which hosted an online version of the postcard. We also ran a social media campaign encouraging people to tweet or comment on Facebook with their vision of a sustainable university
	Discussion groups were created online through the LinkedIn website and 2° network, as well as across alumni groups on various social media sites
	In total 22 responses were received, including responses from alumni, again these responses where collated and included in the thematic review
Focus groups	Based on the information that was collected through the feedback mechanisms above we facilitated four focus groups open to staff, students and members of the community. We facilitated discussions driven by the themes that had emerged in the engagement process so far, and gave an

(continued)

Table 3 (continued)

Stakeholder engagement	
	opportunity for participants to identify any gaps in the vision of a sustainable university
	A graphic artist was present at all groups, and produced images which represented the ideas discussed by participants. We were given a variety of images from each focus group, which contributed to a final image which combines these drawings with commentary, and has incorporated these into illustrations of prominent buildings and landmarks on campus
	The focus groups received 100 attendees
Exhibition	An exhibition was held on campus to overview the engagement process and to ask for feedback on the suggested outline of the final strategy. All illustrations and postcards were displayed, as well as a video which detailed how the engagement process had evolved into an outline of the strategy
	We also provided another opportunity for stakeholders to fill in a postcard and identify any gaps in the suggested themes
Direct consultation	The final strategy will be reviewed with key internal and external stakeholders, and will additionally be sent to all focus group participants to ensure it is an accurate representation of the feedback we collated

5.3 Reflections on the Engagement Process

Providing feedback mechanisms on various platforms, including online, face to face discussions, postcards and visual displays allowed us to obtain a wealth of feedback from various sources, which became the foundations for the strategy themes. The anonymity of the postcards sent to us through the internal mail system ensured we acquired honest feedback and often radical visioning, without the potential restraints stakeholders may feel speaking in their capacity as employees or students of the University.

The physical displays also allowed us to target areas in the University with employees who do not use computers, such as cleaners and maintenance staff. This ensured we presented staff with as accessible an opportunity to contribute as possible.

By recording feedback and organising it thematically we were able to clearly identify areas and work streams which stakeholders considered important, and these were reiterated through the images created in the focus group discussions. The images and commentary which now form the final strategy image provide a visual representation of the engagement process, having been created through stakeholder discussions during a focus group. This has ensured the strategy development is explicitly stakeholder driven, transparent, and underpinned by the contributions of staff, students, and members of the local community.

6 Outcomes

It is important to discuss the outcome of the strategy engagement process from two perspectives; how it has informed the strategy itself and the wider impact of the process.

6.1 An New Integrated Sustainability Strategy

It became clear throughout the engagement process that our stakeholders want us to be leaders in sustainability and in order to do this we should be brave and inspirational and fundamentally, we should be having a positive impact on society. It was important that we encapsulated this aspiration at a high level before we started developing the detailed elements of the strategy. Therefore we set out a vision to 'inspire collaboration for a positive future' within which we would have a positive impact on society by addressing economic, social and environmental well-being. To do this, we would aim to ensure that all our graduates are responsible and aware of sustainability, undertake research that matters and is accessible, and develop an organisation which is knowledgeable, responsible and efficient.

To support this vision and aim, and to reflect the sentiments of the engagement process, five main themes were developed to form the strategy round. These are:

1. Embedding sustainability

Embedding sustainability at the heart of all decision-making is key to successful organisational change, this should be based on clear processes which promote collaboration, ensuring sustainability is considered in all decisions and processes, as well as ensuring solid and simple management systems, coherent policies and procedures, clear accountability and transparency, integrating our core activities and encouraging a living lab approach.

2. Being a positive partner in the community

The community—university nexus covers a range of activities and impacts, and addresses the University and wider society, integrating the University and the Leeds City Region and linking students with the community. This was a strong theme that came through the engagement process and will cover areas such as actively linking research to social and policy needs, linking research and skills to local needs, being a positive and proactive partner in the Leeds City Region, and engaging in policy-making at all levels. Additionally, it will include providing a positive message to young people, having a campus which is open, accessible and welcoming and one which provides cultural experiences for all in the city, ethical investment, reducing and managing our physical impacts and encouraging students to be active and good members of the local community. In addition to looking at our relationship with the local community, we are committing to working to have a positive impact on community in its widest sense to include the global community.

3. Developing knowledge and capacity

As an organisation we are committing to developing sustainability knowledge and capacity, this includes developing sustainability literacy across our student body, equipping students with skills which link to employability and wider post-graduation activity, linking local projects with operations within the organisation and with relevant staff and students, developing knowledgeable staff, working with our partnerships to develop sustainable skills in the wider community including our supply chain, increasing visibility and access to research and ensuring our organisation becomes a living lab.

4. Resource management

Resource management encompasses the efficient use of resources (encompassing energy, water, paper and procurement), impacting upon travel choices both commuting and business, dealing with waste generation and disposal as well as the generation of energy, and working to increase the sharing of equipment across the organisation. Resource management will be based on the principles of circular economy and whole life analysis.

5. Building a collaborative organisation

A collaborative organisation is one which fundamentally works together for common goals through sharing space, equipment and resources, reusing resources before buying more, buying together and sharing best-practice and skills. We are also committing to ensuring we enhance collaboration externally as well as facilitating internal collaboration. Collaboration is key to achieving real change for sustainability.

Within each one of these themes are commitments and indicators which make up the implementation plan, and also ensures we are accountable for our actions and are able to monitor and report progress against our strategy.

By following a participatory engagement process the content of the strategy has been developed through a rigorous process with sense checking and mechanisms to ensure transparency included at every stage, by the team and our stakeholders. In addition, it has meant that the language used to populate the strategy is appropriate for the setting, has been organically formed and reflects a broad range of perspectives and ideas.

6.2 The Engagement Process

There have been a number of key outcomes from an institutional perspective of the using a participatory engagement process;

Creating increased buy-in and ownership

At the outset of the engagement process a key aim was to create buy-in and ownership of the strategy, as outlined in earlier discussions this is a key success factor in any strategy development. The process successfully achieved this, with feedback from participants saying they welcomed the open approach to involvement and to the content development. The use of the graphic artist and then the subsequent use of the imagery in the final documentation also added to the sense of inclusion and explicitly demonstrated that the content developed in the engagement process has been the basis of the strategy.

Increasing knowledge and understanding of sustainability
A significant outcome of the engagement process is that there is now an increased understanding of the true nature and depth of sustainability; through this process there is a much greater understanding that sustainability is not solely about 'green issues' and includes more than energy efficiency and recycling. This process has overcome the problem of fragmented approaches to sustainability identified in previous literature as discussed earlier in this paper. Likewise, the process also encouraged people to think beyond their own internal organisational boundaries and encouraged them to start to think about how joint projects and working together across the different activities of the organisation can be beneficial.

Raising the profile of the sustainability service and the issues
In addition to specifically supporting the development of a new integrated sustainability strategy, carrying out such a participatory engagement process has led to an increase in the profile of the sustainability service and sustainability in general across the University and with some of the local community. This has led to more proactive involvement in other sustainability initiatives and more requests for support in making change within individuals' areas of work. This provides a positive platform to create change through the new integrated strategy.

7 Conclusions

The experience of this process has reiterated the understanding outlined by Hoover and Harder (2014) that HEIs are complex organisations and the University of Leeds echoes this, however, by employing such a participatory approach we have learnt that we can unpick some of these complexities and understand where things can be simplified, as well as acknowledging where complexity is required. We now have a great understanding of the different concerns and challenges which exist across the organisation. By embarking on a participatory engagement process we have managed to develop a more robust and rounded strategy which has buy-in from the earliest point. Although this will not make implementation simple it will make it a more positive experience and will ensure that all elements of the institution are involved. The engagement process has become a positive mechanism for organisational change at the university by creating a positive sense of inclusion and ownership.

We have also experienced secondary benefits of undertaking a participatory engagement approach include the raised profile and boarder understanding of sustainability across the institution, which has helped lay the ground work for organisational change, and will allow greater collaboration, increased knowledge and capacity, a sense of ownership and fundamentally link our research and innovation, student education and operations through sustainability. The process has helped create an understanding and an institutional desire for sustainability to be integrated throughout the University's operation, adding value to our core aims and ensuring sustainability is no longer a green add-on.

A key lesson learnt in this process is that is takes more time and planning to develop a strategy through participatory engagement however the value created both in terms of the strategy outcome and wider organisational benefits significantly outweigh the increase in time required. We also acknowledge that engagement doesn't stop with the 'publication' of the strategy, engagement is key to driving forward sustainability and this is merely the beginning.

In a broader, sector sense, if we encourage more participation at a strategic level we not only will create a better understanding of the sector in term of sustainability but also have more of a change of creating transformation change for sustainable development.

References

Andriof J, Waddock S, Husted B, Rahman SS (2002) Introduction. In: Andriof J, Waddock S, Husted B, Rahman S (eds) Unfolding stakeholder thinking: theory, responsibility and engagement. Greenleaf Publishing Limited, Sheffield, UK, pp 9–17

Andriof A, Waddock S (2002) 'Unfolding stakeholder engagement' introduction. In: Andriof J, Waddock S, Husted B, Rahman S (eds) Unfolding stakeholder thinking: theory, responsibility and engagement. Greenleaf Publishing Limited, Sheffield, UK, pp 19–43

Blackburn WR (2007) The sustainability handbook: the complete management guide to achieving social, economic and environmental responsibility. Earthscan, London

Carroll A, Buchholtz A (2008) Business and society: ethics and stakeholder management. Cengage Learning, Ohio

Desha CJ, Hargroves KC (2012) Fostering rapid transitions to education for sustainable development through a whole-system approach to curriculum and organisational change. In: Leal Fihlo W (ed) Sustainable development at universities: new horizons. Peter Lang, Frankfurt, pp 29–47

Denman B (2005) What is a university in the 21st century? High Educ Manage Policy 17(2):9–28

Disterheft A, Ferreira da Silva Caeiro SS, Manuel de Miranda Azeitiero U, Leal Filho W (2012) Implementing sustainability at the campus—towards a better understanding of participation processes within sustainability initiatives. In: Leal Filho W (ed) Sustainable development at universities: new horizons. Frankfurt, Peter Lang, pp 345–363

Donaldson T, Preston LE (1995) The stakeholder theory of the corporation: concepts, evidence and implications. Acad Manage Rev 20(1):65–91

Freeman RE (1984) Strategic management: a stakeholder approach. Pitman, Boston

Grayson D, Hodges A (2004) Corporate social opportunity. Seven steps to makes corporate social responsibility work for your business. Sheffield, Greenleaf

Hitchcock D, Willard M (2008) The step-by-step guide to sustainability planning: how to create and implement sustainable plans in any business or organisation. Earthscan, London

Hoover E, Harder MK (2014) What lies beneath the surface? The hidden complexities of organizational change for sustainability in higher education. J Cleaner Prod, pp 1–14 (in press)
Lozano R, Young W (2013) Assessing sustainability in university curricula: exploring the influence of student numbers and course credits. J Cleaner Prod 49:134–141
Manktelow R (2005) Stakeholder analysis and stakeholder management—winning support for your project. www.mindtools.com
Marans RW, Shriberg M (2012) "Creating and assessing a campus culture of sustainability: the university of Michigan experience. In: Leal Filho W (ed) Sustainable development at universities: new horizons. Peter Lang, Frankfurt, pp 557–569
Phillips R (1997) Stakeholder theory and a principle of fairness. Bus Ethics Q 7(1):51–66
Pohl M, Tolhurst N (2010) Responsible business: how to manage a CSR strategy successfully. ICCA, Chichester
Russell Group (2014) Russell Group. About us http://www.russellgroup.ac.uk/about-russell-group/ . Accessed on 11 Apr 2014
Stephens JC, Graham AC (2010) Towards an empirical research agenda for sustainability in higher education: exploring the transition management framework. J Clean Prod 18:611–618
Wals AEJ (2014) Sustainability in higher education in the context of the UN DESD: a review of learning and institutionalization processes. J Clean Prod 62:8–15
Wals AEJ, Blewitt J (2010) Third-wave sustainability in higher education: some (inter)national trends and development. In: Jones P, Selby D, Sterling SR (eds) Sustainability education: perspectives and practice across higher education. Earthscan, London, pp 55–74

Author Biographies

Dr. Louise Ellis is Head of Sustainability at The University of Leeds. Prior to this position she held a lectureship in organisational sustainability at the University, and was programme manager of the M.Sc. Sustainability (Environmental Consultancy and Project Management). She completed her Ph.D. in social sciences at the University, and has since conducted research in organisational sustainability, corporate responses to climate change, and the relationship between business and government policy.

Jessica Martin is the research and communications assistant for the Sustainability Service at the University of Leeds. She has just completed her master's degree in Political Theory at the University, with a particular focus on gender relations.

It's Contagious! Developing Sustainability Perspectives in Academic Life at a UK University

Stephen Scoffham and Nicola Kemp

Abstract

Finding ways to bring sustainability into the curriculum in meaningful ways is a challenge faced by universities worldwide. This paper explores how one UK university is responding by developing a futures perspective which can be applied both within and across disciplines. A variety of small scale projects have been initiated across the university including course innovation, one-off conferences, community engagement and international links. This approach is underpinned by clearly articulated principles and aims to build the capacity of staff and students. The examples presented illustrate the richness of the responses from different discipline areas. It is argued that building relationships has a key role in promoting long term institutional change and that staff development is crucial in initiating change. Deeper analysis using ideas drawn from social learning theory suggest that structures need to be able to embrace multiple perspectives and be organically responsive. This paper will be of particular interest to those who are interested in understanding processes which have the potential to embed sustainability thinking on a lasting level, rather than simply meeting short term targets and outcomes.

Keywords

Capacity · Change · Futures · Sustainability · University · Values

S. Scoffham (✉) · N. Kemp
Faculty of Education, Canterbury Christ Church University, North Holmes Road, Canterbury, Kent CT1 1QU, UK
e-mail: stephen.scoffham@canterbury.ac.uk

W. Leal Filho et al. (eds.), *Integrating Sustainability Thinking in Science and Engineering Curricula*, World Sustainability Series, DOI 10.1007/978-3-319-09474-8_7

1 Introduction

There is growing public and political concern about sustainability and the state of the environment. Environmental issues such as land degradation, species loss, pollution and climate change have come to dominate news stories and media headlines in recent years. Numerous scientific studies also confirm that the planet is under increasing ecological and environmental stress. One key report (WWF 2012) shows that there has been around a 30 % decline in global biodiversity health since 1970 and that the human ecological footprint is currently exceeding the Earth's carrying capacity by more than 50 %. Such pressures are manifestly unsustainable. Current levels of biodiversity loss have only been matched by the mass extinctions that occurred in geological history. Furthermore, Lovelock (2007) argues that we may already have reached a tipping point beyond which a vicious cycle of positive feedback will fuel the environmental changes which have already been set in motion. Whether or not this is the case, there is a compelling argument for making wise choices and for placing the natural world at the centre of decision making processes. The environment is the meta-narrative of our time and we ignore it at our peril.

As concern about the state of the natural environment has gathered pace, the discussion surrounding the issues has deepened and matured. When environmentalists first sounded the alarm about ecological degradation in the 1960s and 1970s there was a tendency to view environmental problems as isolated and discrete. Now it is increasingly recognised not only that they are linked but that human intervention is upsetting the balance of nature. Instead of an environment crisis there is actually a human crisis—a crisis of values and the way that we lead our lives (Booth and Ainscow 2010). How we see ourselves, how we relate to others and the principles which guide our actions have profound implications on the way we treat the planet. Social, economic and political dimensions have been introduced alongside environmental and ecological perspectives. This has broadened the agenda and shifted attention towards underlying principles such as equity and justice. Sustainability is more than just connecting with the environment. It is about connecting to others and understanding what it means to be human.

Universities are in a key position in society and have a central role to play in building social capital. The scale and the potential impact of the current environmental challenges require an informed response. Understanding environmental issues has become part of what it means to be educated in the modern world. Furthermore, the students who are at university today will be the leaders of tomorrow and are liable to have a significant role in forming opinion in a participatory democracy. However, universities along with other educational institutions have been surprisingly slow to recognise the importance of the sustainability agenda. A UNESCO (2005, p. 29) report on sustainability and teacher education noted that although many nations had embraced the need for sustainability education 'only limited progress' had been made in implementing it. More recent evidence from Australia and the UK suggests that the reform of the curriculum remains problematic, although sustainability policies and new approaches to estate management are now increasingly in place (Jones et al. 2010).

There are multiple reasons why universities have been so slow to embrace sustainability. One possible explanation is the inertia inherent in all large institutions. Rigid organisational structures may also play a part, along with a lack of funding. The pressures on the curriculum and the division of knowledge into academic disciplines add to this inertia. In their attempt to understand the forces which are restricting sustainability education in higher education institutions Jones et al. (2010) identified three principal inhibitors. Firstly, staff seek to safeguard their academic freedom and sometimes see sustainable development as extraneous to their discipline. Secondly, many staff believe that they lack the necessary knowledge and expertise to introduce it. Thirdly, the ethos of the institution and its organisational structures often tend to be antithetical to sustainability related teaching and learning. Such considerations chime neatly with the conclusion reached by Cotton et al. (2012) that 'key to addressing the issues and barriers that ESD presents, and to tapping its potential to enrich curricula and stimulate innovation, is staff development' (p. 10). The way in which one UK university is seeking to build capacity through a wide ranging staff development and change programme called the *Futures Initiative* is the subject of this paper.

2 Background to the *Futures Initiative*

The *Futures Initiative* was launched at Canterbury Christ Church University in September 2011. Until that time, Education for Sustainable Development (ESD) had been restricted to a few 'bright spots' and depended largely on the enthusiasm of a small band of committed individuals who had had a very modest impact. The appointment of a new Vice Chancellor and increasing pressure from central government heralded a change of thinking. At the same time there was a period of extensive internal debate regarding the formulation of a new strategic plan. The document which eventually emerged made clear references to sustainability and included a section on environmental responsibility. Furthermore, in spring 2011, the university made a successful bid to participate in the 'Green Academy' (an initiative run by the UK Higher Education Academy) which further raised the profile of ESD and created space for detailed discussions about underlying principles and values.

The growing support of senior management and the enthusiasm and expertise of academic staff made it possible to move forward with the ESD agenda. It was decided that given the constraints of the budget it would be best to work with colleagues who were already sympathetic and interested in environmental issues. A small pump priming fund was set up to enable staff to set up new programmes and embark on initiatives relating to ESD which they would not otherwise undertake. These would serve as examples to other colleagues and might become magnets around which new ideas could coalesce. An associated programme of seminars and residential staff development was devised to enhance and build capacity. There was a conscious decision to steer away from 'green' issues which tend to carry a considerable amount 'baggage' and are all too easy to pigeon hole and dismiss. By calling the programme the '*Futures Initiative*' it was hoped to suggest a forward-looking and open-ended approach which would be relevant to all disciplines and which would be manifestly relevant in equipping students for an uncertain and fast-changing world.

3 *Futures Initiative* Projects

Over the past 3 years the *Futures Initiative* has supported around 50 projects. These have varied enormously in both their scope and application. One tutor in Media, Art and Design decided to purchase a yurt to engage students with traditional methods of light and image capture. In sports science, the funding was used to purchase an exercise bike which generates power while it is being pedalled. In education, students who were training as teachers were brought together to design a student led module on 'sustainable schools for the future'. There are currently plans for a 'sound garden' on the main university campus where students, trainee teachers and very young children can interact with their surroundings on a sensory level. These projects, although very different, are united by a single ambition. They all seek to revitalise the curriculum through sustainability perspectives. At the same time, they illustrate how colleagues, once they feel they have been given permission, can bring a remarkable range of creativity and inventiveness to their teaching.

Most *Futures Initiative* projects have multiple aims and multiple purposes. They can, however, be sorted into groups. The most direct approach is to focus on the formal curriculum. Some tutors have favoured small scale changes and modified a single module by adding new case studies or examples which highlight environmental issues. Others have concentrated more on the mode of delivery, emphasising activity based learning (as in the Business School) or the use of electronic communication (as in the Department of Media, Art and Design). Projects which focus on the co-curriculum (the curriculum beyond the confines of validated programmes) have also proved popular. These include one-off conferences and events such as training in digital story telling for health professionals and activities which involve the local community such as work-based placements in archaeology. A third group of projects focuses on campus based activities of one kind or another. Foremost amongst these have been the creation of an allotment with planter spaces on the university campus, setting up 'insect hotels' to support wildlife and a devising a digital marketing website in conjunction with local community groups.

Recently the *Futures Initiative* has begun to have an impact overseas. Christ Church tutors working in both India and Malaysia have encouraged colleagues to include environmental perspectives in their own programmes. Meanwhile international students from around the world have participated in a new module on sustainability and business ethics launched in northern France. Reporting on sustainability initiatives at conferences and writing about them in academic journals has given added impetus, not only to individual tutors but also to the *Futures Initiative* as a whole. They also serve to broaden its base and raise its status. Three years after it was launched, the *Futures Initiative* is growing in maturity. Many academic staff are now aware of its activities and over a hundred have actively engaged with it.

4 Project Models

The *Futures Initiative* was originally envisaged as a set of one-off projects supported by staff development and student activities. It was acknowledged from the outset that there can be severe limitations to piecemeal change of this kind. Such approaches are fragmented, uncoordinated and difficult to sustain. Furthermore, as Stirling (2000) points out, the idea that just providing more environmental education will have the desired impacts is simplistic and ignores larger realities. The dynamics of social, economic and cultural systems serves both to circumscribe and to limit individual initiatives. On the other hand, the advantages of show-casing exciting developments in different faculties and departments across the university seemed a powerful way to reach new colleagues. Pragmatic considerations also came into play. Attempting to change the culture of an entire university is a huge ambition. Tackling it opportunistically seemed much more realistic.

After 3 years of implementation, it is now possible to begin to assess how the *Futures Initiative* has developed in practice. As well as a growing number of individual projects, 'hot spots' can be identified where sustainability activity is clustered either around a theme or a group of individuals. Four different models have emerged each with its own characteristics.

4.1 Stand-Alone and Static

Some of projects have been highly successful within their own terms of reference but have (necessarily) not developed beyond their original conception. A typical example of this is the physic garden which was developed on an under-used part of the university site as a collaborative sustainability venture between the Department of Geographical and Life Sciences and Allied Health Professionals. Including a selection of medicinal plants from the past and present day, the garden represents an important teaching resource and highlights the importance of maintaining biodiversity for the continuing progress of medical science. Now established, the garden is a valuable resource with considerable potential for future activity. However, the project itself has been completed. Projects of this type which are isolated and fully accomplished fall into the category of 'stand-alone and static' (Fig. 1).

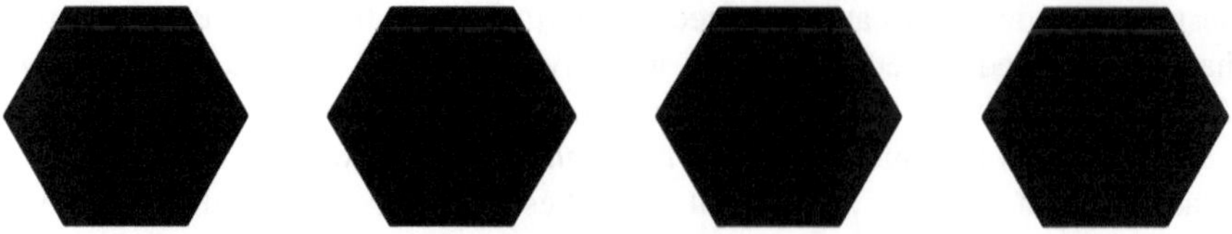

Fig. 1 Stand-alone and static projects are isolated and fully completed

Fig. 2 Some projects have potential for growth and change while remaining isolated

4.2 Stand-Alone and Growing

In contrast, there is a group of projects which have continued to grow beyond the scope of their initial conception. In some cases, the same activity is being refined, consolidated and repeated. The spring festival is a good example. The first festival was organised in 2013 by level 5 events management students and aimed to promote sustainability to local schools and community organisations as well as fellow students and staff. It proved a modest success but the students were hampered by their lack of knowledge about sustainability and environmental issues and the publicity reached a rather limited audience. A similar event in 2014 was informed by this experience. The students had a much better idea of what they were seeking to achieve and there was a much broader range of participants. Further refinements are already envisaged for 2015 but the project remains an isolated event involving specific tutors and a discrete group of students (Fig. 2).

4.3 Loose Clusters

In some cases projects fall within loose disciplinary clusters but are not themselves connected. Such clusters appear to reflect existing institutional structures. For example, within the Department of Childhood Studies six separate projects have been developed over the last 3 years. The projects include an introduction to Forest Schools, plans to develop a dedicated 'thinking space' for staff and students and a proposals for a new undergraduate programme on sustainability and the environment. Taken together these projects represent a 'hot spot' of *Futures Initiative* activity. Such clustering is also evident, to a lesser extent, in other departments where a number of projects have been developed and there is considerable interest

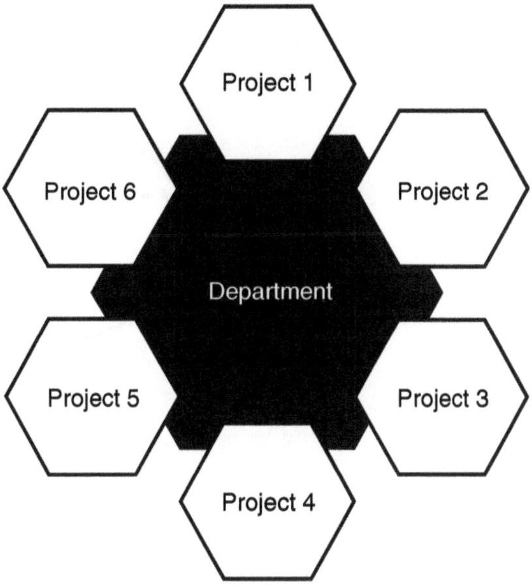

Fig. 3 Projects that have developed in a single department do not always connect

from different academic staff. It is interesting to note the way that these projects are connected through the head of department (who must support the application) but are not yet internally linked. There appears to be latent potential for future synergies and collaboration which has not yet been developed, perhaps due to conflicting or competing demands on staff time and energy (Fig. 3).

4.4 Clustered and Interconnected

There are also clusters of activity which have developed or coalesced around specific themes. Here there is a sense that 'one thing leads to another' and that there are many and varied connections between the projects and those involved in them. One such interconnected cluster is broadly focused around 'the outdoors' or more specifically 'the natural environment'. The starting point was an initial project funded by the *Futures Initiative* that enabled 21 colleagues from across the university to take part in an intensive 1 day Forest School event. Many of the staff who participated found not only that their interest in the natural environment had been rekindled but they were working alongside other colleagues with similar enthusiasms. This has resulted in a number of new *Futures Initiative* projects. At the same time a cross faculty network within the university (Connecting Children and Nature) has been established and strong collaborative relationships developed with a range of local community organisations resulting in collaborative research and knowledge exchange. Examples include an academic business partnership with *Kent Wildlife Trust* to evaluate their Forest School programme and an outdoor arts programme for children and young people supported by network funding from an

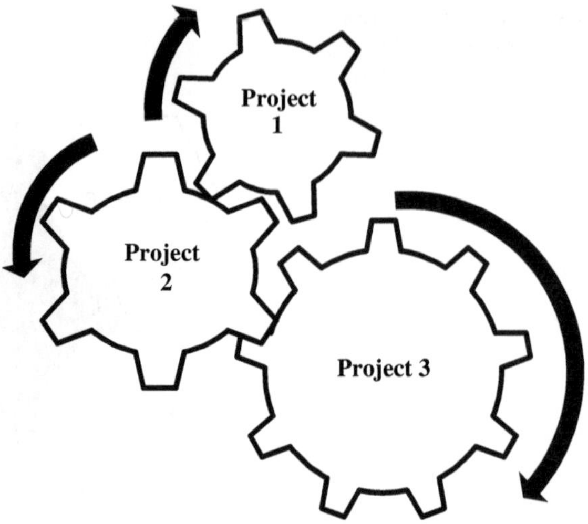

Fig. 4 Interconnected clusters have the potential to drive wider change because they feed off and re-inforce each other

independent community agency called *Artswork*. Large-scale funding is now being sought to formalise the partnership as a university-led hub which will continue to drive research and knowledge exchange as well as teaching and learning. This evolutionary model can be represented diagrammatically as a set of interlinking cogs in which the spurs represent the connections and inter-relationships between people. As a project develops momentum it begins to impact on other projects and new ideas emerge. This in turn has a knock on effect and starts to build a wider network (Fig. 4).

These four models are a way of understanding the impact of *Futures Initiative* projects. They represent different approaches all of which have a role in initiating institutional cultural change. Inevitably some projects have flourished. Others have proved less successful than anticipated. A few have failed to get going at all but might be activated in the future. The projects also overlap. For example, one of the'stand-alone and static' projects involved creating a website about sustainability for staff and students. Now this has been completed it represents a resource which can be used in all manner of ways. The project itself was free-standing but it will have an impact which promises not only to be dynamic but which will also to contribute to capacity building across discipline areas. Thinking about people and the way their ideas are changing focuses attention on one of the central aims of the *Futures Initiativ*—initiating new ways of thinking and being.

The most significant changes appear to be happening where there is connectivity between projects. Here the interaction between colleagues is generating outcomes which are more than the sum of the parts. Much depends on the attitudes of those who are leading the projects. Mutual trust and a willingness to share ideas are crucial. There also needs to be a genuine sense of professional equality in order to generate a culture in which risk taking and innovation can occur. Staff development events of all kinds can do much to facilitate this process. The leadership of senior managers and the commitment of the university also plays an important role as without validation new initiatives can easily fall by the wayside or be obscured by other pressures.

5 Building Connections

Over the past 3 years an increasing number of staff have participated in running *Futures Initiative* projects or come to associated publicity events. They have been invited to participate in a flexible programme of lectures, discussion groups, seminars and activities designed to help build networks and deepen understanding of key issues. One of the most important parts of this programme has been an annual 2 day residential session in which 16 staff and students have shared and developed ideas about sustainability. This event has been run as a joint venture between the university and a local environmental and educational organisation called *Commonwork*. Part of the appeal of this joint programme relates to 'place'. *Commonwork* is based in an ancient manor house with an organic dairy farm and gardens. The opportunity to witness and experience sustainable living in practice has been a significant element in the course. However, the wider advantages of being in a space that is well away from the everyday pressures of academic life have also been apparent. Furthermore, and perhaps crucially, *Commonwork* staff have been able to draw on their experience and expertise in helping to plan and deliver a practical programme which focuses on what it means to be human in a global world. Many participants have found this approach, which is deliberately distanced from the details and realities of managing academic programmes, both liberating and affirmative.

Assessing the impact of the residential courses and staff development events is notoriously difficult. Questionnaires, surveys and formal evaluations tend to focus on surface level impacts; the lived experience and deeper effects are much harder to evaluate partly because they are intangible. The *Futures Initiative* team has made various attempts to understand more about way in which the programme can help individuals transform their thinking. The data suggests that most participants reported that their understanding has matured and deepened. This is neatly summarised in the following response:

> I leave with more questions than answers. But with increased confidence in formulating the questions and a developing framework from which to develop a response. I am seeking time to reflect on my own practice and curriculum development. I haven't done this yet but I have considered broader ideological institutional positions during the programme. This has been hugely beneficial.

Reflecting on personal values and professional practice are fundamental stages in developing new frameworks and models. One of the other respondents from the 2012 cohort made a similar point but also went to talk about how their thinking had deepened:

> Actually I am coming away with something which I don't have. It's almost like a hue, almost like a shade of difference. And nothing negative can come out of that, only positives can come from that. But it's difficult to talk about in hard language. How one measures it I'm not sure.

Others also talked about new layers of understanding. '*My thinking is richer, deeper, wider,*' one said. My notion of sustainability has become '*much more personal,*' said another. The chance to meet colleagues with similar interests and to share ideas was widely acknowledged and highly prized by many. A determination to move forward positively and to have a real impact in the future also shines through in many of the responses. A more detailed analysis of the short term impact of the residential programme appears elsewhere (Scoffham et al. 2014). The longer term benefits have yet to be documented and explored more fully from a critical perspective.

6 Theoretical Considerations

The *Futures Initiative* is generally regarded positively and staff tend to welcome the opportunities that it offers. To understand this further it is illuminating to consider the notion of communities of practice developed by Wenger (1998). According to Wenger, a community of practice exhibits the following three characteristics:

(a) it has a shared common interest or domain,
(b) members discuss ideas, share information and engage in joint activities which builds relationships and enhances learning as they pursue their common interests,
(c) the shared repertoire of experiences, stories, tools and ways of addressing recurring problems which the community builds up further extends its professional expertise.

The embryonic and small networks which have developed as a result of *Futures Initiative* projects show a significant number of these features. Many of those involved talk positively about the experience. They enjoy making links with other colleagues and appreciate the way that they are drawn into reflecting on their personal experiences and values. Some colleagues find this deeply meaningful. Others see it as an opportunity to refresh and reinvigorate their teaching. It would be unrealistic, however, to expect such synergies from the *Futures Initiative* as whole. When staff are divided into different faculties and departments and spread between campuses there are few opportunities to meet. Even small groups find maintaining contact on a regular basis a significant challenge. Furthermore, the scale and complexity of the sustainability agenda makes it hard to achieve the necessary unity. At best colleagues are able to exchange ideas.

As well as supporting the development of small communities of practice, the *Futures Initiative* also respects the professional expertise and competence of colleagues. There is an underlying premise that subject knowledge is held by the academic staff within each discipline and that it is they who hold the key to embedding sustainability within the curriculum. This has meant that there is no unified, top-down check list of whether a particular module or programme is 'sustainable.' Rather, through the *Futures Initiative* process, individual members of staff have engaged with sustainability and have brought their own academic disciplinary knowledge and understanding to the challenge. As one project leader explains, sustainability is not an '*add-on project*' but something which *'arises out of what you're teaching'*. This is an important observation because it shows how a sustainability mind-set can help tutors to circumvent two of the three principal inhibitors identified by Jones et al. (2010) which were referred to earlier. It also highlights the advantages of alternative ways of thinking. Deleuze and Guattari (1999), for example, use the rhizome as a metaphor for representing approaches which validate multiple and parallel perspectives. Not only is rhizomatic thinking much more flexible than linear thought processes, it is also organic. Deleuze and Guattari point out that, along with other natural organisms, a rhizome has neither an organising memory nor a central control mechanism. It is, as they put it, like 'a system without a general' (p. 22). Such ways of thinking resonate with the sustainability agenda which is characterised by feedback, loops, interconnections and complexity.

One other idea which has informed the *Futures Initiative* from a very early stage is the notion of 'wicked problems'. The idea of a 'wicked problem' derives from social policy and was first formulated by Rittel and Webber (1973) who noted that there were a whole range of social planning issues which cannot be successfully treated with traditional, linear, analytical methods of thinking. In contrast to 'tame problems' which are well-defined and stable, 'wicked problems' are ill-defined, ambiguous and often associated with strong moral, political or professional dimensions. The banking crisis which developed in 2008–2010 is a good example of a wicked problem as is the on-going issue of climate change. Rather than having a single or definite solution, wicked problems have multiple causes and can only be addressed through a range of strategies, none of which can be shown to be better than another. A further feature of a wicked problem is that you don't really understand the issues until a solution emerges, but the solution itself changes the nature of the problem. This means there are no stopping rules and the way the problem is defined is itself part of the solution. Sustainability and environmental issues show many of the characteristics of wicked problems. Coming to terms with them implies the ability to tolerate uncertainty and a willingness to accept multiple perspectives—exactly the same qualities that characterise rhizomatic thinking.

7 Conclusion

The *Futures Initiative* was evaluated in 2013 by an independent evaluator (McCoshan 2013) who highlighted some of its distinctive features. Foremost amongst these were:

(a) a very open, grass roots approach to formal curriculum development,
(b) a focus on long term staff development and transformation within an inclusive culture,
(c) the development of a website to introduce staff and students to sustainability thinking,
(d) a recognition (shared by other universities) that sustainability involves persuasion and is best embedded by changing people's hearts and minds.

Ensuring that the *Futures Initiative* was underpinned theoretically and supported by clearly articulated values and principles has been crucial to its development. The ideas behind it were developed over a number of years, first by the ESD committee and then by the Strategic Plan Group followed by the Green Academy team. The emerging principles were then confirmed in conversations, rehearsed in reports and amended in committees.

Developing the sustainability agenda has involved grappling with many contradictions and tensions. There are those who, like Knight (2005), argue that universities should confine themselves to academic reflection and analysis and that it is not their job to become involved in contemporary issues and affairs. Striking a balance between dispassionate analysis on the one hand and engagement and motivation on the other can be challenging, particularly in classroom situations. The dangers of activism, propaganda and bias also need to be recognised and acknowledged as part of a mature understanding of the domain. Tokenism and eco-guilt present further threats. However, sustainability and environmental problems need to be addressed in a positive frame of mind and in a spirit of hope. Critical examination of the issues is essential, but so too is innovation and creative thinking. It is not the role of education to make students into victims; rather it should seek to empower and liberate. Anticipating the challenges which lie ahead is an impossibly difficult task. Equipping students with the capabilities which they are likely to need in an uncertain world and introducing them to the knowledge and understanding which will help them to understand what is happening round about them, offers a much more constructive way forward.

Whilst there is no expectation of orthodoxy or compliance, the philosophy which underpins the *Futures Initiative* is spelt out in guidance notes which accompany the bid application form. Here there is a clear acknowledgement that ESD involves a commitment to social and economic justice and respect and care for life in all its forms. The value of cultural diversity set within a framework of tolerance and peace, and the responsibilities which the current generation has to people of the future generations are also highlighted. The most direct and effective way universities can honour these commitments is by working with students at a formal and informal level. The ultimate test for *Future Initiative* projects is their impact on the student body. It was recognised from the outset that this is not easily measurable, that it is a long term process and that there is no single or direct route to achieving these goals.

There are encouraging signs that the *Futures Initiative* is beginning to gain momentum—that it is indeed contagious. We are now seeking to understand more about its impact and to identify those approaches which are likely to be most effective in the years ahead. One question, yet to be answered, is whether inter-related thematic clusters represent a mechanism that has the potential to drive wider institutional change. Another question is whether the *Futures Initiative* has reached a critical tipping point beyond which it becomes embedded in the university thinking and takes on a momentum of its own. What we can confirm is that where colleagues have similar interests, a commitment to shared values and the opportunity to be innovative, their ideas are likely flourish. By facilitating the development of small communities of practice and, as it were, validating colleagues' enthusiasm, the *Future Initiative* has derived a strength which goes far beyond the resources that have been directly committed to it. This bodes well for the future.

References

Booth T, Ainscow M (2010) Index for inclusion: developing learning and participation in schools. Centre for Studies on Inclusive Education, Bristol, UK, p 190

Cotton D, Sterling S, Neal V, Winter J (eds) (2012) Putting the 'S' into ED—education for sustainable development in educational developments. Staff and Educational Development Association, London, UK, p 38

Deleuze G, Guattari F (1999) A thousand plateaus: capitalism and schizophrenia, London, UK, Athlone, p 687

Jones P, Selby D, Sterling S (2010) Sustainability education: perspectives and practice across higher education. Earthscan, London, UK 364 p

Knight P (2005) Unsustainable developments, The Guardian, 8 February. Available at http://www.theguardian.com/education/2005/feb/08/highereducation.administration (Last Accessed 29 Mar 2014)

Lovelock J (2007) The revenge of gaia, London, UK, Penguin, p 222

McCoshan A (2013) Reflections on the impact of green Academy at canterbury christ church, unpublished internally commissioned report, p 8

Rittel HWJ, Webber MM (1973) Dilemmas in a general theory of planning. Policy Sci 4:155–169

Scoffham S, Buckley Sander J, Bracewell J (2014) Developing a sustainability mindset: a collaborative approach to staff development. In: Education for sustainable development and global citizenship: good practice case studies in higher education, Higher Education Academy Available at http://bit.ly/14Yyggc (Last Accessed 29 Mar 2014)

Stirling S (2000) Sustainable education: revisioning learning and change. Green Books, Totnes, Devon, UK, p 96

UNESCO (2005) Guidelines and recommendations for re-orientating teacher education to address sustainability, Available at http://unesdoc.unesco.org/images/0014/001433/143370e.pdf (Last Accessed 29 Mar 2014)

Wenger E (1998) Communities of practice: learning, meaning, and identity. Cambridge University Press, Cambridge, p 318

WWF (2012) Living planet report 2012, Available at http://wwf.panda.org/about_our_earth/all_publications/living_planet_report/2012_lpr/ (Last Accessed 29 Mar 2014)

Authors Biography

Dr. Stephen Scoffham has been involved in sustainability education for the past 35 years. His first book, *Using the Schools Surroundings* (Ward Lock Educational 1980) offered schools and teachers a range of practical strategies for outdoor work derived from research. His latest publications, *Collins Primary Geography books 1–6* (Harper Collins 2014), bring environmental perspectives to classroom settings in the UK and around the world. Stephen is currently Honorary Publications Officer for the UK Geographical Association and a long term member and trustee of his local development education centre (WEDG). His research interests focus on primary school geography, creativity, the environment and global learning. Stephen is a principal lecturer in the Faculty of Education, Canterbury Christ Church University, UK and Director of the University's *Futures Initiative* which promotes sustainability and futures perspectives in academic life.

Dr. Nicola Kemp currently works in the Department of Childhood Studies at Canterbury Christ Church University, UK and is a leading member of the '*Futures Initiative*' team. With an academic background in rural and environmental geography, she worked in the rural regeneration team at Kent County Council for 10 years before moving to Canterbury Christ Church. She has a particular interest in children's experiences of the natural environment and has developed the university's 'Connecting Children and Nature' network which brings together staff, students and a range of local organisations to explore and develop this agenda through research, knowledge exchange and curriculum development. Her research interests also include home education, alternative curricular and outdoor learning.

A Strategic Framework for Developing Interdisciplinary Minors on Climate Change and Sustainability Policy: The CLIMASP-Tempus Example

Vassilios Makrakis and Nelly Kostoulas-Makrakis

Abstract

Higher education plays a unique and critical role, one often underestimated and/or neglected in making a more sustainable society. This paper deals with the development of an interdisciplinary minor programme on climate change and sustainability policy across 10 universities in Egypt, Jordan and Lebanon. It is a European Commission funded programme led by the University of Crete, Greece. Climate change is one of the most pressing issues affecting all regions in the world and especially the Middle East and Mediterranean region which is the focus of this paper. The climate change topic should shift from a strict disciplinary orientation focused on natural sciences to a science, educational, economic and sociological orientation. The evolved CLIMASP initiative is designed to develop, implement and assess interdisciplinary minors that aspire to meet the needs of climate change education and sustainability policy.

Keywords

Interdisciplinary studies · Minor · Climate change · Sustainability policy

V. Makrakis (✉) · N. Kostoulas-Makrakis
Department of Primary Education, University of Crete, 74100, Rethymnon, Crete, Greece
e-mail: makrakis@edc.uoc.gr

N. Kostoulas-Makrakis
e-mail: nkostoula@edc.uoc.gr

W. Leal Filho et al. (eds.), *Integrating Sustainability Thinking in Science and Engineering Curricula*, World Sustainability Series, DOI 10.1007/978-3-319-09474-8_8

1 Introduction

We are increasingly confronted with complex social, economic and environmental problems locally and globally such as climate change, which is seen as the most critical problem facing humanity in recent times. The concentration of CO_2 in the atmosphere that has contributed to greenhouse effect known as 'global warming' has been increased significantly since the industrial revolution. More specifically, the CO_2 concentration in the atmosphere is currently about 370 parts per million (ppm): an increase of more than 30 % since 1750 (Osman 2010). According to the IPCC (2007) global surface air temperature increased by 0.76 °C from 1850 to 2005. The linear warming trend over the last 50 years is recorded by 0.13 °C per decade, with a full range of projected temperature increase of 1.1–6.4 °C by the end of the 21st century (ibid.). The increase is largely due to anthropogenic emissions of CO_2 from fossil fuel combustion and to a lesser extent land-use change, some industrial processes, and biomass combustion (IPCC 2001). As it is pointed by Rahmstorf (2008), "other possible causes, such as solar activity, volcanic activity, cosmic rays, or orbital cycles, are well observed, but they do not show trends capable of explaining the observed warming" (p. 47). In his study "the evidence for the anthropogenic increase in atmospheric CO_2 concentration and the effect of CO_2 on climate is proven beyond reasonable doubt and that a mass of evidence points to a CO_2 effect on climate of 3C +(-) 1.5C global warming for a doubling of concentration" (p. 48).

Increased energy use by individuals and industries has long been associated with higher levels of development and Gross Domestic Product (GDP). In 31 countries, less than 20 % of the population controls more than 50 % of the national wealth, so an enormous economic wealth has been accumulated almost exclusively in the developed world, while the world's poorest nations have grown even poorer (UNDP 2000). Presently, about 20 % of the world is consuming 80 % of the world's resources, while those consuming less are trying to catch up by following existing models (UNEP 2011). These global economic disparities are associated with a global "carbon disparity" that reflects different levels of consumption, production, and wealth: "Twenty percent of the world's population is responsible for 63 % of the emissions, while the bottom 20 % of the world's population is only releasing 3 %" (Roberts 2001, p. 503). As depicted in Table 1 that shows the development of cumulated CO_2-emissions per capita in selected regions according to GDP, the United States with a 5 % of world's population releases more carbon than all the other regions included in the table (Oberheitmann 2013).

Climate change presents a real threat that has environmental, social, economic and political impacts worldwide and in the Arab region, despite its low contribution to global CO_2 emissions, and even lower if the oil-producing Middle Eastern countries are excluded. The region's emissions of Greenhouse Gases (GHG) are generally small in absolute terms—less than 5 % of the world's total (Osman 2010).

By the end of this century, this region is projected to experience an increase of 3–5 °C in mean temperatures and a 20 % decline in precipitation (IPCC 2007). Due to lower precipitation, water run-off is projected to drop by 20–30 % in most of Middle

Table 1 Development of cumulated CO_2-emissions per capita in the world (2010–2050, in *t* CO_2, cum)

Time period	2010	2020	2030	2040	2050
Africa	38	49	61	73	84
Asia Pacific	86	133	191	263	344
Middle East	200	287	387	497	617
Central and South America	88	113	139	165	192
Europe and Eurasia	614	720	835	959	1,088
North America	985	1,180	1,385	1,596	1,809
USA	1,290	1,537	1,798	2,068	2,343

East & North Africa (MENA) by 2050 (Milly et al. 2005). Water scarcity also threatens food security by reducing agricultural productivity, as well as hindering human health and economic development; water scarcity can also lead to additional environmental stress, as well as increase tensions within and between nations sharing water resources (El-Fadel and Bou-Zeid 2001; Osman 2010). Climate change is associated with other global, national and local environmental problems such as loss of biodiversity, deforestation, desertification and resource depletion.

It is also widely discussed that the poor, women and children are expected to be the most vulnerable to the detrimental impacts of climate change as these groups possess the least assets and resources and has less power to adapt to the climate change impacts (Makrakis et al. 2012a, b). It has been reported that by 2010, an estimated 26 million people had been displaced by climate change; 20 million of them are women (Loftus-Farren 2013). Increasing the capacity of the disadvantaged social groups to adapt to climate change, in the coming years, is extremely important. While the scientific evidence for climate change grows, the policy responses have so far had little or no impact on the production of emissions (Helm 2008). Policy interventions need to be considered to counteract the existing social, political and economic obstacles to the effective and efficient tackling of climate change (Haigh 2011; Dubash et al. 2013; Figueres 2013). Failing to take policies and actions may result in significant costs and irreversible situations, especially on disadvantaged groups that will possible lead to the worsening of existing social and economic disparities worldwide (ibid.). To this end, Higher Education Institutions (HEIs), especially in the Mediterranean and Middle Easter region, that is going to face more severally the climate change impacts, should play a considerable role.

2 The Role of Higher Education

Higher education while being criticised for leading to the current sustainability crisis and climate change (Corcoran and Wals 2004), it can play a unique and critical role that is often underestimated or even neglected (Leal 2010; Makrakis 2012; Tilbury 2012). Although the Middle East region's contribution to the damage of the global climate is less when compared to developed regions, there is broad recognition of the urgency for meeting the climate change challenge since this region is one of the most vulnerable to climate change (Makrakis and Kostoulas-Makrakis 2013). As it has been revealed in the introduction, significant changes in climate are present worldwide and the debate is no longer about: 'Will our climate change?' but rather about 'How will it change, how can we cope with its impacts (adaptation), and how can we limit future climate change (mitigation)?' These questions generate a range of new challenges to higher education institutions, such as: How curricula will be changed to address climate change and sustainability policy? What can we do about climate change? How can we empower disadvantaged groups to face climate change impacts? What are the solutions with climate change? What climate change policies have been formulated locally, regionally and globally? Why is there a conflict and huge variation on climate change issues?

Climate change education is thought to depend heavily on interdisciplinary knowledge and practice. If students are only taught to think in particular disciplines without integrating various disciplines into a coherent framework to study sustainability issues such as the climate change phenomenon in its social, environmental, economic, cultural and political dimensions, it is unlikely that they will truly understand the need for personal and social transformation towards a sustainable society. A world climate change survey revealed that climate change matters are mostly taught in the natural and social science fields, with a lesser emphasis in engineering and humanities (Leal 2010). It may not only be used as a tool towards greater understanding of different variables associated with climate change, but it is also suitable for teaching the English language (Leal Filho and Manolas 2012). Interdisciplinary teaching and learning is a challenge for both teachers and students (Woods 2007; Bleakly and Brennan 2011). There is thus need for interdisciplinary study programmes, either major and/or minor in climate change that aims to link various disciplines. Although, interdisciplinary teaching and learning is highly prioritised in the participating institutions, in practice, there is lack of interdisciplinary perspective and motivation among teaching staff (Makrakis and Kostoulas-Makrakis 2013). Interdisciplinary collaboration is essential for modernizing higher education and it is a necessary condition for any transformation to meeting the challenges of climate change and sustainable development in the region.

It is also important to recognize that not all teaching and learning needs to be interdisciplinary, but that integrating knowledge from multiple disciplines and perspectives is of paramount importance, especially when dealing with climate change and other sustainability issues. Climate change requires educated individuals who are able to bridge the isolated academic fields in order to develop sustainable alternatives and to facilitate policy choices toward a path of ecologically sound and socially equitable solutions. There is a critical need for Middle East universities in light of the climate change challenges their region is facing to cultivate interdisciplinary expertise among their future leaders across many of the societal sectors.

3 The CLIMASP Tempus Initiative

Taking into consideration the documented need for establishing interdisciplinary programmes on climate change and sustainability policy, the UNESCO Chair ICT in Education for Sustainable Development at the University of Crete and RCE Crete invited a number of institutions from its international network to participate in its proposal for the development of such programmes. In total 16 institutions, among them three Institutions from European Union (University of Crete, Greece, Frederick University, Cyprus and Leuphana University, Germany), 10 universities from Egypt (Aswan University, Heliopolis University, Port Side University and Suez Canal University), four from Jordan (Jerash University, Hashemite University, Madaba University and University of Jordan), two from Lebanon (Lebanese American University and Notre Dame University) and three NGOs, namely the Egyptian Business Women Association (EBWA)-Egypt, the Jordan Environment Society (JES)- Jordan and the Education for Employment Foundation (EFE) stationed in Egypt and operating in various Middle East and North African (MENA) countries joined the CLIMASP Consortium. The proposal was approved and financed by the European Commission Tempus programme.

The rationale for selecting the 16 partners in the CLIMASP Consortium lies on the capacity and will of these institutions to carry out the objectives set. In particular, the three EU Universities have established a close collaboration through the UNESCO Chairs they hold (University of Crete and Leuphana University Luneburg) and previous programmes. The University of Crete has also collaborated with five partners institutions in a previous Tempus project in its attempt to turn it into a sustainable university (Kostoulas-Makrakis and Makrakis 2012). With respect to three NGOs, Jordan Environmental Society (JES) with 11 branches in the region is devoted in raising public awareness and promotes community action. As women and young people are the groups who are going to be largely affected by climate change, we have selected the EBWA that aims to empower women, advocate their climate change rights, and build young women entrepreneurship. As the CLIMASP program besides promoting climate change, aims to increase the employability of graduates, we have selected the EFE, an affiliate of EFE Middle East dedicated to youth training, employment, entrepreneurship and civic engagement. The role of these NGOs will be very instrumental in the process of defining the needed competence to be integrated into the CLIMASP course curriculum. All of them showed increased interest on CLIMASP; are committed to involve their faculties in the project; are geographically distributed (centre-periphery) and aspiring to be "role models" in the wider Arab region.

- The wider CLIMASP objectives are to transform current unsustainable practices that prevent interdisciplinary collaboration and promote sustainable leadership in the partners' countries universities and its graduates. The specific objectives are to:
- Develop capacity-building programmes to train university teaching staff and key administrators for interdisciplinary collaboration and building partnerships with local/national/regional partners;

- Involve university staff and other key stakeholders (e.g., students, professionals) in the development of an undergraduate interdisciplinary programme on CLIMAte change and Sustainability Policy (CLIMASP) in each partner country university;
- Integrate and implement the CLIMASP programme as an integral part to existing undergraduate academic degrees in disciplines such as education sciences, applied sciences, technical sciences, economics/business sciences, and social sciences; and
- Monitor, evaluate and review the CLIMASP programme in each partner country institution.

These objectives entail 10 key components:

1. Identify and analyse the barriers and drivers of stakeholders' inputs to interdisciplinary CLIMASP curriculum development.
2. Develop a survey tool to be administered to nearly 400 professionals working in the field of CLIMASP to identify what competences are needed.
3. Develop guidelines for strengthening stakeholders' inputs in the CLIMASP curriculum development and implementation.
4. Organise and carry out three national and two regional training workshops for preparing teaching staff in the CLIMASP curriculum development and implementation.
5. Develop the CLIMASP online platform to provide continuous training support and resources in the Arabic and English language.
6. Develop, validate and accredit 240–300 interdisciplinary course syllabi and course modules of 6 ECTS each related to CLIMASP across the 10 partner universities. Course syllabi and modules will be both in Arabic and English.
7. Develop a Centre for Integrative and Interdisciplinary Studies with an ICT Laboratory in each of the 10 partner Universities to institutionalise and support the CLIMASP programme (N = 10).
8. Carry out formative and summative assessments of the CLIMASP pilot and full implementation, involving more than 100 teaching staff, 120 courses and 3000 undergraduate students.
9. Introduce the Euro-Arabpass Diploma in the CLIMASP undergraduate program adapted from Europass.
10. Develop an International Network of CLIMASP Studies.

In particular, the minor in Climate Change and Sustainability Policy could offer students a unique, inter/multidisciplinary understanding of climate change. The climate change topic should shift from a strict disciplinary orientation focused on natural sciences to a science, educational, economic and sociological issue. The evolved CLIMASP initiative is designed to develop, implement and assess a broad interdisciplinary curriculum to reflect this expanding focus. Choosing a minor combined with a major degree it gives students the opportunity to pursue a second area of interest that can open significant career opportunities, besides building knowledge and skills in an area that fulfils their personal interests. The interdisciplinary CLIMASP courses will be integrated into a regular undergraduate degree in disciplines like education sciences, technical sciences, economics/business sciences and social sciences. It will consist of core courses, elective courses, and the required capstone course in the three concentration areas, namely, (1) Climate Change, Environment and Society; (2) Climate Change, Economics and Public Policy; and (3) Climate Change, Science and Technology. Each of the core and elective courses will be of 6 ECTS and the capstone course of 10 ECTS. The capstone course will be based on an internship that provides a strong mechanism for integrating academic coursework with practical experience. The amount of the minimum courses to be taken by undergraduate students to qualify for the CLIMASP minor to be awarded in addition to their undergraduate major degree is expected to be around 60 ECTS. This will provide students a formal credential through transcript documentation (EuroArabpass) adapting the Europass diploma to certify that they have developed leadership in the field of climate change and sustainability policy.

4 The CLIMASP Approach

To reach the objectives set and outcomes, we have developed a conceptual and methodological framework depicted in Fig. 1. According to this framework, the curriculum development process for the interdisciplinary CLIMASP programme

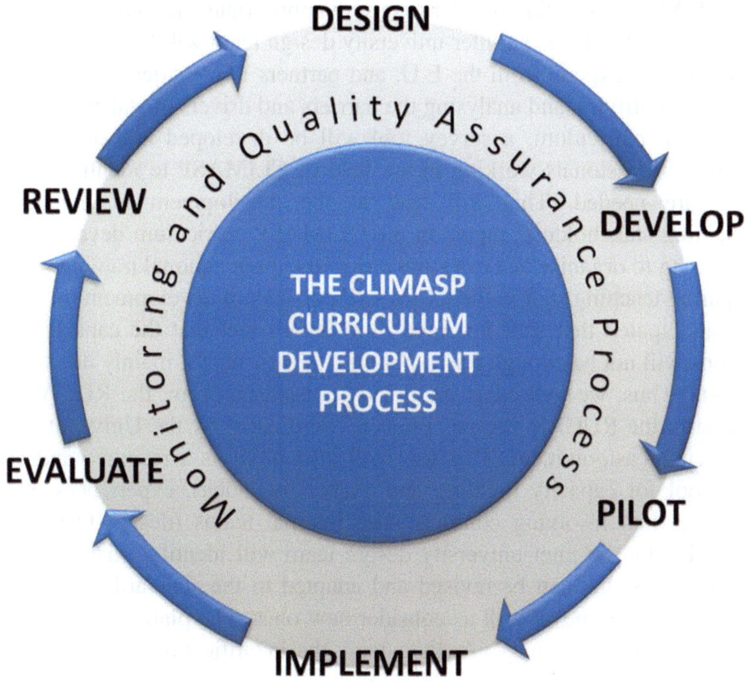

Fig. 1 The CLIMASP curriculum development process

integrates six stages: (1) design, (2) develop, (3) pilot, (4) implement, (5) evaluate and (6) review through a cyclical and interlined process. The monitoring and quality assurance process is planned to run across all the six stages.

5 Design (December 2013–June 2014)

This is the stage in which all stakeholders become part of the process to investigate compatibilities and conflicts of interest related to climate change and strengthening their contribution to operationalise the curriculum. We will focus on stakeholders at the level of interest groups and interest organisations, local trade unions, and civic initiatives and societies.

For monitoring and quality assurance, in each partner university, a small design team consisted of members from each academic discipline involved will be set up to get engaged with: (a) creating a conceptual framework for the interdisciplinary CLIMASP undergraduate programme; (b) engaging and involving stakeholders in curriculum design; and (c) working with individual faculties and university high level administration. The design team will set up and lead three working groups (WG); one for each of the three CLIMASP concentration areas. The WGs will be charged with: (i) mapping of partner university courses that could be reviewed and modified appropriately for the programme; (ii) brainstorming ideas for courses and course content; and (iii) developing of the pedagogical and structural principles that will guide development of study units/packages. Coordination among all design teams and WGs is essential in order to ensure appropriate and timely preparation of all assigned works. Thus, an inter-university design team will be established which will include key experts from the E.U. and partners universities.

Besides identifying and analysing the barriers and drivers of stakeholders' inputs to CLIMASP curriculum, a survey tool will be developed and administered to nearly 400 professionals working in the field of CLIMASP to identify what competences are needed. This will lead to the development of guidelines for strengthening stakeholders' inputs in the CLIMASP curriculum development. We are also going to organise and carry out one of the three national training workshops for preparing teaching staff in the CLIMASP curriculum development (objective 1).

It is anticipated that staff involved in WG will feel that the capacity-building workshops will not be enough for responding to their needs, mainly due to the time constraints. Thus, we will adapt an already existed platform, the RUCAS Toolkit developed in the RUCAS-Tempus project, coordinated by the University of Crete (http://www.rucastoolkit.eu). This tool will also serve as a resource and communication tool for capacity building and exchanging ideas, experiences, problems encountered, and resolving conflicts. The internal teams (design team and WG team) guided by the inter-university design team will identify existing courses in various faculties that can be revised and adapted to the standards needed for the CLIMASP programme as well as consider new ones. The platform will be both in Arabic and English language. At this stage, the identification of suitable existing courses in various faculties that can be revised and adapted to the CLIMASP program will be carried out. We are also going to define the technical specifications for the ICT Laboratories and the functions of the Centres for Integrative and Interdisciplinary Studies to be developed in the next stage.

6 Develop (July 2014–June 2015)

In this stage, every WG led by the institutional design team in each partner university will proceed to the development of 24–30 course syllabi and course modules in the CLIMASP three concentration areas on the basis of stakeholders' inputs (Objective 2). The methodological approach will be based on a modular scheme, especially developed to meet the objectives of the CLIMASP project. The basic idea underlying the CLIMASP modular design is to organize its programme as a set of courses that can be developed independently and then plugged together to form a coherent but flexible programme. The modular characteristic of the programme and its course modules will provide flexibility that enables adapting the different courses and individual course modules to a wide range of academic requirement and enable each partner institution to tailor CLIMASP according to its specific needs. The inter-university design team will identify and promote coherence of a number of courses with other partner and E.U. universities so to increase opportunities for inter-university student and staff mobility and cooperation/co-teaching.

The 240–300 course modules will be developed in Arabic and English across the 10 partner universities will go through validation and accreditation by stakeholders and external evaluators before integration into the respective undergraduate disciplines (Objective 2). The inter-university design team in cooperation with the institutional design teams will define the minimum requirements that undergraduate students have to fulfil in order to qualify for the supplementary interdisciplinary CLIMASP minor in addition to their major disciplinary degree (Objective 3). To better qualify teaching staff for the implementation process, the 2nd national training workshop followed by the 1st regional training workshop will be carried out in each partner country as well as the establishment of the Centres for Integrative and Interdisciplinary Studies and the ICT Laboratories on the basis of the specifications set in the previous stage.

7 Pilot and Implement (July 2015–December 2016)

For a smooth transition from disciplinary to interdisciplinary curricula, a piloting stage will start on a small number of CLIMASP courses. Piloting creates the opportunity to demonstrate what interdisciplinary learning and teaching looks like throughout the semester and allows faculty, students and evaluators to observe processes, methods and practices. Based on the pilot assessment, proper interventions to enhance content and methods will be applied to all the 240–300 CLIMASP course modules across the 10 partner universities. The teaching staff involved in the implementation will participate in the 3rd national workshop followed up by the 2nd regional training workshop to help them review the implementation process. Simultaneous support will be provided online through the web-based platform dedicated to that. As a result of these workshops, proper "on-the-job" interventions will be done to ensure quality (formative evaluation). During the implementation process, courses will be also assessed both internally (peer review) and externally assigned team.

8 Evaluate and Review (Whole Period of the Project)

Evaluation is being integrated into the CLIMASP program *before*, *during* and *after* implementation in the following way: **Evaluation before project implementation includes the activities performed in the design and development stages** of capturing stakeholders' expectations, competences needed and aversions as well as validating/accrediting the course modules. **Evaluation during implementation (formative evaluation) will be based** on four key processes: (1) record; (2) reflect; (3) react and (4) review, which will enable to progressively review the project strategies according to the changing circumstances to attain the desired outcomes. Teaching staff and students will be engaged in collecting relevant data and key experts to analyze data collected for identifying possible deficiencies and root causes that can lead to corrective action (Objective 4). The evaluation methods include: self-assessment, peer/external review, questionnaires etc. Internal and external assessment can provide data that demonstrates how the CLIMASP interdisciplinary courses can foster the objectives set. **Evaluation after project implementation (summative evaluation) which will involve both, teaching staff and students as well as other stakeholders** will demonstrate whether CLIMASP has achieved its aims and objectives.

9 Concluding Remarks

As pointed previously, higher education institutions are being challenged to respond to the sustainability crisis, in which climate change is one of its most cited issues. The CLIMASP initiative adopts an interdisciplinary and systemic approach that at a wider level aims to transform current unsustainable practices that prevent interdisciplinary collaboration; and promote sustainable leadership in the partners' countries universities and its graduates. Specific objectives, within the aforementioned wider ones, include: (1) to develop capacity-building programmes to train university teaching staff and key administrators for interdisciplinary collaboration and building partnerships with local/national/regional partners; (2) to involve university staff and other key stakeholders (e.g., students, professionals, employers) in the development of undergraduate interdisciplinary programmes (minors) on climate change and sustainability policy in each partner country university; (3) to integrate and implement the interdisciplinary programme as an integral part to existing undergraduate academic degrees in disciplines like education sciences, applied sciences, technical sciences, economics/business sciences, and social sciences; and (4) to monitor, evaluate and review the interdisciplinary programmes on climate change and sustainability policy in each partner country institution. Such a programme will lead to high-quality teaching, learning, research and curriculum modernisation at the undergraduate level, which will pave the way for establishing stand-alone undergraduate degrees in the field of climate change and sustainability policy. It will also be harmonised with EU standards and the contextualisation of the Europass as a supplement diploma to students who fulfil the requirements of the interdisciplinary programmes on climate change and sustainability as an integral part of their undergraduate degree. Curriculum development in the new field of climate change and sustainability policy raises a number of risks in cases that those involved in this process should take into consideration. Developing interdisciplinary programmes, overcoming departmental barriers is probably the biggest challenge. Removing departmental barriers will be tackled through strengthening stakeholders' inputs, dialogue and capacity building provided. Capacity building among the academic staff and other stakeholders is an effective response to concerns about competences needed for CLIMASP curriculum development. Concerns regarding linguistic issues should be tackled. To this end, the colloquia will be self-organised in Arabic in each partner university. Coordination of E.U. and partner countries in the Arab region could sometimes lead to misunderstanding due to different perceptions of time. There is need to develop a shared understanding and acceptance of managerial principles among all Consortium members and key staff involved. Concerns may be raised with respect to community-university partnerships in the curriculum development process due to possible misunderstandings about what an institutional culture of multi-stakeholder engagement entails. Obtaining constructive feedback from all stakeholders is critical. Strengthening the inputs of the key stakeholders needs to identify the barriers and drivers related to their potential contribution and turn their involvement meaningful to them.

Change from traditional disciplinary teaching and learning, especially in the field of climate change and sustainability policy would need changes in people's mindsets and confidence among the faculty. Thus, besides previous activities in building necessary capacities, it is necessary to run a few pilot courses before full scale of implementation. Concerns could be raised due to possible insufficient monitoring of the quality and output of the CLIMASP initiatives, and difficulties to establish what can be the impacts. This risk can be tackled by setting up an internal project monitoring and evaluation arrangement to check progress/achievement of milestones, identify problems, recognise the need for change/amendment/ development and ensure quality. Neglect to plan from the very beginning for what will happen to the outputs at the end may risk the whole work. Implementation of interdisciplinary programs requires an amount and kind of attention to university policy and infrastructure that most project teams did not anticipate. This will be tackled through the creation of appropriate infrastructure to promote interdisciplinary curricula, learning and teaching in general and in particular to the CLIMASP program.

Giving consideration to the outputs most likely to be sustainable in the long term is of paramount importance. Demonstrating that the work carried out was useful and has benefits for the students and the community is an effective strategy for maintaining and extending the outcomes of the CLIMASP project.

Acknowledgments This work has been developed within the framework of the CLIMASP-Development of an interdisciplinary programme on climate change and sustainability policy project that has been funded from the European Commission (European Commission, 543879-TEMPUS-1-2013-GR-TEMPUS-JPCR). The content of the paper reflects the views of the authors, and the Commission cannot be held responsible for any use which may be made of the information contained therein.

References

Bleakly A, Brennan N (2011) Does undergraduate curriculum design makes a difference to readiness to practice as a junior doctor? Med Teach 33:459–467

Corcoran PB, Wals AEJ (eds) (2004) Higher education and the challenge of sustainability: problematics, promise and practice. Kluwer, Dordrecht

Dubash N, Hagemann M, Höhne N, Upadhyaya P (2013) Developments in national climate change mitigation legislation and strategy. Clim Policy 13:649–664. doi:10(1080/14693062)(2013, 845409)

El-Fadel M, Bou-Zeid E (2001) Climate change and water resources in the Middle East: Vulnerability, socio-economic impacts, and adaptation. Nota Di Lavoro 46. The Fondazione Eni Enrico Mattei Note di Lavoro Series Index: http://www.feem.it/web/activ/_activ.html. Social Science Research Network. http://papers.ssrn.com/abstract=278514 (Last accessed 9/10/2013)

Figueres C (2013) Climate policy: a new foundation of stability and prosperity. Clim Policy 13 (5):538–540. doi:10.1080/14693062.2013.822736

Haigh M (2011) Climate policy and financial institutions. Clim Policy 11(6):1367–1385. doi:10.1080/14693062.2011.579265

Helm D (2008) Climate-change policy: why has so little been achieved? Oxford Rev Econ Pol 24 (2):211–238. doi:10.1093/oxrep/grn014

IPCC (2007) Climate change 2007. In: Pachauri RK, Reisinger A (eds) Synthesis report: contribution of working groups I, II and III to the fourth assessment report of the intergovernmental panel on climate change. Geneva, Switzerland

Kostoulas-Makrakis N, Makrakis V (2012) Processes, strategies and practices for turning the University of Crete into a sustainable university. Discourse Commun Sustain Educ 3:5–22. doi:10.2478/v10230-012-0001-2

Leal Filho W (2010) Climate change at universities: results of a world survey. In: Leal Filho W (ed) Universities and climate change. Springer, Berlin, pp 1–19

Leal Filho W, Manolas E (2012) The challenge of climate change. In Leal Filho W, Manolas E (eds) English through climate change, pp 1–12. Department of Forestry and Management of the Environment and Natural Resources, Democritus University of Thrace. http://www.solon.org.gr/downloads/english_through_climate_change.pdf (Last accessed 1/10/2014)

Loftus-Farren Z (2013) Why are women left out of climate change policy making? Earth Island J. Latest News. http://www.earthisland.org/journal/index.php/issues/current/. (Last accessed 9/10/2013)

Makrakis V (2012) Scientific and technological progress, political beliefs and environmental sustainability. Discourse Commun Sustain Educ 3:63–74

Makrakis V, Kostoulas-Makrakis N (2013) Sustainability in higher education: a comparative study between European Union and Middle Eastern universities. Int J Sustainable Hum Dev 1(1):31–38

Makrakis V, Gkotzos D, Larios N (2012a) ICT-enabled climate change education and children's rights. J Teach Educ Sustain 14(2):89–110

Makrakis V, Larios N, Kaliantzis G (2012b) ICT-enabled climate change education for sustainable development across the school curriculum. J Teach Educ Sustain 14(2):54–72

Milly PCD, Dunne K, Vecchia A (2005) Global pattern of trends in stream flow and water availability in a changing climate. Nature 438:347–350

Oberheitmann A (2013) Some remarks on the individual contribution to climate change. Am J Clim Change 2:198–202. http://dx.doi.org/10.4236/ajcc.2013.23020. (Last accessed 9/10/2013)

Osman B. (2010) "Elashamapping of climate change threats and human development impacts in the Arab Region. Arab Human Development Report Research". Paper Series United Nations Development Programme. Regional Bureau for Arab States Arab Human Development Report Research Paper Series

Rahmstorf S. (2008) Anthropogenic climate change: Revisiting the facts. http://www.pikpotsdam.de/~stefan/Publications/Book_chapters/Rahmstorf_Zedillo_2008.pdf. (Last accessed 10/5/2013)

Roberts JT (2001) Global inequality and climate change. Soc Nat Resour 14(6):501–509

Tilbury D (2012) Higher education for sustainability: a global overview of commitment and progress. http://insight.glos.ac.uk/sustainability/Education/Documents/GUNI%20HE%20in%20the%20World%204%20HE's%20Committment%20to%20Sus.pdf. (Last accessed 9/10/2013)

UNEP (2011) Towards a green economy: pathways to sustainable development and poverty eradication—a synthesis for policy makers. http://www.unep.org/greeneconomy. (Last accessed 9/20/2013)

United Nations Development Program (UNDP) (2000) Human development report 2000. Oxford University Press, New York

Woods C (2007) Researching and developing interdisciplinary teaching: towards a conceptual framework for classroom communication. High Educ 54:853–866

Authors Biography

Vassilios Makrakis, Ph.D is Professor of ICT in Education for Sustainable Development and Chairholder for the UNESCO Chair "ICT in Education for Sustainable Development, University of Crete, Greece". Prof. Makrakis is also the Director of the Regional Centre of Expertise (RCE Crete) on Education for Sustainable Development. He has led 10 national, European and international projects in the field of ICT and education for sustainability. Currently, he is leading the European Commission funded Tempus project CLIMASP on developing interdisciplinary programmes on climate change and sustainability policy.

Nelly Kostoulas-Makrakis, Ph.D is Associate Professor in Teacher Education Department at the University of Crete, Greece. She graduated from the University of Athens, Greece and did her doctoral dissertation at the University of Stockholm, Sweden. Her teaching and research interests focus on education for sustainable development pedagogy.

Sustainable Development Integration Strategies in Higher Education: Case Study of Two Universities and Five Colleges in Quebec

Véronique Bisaillon, Mourad Ben Amor and Alain Webster

Abstract

With financial support from the provincial ministry for education, two universities and five colleges in Quebec are developing a project towards integrating sustainable development (SD) in the curriculum. This project is not only dedicated to teaching but is also expanded to: (1) professors' support using various activities (conferences, workshops and short lectures); (2) program integrations; (3) the institutional level, as some institutions have adopted various politics and actions to foster SD in the curriculum and regarding their overall missions of research, community outreach and operations; and (4) regional collaboration between the seven academic partners, which is a unique characteristic of this project. This work is advocating an integrated strategy of variable geometry, as on one hand, it recognises the expertise of professors in their field while, on the other hand, it implements structuring actions throughout programs and institutions. As an example, a 2-day workshop is offered on a regular basis gathering teachers from various disciplines and brainstorming strategies beyond "binding" administrative boundaries. More recently, these practices helped in building a critical mass of practitioners in the field and launched a community of practice. Preliminary results show that the integration of SD in student education is not only accomplished through pedagogical and

V. Bisaillon (✉) · A. Webster
Office of the Vice-President, Sustainable Development and Government Relations, Université de Sherbrooke, 2500, Boul. de l'Université, Sherbrooke, QC J1K 2R1, Canada
e-mail: veronique.bisaillon@usherbrooke.ca

M.B. Amor
Faculty of Engineering, Civil Engineering Department, Université de Sherbrooke, 2500, Boul. de l'Université, Sherbrooke, QC J1K 2R1, Canada

W. Leal Filho et al. (eds.), *Integrating Sustainability Thinking in Science and Engineering Curricula*, World Sustainability Series, DOI 10.1007/978-3-319-09474-8_9

extracurricular activities but is also embodied in the overall student experience on campus. SD's main stake is to grow into an institutional culture in which teaching, research and management activities mutually reinforce each other.

Keywords
Higher education institution · Strategies · Practices · Teaching · Collaborations

1 Introduction

In Quebec, as in most places around the world, the integration of SD in higher education has mostly been rendered by a steep increase in specialist training in the environment and SD fields. Over the last 40 years, the Université de Sherbrooke has developed a solid expertise in this field with its Master's program in Environment offered by the University Center for Environmental and Sustainable Development Studies (CUFE—Centre Universitaire de Formation en Environnement et développement durable). In addition, the CUFE developed several short programs related to the environment and SD over the last decade, namely a new Bachelor's degree in Environmental Studies as well as in-service training. In parallel to this approach, aiming to answer requests from specialists in the environment and SD fields, the Université de Sherbrooke and several other higher educational establishments around the world question themselves on their responsibility in training responsible citizens and, especially, professionals of all domains that integrate the principles, values and practices of SD in their profession. The challenge, complementing the specialist training, thus consists in integrating SD to all the general or professional training courses offered by our higher education institutions.

The Université de Sherbrooke, Bishop's University, as well as the Cégeps (Colleges) of Sherbrooke, Drummondville, Granby, Victoriaville and the Lennoxville campus of Champlain Regional College joined forces in order to take on this challenge of integrating SD into higher education. With financial support from the provincial government, these institutions collaborate to create the PACTE 2D project (Partenariat, Apprentissage, Collaboration et Transfert en Éducation au Développement Durable—Partnership, Apprenticeship, Collaboration, and Transfer in Education towards Sustainable Development Education). PACTE 2D aims to foster the integration of SD into the higher education programs of all seven institutions involved, and to increase the expertise in education for sustainable development (ESD).

The present article aims to lay out the different strategies used as part of this approach: (1) the first section focuses on the support and assistance directly provided to professors; (2) the second deals with programs and the overall curriculum; (3) the third highlights the challenges faced by institutions in steering these actions towards a global and coherent approach to integrate SD into their research and education mission as well as their management activities; (4) and the fourth explains how PACTE 2D, with its regional and collaborative approach, represents a key lever for the integration of SD into higher education. In a distinct section, these four strategies will be used as the basis for the presentation of project results. Finally, the article will open onto potential avenues for reflection and action to further advance the integration of SD into higher education.

2 Strategies Used as Part of PACTE 2D

2.1 Assist and Support Professors

The question of assistance and support offered to the teaching staff is fundamental in any approach aiming to integrate SD into higher education, in a context where the programs are already overloaded and the professors very often feel overwhelmed or insufficiently competent regarding SD (Barlett and Chase 2012). The creation of a community or platform allowing interested teachers and professors to exchange on the subject of ESD seems to be the strategy offering the best potential (Gayford 2001; Chase 2009). The Piedmont project at Emory University in Atlanta, launched in 2001, and the Ponderosa project (1995–2002) at Northern Arizona University are among the first initiatives pursuing specifically to engage and support professors interested in integrating SD in their classes. These projects are based on learning from peers and seek to create a community of teachers and professors interested in SD (Barlett and Rappaport 2009).

Although several teachers and professors recognise the importance of SD issues and the relevance to incorporate these notions into student education, moving on to action comes with its own set of challenges. For many, the integration of SD raises questions as to the pedagogical approaches used (Stubbs and Schaper 2011) and is associated with more participative educational efforts (Brunetti et al. 2003). This can include student engagement with initiatives taking place on their campus (Davis et al. 2003), experiential learning (Hegarty et al. 2011; Dieleman and Huisingh 2006) or, more generally, a teaching approach more centered on students (Stubbs and Schaper 2011; Cotton et al. 2009; Barth et al. 2007). Incidentally, the integration of SD into higher education, which can represent a strictly technical change at first (with added content in sustainable development), is actually seen by most as a cultural change to set up in our higher education institutions (Peet et al. 2004; Juárez-Nájera et al. 2006; Le Grange 2011).

2.2 Approaching the Integration of SD at the Curriculum Level

Another level of integration invested as part of the collaborative project is the integration of SD at the level of curriculums and programs. This intervention strategy echoes changes in professional practices inherent to the context and challenges of sustainable development. Therefore, in 2011, the Canadian Engineering Accreditation Board added SD among the qualities required from graduates in its accreditation standards and procedures (CEAB 2013). The Barreau du Québec also recognises the importance of SD within the lawyer profession's framework (Barreau du Québec 2011). In the field of administration, ethical, social responsibility and environmental questions can no longer be ignored. Several initiatives to integrate SD in the curriculum have been identified, particularly in the fields of engineering (Desha and Hargroves 2010; Hussmann et al. 2010; Van Biesen et al. 2009; Svanstrom et al. 2008a, b; Boks and Diehl 2006; Kamp 2006; Fenner et al. 2005) and administration (Wright and Bennett 2011; Sidiropoulos 2011; Rands 2009; Rusinko 2010). The collective work by Jones et al. (2010) compiles the reflection of various authors on the integration of SD and specific actions in the perspective of areas such as nursing care, theology, economy, geography and earth science, law and arts.

2.3 Defining a Global Approach for SD Integration at the Institutional Level

A third category of initiatives concerns higher education institutions, their first responsibility being to adopt and commit to a vision, objectives and actions in ESD within the institutional community. Several initiatives targeting the integration of SD in higher education already exist, ranging from very specific teaching-oriented actions to faculty- or institution-led strategies. Given the diversity of possible strategies and actions, the question is no longer about whether or not, at the level of a student's program, SD should be integrated into a specific course or throughout the whole curriculum (Barth et al. 2007; Hegarty et al. 2011). The issue resides instead in the integration of SD in the student's journey and experience based not only on the expertise of teachers and professors, but also on research involving the contribution of all institutional strategies related to sustainable development. It is also acknowledged that the integration of SD into the student's journey and experience needs to focus on the academic pathways as well as on peri-academic and extracurricular activities, and even on the whole lifestyle on offer in the higher education institution within which the student progresses in his studies. This is consistent with the 4Cs framework developed at Plymouth University and identifying the Curriculum, Campus and Community as forming an integral part of the institutional Culture. These four elements (curriculum, campus, community and culture) are interdependent and complementary targets in the integration of SD in mutually reinforcing universities (Jones et al. 2010). In a second approach, Chase (2009) identifies four elements to the integration of SD into higher education, being curriculums, research, university operations and relationships with the community, a model onto which rest all the actions of the Association for the Advancement of Sustainability in Higher Education (AASHE) (Chase 2009). Figure 1 shows a model of integration of SD in higher education institutions adapted from both approaches.

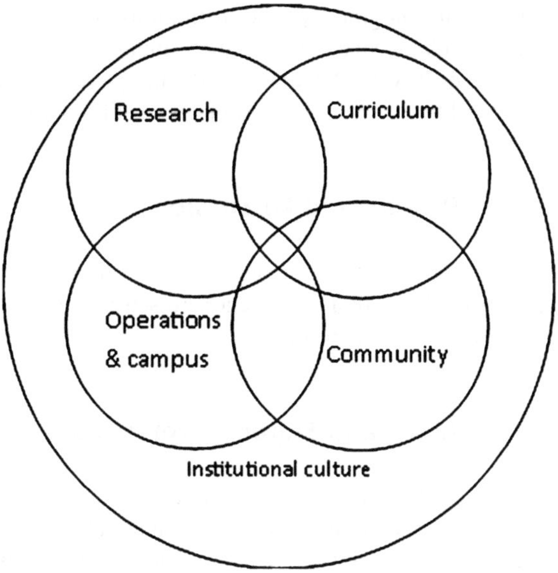

Fig. 1 Model of integration of SD in higher education institutions adapted from Jones et al. (2010) and Chase (2009)

Therefore, each of the four targets for the integration of SD in the institution (research, curriculum, operations and life on campus, as well as community) can be fed by or feed the other three, whilst contributing to an institutional culture integrating SD more efficiently. Moreover, this approach is coherent with an effort to streamline and open up faculties, departments and services making up colleges and universities.

2.4 Collaborating to Integrate SD into Education

Complementing the three scales of the aforementioned strategies, i.e. supporting professors in ESD, integrating it into curriculums and institutional approaches, the PACTE 2D approach opens up the way to a fourth kind of strategy involving the collaboration of colleges and universities. This is the foundation of this project, which sets it in a large collaborative vision such as described by Huxham and Vangen (2005); Balantzian (1997) and Beyerlein (2003). With core values such as transparency and coordinated dialogue, the novel collaboration of PACTE 2D does not follow a predetermined formula but focuses instead on a variable-geometry model, depending on the respective needs of stakeholders, according to logic based on progressive collective construction. PACTE 2D therefore integrates the main three types of collaboration founded on the field of intervention as identified by Margerum (2008). As far as operational collaborations are concerned on the ground, a connection is made between teachers and professors as part of pedagogical days or workshops on professional development, in order to exchange on their teaching practices. At the institutional level, the collaboration tends to focus more on the policies and programs of the institutions concerned and on their interrelations in order to adopt coherent strategies. Finally, as far as political collaborations are concerned, PACTE 2D aims to influence governmental policies in ESD in a context where public policies on the subject are still relatively discrete or inexistent.

3 Results and Discussion

This section reports on project results in the four approaches presented in Sect. 2, and concludes with a more general discussion on avenues for reflection and action to follow up on the project.

3.1 Assist and Support Professors

As part of PACTE 2D, support is offered to interested teachers and professors to assist in a personalised way with the integration of SD in their teaching. Resources have also been made available on the website www.pacte2d.ca. The section "En action!" presents case studies from a variety of courses, programs and fields of study. A database presenting a range of publications (scientific articles, reference material, websites, etc.) has also been built. Assistance offered to teachers and professors also includes lunchtime presentations and discussions, as well as academic workshops organised in several institutions, allowing the teaching staff to share their respective experience and projects related to ESD. An online newsletter is also published periodically so as to foster and feed this community. In November 2013, PACTE 2D more formally launched a community of practice (or working group) in SD education. A specific distribution list has also been recently created to support communication among members in between formal activities.

For the last 3 years, a 2-day workshop has been organised as part of the PACTE 2D project. It targeted two objectives, being to allow teachers and professors to position themselves with regards to ESD, and to adapt their teaching practices in this area. This workshop opened up a space where teachers and professors from wide ranging fields could connect, exchange and develop tangible strategies of action. This workshop is attended annually by 20–30 participants from a range of fields.

The following five subsections present the applied strategies for SD integration into higher education as part of the PACTE 2D project.

3.1.1 Collaboration with Resources from the Field

Hiring a key speaker, visiting a site, watching a documentary or short clips available online are all practical solutions to approach subjects with which the teaching staff is less familiar (Barlett and Chase 2012; Kurland 2011). An increasing amount of resources is available in that respect. For instance, at Cégep de Drummondville, a number of biology laboratories take place in an adjoining forest, giving students an opportunity to use resources from the field to apply in their learning program. They can measure the impact of human activities on their environment in a very tangible way (diversity and productivity of ecosystems, impact of anthropogenic activities on water quality, etc.), predict their long-term consequences and find possible solutions to the problems observed.

At the Engineering faculty of the Université de Sherbrooke, the course on Project design in civil engineering provides students with the possibility to integrate their knowledge previously acquired during their undergraduate courses through a civil engineering project design affected by real problems and constraints. In the first part of the course, students grouped in teams need to find solutions to a real problem submitted by their client (civil engineering firms or others). Environmental issues used to be taught more generally as part of the literature review. Now, following a simplified method of Life Cycle Analysis (LCA), an LCA professor joins this activity and accompanies teams of students so that they can apply it in the proposed range of solutions. The final report submitted to their corresponding client integrates results obtained from the simplified LCA.

3.1.2 Contextualise

SD can become the context surrounding the teaching content. Several teachers already apply this by dipping into the news, filled with SD issues, to introduce these notions. In the Natural Science program at Cégep de Drummondville, the courses' content is presented in a perspective of energy and sustainable development. Indeed, the potential hydrogen represents as a green combustible, as well as biomethanisers and biofuel issues are examples used in the classroom to explain a number of theoretical notions. The use of SD to contextualise such notions can also have a positive impact on teaching and learning by improving the perceived professional relevance and usefulness of acquired knowledge (Bédard 2006).

3.1.3 Involving Students, Inspiring Them to Take Action

Considering the end goal of ESD to train actors capable of not just thinking but also taking action (Svanstrom et al. 2008b; Tilbury 2011; Lundqvist and Svanstrom 2008; Tilbury 2011), the integration of SD into higher education tends to require a more active academic approach centered on students or more closely linked with the community (Rands 2009). The Experiential Learning Program through Community Involvement (or PAEIC—Programme d'apprentissage expérientiel par l'intervention communautaire) at the Université de Sherbrooke pursues two objectives, i.e. to encourage members of the university community to get involved, and to develop the capacity of community groups to fulfill their social mission. PAEIC and other similar initiatives thus make it possible to merge the learning objectives set in academic courses with the needs of a group or organisation, as part of a project generally running over a semester. This interconnection allows students from various fields of study (administration, law, and engineering) to get to know community stakeholders and to better understand their reality. At Cégep de Victoriaville, as part of the geography course Quebec space and methodology, students characterise sections of the riparian zone to determine priority areas for action and offer solutions. Diagnoses are then passed on to the city professionals.

3.1.4 Encourage Students to Get Involved with and Compare Their Views on SD

Various examples of integration of SD into higher education insist on the importance to question the learning process and not just the content (Stubbs and Schaper 2011). According to this view, teachers are asked to act more as guides in the reflective learning process than as mere "knowledge providers". In this perspective, activities have been suggested where students have to run independent research to make up their own opinions, develop their own critical thinking, and compare their views with their peers (Stubbs and Schaper 2011; Hegarty et al. 2011). This position taken by teachers allows them to neutralise the feeling that they may not be qualified or specialised enough to tackle SD issues with their students.

In practice, homework, lab work, projects and other activities can be used as a stepping stone towards the integration of a question or section requiring that the students lead their own reflection with regards to sustainable development. For instance, criteria related to energy efficiency and social acceptability of the project, cost analysis and life cycle analysis are increasingly integrated into projects on Design Engineering at the Université de Sherbrooke. In their internship reports, students can also be invited to reflect on SD in relation to their future professional practice or experience in their internship workplace.

3.1.5 Using Non-intentional Learning and Managing the Class in a SD Perspective

Also called hidden curriculum, non-intentional learning is used in a context that enables knowledge acquisition but without being explicitly targeted as such. This strategy seems particularly interesting in courses where teachers may not necessarily see a direct link between SD and their field of study (Barlett and Chase 2012). For instance, in a translation or reading comprehension course, texts or works to be translated or studied can be chosen in relation to SD topics. At Cégep Beauce-Appalaches, in order to better illustrate the application of a variety of mathematical concepts and operations to be mastered as part of the course, a teacher used social and environmental case studies. For instance, to solve an optimisation problem, students have to minimise greenhouse gases emissions in the delivery of organic vegetables according to a number of set criteria. This approach helps to improve the motivation of students since they are more likely to see the usefulness of the content taught in their courses. Moreover, this approach helps to raise awareness about several issues and introduces a number of notions related to SD while meeting the achievement of mathematical skills development.

However, non-intentional learning tends to be approached in a much wider perspective. It refers to all the messages passed on from a teacher or institution to their students, often in an unconscious, tacit or unplanned manner (Cotton and Winter 2010). According to Orr (quoted in Jucker 2002), the institutional framework as a whole, the studying environment as well as the daily management operations of an institution also form an integral part of non-intentional learning. By extension, the way teachers manage their class can also be included in the non-intentional learning available to students and can represent a first step in the integration of SD within the course. For example, several teachers choose to offer their course in a paperless version, with technological support from Internet platforms for their classes (e.g. Moodle) or specialised correction software.

3.2 Approaching the Integration of SD at the Curriculum Level

In Quebec, the Tourism program at Cégep de Granby has integrated SD progressively over the last 10 years. Also worth a mention, a new project is underway to integrate SD tools according to a life cycle approach in the undergraduate Engineering curriculum at the Université de Sherbrooke. Without going into too much detail for each of these initiatives, it should be pointed out that their main commonalities are the angle from which the question of training future professionals is addressed, and the skills related to SD that can be developed at various stages throughout students' progression in the curriculum. At the program level, SD can be more efficiently integrated and transposed to a professional context, thus building stronger links with the future professional practice.

3.3 Defining a Global Approach for SD Integration at the Institutional Level

At Cégep de Sherbrooke, SD is seen as a skill in the institution's educational project. The will of Cégep de Victoriaville is to offer students the opportunity to steer their project reports required for successful program completion or end of semester projects towards SD themes, with links to local resources and expertise. Therefore, the City of Victoriaville and the college Foundation awarded grants to students' projects dealing with sustainable development; the college also acknowledges SD projects as part of its annual exhibition Expo Sat. Champlain Regional College recently adopted a SD policy. Its Lennoxville campus chose to encourage the Brundtland Green Institutions approach, which is based on four reference values: ecology, pacifism, solidarity and democracy. Activities to mobilise and raise awareness to the population will continue and a teaching staff survey will be conducted to identify academic activities that integrate sustainable development.

For its part, Bishop's University led a vast reflection campaign on liberal education, an issue at the heart of this university's mission. Among the skills that need to be developed throughout the curriculums and programs, one is related to sustainable development, i.e. Knowledge of Environmental, Economic and Social Sustainability. A coherent modification of the recurrent program assessment will follow. At Université de Sherbrooke, the integration of SD to the institution's teaching and research mission features in its 2010–2015 Strategic Plan as well as in the related policy and action plan. These guidelines led to the consideration of a new concern related to the integration of SD into higher education as part of the recurrent program assessment. Furthermore, another guideline adopted by Université de Sherbrooke aims to simplify, open up and develop the synergy between teaching, research and community relations (Strategic Plan Réussir 2010–2015). As far as SD is concerned, Université de Sherbrooke reaches for a synergy between teaching, research, management and relations with the community so as to gradually transform the campus into a SD laboratory.

3.4 Collaborating to Integrate SD into Education

Although collaboration is both time and energy consuming, as highlighted by various authors (Huxham and Vangen 2005), and even though keeping these collaboration mechanisms going in the long-term is a real challenge, this approach is also a powerful lever and springboard in the current project. A number of observations can be compiled concerning the relevance of the collaborative approach of this project. PACTE 2D is a very timely project for several higher education institutions; it has helped reinforce initiatives aiming to integrate SD within partner institutions. The greatest added value brought about by this collaboration can be summed up in the following observations. No matter what the institutional level of commitment towards ESD or the strategies adopted to integrate it to institutions or programs might be, the same fundamental questions arise (and the present article attempted to provide elements of response): how can SD be integrated into higher education, curriculums and programs? How to support teachers and professors? What are the targeted learning objectives?

The collaborative approach made it possible to pool and share resources, and to feed the reflection and elements of response relative to these questions in a more integrated and exhaustive manner. From a political point of view, it is interesting to note that PACTE 2D is as much a project belonging to each one of these member institutions as it transcends the boundaries of their individual organisation. This dual position can lead to an effective engagement from institutions and teachers alike. Proactive institutions bring forward an additional argument to engage their staff towards the integration of SD into higher education. Teachers or program committees wishing to see their institutions becoming more engaged with ESD also find in this project a network with which to get involved. It helps them pursue and go deeper in their engagement in this field and expand it outside of their organisation.

4 Conclusion: Avenues for Reflection and Project's Follow up Actions

So far, the first 3 years of PACTE 2D have mostly focused on establishing collaborative mechanisms between the various institutions involved in order to create and strengthen a community of practice (or working group) amongst their teaching staff. In the fall 2013, PACTE 2D conducted a student survey among partner organisations to assess their interest and expectations in the fields of SD and education. Some 4,000 students completed the survey, which represents a 14.7 % participation rate. Students were invited to express their views on sustainable development, their behaviour in that field, as well as their interest towards topics related to ESD and pedagogical approaches.

The results still need to be closely analysed, but the initial analysis already draws new avenues of reflection and action in order to integrate SD into curriculum. For instance, students were questioned on their idea of SD and on their opinion about it. While only one-fifth identify SD with environmental protection, the three-fifth of the students have a more complex conception of sustainability integrating the social, environmental and economic issues. More than three-quarters of students consider SD as an opportunity for brand new alternatives and solutions to actual crisis, or as a mean of revisiting our ways of living. There thus seems to have a broadly positive opinion of sustainable development. According to their answers concerning their interest to deal with new approaches related to SD within their training, students seem more preoccupied by everyday life issues, by social issues such as health and quality of life, or by human development issues rather than cultural diversity issue for example.

In its first 3 years of existence, the PACTE 2D project brought about positive results as far as tangible actions are concerned, though much is yet to be done. Considering SD as a new way to see the world and take action, it seems that the integration of SD into higher education requires a profound cultural change more than a technical shift. Therefore, the approach should not only be to target the content but also the pedagogical approach and the sometimes compartmented disciplinary structure of our institutions. The SD issue represents an opportunity for our higher education institutions to develop an institutional culture where teaching, research and institutional management activities become mutually reinforcing and create stronger links with the community. SD represents an audacious challenge that universities and colleges, on the strength of their expertise supported by students who make it their priority, have to meet. Who, indeed, will be capable of taking on this challenge if our higher education institutions fail to do so?

Acknowledgments This article was written as part of PACTE 2D, a project funded by the University-College Collaboration Program of Quebec's Ministries for education and higher education. We thank all the people involved in this project within the seven partner institutions involved in this project, i.e. Cégep de Drummondville, Cégep de Granby, Cégep de Sherbrooke, Cégep de Victoriaville, Champlain Regional College (Lennoxville campus) as well as Bishop's University and the Université de Sherbrooke.

References

Balantzian G (1997) L'avantage coopératif: le partenariat, la coopération, l'alliance stratégique. Éditions d'Organisation, Paris 190p

Barlett PF, Chase G (2012). Sustainability and curriculum: AASHE workshop for campus leaders, Emory University, Atlanta, 10–11 Jan

Barlett PF, Rappaport A (2009) Long-term impacts of faculty development programs: the experience of Teli and piedmont. Coll Teach 57(2):73–82

Barreau du Québec (2011) Les avocats de pratique privée en 2021. Rapport du Comité sur les problématiques actuelles reliées à la pratique privée et l'avenir de la profession, 119 p

Barth M, Godemann J, Rieckmann M, Stoltenberg U (2007) Developing key competencies for sustainable development in higher education. Int J Sustain High Educ 8(4):416–430

Bédard D (2006) Enseigner autrement, oui mais pourquoi et comment? Le cas d'un cours universitaire de premier cycle. In: Rege Colet, N. et Romainville, M., La pratique enseignante en mutation à l'université (pp 83–101). Bruxelles, De Boeck

Beyerlein MM (2003) Beyond teams: building the collaborative organization. San Francisco, Calif., Jossey-Bass/Pfeiffer, 247 p

Boks C, Diehl JC (2006) Integration of sustainability in regular courses: experiences in industrial design engineering. J Clean Prod 14(9–11):932–939

Brunetti AJ, Petrell RJ, Sawada B (2003) Team project-based learning enhances awareness of sustainability at the University of British Columbia, Canada. Int J Sustain High Educ 4(3):210

Canadian Engineering Accrediation Board (2013) Accreditation Criteria and Procedures. Engineers Canada, 115 p. http://www.engineerscanada.ca/sites/default/files/sites/default/files/accreditation_criteria_procedures_2013.pdf. Last accessed 26 Mar 2014

Chase G (2009) Engaging faculty members in sustainability: addressing barriers to curriculum change. In: CAUBO, leadership for sustainability in higher education, Toronto, 26–27 Mar

Cotton D, Winter J (2010) It's not just bits of paper and light bulbs: a review of sustainability pedagogies and their potential for use in higher education. In: Jones P, Selby D, Sterling S (eds) Sustainability education. Perspectives and practice across higher education. Earthscan, London, UK, pp 39–54

Cotton D, Bailey I, Warren M, Bissell S (2009) Revolutions and second-best solutions: education for sustainable development in higher education. Stud High Educ 34(7):719–733

Davis JA, Edmister JH, Sullivan K, West CK (2003) Educating sustainable societies for the twenty-first century. Int J Sustain High Educ 4(2):169–179

Desha C, Hargroves KC (2010) Engineering education and sustainable development. A guide to rapid Curriculum Renewal. The Natural Edge Project, Earthscan, London

Dieleman H, Huisingh D (2006) Games by which to learn and teach about sustainable development: exploring the relevance of games and experiential learning for sustainability. J Clean Prod 14(9–11):837–847

Fenner RA, Ainger CM, Cruickshank HJ, Guthrie PM (2005) Embedding sustainable development at Cambridge University engineering department. Int J Sustain High Educ 6(3):229–241

Gayford C (2001) Education for Sustainability: an approach to the professional development of teachers. Euro J Teach Edu 24(3):313–327

Hegarty K, Thomas I, Kriewaldt C, Holdsworth S, Bekessy S (2011) Insights into the value of a 'stand-alone' course for sustainability education. Environ Educ Res 17(4):451–469

Hussmann PM, Trandum C, Vigild ME (2010) How to include sustainability in engineering education? The "Green Challenge" at DTU is one way. In CDIO. Document

Huxham C, Vangen S (2005) Managing to collaborate: the theory and practice of collaborative advantage. Routledge, London, New York 271p

Jones P, Selby D, Sterling S (2010) Sustainability education, perspectives and practice across higher education. Earthscan edition, London and Washington DC

Juárez-Nájera M, Dieleman H, Turpin-Marion S (2006) Sustainability in Mexican higher education: towards a new academic and professional culture. J Clean Prod 14(9–11):1028–1038

Jucker R (2002) Sustainability? Never heard of it! Some basics we shouldn't ignore when engaging in education for sustainability. Int J Sustain High Educ 3(1):8–18

Kamp L (2006) Engineering education in sustainable development at Delft University of Technology. J Clean Prod 14(9–11):928–931

Kurland NB (2011) Evolution of a campus sustainability network: a case study in organizational change. Int J Sustain High Educ 12(4):395–429

Le Grange LLL (2011) Sustainability and higher education: from arborescent to rhizomatic thinking. Edu Philos Theory 43(7):742–754

Learning for a Sustainable Future (S.d.) ESD learning outcomes. http://www.lsf-lst.ca/en/what-is-esd/esd-learning-outcomes. Last accessed 15 Feb 2014

Lundqvist U, Svanstrom M (2008) Inventory of content in basic courses in environment and sustainable development at Chalmers University of Technology in Sweden. Eur J Eng Educ 33 (3):355–364

Margerum RD (2008) A typology of collaboration efforts in environmental management. Environ Manage 41(4):487–500

Peet D, Mulder KF, Bijma A (2004) Integrating SD into engineering courses at the delft University of Technology: the individual interaction method. Int J Sustain High Educ 5(3):278–288

Rands GP (2009) A principle-attribute matrix for environmentally sustainable management education and its application: the case for change-oriented service-learning projects. J Manage Edu 33(3):296–322

Rusinko CA (2010) Integrating sustainability in higher education: a generic matrix. Int J Sustain High Educ 11(3):250–259

Sidiropoulos L (2011) Navigating the journey to sustainability: the case for embedding sustainability literacy into all tertiary education business programs. Int J Environ Cult Econ Soc Sustain 7(3):247–274

Stubbs W, Schaper J (2011) Two approaches to curriculum development for educating for sustainability and CSR. Int J Sustain High Educ 12(3):259–268

Svanstrom M, Lundqvist U, Arehag M, Holmberg J (2008a) Addressing the quality of engineering education for sustainable development—experiences form Chalmers University of Technology. In: Anonyme, engineering education in sustainable development 2008 conference Graz, Austria, 22–24 Sept

Svanstrom M, Lozano-Garcia F, Rowe D (2008b) Learning outcomes for sustainable development in higher education. Int J Sustain High Educ 9(3):339–351

Tilbury D (2011) Education for sustainable development. An Expert Review of Processes and Learning. UNESCO, Paris

Van Biesen LP, Rahier H, Vanherzeele H, Willem R, Hubin A, Veretennicoff I, Deblauwe N, Ponet M (2009) Engineering skills education: the bachelor of engineering programme of the "Vrije Universiteit Brussel" as a case study. Eur J Eng Educ 34(3):217–228

Wright NS, Bennett H (2011) Business ethics, CSR, sustainability and the MBA. J Manage Organ 17(5):641–655

Authors Biography

Véronique Bisaillon is officer in education for sustainable development for PACTE 2D's project and at Université de Sherbrooke. For the last 4 years, Véronique has developed various activities fostering ESD in higher education. She holds a master degree in environmental sciences.

Professor Mourad Ben Amor is an assistant professor at the department of civil engineering in the Université de Sherbrooke. His research and teaching activities focus on sustainable engineering and life cycle assessment (LCA). Since 2013, Ben supervises master and Ph.D. students working in LCA methodological developments and its application in different sectors such as energy, buildings and materials. Ben is also a member of different national and international scientific committees, such as the UNEP/SETAC life cycle initiative.

Professor Alain Webster has been a member of Université de Sherbrooke's governing committee since 2007. After his tenure as Vice-President, Administration and Sustainable Development, between 2007 and 2009, he became the institution's Vice-President, Sustainable Development and Government Relations, and Vice-President, Longueuil Campus.

Making Money and Saving the World: Empowering Students for Sustainability Through Social Enterprise

Kate Arnold

Abstract

In 2013, The Students Green Fund was awarded to 25 Student Unions across England for projects that aim to position students as active agents for sustainability across their institutions and into their communities. In an effort to engage students more deeply with a holistic view of sustainability, The University of Gloucestershire Students' Union is encouraging students to start their own social enterprises. Students gain business skills, and actively apply principles of sustainability to their enterprising activities. This takes sustainability beyond the understood environmental sustainability and teaches sustainability as social and economic in a practical, real life way. The small amount of evidence gained so far is showing that students who become involved in social enterprise are transforming the way they relate to business ethics, gaining a broader understanding of corporate social responsibility and the possibility that business can be a vehicle for positive change towards a better world. These are skills that will follow them into their future careers.

Keywords

Sustainability · Social enterprise · Business · Student skills · Transformative learning

K. Arnold (✉)
The University of Gloucestershire Students' Union, The Park Campus,
Cheltenham, Glos GL50 2RH, UK
e-mail: karnold@glos.ac.uk

W. Leal Filho et al. (eds.), *Integrating Sustainability Thinking in Science and Engineering Curricula*, World Sustainability Series, DOI 10.1007/978-3-319-09474-8_10

1 Social Enterprise

The number of social enterprises in the UK, and indeed, the world, is on the increase. Faith in the willingness of regular businesses to contribute towards the much needed solution to our social, environmental and economic problems has declined in the wake of the 'credit crunch' following 2008.

Since the global economic crash, the UK government has stressed the role of charity and volunteering organisations, commonly known as the third sector. Looking back to 40 years ago, the voluntary sector was greatly overshadowed by public provision of welfare, and the private sector was virtually non-existent (Billis 2010, p. 5). As the third sector became of growing importance, 'third sector staff—along with their colleagues in the public sector—became increasingly subjected to the virtues of concepts originally developed for the private sector', incorporating market logics in the running of every kind of social organisation (Billis 2010, p. 7). With New Labour's 'Third Way', a 'mish-mash' of policy initiatives blurring the boundaries between state, private provision and volunteerism emerged, a 'bewildering array' including 'compacts, partnerships, social enterprise, quasi-markets, networks, transfer associations, community interest companies, foundation trusts, city academies' (Billis 2010, p. 7). It is clear that straightforward public provision of welfare is not on the government agenda, and hybrid organizations and the third sector are increasingly relied upon to provide vital services (p. 11).

Clearly, Social Enterprise is one practical solution to the lack of government funding currently available for positive social projects and interventions—but it goes further still. Increasingly, the movement for social enterprise is not just focusing on how business and management skills can be applied for social ends—how not-for-profits could instead run for-profit ventures to generate revenue—but the social enterprise movement is a 'transformative force' and a call to action to fundamentally change the way we do business (Bornstein 2007, p. 1).

The third sector, including Social Enterprises—or the 'citizen sector' as Bornstein prefers—is growing worldwide. Although the citizen sector has been around for centuries, Bornstein points to several factors that are new and remarkable about the developments in recent years:

1. Mobilisation of citizens through the citizen sector is occurring on a scale never before seen
2. Organisations are more globally dispersed and diverse
3. They are increasingly looking for systemic solutions to problems, rather than mitigating them ("offering better recipes, not just more cooking")
4. Citizens organisations are more independent and exerting more political pressure
5. They're forming increased partnerships with different sectors—businesses, academic institutions, and governments, building new markets and modes of enterprise, and influencing government agenda and operation
6. The sector 'boom' is leading to further innovation and entrepreneurialism (Bornstein 2007, p. 5).

The energy behind the Social Enterprise movement is growing, and so is the belief that Social Enterprise can be a real challenge to existing systems that lead to inequality and environmental degradation. Not only does it revolutionise the role of profit in business, but it's attitude towards cooperation and partnership challenges dominant business paradigms. In doing so "the whole movement toward a sustainable enterprise economy tackles [...] wicked problems, simply because moving in this direction necessitates that multiple considerations be built into enterprises of all sorts" (Waddock and Mcintosh 2011, p. 306). Joe Hsueh, a founding member of the Academy for Systemic Change, also sees sustainable business as central to creating real and lasting change, that "big, complex problems cannot be solved by any one technological improvement or any one organization acting alone," (in Gunther 2014).

Daniella Tilbury, Head of Sustainability at the University of Gloucestershire, has asserted that innovation not integration lies at the heart of Education for Sustainable Development (Tilbury 1998). This is exactly what Social Enterprise is bringing to the world of business.

2 Sustainability in Higher Education

Research from NUS in recent years has been clear about the appetite of students across the UK to gain skills related to sustainable development. Having run research from 2010 to 2013—across the entire period of a student's career—over eight in every ten students surveyed by NUS 'consistently believe that Sustainable Development should be actively incorporated and promoted by universities, and this increases as respondents progress through their studies' (Drayson et al. 2013, p. 4). Furthermore, their evidence over repeated years suggests that students become increasingly focused on employability throughout their time at university, with third years, unsurprisingly, being the most enthusiastic about furthering their employability (Drayson et al. 2013, p. 6). The idea that employability is high on our students' agenda is supported by findings within our own institution, with the overwhelming majority of respondents in our 2012 annual SU survey telling the Union that they believe our main function should be providing them with part time jobs and volunteering opportunities.

Another interesting revelation of the NUS skills for sustainable development survey showed that 88 % of first years and 86 % of third years believe that 'acquiring skills enabling them to act as responsible citizens both locally and globally' is important to future employers (Drayson et al. 2013, p. 6).

How best to deliver the skills for sustainability which students so clearly desire is another question. Although modules and courses on sustainability have been running for some time, groups such as Forum for the Future have been leading the way in ensuring sustainability is ubiquitous throughout the university curriculum. They assert the importance of integrating sustainability into courses teaching 'functional specialities', rather than offering 'stand alone' modules such as environmental management or social responsibility, asserting that this approach lends "increasing overall perceived legitimacy [...] than if it is presented in electives or courses 'peripheral' to the core" (Forum for the Future 2000 in Thomas 2005, p. 190).

This has led to a further integration into courses such as business management—although this integration has not been without resistance. Tokenistic gestures towards sustainability—such as a brief discussion of Corporate Social Responsbility, raises the question as to whether sustainability has yet "achieved legitimacy among students and faculty as a topic worthy of serious consideration" (Springett and Kearins 2001 in Thomas 2005, p. 186).

This is where Social Enterprise may again rear it's head. Thomas stresses that "Education for sustainability requires a commitment to a *transformative* learning process in which students are challenged and given the opportunity to develop entirely new ways of understanding the role and behaviour of the business enterprise" instead of *transmissive* learning that rewards students for adhering to narrow 'accepted' models and modes of business (Thomas 2005, p. 186). By taking a role in an operating Social Enterprise, students of business—and outside of the business school—are able to actively practice skills for sustainability, such as building partnerships, long term thinking, working to make a positive difference in the community. These are the skills that will assist "learners [to] develop the skills to influence change within a system, organisation or wider society" (Tilbury 2007, p. 119).

3 Social Enterprise, Students, and Understanding Sustainability at the University of Gloucestershire

When the National Union of Students (NUS) announced the availability of £5,000,000 for student led, transformational sustainability projects in April 2013, the University of Gloucestershire Students' Union was presented with a challenge. Coming from a University well-known for its strategic focus on sustainability and proven record for incorporating sustainability into the curriculum, and a Students' Union that had consistently received the Green Impact Gold award for pro-sustainability behaviour; in many ways the infrastructure for engaging students with sustainability was already in place.

Despite this, however, levels of student engagement in and understanding of sustainability were still lacking. For many students with whom we were working on a day-to-day basis, the idea of 'sustainability' still meant little more than turning off lights and recycling—an attitude that is not easily overcome with changes to our carbon footprint and waste management plans.

How, then, were we to engage those students whose interest in sustainability was latent, but who may never see themselves as environmental or social campaigners? To reach the students who would never actively seek to be involved in 'green' activities?

Our own annual survey and research from NUS, explored above, clearly demonstrated to the SU that there is a demand for employability opportunities, but with a sustainability theme. Fortunately, this also meant that we had positioned the project to integrate key University of Gloucestershire priorities of sustainability and business development, tying neatly into the overall focus of our parent institution. With this in mind, we decided to approach the Students' Green Fund with those hard-to-reach and employability-focused students at the fore. Our project takes on the challenge of reaching out to students who are yet to engage with sustainability and helps reposition the role of students as active partners shaping institutional and regional change. The result will aim to provide activities for students that offer a range of practical ways for them to develop competencies and experiences in this area as well as improve formal learning opportunities.

The means to unite business and sustainability, we believe, is encapsulated in the social enterprise model of business. By equipping students with the applied knowledge of how to run a business that is economically, environmentally and socially sustainable, they will not only gain crucial professional acumen, but skills for sustainability that will follow them into their future careers, and the means to challenge business-as-usual by being part of a movement that revolutionises the way we relate to profit and competition. According to Social Enterprise UK, there are approximately 70,000 social enterprises in the UK today, contributing £18.5 billion to the UK economy every year. Demonstrably then, social enterprise cannot be considered a 'fad' or insignificance, but a viable and potentially powerful up-and-coming development to the world of business.

We are not the only Students' Green Fund project to be engaging with the Social Enterprise model. Lancaster, Leicester, Bedfordshire, Leeds and City Universities have all developed projects that are geared towards creating more student led enterprises that directly address issues of sustainability.

The most exciting aspect of introducing students to social enterprise is to ask them to explore what they are passionate about, what makes them angry, or what changes they would make to the world if they could. As well as getting them thinking about how business can contribute to the solution instead of the problem, this also engages students in the world around them. It's a call to action to tackle social and environmental problems, and whether it results in their involvement in social enterprise is almost irrelevant. Issues such as food poverty, global inequality, inclusivity in sport, mental health and waste reduction were all suggested as potential 'social missions' around which an enterprise could be based in order to tackle them. It engages them in that crucial act of systems thinking—an element of effective Education for Sustainability, which must "encourage learners to develop critical and systemic thinking skills, enabling them to get to the core of the issues" (Tilbury 2007, p. 120).

For one student in her first year who has designed a social enterprise around the promotion and protection of bees and local honey, the benefits of being involved in a social enterprise were multifaceted:

> For some years I have wanted to keep bees, but now I am heading a project that will provide so much more than a few pots of honey. It has enabled me to **get involved in a community** that I am new to, **learn about sustainability** beyond an environmental perspective and **learn skills** that I never expected to learn as part of my course. While I came to University to put my head down and get a degree, I realise now the fun to be had by

getting involved with a social enterprise scheme such as this. I would strongly encourage anyone with even a glimmer of an idea to give it a go.

—Charlotte Powell, First Year Biology student and Head of the Beekeeping Project.

It's clear that Charlotte sees beyond merely making money through her beekeeping business. The involvement in a social enterprise has given her deeper understanding and involvement in sustainability, beyond environmental factors. It has linked students and staff with the local community through the beekeeping activities, and presented the opportunity for many students to pick up a new and unusual skill. Furthermore, she has managed the entire project independently—designing risk assessments and Gantt charts, budgeting and sourcing, held steering groups, liaised with stakeholders, applied for planning permission and created products to sell in order to sustain the enterprise and ensure expansion and legacy. Undoubtedly, these are skills transferable to any business situation.

Students are empowered to pursue their passion and make things happen and although the funding is initially provided by the SU, this is only for start-up costs. In the long term, the businesses will be able to independently fund their projects through enterprising activities, something that, in the current economic climate, is more pertinent that ever. Third sector funding is increasingly squeezed by the drive for austerity but by introducing students to ways and means of achieving social change without a reliance on grant funding, we are helping them to think creatively and overcome financial obstacles. From the Students' Union point of view, it means that every year more students will be able to get involved and improve their skills with long-term sustainability projects, such as the on-campus apiaries, without worrying about sourcing continued funding for the projects.

It can certainly be said that The University of Gloucestershire did not have a culture of student engagement. However, the creation of social enterprises such as the Cheltenham Chilli Company and Core Blimey Juice and Cider have been challenging this tradition and we feel the increased engagement of students is due to linking sustainability with employability and enterprise.

You cannot solve a problem from the same consciousness that created it.
You must learn to see the world anew.

—Albert Einstein

References

Billis D (2010) From welfare bureaucracies to welfare hybrids. In: Billis D (ed) Hybrid organizations and the third sector: challenges for practice, theory and policy. Palgrave Macmillan, Basingstoke, p 3–15

Bornstein D (2007) How to change the world: social entrepreneurs and the power of new ideas. Oxford University Press, New York

Drayson R, Bone E, Agombar J, Kemp S (2013) Student attitudes towards and skills for sustainable development, Higher Education Academy. Available at: https://www.heacademy.ac.uk/sites/default/files/ESD_student_attitudes_2013_v4.pdf. Accessed 5 Aug 2014

Forum for the future (2000)

Gunther M (2014) The art and science of systems change. In: Guardian sustainable business (online). Available at: http://www.theguardian.com/sustainable-business/systems-change-joe-hsueh-secondmuse?CMP=twt_gu. Accessed 20 Apr 2014

Thomas TE (2005) Are business students buying it? A theoretical framework for measuring attitudes toward the legitimacy of environmental sustainability. Bus Strateg Environ 14 (3):186–197

Tilbury D (1998) The role of research in initiating and sustaining developments in teacher education. Int Res Geograph Environ Edu 7(3):239–264

Tilbury D (2007) Learning based change for sustainability: perspectives and pathways. In: Wals AEJ (ed) Social learning towards a sustainable world: principles, perspectives, and praxis. Wageningen Academic Publishers, Netherlands

Waddock S, McIntosh M (2011) Business unusual: corporate responsibility in a 2.0 world. In: Business and Society Review 116(3):303–330

Author Biography

Kate Arnold Green Fund Project Manager, University of Gloucestershire Students' Union. After graduating from The University of Warwick in 2013, sociology graduate and sustainability activist Kate Arnold join the team at The University of Gloucestershire Students' Union as Project Manager of The Students' Green Fund, which aims to drive student-led sustainability across campus life and into the local community. Kate oversees the delivery of the 'Greener Gloucestershire' project's five strands, including encouraging students to start up their own social enterprises.

Promoting Education for Sustainable Development at the University of Prishtina, Kosovo

Arlinda Beka and Adriano Ciani

Abstract

Kosovo Education for Sustainable Development—KESD is a local NGO which is one of the first organizations dealing with matters of education for sustainable development in Kosovo. The purpose of KESD is to help Kosovo in sustainable development in all aspects including education. KESD has played a major role not only in increasing citizens' awareness about Education for Sustainable Development, but has also influenced the Faculty of Education at the University of Kosovo. Bearing in mind that the Faculty of Education is the only institution that prepares future teachers; it is a great way to influence in general, with ESD through professional development, future teachers about the role and importance of the education for sustainable development and the use of the teaching methodologies that support ESD. In this paper we will present current situations about ESD, our country, achievements, challenges and obstacles, KESD's contribution in raising awareness of Kosovo citizens for ESD, in order to understand their role in developing ESD. The methodology used in this paper is a case study methodology. Through different activities which are mainly organized on a voluntary basis, KESD has been able to promote Education for Sustainable Development through various projects from preschool education to

A. Beka (✉)
University of Tirana, Tirana, Albania
e-mail: arlinda.beka@uni-pr.edu

A. Beka
University of Prishtina, Rr. Rrustem Statovci 31/20, Prishtina 10000, Republic of Kosovo

A. Ciani
Department of Agricultural, Environment and Food Science, University of Perugia, Borgo XX Giugno 74, 06121 Perugia, Italy
e-mail: adriano.ciani@unipg.it

W. Leal Filho et al. (eds.), *Integrating Sustainability Thinking in Science and Engineering Curricula*, World Sustainability Series, DOI 10.1007/978-3-319-09474-8_11

University level. However one of the greatest achievements is involvement of a course which is called Education for Sustainable Development and which will be part of elective courses taught at the Faculty of Education of the University of Prishtina. It is the first time that the Faculty of Education has organized a course like this which is very important for the future generations, particularly the future teachers. Through the education of students, KESD aims to increase the role and the importance of the sustainable education to new generations in our country and beyond. Increasing the awareness means more attention will be given to teaching methodologies, and teaching contents, climatic change, economic development, and protection of the environment inside of the Framework of reference of Sustainable Development Goals—SDG's 2015–2030 and Green Economy.

Keywords
Education · Sustainable development · Social inclusion · Curricula · Green economy

1 Introduction: Education for Sustainable Development

The UN Decade of Education for Sustainable Development (2005–2014) seeks to mobilize the educational resources of the world to help create a more sustainable future. Many paths to sustainability (e.g. sustainable agriculture and forestry, research and technology transfer, finance, sustainable production and consumption) exist and are mentioned in the 40 chapters of *Agenda 21*, the official document of the 1992 Earth Summit. Education is one of these paths. Education alone cannot achieve a more sustainable future; however, without education and learning for sustainable development, we will not be able to reach that goal (ESD-UNESCO 2005).

The vision of Education for Sustainable Development (ESD) is a world where everyone has the opportunity to benefit from quality education and learn the values, behavior and lifestyles required for a sustainable future and for positive societal transformation. ESD is for everyone, at all stages of life and in all possible learning contexts. ESD employs a partnership approach that engages multiple sectors and stakeholders—including media agencies and the private sector—and utilizes all forms and methods of public awareness raising, education and training to promote a broad understanding of sustainable development. ESD equally addresses all three pillars of sustainable development—society, environment and economy—with culture as an essential additional and underlying dimension. By embracing these elements in a holistic and integrated manner, ESD enables all individuals to fully develop the knowledge, perspectives, values and skills necessary to take part in decisions to improve the quality of life both locally and globally on terms which are most relevant to their daily lives (ESD-UNESCO 2005).

"Education for Sustainable Development (ESD)", is a concept that goes far beyond environmental education. ESD is the educational process of achieving human development ("the three pillars of human development" proposed by UNDP: economic growth, social development, and environmental protection) in an inclusive, equitable and secure manner. It thus includes education for poverty alleviation, human rights, gender equality, cultural diversity, international understanding, peace and many more. UNESCO proposed that the vision of education for sustainable development is a world where everyone has the opportunity to benefit from quality education and learn the values, behavior and lifestyles required for a sustainable future and for positive societal transformation (UNESCO-ESD).

Even though "Sustainable Development" has become a phrase with wide use in public speaking of scholars, politicians and other public figures, often the general sense when speaking about "sustainable development"is considered only the issues related to the environment protection (Odile and Buchs 2014). Or there are others who relate this phrase with economic development only (Waas et al. 2011). However there have been other schoolars who made it clear the interdependent relation between Education and Sustainable Development (de Haan 2006). He goes even further in his reasoning, that any effort toward sustainable development should start with education as a starting point. Furthermore in this direction goes David W. Orr in his book "Earth in mind" when he suggests his six principles of education related to sustainable development: Education as environmental education, Education as a way of mastering one's personhood, Education as responsible way of carrying for the world, Education as a way of understanding the impact of the knowledge to the people and communities in real life, Education as exemplary way of teaching, and Education in connection with the real world (Orr 2004). In a line with this thinking is also Michael Fullan in his book "Leadership & Sustaiability", under chapter five, where he describes the changing of "school cultures", he suggests that sustainable education reforms should bring out a process that produces both good ideas and social cohesion. He goes on to describe sustainability as much changes in culture, powerful strategies that enable people to question and alter certain values and beliefs as they create new forms of learning across all levels of education system (Fullan 2005). Therefore if we want social cohesion, poverty alleviation, gender equality or other MDG's, education is the vehicle for that. But changes in education, in terms of reforms and improvements should address sustainable development as a main goal or a core value. Otherwise as Geoff Fagan states: "If we do not change the education system in a fundamental way, we will, by making it more efficient and fit-for purpose, support and encourage unsustainable practices" (Faggan 2009).

Education for Sustainable Development is a new idea to Kosovo. Kosovo declared its independence in February 2008, thus becoming the newest state in Europe. Immediately after independence, the government and the country's parliament took steps and actions to include the European agenda as a strategic goal aiming to integrate Kosovo into the European Union. To start with they began drafting legislation and other documents with EU standards as well as the harmonization of existing ones. Although Kosovo is not yet a member of the United Nations and has no right or obligation to accept international agreements, unilaterally Kosovo parliament has approved/ratified various international conventions by including them in its constitution and other relevant laws. In the field of Education for Sustainable Development, Kosovo's Parliament adopted the Millennium Development Goals (MDG), pledging to commit to achieve these targets by 2015 (MDG Kosovo 2008). This marks the beginning of involving Education for Sustainable Development in the policies and actions of the institutions in Kosovo. ESD is also included in the Kosovo Environmental Strategy (KES 2003). This strategy has several priorities regarding the environment:

1. Completion of environmental protection legislation, in accordance with existing conditions in Kosovo, gradual fulfilment of EU standards, and efficient implementation of the existing ones.
2. Building competent and capable institutions, equipped with human resources and instruments for the implementation of environmental policies.
   (Both priorities are in line with UNESCO ESD principles: "deals with the well-being of all four dimensions of sustainability—environment, society, culture and economy; and accommodates the evolving nature of the concept of sustainability) (UNESCO 2005).
   First of all this impacts:

   a. Decision making institutions, (*reflecting that all decisions that are to be taken should be in accordance with ESD principles*)
   b. Monitoring institutions (*to assure that implementation of the legislation is happening and all law and other regulations are respected*)
   c. Supervisory and inspecting institutions (*to ensure accountability and provide other measures if needed*)

3. Provision of necessary financial and efficiently economic instruments (Eco-founding), for environmental protection, in harmony with economic development. This is in line with UNESCO ESD principle; "builds civil capacity for community-based decision-making, social tolerance, environmental stewardship, an adaptable workforce, and a good quality of life" (UNESCO 2005).
4. Establishment and operation of environmental monitoring network at the national level by extending it with priority to major industrial polluter's "hotspots "in Kosovo.
5. Gradual increase the population's access to safe drinking water, sewerage system, the waste treatment system and support recycling programs and waste sewage.
6. Rational use of natural resources: land, water, minerals, and forests. Special attention will be given to the use of endangered areas and orientation towards renewable resources.
7. Increasing the area and the protection of natural heritage areas and those with special natural values along with increased capacity for efficient management in accordance with the principles of the Rio Declaration. (The last three principals

are in line with the UNESCO ESD principle; deals with the well-being of all four dimensions of sustainability—environment, society, culture and economy) (UNESCO 2005).
8. Development of long-term programs of education, awareness campaigns and support of scientific projects focused on the environment.
9. Support of clean technologies behaviour in Kosovo in the starting phase of new industries and adopting the existing industries.
10. Applying the concept of energy efficiency in all sectors of energy use (KES 2003).

(The last three priorities are in line with UNESCO ESD principals; uses a variety of pedagogical techniques that promote participatory learning and higher-order thinking skills; promotes lifelong learning; is locally relevant and culturally appropriate) (UNESCO 2005). Some of these priorities give utmost importance and speak directly to an educational approach based on the principles of Education for Sustainable Development.

On the other hand the Ministry of Education, Science and Technology has made some moves towards integration of Education for Sustainable Development within the Strategic Plan of Education in Kosovo 2011–2016 (KESP 2011). This strategic plan is based on the achievement of the Millennium Development Goals aiming that "inclusive education strategies, priorities and planned actions, will be crucial to achieving the Millennium Development Goals (MDGs) and comprehensive policy of equal access by 2015" (KESP 2011). This is emphasized even more clearly in the KESP's vision of education:

The vision of education is to create a comprehensive system of education, which will provide all citizens of Kosovo with equal access to quality education at all levels, a system that provides for all the people of Kosovo, versatility throughout life for an advanced knowledge integrated into European society and contribute to long term sustainable development of the country through job creation and social cohesion advanced (KESP 2011). A step forward in the promotion and integration of the principles of Education for Sustainable Development is done with the approval by the Assembly of Kosovo of the National Council of Science. The Kosovo Assembly adopted in 2010 the National Research Program, developed by the National Research Council (NRC), which defines priorities for scientific research in view of social and economic development of the Republic of Kosovo. Research priorities were identified after a comprehensive consultation process which involved all stakeholders and relevant local and international factors in the research sector, higher education and in the area of economic cooperation. Initially, 33 fields were analyzed to design a narrower list of 16 priorities areas for research (NRP 2010).

Further discussions resulted in the following five priorities:

1. Natural resources, energy and environment;
2. Agricultural production and food security;
3. Medical research;

4. Social and economic studies;
5. Linguistic, cultural and historical research (NRP 2010).

Most of these priorities are fully consistent with the characteristics of Education for Sustainable Development as they are stated in UNESCO/ESD website:

Education for sustainable development:

- is based on the principles and values that underlie sustainable development;
- deals with the well-being of all four dimensions of sustainability—environment, society, culture and economy;
- uses a variety of pedagogical techniques that promote participatory learning and higher-order thinking skills;
- promotes lifelong learning;
- is locally relevant and culturally appropriate;
- is based on local needs, perceptions and conditions, but acknowledges that fulfilling local needs often has international effects and consequences;
- engages formal, non-formal and informal education;
- accommodates the evolving nature of the concept of sustainability;
- addresses content, taking into account context, global issues and local priorities;
- builds civil capacity for community-based decision-making, social tolerance, environmental stewardship, an adaptable workforce, and a good quality of life;
- is interdisciplinary. No single discipline can claim ESD for itself; all disciplines can contribute to ESD (UNESCO 2005).

Kosovo has made concrete steps in creating legislation for developing into a sustainable country, but implementation of this legislation is moving slowly in comparison with developed countries.

Kosovo is a developing country and doesn't have a sustainable economy yet. This is shown by the fact that Kosovo still does not base economic development in production, since the productions are still low and the process of privatisation of state owned enterprises is going slowly. This is best shown through the high rate of unemployment in Kosovo society, which is 30.9 % according to the Statistics Agency of Kosovo (SAK 2013) or 35.1 % according to UNDP in Kosovo and some other sources (UNDP-KS 2013). Although the average percentage of the population of Kosovo is young, 25 years old, the majority of them have completed higher education but still remain unemployed or they look for employment outside their profession (60.2 %) (SAK 2013). The high percentage of unemployment and lack of production in the country shows that Kosovo is far from reaching sustainable development. Equality in Kosovo still continues to be among the biggest challenges of society. Regardless of the great efforts that have been made to raise awareness regarding gender equality from various government and non-governmental institutions, the position of women in society is still unenviable. Female representation in the Parliament of Kosovo is quite high compared to other countries in the region but that's because of the electoral quota that is constitutionally guaranteed, 31 % participation in parliament (Gazette 2008). Managerial positions are mainly held by men.

Also, female employment is lowest in the region. According to the UNDP office in Kosovo 28.4 % of women who are part of the workforce only 12 % have a regular job (UNDP-KS 2013) and under Statistical Agency of Kosovo only 6 % of women are business owners (SAK 2013). Another issue in this matter is that the younger generation of women don't have more than 1 or 2 children. This happens because of existing legislation and laws that are not in favour of maternity leave. Kosovo, therefore, still leaves much to be desired in terms of gender equality which is a very important component of sustainable development in the country.

Regarding environmental protection, Kosovo as a country has had legislation drafted according to international standards, but the implementation of this legislation is extremely slow. Kosovo has turned into a place of high scale polluted air. Air pollution mainly comes from Thermo Power Plants that are almost the only source of energy in the country. Although our country has numerous opportunities for the production of renewable energy as we are mountainous country, this still remains to be realized in the future. Electric power is currently fuelled primarily by coal, which makes life harder for all citizens, especially in the settlements around the plants. In addition to the thermo power plants another cause of air pollution has been the war and weapons and bombs used during the war in 1999. Additional to this is the lack of control on gas (CO_2) emissions from cars. As a result of air pollution Kosovo is facing drastic increases in cases of children suffering from diseases like leukemia or other kinds of cancer and fatal diseases. Air pollution also affects the cultivation of agriculture and farming.

Therefore, despite the progress Kosovo has made in developing legislation for sustainable development, in practice it still leaves much to be desired. This situation makes sustainable development in the country a big challenge.

The overall aim of this paper is to present the situation of the Education for Sustainable Development in Kosovo. Special attention will be given to the inclusion of ESD in higher education institutions. We will also present in more details the role of local NGO Kosovo Education for Sustainable Development in promoting and implementing ESD projects and activities.

2 Education for Sustainable Development in University of Prishtina

The Brundtland Commission defined sustainable development as a pattern of resource use that "meets the needs of the present without compromising the ability of future generations to meet their own needs." In order to preserve the natural world, economic, social and environmental factors must be jointly considered and harmonized. Formal and informal learning, through raising awareness and influencing behavior, has a pivotal function if sustainable development is to be achieved. This role is especially pronounced in the realm of higher education (HE) because at this level students are being prepared to enter the labor market and emerge with skills to support green economies and as messengers of ideas. Progressively, universities and other higher education institutions (HEIs) have been incorporating sustainable development values and practices into their core activities of teaching and research, institutional management and operational systems. However, the debate thus far has focused primarily on the rationale and reasoning for why sustainable development needs broad adoption. The international discussions, however, have failed to specify the various actions that higher education institutions can adopt (OECD 2007). While these kind of discussions are happening at the international higher education area, the University of Prishtina as a leading institution of higher education in Kosovo also sought to include the Millennium Development Goals and the principles of the Education for Sustainable Development in their curricula's and teaching. Nevertheless ESD is still considered, within the University circle, as something related mainly to environmental preservation or development of ecology in the country.

ESD is mainly introduced in curricula and teachings in natural science faculty's courses/subjects in which they are treated as topics for the development of agriculture and animal husbandry or for the environment protection subject/courses. Given that Education for Sustainable Development is more than just the development of agriculture, animal husbandry or environmental protection, there have been efforts within the University, to develop activities which will include students from the Faculty of Education who are preparing to be future teachers in undergraduate institutions in the country. Training of future teachers with ESD principles is a great asset towards influencing future generations.

Education for Sustainable Development as a University course was organized for the first time at University of Prishtina during the Prishtina International Summer University in 2011 (PISU 2011). The course was lead by American and Kosovar co-professors and international students. Students of this course became the first ambassadors of Kosovo Education for Sustainable Development, a local organization established in 2011. These students, a number of whom are now teaching in various schools in Kosovo, still continue to be active in organizing various activities not only for their students but beyond their schools, by which they want to influence the growth of society's awareness about the importance of Education for Sustainable Development in the country.

The organization of this course within PISU was a small initiative to promote ESD within the University. From this period, Education for Sustainable Development was given more space by trying to engage in more subjects and in different profiles of studies in other faculties. Due to the high interest of student for ESD related courses, Prishtina International Summer University launched another course on ESD during the 2012 session. This was a great opportunity to enhance and further promote ESD within the University of Prishtina.

KESD organization's representatives, as part of the Faculty of Education, gave the idea that within the courses taught at the Faculty of Education to develop new curricula that will give all students the opportunity to learn more about EDS and its role in developing the country. Since the idea of Education for Sustainable Development is still treated as a problem only of natural sciences, there were major objections to include an ESD course in the plans and programs of the Faculty of Education. There was a need for additional information and explanations of the important connections of ESD and Education like the fact that teacher training and curricula reform and reorientation are the core values and a very important part of ESD, as de Haan would say: "since where a major mental shift is primarily involved, we must encourage the processes for changing awareness among individuals—and this can only be accomplished through learning" (de Haan 2006).

Since students of this faculty will be future teachers they can use Education for Sustainable Development from pre-school education, primary education up to secondary level. This investment in the education of children of a preschool age will enable these children to grow up with the idea of the importance of sustainable development. Education of children from preschool age up to the middle school will indirectly influence the awareness of these children and their families and in this way will lay down solid foundation of ESD to the wider society.

Besides organizing the course called "Education for Sustainable Development", at the Faculty of Education, other results saw an increased network of students who become active in promoting ESD, through various activities which are organized by KESD. The number of young volunteers from day to day is growing. At the beginning there was only one representative of KESD at University, today there are over 700 students who voluntarily contribute to the planning and implementing of training and other educational activities about the necessity of Sustainable Development in the country. KESD aims that other Departments of the University of Prishtina will take concrete steps to include Education for Sustainable Development within their teaching syllabuses. Full integration of the principles of Education for Sustainable Development remains an on-going challenge in higher education in the country; especially inclusion of these principles in the context of syllabuses of all courses/subjects and study programs in various faculties. It is not only the level of higher education that should include ESD, but also other levels of education in the country.

So, there is much more we should do to in rising awareness of all relevant factors in Kosovo so that Education for Sustainable Development will be starting point of policy changes and education for future generations, or even in the framework of activities for adult education. Further work needs to be done in advocating the inclusion of Education for Sustainable Development and the promotion of the further benefits arising from this.

3 Kosovo Education for Sustainable Development—KESD

Kosovo Education for Sustainable Development was established as a non-governmental organization in June of 2011. It is as a civic initiative that will promote and integrate the Principles of Education for Sustainable Development in Kosovo, through providing professional training, projects and different activities about ESD.

KESD adopted as its mission and vision, UNESCO/ESD mission: "*Integrate the principles, values and practices of sustainable development into all aspects of education and learning. This educational effort will encourage change in behavior that will create a more sustainable future in terms of environmental integrity, economic viability and fair society for generations present and future"*

UNESCO/ESD vision is: *"Education for Sustainable Development in Kosovo (KESD) aims to help people develop their views, skills, perspectives and knowledge to make decisions by being well informed and act on their benefits and others, now and in the future"* (*DESD*).

KESD so far has developed different projects and activities in several areas:

3.1 Promotion and Awareness of the Importance of Sustainable Development

The first project that KESD organized was greening the outdoors of preschools in Prishtina. In this event they invited teachers, children, parents and the community, to plant flowers together and make the outside of the preschool more green. The purpose of this activity was to increase the awareness of society to the importance of Education for Sustainable Development, which should start with children of a young age. Also the involvement of parents and the community gave a message that to the country's sustainable development; everyone should be engaged and collaborate without exception. Being part of society means also contributing as an individual volunteer, to provide a better life for all.

Another KESD activity worth mentioning was the distribution of over 3,000 copies of books, that where distributed in different places, mainly in poor areas where people of different nationalities live, in order to motivate them to learn how to read and write. Many Roma children, although in grown age (10, 15, 17 years old), began for the first time, learning literacy through these books. Now these children are supervised and supported by various organizations in their learning activities. This activity was used as a form for the eradication of illiteracy in the country (MDG 2).

Another project that KESD has participated was national clean-up campaign: "Let's do it Kosovo", with over 400 volunteers who cleaned the different areas of the country and created cleaner space for children and adults.

The cycling campaign was another campaign which was aimed to increase awareness of society to the use of means of transport that does not pollute the air. This campaign consisted of inspiring people to use bicycles to travel instead of motor vehicles. About 250 people of different ages joined this campaign. KESD volunteers motivated young people to use bicycles as a mode of transportation that helps health not only physically, but also does not pollute the air we breathe. Such activities are organized by KESD constantly and the implementation of all activities performed by volunteers and active citizenships activism. Work is supported by donations to KESD from various local and international institutions. KESD's motto is "Sustainability begins with you", which promotes the organization's activities without a big financial budget.

3.2 Training and Capacity Building

One important contribution from KESD is the training of young people for sustainable professional development. Training is mainly focused on professional development of future teachers. Training in the way they should be working with children is the strategic point for promoting ESD in country. KESD provides training on child protection policy, indoor and outdoor activities at educational institutions and after school activities. KESD training also focuses on promoting voluntary work for the community. Developing countries rarely have applied volunteering as a tool for development. They are more concerned with jobs from which they benefit and very little in their contribution to community volunteering in the place where they live. Initially this training was not of interest to young people. KESD's initial challenge has been how to motivate them to be involved in various volunteer activities. In order to motivate young people, for their engaging as youth volunteers for one-year period, KESD provides them with a reference letter, which they can use in their CV. Besides the local training KESD has provided training at regional and international level through various projects, where participants were able to be trained in different topics and areas. Now the number of students who want to engage in KESD's trainings is very high, and their interest to participate in voluntary work has reached the point that KESD initially aimed for.

3.3 Representation and Support

Representation and support is another field where KESD as developed several activities. From the beginning KESD has played an active role in representing the Education for Sustainable Development in any activity that is related to the topic. Therefore KESD's role has been recognized by Kosovo's institutions and other organizations and we are an active partner and participant to all activities related to ESD. Something worth mentioning here is Working Group for inclusion of ESD. This group is formed from Ministry of Education Science and Technology, Ministry of Environment and Spatial Planning, Regional Environmental Centre-Kosovo and Kosovo Education for Sustainable Development. The aim of the group is institutional promotion and inclusion of ESD into our planning and regulations. In terms of supporting, KESD has provided different kinds of support bona fide to individuals and grass root organizations that had any activity related to ESD. This support consisted of consultancy, volunteering, expertise or access to our researches and other tools. We will continue to provide such support in hope of establishing a wider network of ESD promoters in Kosovo.

4 Conclusion

Kosovo Education for Sustainable Development, although only established in 2011, has achieved a great influence in promoting and organizing various activities for sustainable development in the country. The biggest impact of KESD achievements has been at the University of Prishtina, Faculty of Education, where mobilizing of students in organizing various volunteer activities, KESD has managed to influence the creation of the course/subject that will be part of the teaching program at the Faculty of Education, as professional preparation of students who will be future teachers. Investing in students, who will become teachers, is considered the more influential, since these students will work directly with children of pre-school level up to secondary level.

Parental involvement in the institutionalized education of children means that indirectly students-prospective teachers, will have an impact on their awareness of the values and MDG. Inclusion of citizens in various activities helps creation of a culture for sustainable development in the country.

Engaging students in various extra training activities empowers the impact that KESD has on higher education, by encouraging them to give a practical contribution to the development of the country. The idea of volunteering and active citizenship, helps our country not only to develop various activities that improve the quality of life for its citizens, but also promotes the idea that implementation of these activities would not require huge amount of funds. KESD is an example for many other local and international organizations, who are applying similar approach to increase the quality of life for citizens, taking into account the voluntary work that University students are doing. KESD continues to achieve its vision, making investments that are sustainable and long-term, such as higher education.

References

de Haan G (2006) The BLK '21' programme in Germany: a 'Gestaltungskompetenz'-based model for education for sustainable development. Environ Educ Res 12(1):19–32

ESD-UNESCO (2005) (n.d.). http://www.unesco.org/new/en/education/themes/leading-the-international-agenda/education-for-sustainable-development/education-for-sustainable-development/. Accessed Mar 2014

Faggan J (2009) In: Stibbe A The handbook of sustainability literacy, Green Books, UK

Fullan M (2005) Leadership and sustainability. Corwin Press, USA

Gazette O (2008) (n.d.) http://www.gazetazyrtare.com/e-gov/index.php?option=com_content&task=view&id=157&Itemid=56. Accessed Mar 2014

KES K E (2003) (n.d.) http://www.rit.edu/~w-cenr/documents/data/Environ.Strategy-MEM.pdf Retrieved March 2014. Accessed Mar 2014

KESP M O (2011) (n.d.) http://www.masht-gov.net/advCms/documents/KESP_2011_2016.pdf. Accessed Mar 2014

MDG Kosovo D O (2008) (n.d.) assembly-kosovo.orgRetrieved March 2014. Accessed Mar 2014

NRP M (2010) (n.d.) http://www.masht-gov.net/advCms/documents/NRP_FINAL_%20English.pdf. Accessed Mar 2014

OECD (2007) (n.d.) oecd.org. Retrieved from http://www.oecd.org/education/innovationeducation/centreforeffectivelearningenvironmentscele/45575516.pdf. Accessed Mar 2014

Odile B, Buchs A (2014) Teaching sustainable development issues: an assessment of the learning effectiveness of gaming. halshs-00946227 1:1–16
Orr W (2004) Earth in mind. Island Press, USA
PISU (2011) (n.d.) http://uvp.uni-pr.edu/. Accessed Mar 2014
SAK S A (2013) (n.d.) http://ask.rks-gov.net/eng/. Accessed Mar 2014
UNDP-KS C P (2013) (n.d.) http://www.ks.undp.org/content/kosovo/en/home/countryinfo/. Accessed Mar 2014
UNESCO (2005) (n.d.) unesco.org. Accessed Mar 2014
Waas T, Hugé J, Verbruggen A, Wright T (2011) (n.d.) Sustainable development: a bird's eye view. Sustainability 3(10):1637–1661. doi:10.3390/su3101637

Authors Biography

Arlinda Beka is a doctoral student of the Education Sciences, at the Faculty of Social Sciences, University of Tirana. Her fields of study are the education system policies in Kosovo, reforms and curriculum development.

She is a lecturer at the Faculty of Education, of the University of Prishtina and Executive Director of Kosovo Education Organisation Director for Sustainable Development.

She is a member of various professional working groups in the education field, such as group for the development of pre-school developing standards 0–6 years, supported by MEST and UNICEF, member of many international organizations such as: UNESCO/Education for Sustainable Development Network, European Association International Education etc.

She is active with research in education, including teacher professional development, teaching strategies, curriculum development, etc.

Publications which are presented in prestigious national and international conferences and has published works in various scientific journals.

Since 2010, she is University of Prishtina, representative in UNESCO, for ESD.

Adriano Ciani is Professor of Agricultural Economics and Rural Appraisal at the Department of Agricultural Economics, Farm Appraisal and Food Sciences-DSEEA at the University of Perugia. He is Professor of Environmental Economics and Farm Management. His current scientific activity it's addressed in: Sustainable Management of Territory, Common Goods, Water Resource, ICT and Sustainable Rural Development. He wrote several books and 235 publications. He played collaborative activities of research, teaching, technical and dissemination, in over 50 countries. He founded the BIOSPHERA Association and in 2000 has organized the 1^ World Forum on Rural Tourism and Agro-tourism and founded the Association IAERT (International Association of Experts in Rural Tourism and Agro-tourism). From 2001 to 2009 he was commissioned by the MoFA as Scientific Attaché at the Italian Embassy Tirana where has promoted the wide Scientific Program "A Sustainable Development for Albania". He has designed and conducted, as Director, the International Summer School in Sustainable Management and Promotion of Territory. He was by 1980–1990 Adviser at the Provincial Government of Perugia, at SASE Lt. Airport (Perugia), Vice President Umbria-Tuscany Irrigation Authority, Auditor of ERI New Editions. He is honorary member of the Academic Senates of the 7 Universities of Romania.

Increasing Global Environmental Literacy via Faculty Experiential Learning

Antje Danielson and Panagiota Kaltsa

Abstract

Environmental sustainability is a priority area in academia, as evidenced by the recent growth in programs and degrees devoted to this field. However, encouraging communities to address the environmental issues that will shape upcoming generations remains a challenge. This paper presents the value of investing in faculty training and experiential learning as methods for disseminating environmental principles through a wide selection of academic courses and to create social impact. The Tufts Environmental Literacy Institute (TELI) is a long-standing initiative that spurs faculty and, consequently, students to environmental action. Since TELI's inception in 1990, more than two hundred faculty and doctoral students from Tufts and partner universities have come together to exchange ideas, share best practices, and shape curricula. The current work highlights the importance of interdisciplinary collaboration, the value of forming inter-university teaching teams, and the role of social networks in building environmental consciousness. Combining empirical data and statistics, this paper can be useful for universities that seek to effect environmental impact and to establish partnerships with other academic institutions that share the vision of sustainable development.

A. Danielson (✉) · P. Kaltsa
Tufts Institute of the Environment, Tufts University,
Miller Hall, 210 Packard Avenue, Medford, MA 02155, USA
e-mail: antje.danielson@tufts.edu

P. Kaltsa
e-mail: panagiota.kaltsa@tufts.edu

W. Leal Filho et al. (eds.), *Integrating Sustainability Thinking in Science and Engineering Curricula*, World Sustainability Series, DOI 10.1007/978-3-319-09474-8_12

Keywords
Environmental literacy · Exponential knowledge distribution · Interdisciplinary collaboration · Faculty education · Social networks

1 Introduction: The Tufts Environmental Literacy Institute —TELI

The Tufts Environmental Literacy Institute (TELI) is a well-established faculty development workshop aimed at enhancing environmental literacy and assisting participants in incorporating related themes into existing or new courses. TELI's ultimate goal is to increase its participants' exposure to and knowledge of environmental literacy, enabling them to engage effectively with the issues facing incoming generations of students. The Tufts Institute of the Environment (TIE) has run TELI programs for over two decades. Because of its pioneering nature, the program received the Presidential Environment and Conservation Challenge Award in a White House Rose Garden ceremony in 1990.

Since its inception by Anthony Cortese at Tufts University in 1990 TELI has focused on engaging Tufts faculty with sustainable development concepts with the aim of raising environmental literacy across the University (Cortese 2003). Participating professors have been urged to apply what they learn at TELI into their teaching and to change their curricula to include environmental principles. Both the development of the Environmental Studies Program and the creation or updating of approximately 50 courses at Tufts University can be directly attributed to this faculty workshop. It has also been shown that TELI alumni become more collaborative and include environment related content in their research (Barlett and Rappaport 2009).

Each year, TELI addresses an issue of major environmental importance. Previous workshops have covered issues like environmental communication, the urban environment and social justice, sustainable agriculture, and climate change. Each workshop includes interactive presentations by a variety of experts in the fields of environmental science and engineering, public policy, health, and economics. TELI also provides field-based learning experiences, introductions to new tools, and access to a range of resources for course development. Facilitated contextual and non-contextual group experiences (Molineux and Haslett 2007) enhance sharing and problem solving among the participating faculty.

At Tufts University TELI has played an integral part in preparing students to address life challenges with an environmental point of view simply by bringing this context to their professors. More than two hundred Tufts faculty have participated in TELI to date and as a result, thousands of students have learned to synthesize different disciplines and to integrate awareness of complex issues such as climate change and environmental sustainability into their worldviews. Moreover, the interest of several alumni to repeat the TELI experience indicated the workshop was so beneficial to them and their work that it was worth repeating.[1] Today, many other U.S. colleges and universities now offer similar faculty workshops, thereby contributing to the exponential knowledge distribution TELI initiated in the 1990s (Barlett and Rappaport 2009).

Recognizing the tremendous potential of TELI, Antje Danielson and Panagiota Kaltsa (hereinafter "TELI leaders") started working to widen the scope of the workshop in Spring 2012. The goal was to maintain the same structure and address faculty members within and beyond the Tufts community. This approach was based on the hypothesis that such change would foster the creation of inter-university connections and promote sustainable development in academia.

2 TELI 2013

TELI 2013 was the first to foster international participation. In May 2013, 26 academic professionals representing 16 universities from 10 countries[2] in Africa and Southeast Asia gathered in Boston. The goals of TELI 2013 were to (1) promote "One Health Thinking" and expose professionals from the medical, veterinary, public health and environmental sector to this concept; (2) introduce the participants to tools that can be used in cutting edge, creative, interdisciplinary teaching and curriculum development; (3) create and maintain global connections between the participants. The workshop was run with grant support from the United States Agency for International Development (USAID).

Consistent with the hypothesis of the TELI leaders, the participants of TELI 2013 unanimously agreed in the workshop's exit survey, on the value of cross-cultural interaction and the exchange of ideas about sustainable development among experts, as well as the unique positioning of the United States (US) educational system to initiate dialogue and collaboration channels across disciplines.

Both TELI leaders and participating faculty were interested in using Social Network Analysis (SNA) to quantitatively analyze the momentum of the network relationships created at the workshop. TELI leaders needed to evaluate the workshop design's efficacy, while the participants were curious about the application of SNA, a tool they had learned about during TELI. Three months after the completion of the workshop, the TELI leaders invited participants to complete a customized SNA survey. The TELI goals and the results of the TELI 2013 SNA are described here, showing that this type of workshop is suited to strengthening international faculty networks.

[1] Until 2012 TELI participants were given a small stipend at the end of the course. To accurately assess the workshop's value and impact the authors addressed the question of whether this stipend drove the faculty's eagerness to attend. A survey conducted among TELI participants in 2012 showed that this was not the case.

[2] The ten represented countries were the **Democratic Republic of the Congo, Kenya, Ethiopia, Uganda, Rwanda, Tanzania in Africa; Thailand, Vietnam, Indonesia, Malaysia in SE Asia and the United States of America**.

3 Analysis of TELI Goals

(1) One HealthIn the past decade, the world has witnessed increasing numbers of health and health-related problems that are complex in etiology as well as impact, affecting simultaneously human and animal populations and the environment. Examples of these problems include emerging infectious diseases, climate change, respiratory disease from fuel burning, widespread pollution, and natural and manmade disasters and emergencies.

Awareness of these issues has led advocates for human, animal, and environmental health to unite around the concept of One Health. This concept bridges both disciplinary and geographical boundaries by bringing together physicians, veterinarians and public health officials with ecologists, environmental, engineers, and environmental planners, in order to focus on working locally, nationally, and internationally. This definition acknowledges that significant, long-standing barriers to collaboration between health professionals and members of other disciplines have limited our understanding of causation, constraining the development of balanced, effective solutions to One Health problems. The concept also implies that solutions to complex health problems must balance competing interests in the political sphere; it is no longer sufficient to blindly elevate the health of one group over that of others. All constituents are linked in a complex web of health.

"One Health Thinking" is already being taught at many colleges and universities in the US, in addition to traditional undergraduate and graduate courses. For example, Tufts University created an interdisciplinary, team-taught course on One Health through its University Seminar program, a program that aims to "prepare leaders with a rich and textured understanding of the world in all its complexity and diversity." Similarly, the University of Wisconsin created a Certificate on Humans and the Global Environment to prepare students to devise effective solutions to interdisciplinary environmental problems (Kaufman et al. 2008).

Many One Health problems are global in nature, necessitating the establishment of global networks and a global understanding of the One Health concept. Not only did the USAID recognize this idea by granting funds to a University Consortium including Tufts University in 2009, it was also one of the main reasons for opening TELI 2013 to a global audience.

(2) Tools
According to Koehler and Mishra the ability of a teacher to make connections between content, pedagogy, and technology and understand the complex interactions among these three fields in defined contexts is what leads to effective problem solving (Koehler and Mishra 2009). Revisiting academic teaching in this framework, the TELI leaders designed the program to introduce not only the scientific aspects of One Health, but also to give a new perspective on a variety of tools and communication technologies, and demonstrate how those could be applied in *the participants'* teaching. To ensure that the exposure to new tools was not

overwhelming, participants were asked to fill out a survey before their arrival in Boston, where they assessed their knowledge of those technologies. The reading materials and content of these sessions were adjusted to fit the learning needs of the group.

Interactive group activities and case studies introduced the tools, Geographic Information System (GIS) (Koch and El-Baz 1998), crisis mapping (Meier and Leaning 2009), asset mapping (Mor et al. 2013), digital storytelling, and social networks and communication (Tunnard 2003), to the participants. Participants were asked to apply these tools to hypothetical cases from their respective disciplinary backgrounds and to share their ideas with a small group of other participating faculty in a collaborative learning experience. Most of the presented tools were so new to the participants on so many levels, that they were able to ignore their cultural differences and create a common understanding of the problems facing them in their home countries, on a collaborative professional level.

(3) Network Formation

Firestein found that groups that laugh and smile more also performed better creatively (Firestien 1990), therefore TELI 2013 included diverse activities that added value to network formation. One of the highlights of TELI are the environmentally themed fieldtrips that in 2013 brought the participants to the Deer Island Waste Water Treatment Plant, the Tufts School of Medicine in Boston and the Cummings School of Veterinary Medicine in Grafton. Beyond classroom interaction, extra-curricular activities were organized for participants to make social connections during day time group meals on site, dinners at restaurants or at Tufts faculty homes and tours in the Boston Area. These arrangements brought the participants closer and allowed them to bond on both a professional and personal level, which is prerequisite for building lasting networks.

Developing new approaches and concepts for problem solving requires a high degree of creativity. When bringing interdisciplinary collaborators together, group creativity must be enabled through conducive environmental conditions such as psychological safety and freedom, tolerance of ambiguity and individual differences, and support for creative ideas (Sternberg and Lubart 1996). In fact, integration and trust building are necessary conditions for successful interdisciplinary collaboration. The literature acknowledges the importance of social networks to learning performance, including a sense of community, cohesion, and resource exchange (Dawson 2010). The development of a trusting relationship within the interdisciplinary community is the foundation for developing shared mental models, understanding, and fruitful learner-to-learner interactions. Problem-based learning and opportunities to interact socially were integrated into the TELI 2013 curriculum exactly to facilitate such positive group interactions.

4 Analysis of TELI Network

SNA—theoretical framework Hicks and his coauthors used SNA focused on *faculty integration* (Hicks et al. 2010) in order to evaluate the interdisciplinary and impact of disciplines related to the environmental sciences, focusing on journal citations as data input to their study. Network analysis was a useful tool for them in measuring the crossing of disciplinary boundaries and for identifying disciplines that facilitate interdisciplinary roles.

SNA helps analyze the structure of social networks and enrich our understandings of how they work by looking at the flow of information or resources between the participants (Wasserman and Galaskiewicz 1994). Every social group of people (or organizations) characterized by dyadic ties among its members constitutes a social network. Over time it is possible to see developments in network interactions and the effects that these relationships have on the people (Wasserman and Faust 1997). SNA focuses on data describing the ***relationship*** between the individuals such as the exchange of messages, projects worked on together, or number of papers published together (Laat et al. 2007). The two key indicators of SNA are density and centrality. Density describes the number of links in a network over the number of possible connections and is a measure of the overall connections (Scott 1991). Centrality describes the level of engagement of individuals in a network (Wasserman and Faust 1997).

5 Testing the Network of TELI 2013

To test the hypothesis that a wider participant scope for TELI would foster the creation of inter-university connections and promote sustainable development in academia, the question was whether TELI 2013 had managed to bridge international divides and facilitate substantial connections that would allow for international collaboration. In this context it was important to understand who was involved in collaborative tasks, who where the active participants, and who was just participating peripherally.

For this study a SNA questionnaire was developed to (1) capture relational ties specific to this group; (2) identify their common interest areas around the concept of One Health; and (3) gather information regarding the tendency of faculty to use technology for personal and professional purposes. The data were to enable TIE to tailor post-TELI communication with alumni, and to use the right media to share applicable information with alumni sub-groups that share the same environmental interests.

The network community was defined as the whole group of TELI 2013 participants including Antje Danielson the course director. The SNA Survey Questionnaire was emailed to the survey participants in the first week of September, using the Tufts Qualtrics survey tool. In total, TIE asked 27 people to complete the survey and 25 did so (Male = 16, Female = 9) between the 25 and 55 years of age. To evaluate SNA surveys, achieving the maximum number of responses in the target group is essential; the TELI leaders also prompted each participant individually.

6 Results and Discussion

(1) TELI Participant Relational Ties

The SNA survey data reveal that before TELI 2013 there was no direct initial connection between the participants from the African delegation to the SE Asian delegation. While members of each network were connected, the networks were not connected to one another (Fig. 1). When including U.S. participating faculty in the pre-TELI 2013 SNA, it becomes clear that the U.S. delegation represents a bottleneck between the two international faculty clusters. Specifically, only some of the American participants are the nodes that connect both sides of this network (Fig. 2). During the workshop, the outstanding number of 138 new ties were created, bridging the Africa-Asia divide (Fig. 3).

Following the workshop, five ties seem to have remained strong between the African and Asian cohorts (Fig. 4). The decrease in number of connection lines does not necessarily indicate the absence or negation of ties formed during the conference. Participants are expected to use the TELI ties to communicate when they need to (i.e., joint projects, consultations, academic research). In conclusion, while the post-TELI number of expressed connections might seem small, these five ties are channels of information flow between the two continents that did not previously exist. The number is also remarkable in light of the small size of the survey sample.

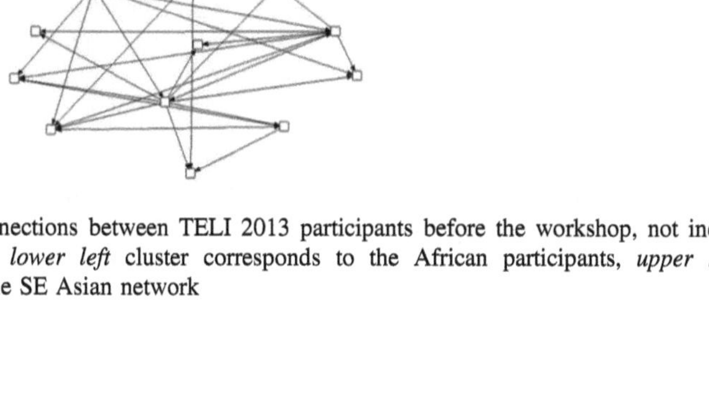

Fig. 1 Connections between TELI 2013 participants before the workshop, not including U.S. participants; *lower left* cluster corresponds to the African participants, *upper right* cluster represents the SE Asian network

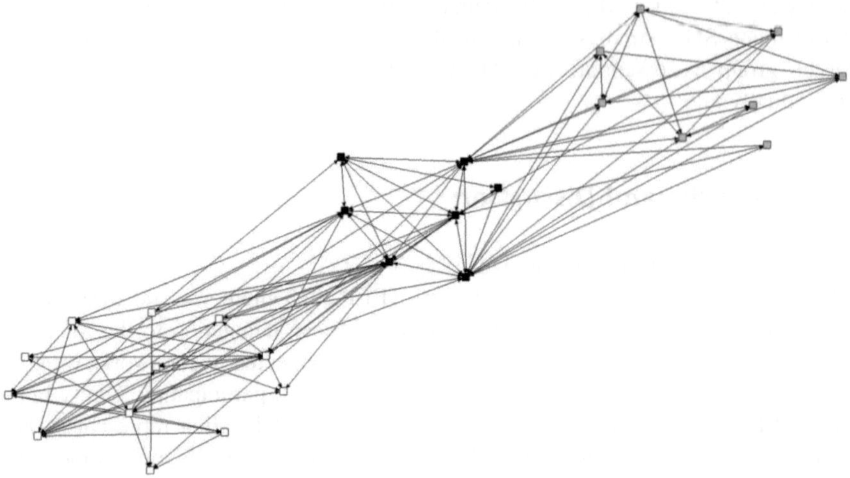

Fig. 2 Connections between TELI 2013 participants before the workshop, including U.S. participants in the *middle*

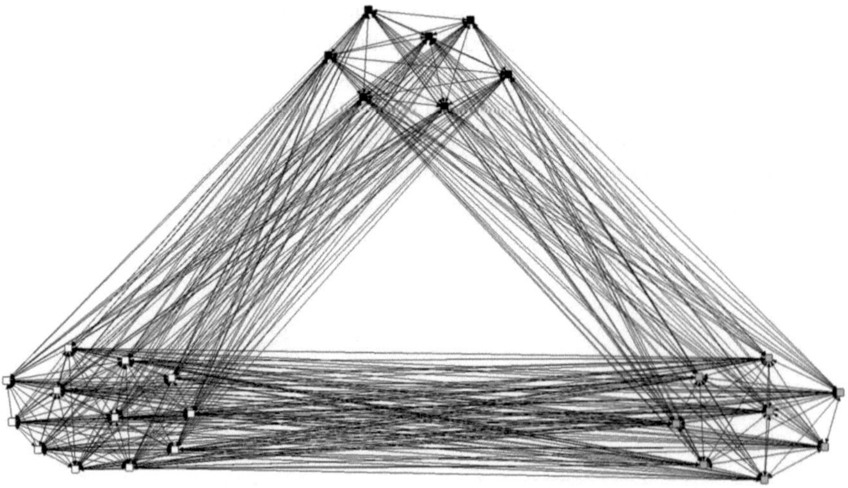

Fig. 3 Connections during TELI 2013; the *top* cluster is the American delegation, the *lower left* cluster corresponds to the African participants, and the *lower right* cluster represents the SE Asian network

The interactive tools introduced and used through the TELI curriculum contributed directly to the number of connections made during the workshop. For example, model-based reasoning over boundary objects has been shown to be a useful tool for fostering cooperation and communication among diverse members

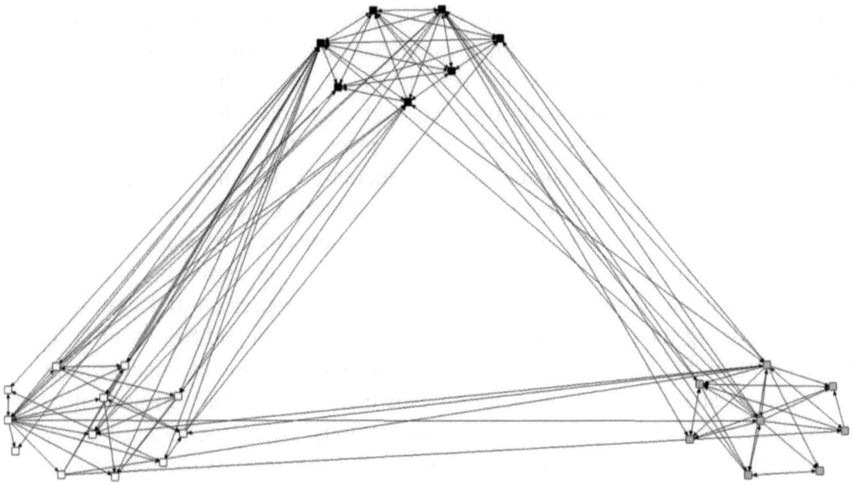

Fig. 4 Connections three months after TELI 2013; the *top cluster* is the American delegation, the *lower left* cluster corresponds to the African participants, and the *lower right* cluster represents the SE Asian network

of heterogeneous working groups (Wilson and Herndl 2007). During two of the days at the workshop participants exercised model-based reasoning using boundary objects during their collaborative exploration of digital story production and asset mapping. Most of the participants were not familiar with these techniques. From the lively exchanges that developed during those 2 days, it was clear how much the participants enjoyed this kind of learning and how much it contributed to the cross-cultural, geography-bridging success of the workshop.

Additionally, bringing a group together face-to-face (Bracken and Oughton 2006) in one physical space and framing the meeting around a specific theme are crucial factors for fostering effective encounters and for developing long term interdisciplinary research relationships. Bridle and co-workers describe how theme, size of gathering, and location of the meeting contribute to fruitful interactions. In particular they emphasize the importance of the availability of quiet break-out spaces and an "inspiring location" (Bridle et al. 2013). During TELI 2013, in addition to the above space features, we also provided an interfaith worship space and invited the participants to the private homes of some of the Tufts host faculty. All of these factors contributed to the 138 ties that where established during the workshop and laid a sound foundation for post-TELI relational ties.

(2) Common Interest Areas of TELI participants

The SNA survey was designed to reveal more than just personal connections within the TELI network. It was important for the TELI leaders to pair participant's interests and to understand whether those who share the same environmental concerns connected during the workshop. Even identifying where such connections were ***not*** formed can be helpful to TIE in steering inter-university collaboration between faculty who share a common experience and a common interest. Networks like TELI can be used to share best practices in teaching and to help the improvement of environmentally oriented courses worldwide.

The analysis of the SNA survey results proves that inter-regional and intra-regional connections with shared interests increased over the course of the conference (Fig. 5). In fact, during the conference, interest based connections increased by 350 %, up to 405 ties, with the majority connecting African and SE Asian delegates (Fig. 6). Post-TELI the interest connections decreased insignificantly by 8 %—7 fewer ties compared with the pre-conference period—but new ties were created bridging the communication gap between African and SE Asian Delegates (Fig. 7). Weak ties like these are ***acquaintances*** that function as crucial bridges for information sharing and collaboration between the separate primary networks represented by TELI alumni (i.e., home institutions, countries, regions).

The thickness of the connecting lines in the network maps, reflect common interest ties and represent the number of common interests shared between participants. This number can be 1, 2, or 3. Table 1 shows how many interests were shared according to the formed ties.

Moreover, using paired interests to introduce potential collaborators to one another is relevant to both faculty and students at different universities. For example, a Tufts PhD Candidate in Water Diplomacy at the Integrated Graduate Education and Research Traineeship (IGERT) program travelled to Ethiopia in March 2014 to conduct field research with the help of a local TELI 2013 alumnus. The alumnus welcomed the student, assisting her in making professional

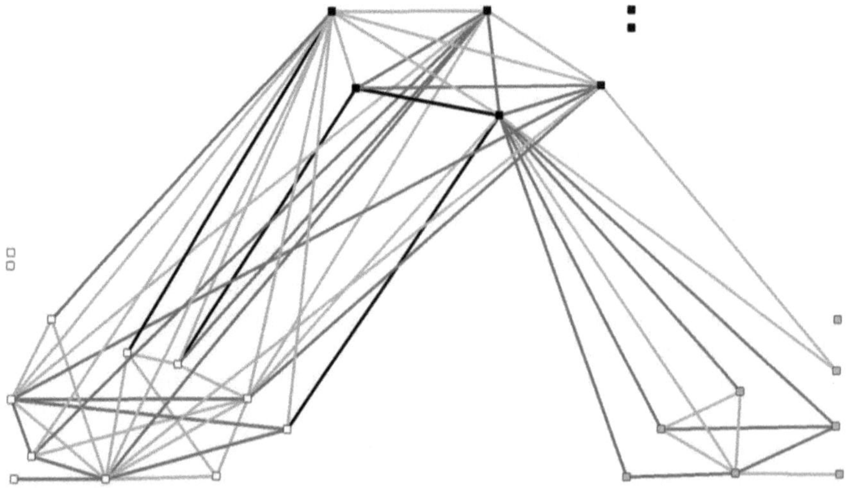

Fig. 5 Shared interest connections before TELI the top cluster is the American delegation, the *lower left* cluster corresponds to the African participants, and the *lower right* cluster represents the SE Asian network. Consistent with previous data, there are no direct common interest ties between the African and SE Asian delegates

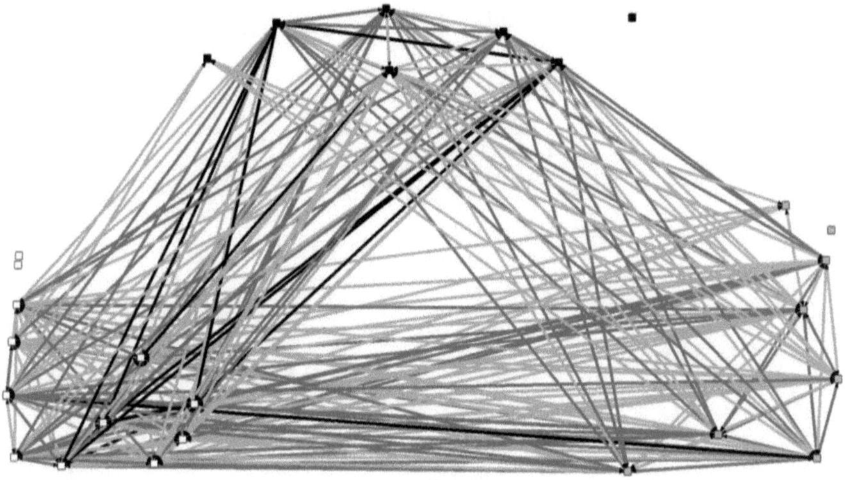

Fig. 6 Common interest connections during TELI 2013, the top cluster is the American delegation, the *lower left* cluster corresponds to the African participants, and the *lower right* cluster represents the SE Asian network. Consistent with previous data, this network map demonstrates the contribution of TELI to the formation of new ties this time—common interest ties

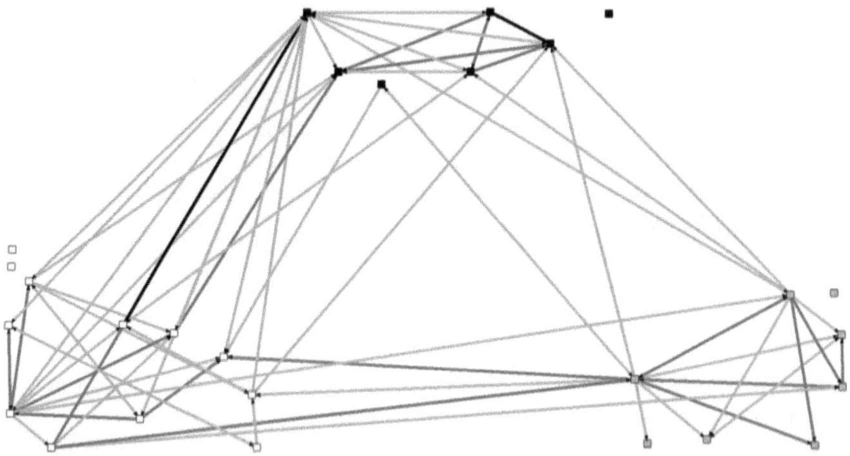

Fig. 7 Common interest connections three months after TELI the *top* cluster is the American delegation, the *lower left* cluster corresponds to the African participants, and the *lower right* cluster represents the SE Asian network. Because of TELI, five new common interest ties were created between African and SE Asian delegation, which did not exist before

connections in the field of climate change adaptation. This case is just one example of how valuable inter-university networks, especially across continents, can be for higher education.

Table 1 Common interests

Pre-TELI	
No. of interest	No. of ties
3	5
2	31
1	54
Total	90
During TELI	
No. of interest	No. of ties
3	24
2	143
1	238
Total	405
Post-TELI	
No. of interest	No. of ties
3	3
2	24
1	56
Total	83

Number of shared interests amongst participants before, during and after TELI

(3) Use of Technology among TELI participantsThe tools presented at TELI, including digital media storytelling, crisis mapping, GIS and SNA, opened a new subject area for discussion among participants. Using social media in education was novel for most faculty participants. If they were at all familiar with different digital media applications, it was only for personal use. Survey respondents named email as their primary tool for professional communication; while standard, this finding allows room for advancement in academic communication channels.

Specifically, according to the SNA survey results, the average TELI participant was not a frequent social media or technology tools user. As expected, Facebook, LinkedIn and YouTube were the top-ranked online media used, but those were primarily used for personal purposes (Fig. 8). Only four of the survey respondents used social media for academic purposes. Of those four, two were from the United States and two from SE Asia, namely Indonesia and Malaysia. This finding corresponds with global social media use trends; social media are indeed more popular and widespread in both the US and certain countries in SE Asia (Winkels et al. 2013).

Considering the rapid growth of technological savvy among younger generations —those known as technology natives—the application of social media and other technology tools in education has become implicit (Prensky 2001). Modern students

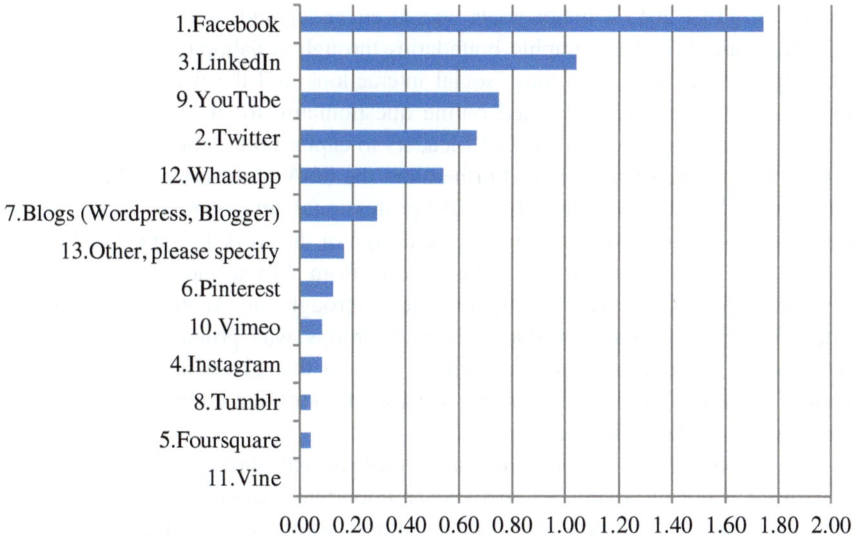

Fig. 8 Social media use of TELI participants primarily for personal purposes

learn differently and therefore education methods should be updated to match their needs and habits. Teaching students how technology can be used to create social impact beyond academia constitutes a large opportunity for growth that must be further explored. Aiming to increase the number of faculty using new technology in their teaching across international universities, experiential workshops like TELI can change teaching practices.

7 Conclusion

Past surveys to assess the impact of TELI on participating faculty indicated that alumni place a high value on building a community within the university (Barlett Rappaport 2009). TELI 2013 aimed to take this sense of community a step further both in a geographical sense and through the aim of broadening the perception of faculty with accessible technology tools such as digital media. It also showed them how such tools can be used to raise awareness of environmental issues within and beyond academia, creating a broader community and a greater environmental impact.

This paper demonstrates the unique aspects of the first international TELI, including the value of diversity among its participants and of the sharing of best practices from their experience. During the seminar it became evident that this was more than an environmental literacy workshop; the participants created an international network of experts on human, animal, and environmental health with tremendous potential to train future leaders in complex, creative One Health problems solving.

One of TELI 2013's major goals was to engender understanding across disciplinary, cultural, and geographic boundaries through collaborative problem based learning, model-based reasoning, social interactions and the use of technological tools. Using a specially designed online questionnaire to survey the TELI 2013 faculty participants allowed the TELI leaders to capture pre-, concurrent, and post-program participant interactions. Furthermore the SNA of the collected data helped us evaluate the efficacy of the TELI 2013 professional development workshop. The described analysis shows that through the design of the workshop the TELI leaders created new lasting ties between the cohorts from Africa and SE-Asia, thereby bypassing the previously existing bottleneck through the American participants. The post-TELI decrease in shared interest groups was probably based on the introduction of additional interdisciplinary content that diversified the knowledge pool. This is a solid indication of our success in increasing environmental literacy amongst the participating faculty.

The infrequent use of social media for professional purposes amongst the participants was not surprising but uncovers an important area of instruction. Teaching students to adopt a synthesis approach to address complex issues that interest them can be supported by the latest technology. Demonstrating that tools like online mapping or even social media can be used for more than one purpose, including raising environmental awareness and engagement, can be the first step towards an active environmental dialogue in the wider community. Workshops such as TELI can create new occasions for constructive dialogue and collaboration in teams or on a one-to-one level depending on the occasion and need. As the Social Network Analysis indicates, TELI has successfully contributed to building an international network of One Health thinkers. During the closing remarks of TELI 2013, participants requested bringing TELI to their home countries, pointing to a tremendous opportunity for developing global educational partnerships.

This study must be seen as a pilot for further research. Following this TELI, the analysis of upcoming TELI participant networks will provide statistically significant data in the future. This first attempt to quantify the efficacy of this workshop will also allow the TELI leaders to refine future survey questions and in order to tease out which aspects of the workshop where more significant for reaching their objectives.

Acknowledgments We would like to thank Professor Christopher Tunnard and his students, Ryo Kato, Chi Guo, and Charlene Hasib, for conducting the analysis of the TELI SNA survey data within the framework of «The Social Networks in Organizations (DHP P212m)» course taught at the Fletcher School of Law and Diplomacy at Tufts University.

References

Barlett PF, Rappaport A (2009) Long-term impacts of faculty development programs: the experience of Teli and Piedmont. Coll Teach 57:73–82

Bracken LJ, Oughton EA (2006) "What do you mean?" The importance of language in developing interdisciplinary research. Trans Inst Br Geogr 31:371–382

Bridle H, Vrieling A, Cardillo M, Araya Y, Hinojosa L (2013) Preparing for an interdisciplinary future: a perspective from early-career researchers. Futures 53:22–32
Cortese AD (2003) The critical role of higher education in creating a sustainable future. Plan High Educ 31(3):15–22
Dawson S (2010) "Seeing" the learning community: an exploration of the development of a resource for monitoring online student networking: Monitoring online student networking. Br J Educ Technol 41:736–752
Firestien RL (1990) Effects of creative problem solving training on communication behaviors in small groups. Small Group Res 21:507–521
Hicks CC, Fitzsimmons C, Polunin NVC (2010) Interdisciplinarity in the environmental sciences: barriers and frontiers. Environ Conserv 37:464–477
Kaufman GE, Epstein JH, Paul-Murphy J, Modrall JD (2008) Designing graduate training programs in conservation medicine—Producing the right professionals with the right tools. EcoHealth 5:519–527
Koch M, El-Baz F (1998) Identifying the effects of the Gulf War on the geomorphic features of Kuwait by remote sensing and GIS. Photogramm Eng Remote Sens 64:739–747
Koehler M, Mishra P (2009) What is technological pedagogical content knowledge (TPACK)?. Contemporary issues in technology and teacher education, 9(1):60–70. AACE. Accessed April 7, 2014 from: http://www.editlib.org/p/29544
Laat M, Lally V, Lipponen L, Simons R-J (2007) Investigating patterns of interaction in networked learning and computer-supported collaborative learning: a role for Social Network Analysis. Int J Comput Support Collab Learn 2:87–103
Meier P, Leaning J (2009) Applying technology to crisis mapping and early warning in humanitarian settings, Working Paper Series, Harvard Humanitarian Initiative Accessed April 17, 2014 from: http://hhi.harvard.edu/sites/default/files/In%20Line%20Images/working%20paper%20-%20applying%20tech.pdf. Accessed 20 March 2014
Molineux J, Haslett T (2007) The use of soft systems methodology to enhance group creativity. Syst Pract Action Res 20:477–496
Mor SM, Robbins AH, Jarvin L, Kaufman GE, Lindenmayer JM (2013) Curriculum asset mapping for one health education. J Vet Med Educ 40:363–369
Prensky M (2001) Digital natives, digital immigrants. On the horizon, MCB University Press, 9 (5):1–6
Scott J (1991) Social network analysis: a handbook. Sage, London
Sternberg RJ, Lubart TI (1996) Investing in creativity. Am Psychol 51:677–688
Tunnard CR (2003) From state-controlled media to the "anarchy" of the internet: the changing influence of communications and information in Serbia in the 1990s. Southeast Eur Black Sea Stud 3:97–120
Wasserman S, Faust K (1997) Social network analysis: methods and application. Cambridge University Press, Cambridge
Wasserman S, Galaskiewicz J (1994) Advances in social network analysis: research in the social and behavioral sciences. Sage, London
Wilson G, Herndl CG (2007) Boundary objects as rhetorical exigence—Knowledge mapping and interdisciplinary cooperation at the Los Alamos National Laboratory. J Bus Tech Commun 21:129–154
Winkels M, Aka V, Cheung M, Enberg J, Franco O, Guimaraes T, Jhun J, Marin-Sharp N, Pearson J, Rengert H, Wililamson D (2013) The global social network landscape, a country-by country guide to social network usage. Report published by Emarketer. Accessed March 25, 2014 from: http://www.optimediaintelligence.es/noticias_archivos/719_20130715123913.pdf

Authors Biography

Dr. Antje Danielson has been a researcher, consultant, entrepreneur, lecturer, and student mentor in environmental geosciences and sustainability since 1993. She is an Assistant Professor and currently leads the Tufts Institute of the Environment. She is also the President of the Council for Environmental Deans and Directors at the National Council for Science and the Environment. She received her PhD in geochemistry in 1989 from the Freie Universität in Berlin, Germany. Between 1991 and 2005 she worked at Harvard University in the Department of Earth and Planetary Sciences, the Center on the Environment, and the Green Campus Initiative on teaching and research related to greenhouse gas emissions reductions from energy consumption in the U.S. From 2005 until 2008 she was the Deputy Director for Sustainability at the Centre for Research into Earth Energy Systems at Durham University in the UK, working on carbon capture and sequestration and creating a campus sustainability program, which was awarded the Green Gown Award in 2008. She joined Tufts University in 2008. She is currently interested in the mechanisms that underpin interdisciplinary research and collaboration. In 1999/2000 she also co-founded the car-sharing company Zipcar.

Panagiota Kaltsa (M.A.) has transitioned between the corporate, education, NGO and international development fields since 2007. She is a Project Manager at the World Wildlife Fund in Greece where she coordinates the establishment of a transboundary Conservation Trust Fund. She is also the Online Education Strategist at the Tufts Institute of the Environment where she previously coordinated the Tufts Environmental Literacy Institute. Panagiota graduated with an MA in strategic management and international communication from the Fletcher School of Law and Diplomacy in Boston, USA. She started her career working as an international manager for HSBC Bank in the UK, Malta and Greece. In banking she held positions in retail, corporate, shipping and marketing sectors and was actively involved in corporate social responsibility initiatives. She serves as a member of the Fulbright-Greece, the Fletcher Women Network and The Fund for American Studies alumni councils. She has a vivid interest in the application of new technologies in education and humanitarian work while she is currently designing an online platform to enable citizen participation against corruption in Greece.

What Does the Concept of Responsibility Contribute to the Teaching of Sustainability in Universities?

Harold Goodwin

Abstract

The realisation that sustainability objectives require inter-disciplinary 'holistic' thinking and the willingness and capacity to take action to address often inter-connected issues. Too often knowledge about the issues does not lead to effective action, the ideas remain inoperative, graduates are unable to effectively challenge denial in others, and lack the knowledge and skills to address the prisoners' dilemma and the tragedy of the commons. They lack the skills and theoretical insights to effect the necessary change. Research conducted with UK tour operators in 1999 into their attitudes to ethics and sustainability suggested that the concept of responsibility might have more traction with business than ethics and sustainability. This paper reviews the development of Responsible Tourism in the UK, and considers the relationship between the development of more sustainable business practices and the evolution of the curriculum in a Responsible Tourism Masters, which has been taught in three universities in the UK. The paper concludes with a broader reflective discussion of responsibility, the Aristotelian concept of phronesis and practical wisdom and the utility of this conceptual approach in providing a framework for students to understand the challenges of sustainability, theoretical approaches to change leading to the realisation of change in practise.

Keywords

Responsibility · Ethics · Sustainable development · Change · Responsible tourism

---

H. Goodwin (✉)
Centre for Responsible Tourism, Manchester Metropolitan University,
Hollings Faculty, Cavendish Street, Manchester, M15 6BG, UK
e-mail: h.goodwin@mmu.ac.uk

W. Leal Filho et al. (eds.), *Integrating Sustainability Thinking in Science and Engineering Curricula*, World Sustainability Series, DOI 10.1007/978-3-319-09474-8_13

Our species confronts a major environmental challenge one which arises from the increasing demands which we make of the environment in a finite world. Much of the language which we use as academics suggests that the environment is something out there, independent of us. In a similar way we refer to traffic as something external to us, although of course we too are traffic. When we teach sustainability, when we design curricula and determine the content of the courses and degrees we teach, we make choices about the knowledge and skills we impart to our students. We choose whether our students will leave with a rounded knowledge of the range of challenges which confront our species—population, food supply, water, energy, waste, climate change and poverty—and the connections between them. We also choose whether to impart the knowledge and skills necessary to a particular career or to extend the range of knowledge and skills we teach to include those necessary to enable our graduates to work with others to secure the change necessary to sustainability. This requires the development of a range of practical and intellectual skills and more than passing familiarity with the thesis of *the tragedy of the commons* (Hardin 1968), *the prisoners' dilemma* (Kreps et al. 1982) and System 1 & System 2 thinking (Kahneman 2011). Sustainability requires change, large scale change by large groups of consumers and producers, it requires politics.

1 The Challenge of Inter-disciplinary

It was over 40 years ago that the first UN conference explicitly addressed the challenge of environment and development. At the Conference on the Human Environment in 1972, Maurice Strong, the Secretary-General of the conference said that,

> the Conference was launching a new liberation movement to free men from the threat of their thraldom to environmental perils of their own making. ... [success was only possible] if there was a new commitment to liberation from the destructive forces of mass poverty, racial prejudice, economic injustice, and the technologies of modern warfare' ...'the physical interdependence of all people required new dimensions of economic, social and political interdependence' ... 'developing countries could ill afford to put uncertain future needs ahead of their immediate needs for food, shelter, work, education and health care. (Strong 1972: 34–36)

The interdisciplinary challenge was recognised by Strong from the beginning as was the importance of tackling the difficulty of maintaining the environment and achieving development. At the Rio Earth Summit in 1992 the agenda was Environment and Development, the use of the conjunction does not resolve the conflict between the two priorities, the phrase is oxymoronic. Post Rio the environmentalists tended to set the agenda. The launch of the Millennium Development goals resulted in more attention being paid to development at the World Summit on Sustainable Development in 2002, where, for example, the World Tourism Organization launched it Sustainable Tourism—Elimination of Poverty initiative. At Rio +20 in 2012 there was renewed commitment to sustainable development and to poverty reduction in the outcome document: **The Future We Want**

> … to ensure the promotion of economically, socially and environmentally sustainable future for our planet and for present and future generations, to sustainable development, and to ensure the promotion of economically, socially and environmentally sustainable future for our planet and for present and future generations (Assembly 2012: 1)

Clear recognition of the importance of the triple bottom line (Elkington 1997) and of the poverty agenda:

> Eradicating poverty is the greatest global challenge facing the world today and an indispensable requirement for sustainable development. In this regard we are committed to free humanity from poverty and hunger as a matter of urgency. (Assembly 2012: 1)

Forty years on, and with the benefit of having seen the utility of the Millennium Development Goals, there was renewed commitment to local action, reassertion of the importance of Agenda 21 "taking into account different national circumstances, capacities and priorities" and calling for Sustainable Development Goals (SDGs) which "should address and incorporate in a balanced way all three dimensions of sustainable development and their inter-linkages." (Assembly 2012: 43)

The realisation of sustainability objectives requires inter-disciplinary 'holistic' thinking and the willingness and capacity to take action to address often interconnected issues. Courses which address sustainability in a holistic interdisciplinary way are particularly stretching for students who are expected to cross the great divide of "The Two Cultures" (Snow 1959) of science and the humanities, not to mention a large body of social scientific literature. Pezzey in an early paper on sustainability pointed to the importance of "ideas from physics, ecology, evolutionary biology, anthropology, history, philosophy, economics, psychology" in developing a "coherent interdisciplinary analysis of the potential for sustaining industrial civilisation." (Pezzey 1992: 321)

Sustainability is researched and taught as a subject in a range of course in universities as well as a dimension of many subjects. Since the latter half of the seventeenth century there has been a rapid growth in knowledge resulting in the "internal differentiation of scholarly activity" into disciplines (Weingart 2010: 5) Academic careers are largely founded on peer reviewed publication in disciplinary journals; there is some evidence that this suppresses interdisciplinary research (Rafols et al. 2012; Bruce et al. 2004). Interdisciplinarity is identified as a challenge in sustainability assessment (Morrison-Saunders et al. 2014) and in developing key competencies for sustainable development (Barth et al. 2007).

There is not space here for a discussion of the purposes of universities in research and higher education, less still for a discussion of the different ways in which knowledge is valued. Sustainability requires significant changes in the way we use our environment and manage production and consumption; this requires both interdisciplinary knowledge and its application using a variety of skills. As Brewer has perhaps provocatively put the problem, "environmental problems require interdisciplinary treatment which the conventional knowledge institutions have been unable, unwilling or slow to provide. Or, as cynics have stated it: 'The world has problems, but universities have departments.' (Brewer 1999: 328)

Interdisciplinarity requires a combination of knowledge from a range of disciplines, where it is effective the impact is greater than the sum of the parts. As Brewer has pointed out

> Much high-quality science illuminates environmental problems, but it is often poorly organized and incomplete. It often does not have an interdisciplinary integration and synthesis that permit problems to be seen in a larger context, especially in an ecologically sensitive and sensible one. It is often not geared to the scale needed to shed light on environmental problems of long-term importance to human well-being. In short, much essential knowledge is not capable of guiding the development of policy, heightening public awareness, or even informing and enlightening political debate (Brewer 1999: 327).

This paper accounts for the development of a programme of post-graduate education and research which has evolved through engagement with the challenges of creating more sustainable forms of tourism and the experience of conducting research, with the tourism sector in the UK and abroad, in four universities over the last 20 years. The evolution of the series of Masters courses briefly described here took place within the tradition of Continuing Professional Development (CPD), focused on the learning needs of a range of practitioners who were mainly in employment, engaged with tourism and who shared an aspiration to make tourism more sustainable. This introduction is an *ex post facto* justification, presented here to contextualise the evolution of the development of a programme of work which in fact grew out of engagement with potential students, students, employers and other stakeholders who shared the aspiration to generate managers and leaders who could change tourism. The approach presented here developed out of praxis, a background in adult education meant a willingness to accept that education and change are intimately related.

> The purpose of a theoretical discipline is the pursuit of truth through contemplation; its telos is the attainment of knowledge for its own sake. The purpose of the productive sciences is to make something; their telos is the production of some artefact. The practical disciplines are those sciences which deal with ethical and political life; their telos is practical wisdom and knowledge. (Carr and Kemmis 1986: 32)

Here the development of the Responsible Tourism CPD curriculum is presented as an iterative process developed with participants and the stakeholders with whom they engage whether in hospitality or tourism, national and local government, conservation organisations, the media, local communities, NGOs or inter-governmental organisations.

In British Colombia, Moore (2005) approached the challenge of introducing sustainability education into the undergraduate university curriculum through 'value focused thinking' using a stakeholder dialogue process arguing that "Given what academics know about the current ecological condition of the planet, there is an obligation for universities to become leaders in the movement to prevent global ecological collapse." (Moore 2005: 326) Moore's research suggested that the university needed to make sustainability a fundamental priority, to encourage collaboration and transdisciplinarity, focus on personal and social sustainability, to address the question about how the university knows that its activities are making a difference, engage with the community through community service in its teaching and research and "injecting inquiry, experience and reflection into the undergraduate classroom." (Moore 2005: 337).

Too often knowledge about the issues does not lead to effective action, the ideas remain inoperative, graduates are unable to effectively challenge denial in others, and lack the knowledge and skills to address the prisoners' dilemma and the tragedy of the commons. They lack the skills and theoretical insights to effect the necessary change. CPD obligations are part of the requirements of many professional bodies, it has more status and less aspiration than the concept of lifelong learning, but lifelong learning is what is required. It is surely no longer sensible to think that a graduate in the twenty-first century has the skills and knowledge for a life time—lifelong learning is not optional, particularly for those engaging with the challenge of sustainability, and it must include formal, informal and reflective learning. (Blewitt and Cullingford 2004)

> lifelong learning is development of human potential through a continuously supportive process which stimulates and empowers individuals to acquire all the knowledge, values, skills and understanding they will require throughout their lifetimes and to apply them with confidence, creativity and enjoyment in all roles, circumstances and environments. (Longworth and Davies 1996: 22)

Learning to do is an important part of lifelong learning and CPD (Blewitt 2004: 27). The lifelong learner needs to be an active creative explorer in the world, a reflexive and self-actualising agent and able to integrate learning (Medel-Añonuevo et al. 2001) and designed to graduate independent learners (Brookfield 1982). One of the privileges of having been involved in working with mid-career professionals through Masters course and research and has been the opportunity to see theory tested in practice, a form of engaged scholarship (Van de Ven 2007) and a major generator of lifelong learning.

2 Responsible Tourism

There has been a plethora of concepts used to describe forms of more sustainable tourism: green tourism, ecotourism, ethical tourism, alternative tourism, soft, appropriate, community-based tourism and many more (Wheeller 1992). In the late nineteen nineties Responsible Tourism emerged as a potential new conceptual framework for thinking about how tourism could be changed. The language used to describe aspirations for tourism, the language we use to discuss tourism embodies a "particular *perspective* of what the 'world' is like." This perspective taking process is value-laden, articulating

> what is 'normal' or not; what is 'acceptable' or not; what is 'right' or not… what is the 'way things are' or not; what is the 'way things ought to be' or not.. and all have deep implications for how we believe or wish potential social goods are or ought to be distributed. (Gee 2013: 2)

Language is used to "enact activities and identities"; "language has meaning only in and through social practices". (Gee 2013: 7, 8) The language of Responsible Tourism and its antithesis was, and is, heavily value-laden.

Responsible Tourism emerged as a new discourse for understanding the social activity of tourism in the nineties from three sources and created the opportunity to hold an International Conference on Responsible Tourism in Destinations in Cape Town as an official side even to the World Summit on Sustainable Tourism in Johannesburg in 2002. The Cape Town Declaration on Responsible Tourism in Destinations recognised the importance and value of the World Tourism Organizations Global Code of Ethics which aims to promote responsible, sustainable and universally accessible tourism and "the importance of making all forms of tourism sustainable through all stakeholders taking responsibility for creating better forms of tourism and realising these aspirations." (ICRTD 2002: 1) Participants in the conference undertook "to work in concrete ways in destinations to achieve better forms of tourism and to work with other stakeholders in destinations (ICRTD 2002: 2). In the Cape Town Declaration Responsible Tourism was embedded in the values of auditable transparency, partnership working and the ethic of responsibility, of responding, of doing, whilst recognising the world's diversity and the importance of addressing local priorities. Responsible Tourism was defined by the Cape Town Declaration with a clear imperative: "to take responsibility for achieving sustainable tourism, and to create better places for people to live in and for people to visit." (ICRTD 2002: 2).

In 1987 Krippendorf's The Holiday Makers had been published in English, a seminal work which established a new paradigm for tourism (Lane 2003). Krippendorf had understood the power of the ethic of responsibility. Krippendorf called for a new form of tourism, one that 'will bring the greatest possible benefit to all the participants—travellers, the host population and the tourist business, without causing intolerable ecological and social damage', a form of tourism he called "soft" or "adapted" (Krippendorf 1987: 106, 107). This required 'rebellious tourists and rebellious locals', to create more 'fulfilling and enjoyable' forms of tourism (Krippendorf 1987: 107, 109). Krippendorf argued that proposals for change needed to be 'infectious'. Codes and advice about improving the quality of travel must not 'degenerate into rules for regimentation and manipulation. They must make the experience of freedom possible.' '[E]very individual tourist builds up or destroys human values while travelling.' 'Orders and prohibitions will not do the job—because it is not a bad conscience that we need to make progress but positive experience, not the feeling of compulsion but that of responsibility'. (Krippendorf 1987: 108–110) In articulating this connection between the idea of the positive experience and the power of the concept of responsibility, couples with the idea of rebellious tourists and rebellious locals Krippendorf identified an approach which has gained traction in generating progress toward sustainability.

In 1998 Voluntary Service Overseas (VSO), a British NGO, launched a *WorldWise* campaign challenging British holidaymakers to get more out or their holiday pointing out that many travelled to the most distant locations on earth and never ate, drank or shopped outside the hotel; and asked whether they would go shopping at home dressed only in a swimsuit. The campaign carried positive messages about how to have a better holiday. VSO commissioned NOP, a commercial polling company, to collect evidence of UK consumer preferences in their random UK omnibus survey. 72 % of respondents said that they 'would welcome a campaign that dealt with the impact that tourists can have on the developing world,' legitimating VSO's WorldWise action to raise awareness about tourism's impact on local people in the developing world. In 1999 VSO ran *Travelling in the Dark*, which surveyed 50 operators who sent tourists to the countries where VSO had volunteers on the ground. Two-thirds of operators, reported VSO, failed to meet even a minimum standard, and many had failed to provide anything at all. The data was presented in league tables. The findings motivated some of the companies which had performed poorly to improve the information which they sent out to their customers, and some to develop formal Responsible Tourism policies. This was the start of the campaign for Responsible Tourism in the UK, subsequently taken up by Tearfund. (Goodwin 2011: 53–57)

The VSO campaign culminated with the funding of research in 1999 into the activities and attitudes of the Association of Independent Tour Operators (AITO) members towards ethical tourism. The research undertaken in the summer of 1999 revealed that 40 % of respondents felt that their company's ethical commitment meant very little to their travellers; 30 % felt that it meant something; 12 % felt that it meant a great deal. The final question canvassed AITO members' views on whether, as members, they should share a stated commitment to ethical tour operations: 52 % felt that they should, 20 % sat on the fence and 27 % said that they should not. The AITO sub-committee concluded that the survey revealed 'unexpectedly extensive good practice in responsible tourism', that members operating within Europe as well as further afield were engaged, and that individual AITO members were making a significant difference. As a consequence of the survey, the AITO Responsible Tourism Committee although unable to commit to the word ethical, which it thought 'carried with it unacceptable implications and inferences' developed a set of advisory rather than prescriptive guidelines adopted by the Association in May 2000. These guidelines acknowledged that "wherever a Tour Operator does business or sends clients it has a potential to do both good and harm, and we are aware that all too often in the past the harm has outweighed the good"; and that all "tourism potentially has an Environmental, Social and Economic impact on the destination involved. We accept, therefore, that we as Tour Operators should aim to be responsible in all our dealings..." (Goodwin 2011: 85–88).

The research conducted with AITO tour operators in 1999, into their attitudes to ethics and sustainability, suggested that the concept of responsibility might have more traction with business than ethics and sustainability.

> The commitment to responsibility, rather than to ethical practice, may be seen by some as a weaker proposition. However, the advantage of the concept of responsibility is that it suggests that members need to respond, to act, rather than standing, or sitting, on their principles and their ethics. Responsibility implies and requires action. AITO was the first trade association to commit to Responsible Tourism, and the recognition that sometimes harmful effects out-weigh the good was important. Critical to creating change is

> acknowledging and owning up to problems, and taking responsibility for making changes. (Goodwin 2011: 87)

Although travel to the destination has significant negative environmental impacts which need to be addressed, and these issues are not ignored in Responsible Tourism, the Cape Town Declaration recognised that it is in destinations, where tourists and locals meet, and where tourism takes place that its sustainability has to be managed. AITO members acknowledged their role in that and larger UK outbound operators subsequently publicly recognised their responsibility too (Goodwin 2011: 93). The first destination to formally adopt a commitment to Responsible Tourism in its national policy was post-apartheid South Africa. The 1996 White Paper was conventionally titled: The Development and Promotion of Tourism in South Africa, but the policy was radical. Responsible Tourism was in the graphics on the cover of the White Paper and used throughout. Responsible Tourism was defined in the White Paper as

> tourism that promotes responsibility to the environment through its sustainable use; responsibility to involve local communities in the tourism industry; responsibility for the safety and security of visitors and responsible government, employees, employers, unions and local communities. (DEAT 1996: vi)

Responsible Tourism was identified "as the key guiding principle for tourism development" in South Africa and defined the roles envisages for all the stakeholders. With funding from the Department for International Development (DFID) the University of Greenwich provided technical assistance to the Department of Environmental Affairs and Tourism which led to the publication of *Guidelines for Responsible Tourism* in 2002, comprising 'three inter-related sets of guiding principles, objectives and indicators' and organised around the triple bottom line of social, economic and environmental responsibility (Goodwin 2011: 138–141).

The opportunity to run the 1st International Conference on Responsible Tourism in Destinations in Cape Town in 2002 came as a result of the conjunction of the hosting of the WSSD in Johannesburg, the desire of the South Africans to share their work on Responsible Tourism and the role which one academic centre had played in the UK., a significant originating market for South Africa, where the same centre had been involved in supporting the development of national Responsible Tourism guidelines. The Cape Town Declaration was the consequence of academic engagement with industry, government and a wide range of stakeholders, an engagement which facilitated the realisation of Krippendorf's vision.

3 The Evolution of Masters Provision

In the Durrell Institute for Conservation and Ecology (DICE) in 1994 while conducting research into the economic, social and environmental impacts of tourism in national parks funded by the Overseas Development Administration (ODA) in order to assess the reality of ecotourism (Goodwin 1998; Chalker 1994). At DICE there were many national park managers from, Africa and South America who received tourists, had to manage them, but understood little of how they arrived there, nor about how they could be managed so as to reduce negative impacts and benefit conservation. Conservationists also sought the knowledge and skills to influence the systems within which they operated and to make change. Modules on politics and tourism and conservation followed, with research findings from the ODA funded research entering directly into the curriculum. An MSc in Tourism and Conservation ran from 1997–1999.

In 1998 DFID requested a paper on Sustainable tourism and poverty elimination (Goodwin 1998), which resulted in the development of a stream of research on Pro-Poor Tourism which again directly entered the curriculum. The research with UK tour operators on their views on ethical responsible tourism, research conducted by Masters students and academic staff on the practices of UK outbound and consultancy on Responsible Tourism policies and their implementation again generated teaching material and secured engagement with the industry, it also assisted in recruiting students into a programme which was seen as relevant by people in managerial roles in the industry and in conservation organisations. The scope of the Masters programme was by now considerably broader than focus on the conservation of habitat and species originally envisaged and a new Masters programme on Tourism, Conservation and Sustainable Development ran 2000–2003 in the School of Earth and Environmental Sciences at the University of Greenwich. With geography within the School there was a far great willingness to accept interdisciplinary work. Consultancy work in South Africa on the implementation of their national Responsible Tourism policy entered directly into the Masters modules.

The International Centre for Responsible Tourism (www.icrtourism.org) was established as an independent virtual network of people committed to implementing the principles of the Cape Town Declaration in 2002 and in 2003 a new MSc in Responsible Tourism Management evolved out of its predecessor with a broader curriculum and some blended learning. The blended learning approach which included distance learning packs with limited evening attendance for classes was designed to meet the needs of employees wanting to study part-time on an applied Masters programme—learning packs ensured that they had all of the essential reading and that study was not constrained by the challenge of library access. A Masters programme taught in a School of Earth and Environmental Sciences, with no undergraduate provision in tourism and staff members that were consultancy active was always going to be an anomaly and in 2006 in a round of redundancies staff was reduced to a fractional post.

Leeds Metropolitan University had appointed Professor Simon Lee at Vice Chancellor in 2003. In his inaugural lecture at Leeds Met, entitled Beyond Boundaries, he quoted a 19th century English college principal as saying that students should be able to drink from a ‘running stream’ of insights, a curriculum informed by research, rather than from a ‘stagnant pool’ and announced a focus on sporting and cultural partnerships, with the Vice Chancellor’s academic background in law, ethics and politics and a clear commitment to the responsibility agenda Leeds Met offered an attractive base for work a post-graduate centre focused on Responsible Tourism. The MSc in Responsible Tourism Management flourished there from 2006 to 2009 with a mix of MSc provision, contract research and consultancy, links with industry, government and inter-governmental organisations flourished. With the arrival of a new Vice Chancellor in 2009 the ethos and structure of the university changed and the Responsible Tourism team at Leeds Met shrank.

A new MSc in Responsible Tourism has now been launched in the Centre for Responsible Tourism at Manchester Metropolitan University with a focus on post-graduate teaching and research, contract research, consultancy and advisory services. There are also opportunities for interdisciplinary work with colleagues in carbon measurement and management conservation biology and accountancy. With a new academic base and a willingness to make fractional appointments and to work closely with practitioners in the tourism sector the curriculum of the MSc has evolved again with a stronger focus on responsible business, leadership and change management. The Continuing Professional Development conferences, engagement in the World Responsible Tourism Awards and the programmes run with advice from the Centre for Responsible Tourism at the World Travel Market trade shows in London, Sao Paulo and Cape Town each year have challenged, educated and inspired the industry to change. Over the last fifteen years the Masters curriculum has evolved, shaped by research and continuing engagement with mid-career students, and former students, and with practitioners in the sector, the people working to make tourism more responsible.

4 Conclusions

It is for others to say whether this approach has anything to offer their teaching or academic practice. Placing responsibility at the core of our approach to sustainability has been very productive in achieving traction with the sector, attracting students and achieving change it has given a strong sense of purpose to our educational and research endeavours and it has ensured that we can engage with industry. In the XIth of the theses on Feuerbach, Marx wrote "The philosophers have only interpreted the world, in various ways; the point, however, is to change it." (Marx 1888: 199) In each of the universities in which this engaged approach has been developed there has been a sense of being at the edge, between the tourism sector and the university. It has sometimes been uncomfortable, but it has been both challenging and productive.

Maurice Strong identified the challenge over 40 years ago, "…to free men from the threat of their thraldom to environmental perils of their own making." (Strong 1972: 34) It is an old problem. Aristotle commented in The Politics, written in 350 BCE: "what is common to the greatest number has the least care bestowed upon it. Everyone thinks chiefly of his own, hardly at all of the common interest" (Aristotle 2.3 Jowett). Neither the tragedy of the commons nor the prisoner's dilemma (Ostrom 2000; Lichbach 1996) is a new challenge, but they are compounded by the realisation that we live in a finite world. Without an understanding of the dilemmas of collective action, the politics of change, psychology and behavioural change we risk continuing to create graduates unable to make the changes which are needed.

The focus on Responsible Tourism in formal and informal teaching has differentiated our approach, as has the insistence that tourism is a social phenomenon, that tourism is what we, producers and consumers make it (Goodwin 2011: 5) The curriculum has been radically revised, regularly. Incorporating new research findings and responding to the changing needs of students and employers, but firmly grounded in the Cape Town Declaration and the commitment to making tourism better. Recognising that local priorities matter and challenging the stakeholders, and our students, to take responsibility in the Aristotelian sense of phronesis, to act with practical wisdom, to determine ends, adopt appropriate means and to secure change. (Goodwin 2011: 34) The knowledge and skills for creating change have become a larger and more significant part of the curriculum. Basing the course on the ethic of taking responsibility and being willing to engage with the issues and all the stakeholders from a committed position, has ensured a relevant curriculum with a running stream of new content from research and the testing of theory in practice; and the graduation of students who go on to contribute to making tourism more sustainable, and who continue to contribute to our work. Sustainability is an abstract aspiration, too often an inoperative idea. Responsibility requires engagement in securing change, and places on us the onus to enable our students to make the changes they know to be necessary.

References

Assembly UG (2012) The future we want. UN General Assembly, New York

Barth M, Godemann J, Rieckmann M, Stoltenberg U (2007) Developing key competencies for sustainable development in higher education. Int J Sustain High Educ 8(4):416–430

Blewitt J (2004) Sustainability and lifelong learning. In: Blewitt J, Cukkingford C (2004) The sustainability curriculum: the challenge for higher education. Earthscan, London, pp 24–42

Blewitt J, Cukkingford C (2004) The sustainability curriculum: the challenge for higher education. Earthscan, London

Brewer GD (1999) The challenges of interdisciplinarity. Policy Sci 32(4):327–337

Brookfield S (1982) Independent adult learning. Publications Unit, Department of Adult Education, University of Nottingham

Bruce A, Lyall C, Tait J, Williams R (2004) Interdisciplinary integration in Europe: the case of the Fifth Framework programme. Futures 36(4):457–470

Carr W, Kemmis S (1986) Becoming critical. Education, knowledge and action research. Falmer, Lewes

Chalker L (1994) Ecotourism: on the trail of destruction or sustainability? A minister's view. In: Cater E, Lowman G (eds) Ecotourism a sustainable option. Wiley, London

DEAT (Department of Environmental Affairs and Tourism) (1996) The development and promotion of tourism in South Africa. DEAT, Pretoria

Elkington J (1997) Cannibals with forks: triple bottom line of 21st century business. Capstone, Oxford

Gee JP (2013) An introduction to discourse analysis: theory and method. Routledge, London

Goodwin H (1998) Sustainable tourism and poverty elimination. In DFID/DETR workshop on sustainable tourism and poverty. Department for International Development, London

Goodwin H (2011) Taking responsibility for tourism. Goodfellow, Oxford
Hardin G (1968) The tragedy of the commons. Science 162(3859):1243–1248
ICRTD: International Conference on Responsible Tourism in Destinations (2002) The Cape Town Declaration, Cape Town http://www.responsibletourismpartnership.org/CapeTown.html
Kahneman D (2011) Thinking, fast and slow. Macmillan, New York
Kreps DM, Milgrom P, Roberts J, Wilson R (1982) Rational cooperation in the finitely repeated prisoners' dilemma. J. Econ. Theor 27(2):245–252
Krippendorf J (1987) The holiday makers. Understanding the impact of leisure and travel. Butterworth Heinemann
Lane B (2003) Jost Krippendorf, Pioneer of sustainable tourism, obituary The Guardian 7 April 2003
Lichbach MI (ed) (1996) The co-operator's dilemma. University of Michigan Press, Ann Arbor
Longworth N, Davies WK (1996) Lifelong learning: new vision, new implications, new roles for people, organizations, nations and communities in the 21st century. Kogan Page, London
Marx K (1888) Theses on Feuerbach, no. 11. Marx and F. Engels, The German Ideology, Pts I &, 3, 199
Medel-Añonuevo C, Ohsako T, Mauch W (2001) Revisiting lifelong learning for the 21st century. UNESCO Institute for Education
Moore J (2005) Seven recommendations for creating sustainability education at the university level: a guide for change agents. Int J Sustain High Educ 6(4):326–339
Morrison-Saunders A, Pope J, Bond A, Retief F (2014) Towards sustainability assessment follow-up. Environ Impact Assess Rev 45:38–45
Ostrom E (2000) Collective action and the evolution of social norms. J Econ Perspect 14:137–158
Prezzey J (1992) Sustainability: an interdisciplinary guide. Environ Val 1(4):321–362 (White Horse Press)
Rafols I, Leydesdorff L, O'Harea A, Nightingale P, Stirling A (2012) How journal rankings can suppress interdisciplinary research: a comparison between innovation studies and business & management. Res policy 41(7):1262–1282
Snow CP (1959) The two cultures. Cambridge University Press, London
Strong MF (1972) Statement to the conference on the human environment at its 1st plenary meeting. United Nations, Paris, 5 June 1972
Van de Ven AH (2007) Engaged scholarship: a guide for organizational and social research: a guide for organizational and social research. Oxford University Press, Oxford
Weingart P (2010) A short history of knowledge formations. In: Frodeman R, Klein JT, Mitcham C (eds) The oxford handbook of interdisciplinarity. Oxford University Press, Oxford
Wheeller B (1992) Alternative tourism—a deceptive ploy. In: Cooper C, Lockwood A (ed) Progress in tourism, recreation and hospitality management, vol 4, pp 140–146

Author Biography

Professor Harold Goodwin is a political scientist with a Ph.D. in government. His initial career was in lifelong learning in adult education. In 1994 he directed a research project at the Durrell Institute of Conservation and Ecology looking at tourism in national park systems and engaged in a campaign for ethical tourism. For nearly twenty years he has been teaching graduate students, many of them employed in the industry, and engaging with them to make tourism better. After periods of course development, teaching and research at Kent, Greenwich and Leeds Metropolitan universities he is now based at Manchester Metropolitan where, with colleagues there, a new M.Sc. in

Responsible Tourism has been launched drawing on twenty years' experience of teaching triple bottom line sustainability to mid-career students engaged in the tourism sector. Harold would like to thank the industry people with whom he has worked and in particular the students with, and from, whom he has learned so much. They have achieved a great deal in making tourism more responsible.

The Road Less (Sustainably) Traveled: A Case Study of Academic Travel at Franklin University Switzerland

Brack W. Hale and Alison Vogelaar

Abstract

Educational travel programs (e.g. study abroad) have long been an integral part of the higher education curriculum. As interest in internationalization and the development of students into global citizens has grown in recent years, so has the popularity of education travel. However, institutions of higher education often find themselves at odds between the desire to internationalize campuses and curricula and the growing movement to make them more sustainable. This paper provides a case study of the practices at one institution where educational travel is an integral and required part of the curriculum, known as Academic Travel. First, it briefly reviews some of the current thinking about educational travel and sustainable development. It then examines the experiences of several professors who integrate themes of sustainability into the curriculum and design of their Academic Travel programs.

Keywords

Educational travel · Sustainability · Internationalization · Curriculum · Logistics

B.W. Hale (✉)
Department of Math and Natural Science, Franklin University Switzerland,
Via Ponte Tresa 29, 6924 Sorengo, Switzerland
e-mail: bhale@fus.edu

A. Vogelaar
Department of Communication and Media Studies, Franklin University Switzerland,
Via Ponte Tresa 29, 6924 Sorengo, Switzerland

W. Leal Filho et al. (eds.), *Integrating Sustainability Thinking in Science and Engineering Curricula*, World Sustainability Series, DOI 10.1007/978-3-319-09474-8_14

1 Introduction

Institutions of higher education (IHEs) face a growing challenge in their attempts to become both more international and more sustainable, as these goals are not always congruent. Educational travel programs, for example, are important means of internationalizing student experiences but can be highly unsustainable (e.g. carbon emissions or negative impacts on host cultures and economies). Student participation is these programs has been increasing steadily and is expected to continue to do so (Institute for International Education 2014), meaning that negative impacts from educational travel programs will also increase if changes in practice do not take place. Surprisingly, campus sustainability efforts rarely address this important tension. Further, some common sustainability reporting tools, such as GASU (Lozano 2006a), STAUNCH (Lozano 2010), and STARS (AASHE 2013) do not to incorporate explicit criteria that examine this aspect of an institution's approach to sustainability in its curricula and operations. The reasons behind this are likely varied and not a topic for this study. Instead, this paper engages the tension by examining how the *Academic Travel* Program at Franklin University Switzerland attempts to negotiate it. It begins with definitions of key concepts and briefly reviews the literature surrounding *sustainability* and *education travel*, and then proceeds with the case study and discussion.

1.1 Definitions

This study defines *sustainability* broadly as a theoretical and practical approach invested in the economic, environmental, and/or socio-cultural health and vitality of ecosystems, both human and natural (Brown et al. 1987). *Sustainable development* is one critical (indeed seminal) aspect of sustainability, however, the terminology of *sustainability* is privileged herein as it captures the diverse approaches to sustainability on university campuses that often do not necessarily involve "development" in the popular sense of the word. *Educational travel* refers to any academically-affiliated program in which participants travel to a destination for the primary purpose of learning about or within that destination (Bodger 1998). As a major tool used by IHEs in the internationalization of campuses, educational travel is also part of a larger project to internationalize student attitudes, beliefs, and behaviors. While study abroad is one form of educational travel, it increasingly takes different shapes and practices; Franklin's Academic Travel Program discussed below is one such example.

1.2 The Tension Between Internationalizing and Greening Campuses

In recent decades, IHEs have placed increasing emphasis on internationalizing campuses and communities (Altbach and Knight 2007; Tarrant 2009). The rhetoric and practice of "global citizenship" is now widespread in IHEs, with increasing numbers of institutions incorporating principles of global citizenship into their missions, visions, and curricula. Rationales range from the practical—increased mobility and diversity in the labor force requires graduates who can work in a global marketplace (Bakalis and Joiner 2004; Sjoberg and Shabalina 2010), to pedagogical—"global" students gain a greater sense of intercultural understanding, social justice and equity, self-awareness, and environmental literacy (Bellamy and Weinberg 2006; Lutterman-Aguilar and Gingerich 2002). Educational travel programs are a key component of the processes of internationalizing. The Institute for International Education (2014) reports that in 2011/12, the number of US students going abroad grew to over 283,000, while the number of foreign students coming to study in the US also grew, to almost 117,000. The EU Erasmus programme also reports growing numbers of students studying outside of their home country, with over 33,000 participants in 2011 (Lifelong Learning Program 2014).

A parallel development in IHEs has been the growing trend towards sustainability, in its diverse forms and interpretations (Long et al. 2014). IHEs are microcosms of real world communities in terms of their resource consumption, energy use, and waste production; however, they are also communities where learning and experimentation are often an accepted part of the culture. This makes IHEs ideal laboratories for sustainability and they have housed initiatives across operations, curricula, research, and outreach that have investigated, experimented with, and implemented sustainable design, practices, and programs (e.g. Rowe 2002; Beringer and Adomßent 2008; Way et al. 2009; Leal Filho 2010). Moreover, these initiatives have become important marketing tools for the IHEs (Mcintosh et al. 2008; Stafford 2011).

Educational travel, however, has been noticeably absent from the sustainability discourse at IHEs. The reasons likely vary and may relate to budgetary, convenience and/or informational issues (Hale et al. 2013). The authors of this paper would also argue that the conflict between the goals of internationalization and sustainability also contribute to this oversight. Reviewed in depth elsewhere (e.g. Long et al. 2014), the possible impacts from educational travel programs are large and span the full range of economic, environmental, and sociocultural aspects of sustainability. A few examples include:

- significant carbon emissions and other air pollution from the use of air travel;
- depletion of local natural resources in host communities due to Western-style consumption by students;
- introduction of disruptive or undesirable behaviors by students;
- support of specific cultural traditions deemed authentic by outside groups at the expense of other, less recognized, local cultural groups; and
- accumulation of economic benefits from travel programs to groups and entities removed from local economy.

While recent research has identified many of these challenges, studies that adequately address program impacts are less common and are only recently emerging. Reilly and Senders (2006) introduced a self-reflexive model ("critical study abroad") that directly challenges IHEs to examine study abroad programs more closely within internationalization goals. Similarly, Palacios (2010) raised concerns about the potential for some study abroad models to reproduce a post-colonial development agenda. Other studies have examined the effects of short-term study abroad programs on host communities (Schroeder et al. 2009), the effects of study abroad on student environmental attitudes and intercultural sensitivity (Rexeisen and Al-Khatib 2009), and the potential incorporation of sustainable development goals into the curriculum of environmentally-focused study abroad programs (Cusick 2008).

Some studies have examined more closely the connections between study abroad and sustainable practices. Dvorak et al. (2011) discuss study abroad programs as a mechanism to achieve goals related to sustainability and internationalization using several examples from the authors' own institutions. They also highlight the concept of *collateral learning*, with which they describe student learning that occurs during a program yet goes beyond the stated learning goals. This phenomenon is reflected in results from Paige et al. (2009) who found the study abroad experience to influence participants' behaviors decades afterwards and in ways that did not relate necessarily to the content of the program. Another recent study (Hale et al. 2013) surveyed current educational travel practitioners to gain insight into how some travel leaders and institutions are thinking about this issue. They found that practice does not always reflect discourse and that programs with intentional sustainability efforts result more from individual efforts than institutional policies.

2 Methodological Approach

In order to contribute to the literature, this study examines practices in an educational travel program (*Academic Travel*) at Franklin University Switzerland, a small liberal arts institution located in southern Switzerland. Franklin may be unique in that it requires its students to participate in these short-term educational travel courses every semester for the students' first few years. Furthermore, Franklin has been explicitly experimenting with sustainability since the inception of its the Center for Sustainability Initiatives (CSI) in 2009. Thus, the Franklin experience provides an important case study to investigate the intersection of sustainability and educational travel.

2.1 Academic Travel at Franklin University Switzerland

Since 2009, CSI has been facilitating conversations about the role sustainability can and does play in Franklin's Academic Travel (AT) program, which stems from the institution's "international imperative" to bring the world into the classroom and to use the world as a classroom. The AT courses are led by faculty members from across Franklin's academic programs and meet on campus during the semester with an embedded 10–14 day travel during the middle of the semester. The courses bring together a geographical site and disciplinary lens. Heading to destinations in Europe and around the world, the courses allow students to engage in site-based study, service learning, and/or cultural immersion. AT courses represent an integral part of the curriculum, fulfilling core, elective, and/or major requirements and by graduation, most students will have completed 12–24 AT credits.

While AT is a powerful tool for immersive, international, and intercultural learning, it is not without complications. The discussions emerging from CSI about sustainability has caused us to reflect on our own practices. The approximately 35 AT program each year account for over half of Franklin's annual carbon emissions (Center for Sustainability Initiatives at Franklin, *personal communication*). Moreover, AT courses likely have both positive and negative consequences for the cultural, economic, and environmental sustainability of the locations visited.

2.2 Data Collection and Analysis

This study features four AT programs that incorporate aspects of sustainability: "sustainability" in Iceland, "folk culture" in Scotland, "sustainable development" in southern Africa, and "travel writing" in the Switzerland. The authors selected these destinations both due to the explicit treatment of sustainability themes within the travels' curricula, as well as the professors' involvement in CSI activities. To examine the four destinations, the authors developed a short set of interview questions based upon a framework from Hale et al. (2013), which suggests a role for sustainability both in the curriculum and the logistics of educational travel programs (Table 1). The authors interviewed several AT professors actively involved in CSI based on this framework and synthesize the responses below.

3 Results and Discussion

The tension between internationalization and sustainability will be a critical issue facing IHEs in the coming decade. Franklin University's *Academic Travel* program provides an interesting case study to examine both the problems and possibilities of this tension. Indeed, much can be gleaned from examining a small institution so explicitly committed to both internationalization and sustainability. The diverse disciplines and destinations involved in Franklin's AT program also make it a

Table 1 Areas of interest regarding sustainability in educational travel programs

Area	Possible components
Curriculum	Examines issues of sustainability
	Focuses specifically on sustainability and travel
Logistics	Intentional choice of destination
	Demonstrates sustainable practices
Other	*Collateral learning*
	Challenges
	Insights

Table 2 Case study program details and importance of place

Program destination (courses)	Sponsoring department	Relevance of place
Iceland (1)	Environmental studies	Iceland provides a good case study for sustainability, in particular sustainable travel. When first developed in 2010, Iceland ranked number one on the Environmental Performance Index (Emerson et al. 2010). Further, Iceland's infrastructure, particularly its energy sources and abundant natural resources (water, food) enables sustainable practices on site. The increasing pressure of tourism permits a firsthand look at the issues involved in tourism
Scotland (1)	Communication and media studies	Scotland as a destination allows an examination of folk culture as made meaningful through Scottish representations, practices, people and sites. As a course interested in representations of national identity, tourism is a critical focus that also naturally turns attention to sustainability—cultural, economic, and environmental. In addition to identifying the form and function of folk in Scottish identity and tourism, students reflect on the sustainability of folk representations and realities on Scottish culture, economies, and environments.
Southern Africa (3):	International relations	The destinations in sub-Saharan Africa offer interesting case studies to examine issues of sustainable development, sustainable agriculture, food security, ecotourism and resource conservation. Franklin is able to draw upon a well-established relationship with Tikwonde Freedom Gardens to allow students a unique opportunity to participate in a sustainable community project
Botswana, Malawi, and Mozambique		
Switzerland (1)	Literature and culture	Switzerland makes an excellent location both to study travel writing, given its literary tradition, and to demonstrate sustainable travel due to its infrastructure such as exceptional public transportation and waste management systems, abundant high-quality food and water resources, and an energy system that uses few fossil fuels

representative (and therefore useful) example of the range of programs in the wider educational travel community.

This case study reveals, first and foremost, the diverse curricular and logistical modalities with which sustainability can be incorporated into educational travel courses across disciplines, reflecting ideas expressed in studies by Blewitt and Cullingford (2004) and Dvorak et al. (2011). The programs presented here (Table 2) represent four different academic disciplines, including humanities, social, and natural sciences, demonstrating the applicability of sustainability in multiple disciplines. As Table 2 also shows, each program has selected a destination that the

Table 3 Sustainability components incorporation into curriculum

Program destination	Themes of sustainability	Themes of sustainable travel	Sample learning tools
Iceland	Renewable energy	Environmental impacts of tourism	Calculation of ecological footprint Reflective journals In-situ research projects Visits to relevant industrial, tourist, cultural sites Participation in local cultural activities Workshops with local experts Low-impact explorations on site "Walking the talk" and resulting discussions
	Climate change	Frontstage vs. backstage travel	
	Natural resource conservation	Sustainable travel practices	
Scotland	Globalization	Commodification of culture	
	Cultural authenticity	Frontstage vs. backstage travel	
	Folk culture/traditions and sustainability	Sustainable travel practices	
Southern Africa	Sustainable agriculture/ food security	Ecotourism	
	Sustainable development		
	Energy resources		
	Wildlife conservation		
Switzerland	Eco-criticism	Travel writing and literature	
		Nature of travel	
		Sustainable travel practices	

highlights and facilitates education about sustainability and sustainable travel. Whether sustainability is an intuitive theme or not for a given course, the destinations are always critical for making sustainability possible. For example, both Iceland and Switzerland allow good demonstration of sustainable practices due to green infrastructure such as reliance on renewable energies (both) or an excellent public transport system (Switzerland). Iceland, Scotland, and Southern Africa all highlight struggles with tourism and sustainability, be it rapid growth in the tourism sector (Iceland), cultural marketing (Scotland), or ecotourism (southern Africa).

Different travel programs and destinations call forth different applications and approaches to how they examine sustainability in theory and practice. Each program, regardless of home discipline, is able to incorporate issues relating both to general themes of sustainability and topics specifically connecting to sustainable travel into its curriculum (Table 3). Although each course naturally focuses on that aspect of sustainability most closely related to its discipline (e.g. cultural sustainability in Scotland, socioeconomic sustainability in southern Africa), each course strives to incorporate broader discussion of all aspects of sustainability into its curriculum.

Connecting students to place and topic takes place through a variety of approaches. For example, the program to Scotland seeks to engage students by having them examine the "ancient wisdom" of folk culture and practices in terms of community building and living arrangements, work and livelihood, food preparation and consumption, entertainment, memory and tradition bearing, and relationship with the land. On location, they visit sites, meet people, and participate in activities (e.g. a ceilidh) that embody and/or problematize each of the dimensions of sustainability and folk culture. The Iceland program employs the student research projects that allow students to engage in literature research about the destination before they travel and field data collection on site during the travel portion of the course. For example, students on a recent Iceland program developed a tool using the *frontstage-backstage* framework (*sensu* Maccannell 1976) to look at tourist impacts in Iceland. The program to Africa uses a long-standing relationship with Tikownde Freedom Gardens where students get hands-on experience with organic, community-based agriculture. While on site, students participate in fieldwork and food preparation and have the opportunity to work with local women's groups. The Swiss program allows students to study travel writing while traveling in the Alps, what better place to engage in Rousseau's *Reveries of a Solitary Walker* both figuratively and literally?

The programs also strive to facilitate collateral learning (*sensu* Dvorak et al. 2011). The Iceland program requires students to estimate the ecological footprints for campus and for travel, giving them an opportunity to explore the environmental impacts of travel within the context of their own lifestyles. The African and Swiss programs require students to keep reflective travel journals, providing students to examine their experiences in a less formal, yet nonetheless potentially powerful way. On all of the travels, attempting to "walk the talk" provides much opportunity to formal and informal discussion about the privilege of travel and what in means in today's world.

As Long et al. (2014) discuss, an educational travel program should do more than just teach about sustainability; it must also, as the saying goes, practice what it preaches. The AT programs examined here attempt to do so, although each destination, discipline, budget requires a different mix of approaches (Table 4). Due to the short-term nature of the AT courses, flying is almost unavoidable to destinations off continent (e.g. Iceland, Scotland, and Africa). However, in recognition of the increase in environmental impacts with each additional leg of plane travel, the three destinations try to limit plane travel to a bare minimum. The Iceland course also budgets money to purchase carbon offsets to attempt to mitigate some of these impacts. Such measures are important given the role that tourism related air travel plays in anthropogenic greenhouse gas emissions (Scott et al. 2010). The Swiss program may provide a more ideal approach with regard to environmental sustainability, as it is able to rely on coach travel and public transportation for its travel.

In terms of sustainable lodging, each program privileges local, small-scale providers (Table 4). The Iceland program takes advantage of several eco-certifications for accommodation that exist in the Nordic countries (e.g. *Nordic Swan* or *Green Hostel)* and intentionally selects facilities that carry these certifications. The Swiss program also takes advantage of the fact that Geneva hotels provide all customers

Table 4 Sustainability components incorporation into logistics of case study programs

Program destination	Transport	Lodging/dining	Challenges
Iceland	Single-leg plane flights with bus/train transport to/from airport	Use locally-owned small hotels and youth hostels with eco-labels	Budget limitations Short length of stay in situ Visit during off-season Information on sustainable alternatives in destinations Lack of literature on outcomes assessment for sustainable initiatives
	Carbon offsets purchased for transport emissions		
	Local-based bus provider	Lodgings used for 2+ nights, when possible	
	City bus passes for students	Use locally-run restaurants that highlights local cuisine using local products	
Scotland	Single-leg plane flights with bus transport to/from airport	Lodgings selected based upon economic and environmental sustainability, where possible	
	Foot travel in cities; bus travel in Highlands	Most dining in local establishments committed to local sourcing/employment and environmental sustainability	
Southern Africa	Air travel limited to two legs in each direction	Basic accommodation in tents or bungalows	
		Many meals self-catered	
	Overland travel limited to two vehicles	Dinners in villages with traditional cuisine	
Switzerland	No air travel	Use of locally-run 2-3 star lodging	
	Locally-hired bus		
	Public transport in all cities	Group meals in local restaurants that offer traditional Swiss cuisine and local products	

with regional public transportation passes. All programs also highlight local food cultures sourcing local products as part of their group meals, avoiding fast food and chain options. Because of the food cultures in each of the chosen destinations, these choices are possible and affordable.

Despite travel leaders' best efforts and intention, a completely sustainable travel is not feasible. Table 4 provides some of the challenges the programs highlighted here face. One of the most common is the budget, a problem also reported elsewhere (Lozano 2006b; Hale et al. 2013). A more sustainable option often is more expensive. Furthermore, expenses such as carbon offsets can be viewed as unnecessary extras by administrators. It is imperative that institutions and not just individual travel leaders recognize the economic value of sustainable measures. A related concern is the season of travel. In Franklin's case, the travel takes place in mid-October and mid-March, which do not represent high season in several locations (including Iceland, Scotland, and Switzerland). Travelling during off-season can be challenge when trying to choose sustainable options, as not all facilities may be open year-round. Furthermore, the off-season might represent a temporal *backstage* for some destinations, which would mean the arrival of a group of student could be intrusive. On the other hand, traveling during the off-season may limit the environmental impacts from the group and can help distribute the income local communities derive from tourism more evenly throughout the year. Each of these points is an intriguing issue that future research needs to tackle.

3.1 Synthesis

Switzerland, Scotland, Iceland and Africa are natural places to examine different aspects of sustainability. Site is indeed one of the most important decisions that can be made when designing sustainable educational travel programs. That said, we caution the oversimplified conflation of specific sites with the study of sustainability. Sustainability issues span the Global North and South, thus so should educational travel programs that examine them. As these courses demonstrate, a balance of destinations and approaches allows students to obtain a variety of insights and experiences. Furthermore, sustainability must be transparent in program planning and execution. In some locations, sustainable logistics are more accessible, convenient and/or affordable, in others they are less so. For the latter cases, finding ways to both offset impacts and take advantage of the educational opportunities is essential

It is also imperative that efforts at sustainability in travel programs connect to larger campus discourses about sustainability. At Franklin, the efforts described above complement and are complemented by other on-campus activities. As a community enmeshed in travel—most students and faculty originate from outside of Switzerland and participate in both the AT program and independent travel—the tension between living globally and living sustainably resonates with the campus culture. The Center for Sustainability Initiatives at Franklin, whose mission is to facilitate sustainability discourse and efforts across all facets of the campus community, has focused outreach efforts to educate the community about travel and sustainability. Further, several regular campus courses take advantage of this topic and have built units into their curricula that use the students' AT experiences to examine the challenges of sustainability and travel.

3.2 Directions for Future Research

The categories of *curriculum* and *logistics* used here to examine the application of sustainability are instructive and important, but lack insight into impacts of these measures on actual student behavior and actions as well as the larger cultural, economic, and environmental impacts on host communities. Short-term gains of knowledge are easy to assess in course assignments or through pre and post-experience surveys (Tarrant and Lyons 2012); however, the real benefit must come from long-term positive impacts for students and host communities. Further, the potential for collateral learning must also be understood. Such outcomes are not immediately apparent and necessary assessment tools are lacking. Thus, an important next stage in the development of sustainable travel programs at Franklin (and elsewhere) must focus on assessment of how these programs influence students' long-term behavior and attitudes towards sustainability. This would be a logical step to follow the development of the many reporting tools that assess the incorporation of sustainability at the institutional level of IHEs.

Another direction learned from the Franklin case study is how to become more deliberate about the incorporation of sustainability into educational travel. Each of the contributors would agree that being asked to evaluate how sustainability is incorporated into their program was useful in terms of making the implicit explicit and in terms of helping travel leaders to imagine additional means of incorporating sustainable measures and curricula. Asking all travel leaders to evaluate the use of and limitations to sustainability in their course could be an important next step for the institution to take. Nonetheless, the interviews show that striving for sustainability is not necessarily easy. Without an institutional imperative, sustainability efforts are largely a result of professor (and sometimes student) initiative. However, professors are also dealing with issues ranging from budget to student reception. Costs certainly influence a program's sustainability, particularly with respect to the trip logistics and site selection. Some sustainable practices result in lower costs (e.g. lodging in a hostel versus a higher-end hotel). Others increase costs, if you choose organic food, require specific flight routes, and purchase carbon offsets. Trying to negotiate sustainable measures with a limited budget is not always possible. Further, the perceived inconvenience of some sustainable practices (e.g. limiting the legs of flights or staying in hostels) can impact student program evaluation.

Without an institutional imperative, access to information is crucial, particularly with respect to travel logistics. The environmental impacts of air travel are well-known, as are options to mitigate them for a travel program. Addressing lesser well-known aspects of sustainability, be it timing of travel, impact of various lodging, dining, or activity choices on student learning outcomes or on host community viability, requires good information about impacts and options. Research is needed to fill these gaps and allow future programs to adapt logistics and curricula accordingly. In turn, this research must be readily available to travel practitioners.

4 Conclusions

The Academic Travel program at Franklin provides important insights into how the tension between goals of internationalization and sustainability can be explored and to some extent mitigated. Travel programs across a diverse range of disciplines are able to address sustainability in a variety of fashions and travel logistics can be explicitly designed to demonstrate best practices. Nonetheless, issues of budget and institutional support remain big challenges, as do instruments of assessment that allow better understanding of long-term student outcomes. Moreover, research is needed to understand the overall impacts of such programs on their host communities. Overall, this tension does not have to mean the end of travel, but rather a more intentional approach to it.

References

AASHE (2013) STARS overview. association for the advancement of sustainability in higher education. Retrieved 03 Feb 2014. https://stars.aashe.org/pages/about/stars-overview.html

Altbach PG, Knight J (2007) The internationalization of higher education: motivations and realities. J Stud Int Educ 11(3–4):290–305. doi:10.1177/1028315307303542

Bakalis S, Joiner TA (2004) Participation in tertiary study abroad programs: the role of personality. Int J Educ Manage 18(5):286–291. doi:10.1108/09513540410543420

Bellamy C, Weinberg A (2006) Creating global citizens through study abroad. Connection: J New Engl Board High Educ 21(2):20–21

Beringer A, Adomßent M (2008) Sustainable university research and development: inspecting sustainability in higher education research. Environ Educ Res 14(6):607–623. doi:10.1080/13504620802464866

Blewitt J, Cullingford C (eds) (2004) The sustainability curriculum. London, Earthscan, p 257

Bodger D (1998) Leisure, learning, and travel. J Phys Educ Recreation Dance 69(4):28–31

Brown BJ, Hanson ME, Liverman DM, Merideth RW Jr (1987) Global sustainability: toward definition. Environ Manage 11(6):713–719

Cusick J (2008) Study abroad in support of education for sustainability: a New Zealand case study. Environ Dev Sustain 11(4):801–813. doi:10.1007/s10668-008-9144-5

Dvorak AMW, Christiansen LD, Fischer NL, Underhill JB (2011) a necessary partnership: study abroad and sustainability in higher education. Front: Interdisc J Study Abroad 21:143–167

Emerson J, Esty CD, Levy MA, Kim CH, Mara V, de Sherbinin A, Srebotnjak T (2010) 2010 environmental performance index. Yale Center for Environmental Law and Policy, New Haven, p 87

Hale BW, Vogelaar AE, Long J (2013) A-broad spectrum: sustainability in educational travel. Int J Sustain High Educ 14(4):349–366

Institute for International Education (2014) U.S. study abroad. Open doors data. Retrieved 11 Mar 2014. http://www.iie.org/Research-and-Publications/Open-Doors/Data/US-Study-Abroad

Leal Filho W (2010) Teaching sustainable development at university level: current trends and future needs. J Baltic Sci Educ 9(4):273–284

Lifelong Learning Program (2014) Erasmus students mobility. Statistics for all. Retrieved 08 Mar 2014. http://www.statisticsforall.eu/maps-erasmus-students.php#

Long J, Vogelaar A, Hale BW (2014) Toward sustainable educational travel. J Sustain Tourism 22 (3):421–439

Lozano R (2006a) A tool for a Graphical Assessment of Sustainability in Universities (GASU). J Cleaner Prod 14(9):963–972

Lozano R (2006b) Incorporation and institutionalization of SD into universities: breaking through barriers to change. J Cleaner Prod 14(9):787–796

Lozano R (2010) Diffusion of sustainable development in universities' curricula: an empirical example from Cardiff University. J Cleaner Prod 18(7):637–644

Lutterman-Aguilar A, Gingerich O (2002) Experiential pedagogy for study abroad: educating for global citizenship. Frontiers: Interdisc J Study Abroad 8:41–82

Maccannell D (1976) The tourist: a new theory of the leisure class. Schocken Books, First, New York, p 214

Mcintosh M, Gaalswyk K, Keniry LJ, Eagan DJ (2008) Campus environment 2008: a national report card on sustainability in higher education. National Wildlife Federation, Merrifield, p 132

Paige RM, Fry GW, Stallman EM, Josić J, Jon J (2009) Study abroad for global engagement: the long-term impact of mobility experiences. Intercultural Educ 20(sup1):S29–S44. doi:10.1080/14675980903370847

Palacios CM (2010) Volunteer tourism, development and education in a postcolonial world: conceiving global connections beyond aid. J Sustain Tourism 18(7):861–878

Reilly D, Senders S (2006) Becoming the change we want to seeȟ: critical study abroad for a tumultuous world. Frontiers: Interdisc J Study Abroad 18:241–267

Rexeisen RJ, Al-Khatib J (2009) Assurance of learning and study abroad: a case study. J Teach Int Bus 20(3):192–207. doi:10.1080/08975930903099077

Rowe D (2002) Environmental literacy and sustainability as core requirements: success stories and models. In: Leal Filho W (ed) Teaching sustainability at universities. Peter Lang, New York, pp 79–103

Schroeder K, Wood C, Galiardi S, Koehn J (2009) First, do no harmȟ: ideas for mitigating negative community impacts of short-term study abroad. J Geogr 108:141–147

Scott D, Peeters Paul, Gössling S (2010) Can tourism deliver its "aspirational" greenhouse gas emission reduction targets? J Sustain Tourism 18(3):393–408

Sjoberg SD, Shabalina OI (2010) More than a sight-seeing trip : enhancing the value in short-term study abroad. Bus Educ Innov J 2(2):46–59

Stafford SH (2011) How green is your campus? an analysis of the factors that drive universities to embrace sustainablity. Contemp Econ Policy 29:337–356

Tarrant MA (2009) A conceptual framework for exploring the role of studies abroad in nurturing global citizenship. J Studies Int Educ 14(5):433–451. doi:10.1177/1028315309348737

Tarrant M, Lyons K (2012) The effect of short-term educational travel programs on environmental citizenship. Environ Educ Res 18(3):403–416

Way T, Matthews C, Rottle N, Toland TR (2009) Greening the American campu. Res Facil Des 40:25–47

Authors Biography

Brack W. Hale has a Ph.D. in land resources from the Nelson Institute for Environmental Studies at University of Wisconsin-Madison. He is currently an associate professor of biology and environmental science at Franklin College Switzerland and co-director of the Center for Sustainability Initiatives at Franklin. His research interests include sustainability issues in higher education, particularly in regard to the environmental impacts of educational travel programs. He also regularly leads educational travels through Franklin's Academic Travel program, having taken students to destinations in Europe and Latin America.

Alison E. Vogelaar has a Ph.D. in communication from the University of Colorado, Boulder. She is an assistant professor of communication and media studies at Franklin College Switzerland where she also co-directs the Center for Sustainability Initiatives. Her research interests include environmental discourses, the rhetoric of social movements, and sustainable educational travel. She also leads a yearly academic travel program to Scotland that focuses upon the themes of folk culture, tourism, representation, and sustainability.

One Metaphor—Several Meanings: An Interdisciplinary Approach to Sustainable Development

Astrid Ouahyb Sundsbø, Benjamin R.K. Runkle, Sarah McMonagle, Kerstin Jantke, Florian Lottermoser, Manuel Gottschick, Sönke Häseler, Juan Miguel Rodriguez Lopez and Marcus Scheele

Abstract

This paper presents reflections on how sustainable development (SD) is defined as a research topic and on differences in references to SD in research. The paper argues that the reference of SD as a research topic must expand to include a broader range of research subjects. During two years of methodical discussions and workshops conducted through the interdisciplinary postdoctoral research group *Sustainable Future* at the Universität Hamburg (Germany), the authors have observed that SD has played a major role in natural sciences (i.e. environmental and earth system sciences) and engineering, and a far lesser role in the social sciences and humanities (SSH). This imbalance in the discourse on SD in research impedes a full understanding of the diverse tasks and challenges that must be addressed in SD. An integrated approach is suggested to unite perspectives from natural sciences, engineering and SSH. The authors therefore encourage a more explicit notion of the concept of SD in SSH research, which builds upon and goes beyond current approaches to SD in SSH. The paper presents some illustrative examples which investigate the implications of expanding or restricting the disciplinary boundaries of sustainability research. The examples advocate an inclusive approach for which the authors describe and clarify several methods for expanding research towards cross-disciplinary methods. To this end, the authors present some practical recommendations for the integration of sustainability into the design and implementation of research academic practice. The paper closes up with an argument that all disciplines

A.O. Sundsbø (✉) · B.R.K. Runkle · S. McMonagle · K. Jantke · F. Lottermoser · M. Gottschick · S. Häseler · J.M. Rodriguez Lopez · M. Scheele
Center for a Sustainable University, Universität Hamburg, Mittelweg 177, 20148 Hamburg, Germany
e-mail: astrid.sundsboe@uni-hamburg.de

W. Leal Filho et al. (eds.), *Integrating Sustainability Thinking in Science and Engineering Curricula*, World Sustainability Series, DOI 10.1007/978-3-319-09474-8_15

should expand their definition of sustainability in order to hasten research outcomes from all the diverse sectors of sustainability research.

Keywords
Sustainability Goals • Discourses on sustainable development • Imbalance in sustainability research • Transdisciplinary research

1 Introduction: One Metaphor—Several Meanings

The framework of sustainability generally comes out of the Stockholm conference (1972) with origins in environmental law and the promotion of intergenerational equity with respect to natural resources (Wright 2002). Since the Brundtland Report (WCED 1987) though, which claims that SD is about meeting "the needs of the present without compromising the ability of future generations to meet their own needs", a broader definition of the concept has been established. According to Theis and Tomkin (2012), the idea of sustainable development is to improve the quality of life for all people and therefore ranks among historical human social movements such as human rights, racial equality, gender equity or labour relations.

This paper examines the manner in which these ideas have been translated into the use of SD as a research topic and the ways in which SD has been used to justify or define research focal points. This work is based on a set of discussions and workshops held at the Universität Hamburg through its interdisciplinary post-doctoral college, *Sustainable Future*. Because the paper comes out of a series of guided and monitored discussions, it is not empirical in a traditional sense, and instead lays out reflective arguments to help push different members of the SD community (and beyond) towards a more expansive notion of SD research.

The first chapter of the paper presents a variety of perspectives and aims of SD as a research topic. The diverse approaches to SD are illustrated through examples of how different disciplines traditionally define the SD challenge and the common questions that are posed in relation to it. The following chapter discusses the imbalance in the approach towards SD in research. In the last chapter it is argued that there is a need for an integrated approach in order to understand the full set of challenges for SD. It concludes with a set of suggestions for advancing a more balanced approach to research within SD. Within each section the critical and relevant literature is reviewed.

2 Different Perspectives on Sustainable Development

In this section, disciplinary perspectives on SD research subjects are described based on the membership of the Universität Hamburg post-doctoral college, *Sustainable Future*. To develop these perspectives the college fellows examined the literature in their individual fields and led critical and focused discussions with their disciplinary peers.

In the field of natural sciences, research in the arena of sustainability typically examines the impacts of anthropogenic pressures on the earth system and the development of materials, procedures or concepts to reduce this impact. Research topics include anthropogenic climate change, ocean acidification, stratospheric ozone depletion, human interference with the biogeochemical nitrogen and phosphorus cycle, freshwater use, land use change, biodiversity loss, chemical pollution and atmospheric aerosol loading (Rockström et al. 2009). The rationale here is that these phenomena are consequences of the non-sustainable use of natural resources. The scientific and subsequent societal and political solutions to these problems pose significant challenges for SD and the survival of humankind as well as fundamental shifts in how the earth's systems behave.

In engineering research, sustainability implies a concern with the interface between societal needs and the natural resources those needs demand. The research topics therefore often involve maximising some needed output while minimising the environmental or resource inputs (Anastas and Zimmerman 2003). These projects are often designed around goals seen as fundamental to a high quality of life or to a society's sense of 'civilisation'. Many engineering disciplines have sustainability components related to food supply, including the biological engineering of crop varieties, watershed engineering of water supplies and decision-making regarding irrigation scheduling, chemical engineering of fertilizer resources and the mechanical engineering of farm equipment. An engineer focused on sustainability would also research technical approaches to climate change mitigation, adaptation strategies to climate change and the development of renewable energy supplies. Finally, concerns about poorer or less powerful communities often arise in engineering research in the development of cost-efficient or very low-cost solutions to societal needs (including light for reading, sunlight for power or water purification or redesigned cities for public transportation).

Economists research SD by analysing the conditions under which an individual's efforts to maximise her own well-being simultaneously promote the welfare of the larger society. In ideal conditions markets ensure that the individual's interests are aligned with those of society. Even if such 'interpersonal' alignment is achieved, however, the outcomes need not be compatible with sustainable development. In fact, markets routinely fail to produce sustainable outcomes due to their limited ability to incorporate 'intertemporal' aspects. Climate change, for example, has been referred to as the greatest market failure in history (Stern 2008), where a market failure is the incorrect and suboptimal assignment of costs and benefits (Mankiw 2012).

Economic theory proposes two general approaches to overcoming the market failures that stand in the way of SD (Hess 2013). Firstly, it may be possible to internalise the external benefits and costs that lead individuals to make unsustainable choices. For that purpose it is necessary to attach prices to those external (future) costs and benefits, which is not an easy task. The second approach consists in regulating potentially unsustainable activity. Both approaches require international cooperation to tackle the global issue of SD as well as new research within and including the field of economics. The economic and business literature thus examines the conditions under which self-interested optimisation by individuals, societies or firms will produce outcomes that are compatible with SD, where the issue of climate change attracts most research attention.

Sociologists have also engaged in the debate on sustainability in terms of the social and cultural aspects in environmental challenges; see the summaries provided by Burns (2012) and Jetzkowitz (2012). For instance, they see the threats posed to the environment on account of consumer behaviour or the ways in which people (have to) manage their lives (Jetzkowitz 2012). There are also numerous opportunities within the sociological approach to intuitional actors in SD (see Box 1). Such approaches share the perspectives on SD as an environmental challenge, as mentioned above.

Box 1: The sociological approach promises much potential in sustainability research

The capability of current sociology goes far beyond previous contributions to sustainability in research. State-of-the-art sociological theory (Kroneberg 2011; Maurer and Schmid 2010) allows for understanding (un-)sustainable phenomena in a holistic way—both as individual action and embedded in collective structures in social and ecological contexts. Sustainable topics, practices and gaps can be analysed on the micro level as resulting from influences in societal structures, norms and institutions. Intentional actors with limited rationality have to solve challenges of social coordination and create aggregated (un-)sustainable solutions on the macro level. In doing so, their intentions vary from ecological and social considerations to (socio-) political claims to power and economic profit. From this perspective, sociological theory and behavioural economics complement and cross-fertilise each other, not least because they build on similar theoretical roots. Combined with deep insights on the mechanisms of institutional path dependence (Beyer 2006), the long-term process of sustainable development with its reinforcements and hindrances should be looked at in its entirety (Lottermoser 2014). Taken as a whole, the sociological approach has much to contribute to sustainability research. Together with economic approaches, it can identify, describe and quantify societal and individual forces, goals, triggers and obstacles, and can subsequently transfer such findings to practical support systems for SD.

There are also other areas of research where sociologists can explicitly engage with SD—for instance on issues of social inequality. Research on social inequality is highly relevant to the wider understanding of SD in terms of providing future generations with the skills they need in order to improve their position in the socio-economic hierarchy (cf. the Brundtland definition). Hence research considering the (unequal) distribution of economic and social resources (including education) and the mechanisms for the (re-) production of social inequality are also to be considered as SD research.

In the report *Urban Future 21, A Global Agenda for 21st Century Cities* provided by Hall and Pfeiffer (2000), social inequality and exclusion are defined as SD challenges. SD is thus not necessarily confined to the ecological framework. It can also address the conditions for social mobility of disadvantaged groups. Based on this wider understanding of SD and the interpretation of its challenges, research on the quality of schools and their ability to compensate for social disadvantage also forms part of the sustainability discourse. However, research on social inequality and obstacles to social mobility is seldom labelled as SD research.

Language, too, is a resource that is necessary for participation in society yet whose marginalisation can lead to social exclusion. Since language is critical to how individuals and societies function, it should play a much broader role in SD research. The special resource status of language and its role in the social sciences and humanities (SSH) contribution to SD is explicated further in Box 2.

Box 2: Language as resource: one field's guide towards a broader concept of sustainability research

A stronger involvement of linguistic scholars in sustainability research would enhance the inclusion of 'language as resource' in SD discourses. The role of language in such discourses is often limited to the preservation of endangered languages which are likened to biological species that must be maintained in the interest of diversity. Stanford and Whaley (2010) propose the discourse of sustainability to conceive of language as a valuable cultural resource whose continued use must be ensured for future generations. 'Language as resource' is both sensible and multilayered, yet it is necessary to move beyond the 'endangered language' phenomenon when considering language in terms of sustainability. Languages are not biological organisms and attempts at comparison to endangered species present language in a rather static fashion, rather than as a dynamic resource drawn upon by speakers in a range of situations.

This argument does not overlook the language endangerment phenomenon; policies for the promotion and protection of lesser-used languages are critical. Rather, it seeks a broader perspective on language in sustainability research in two main respects:

(1) Language *per se* must be considered as an individual and societal resource that does not require biological and environmental metaphors for inclusion in SD discourses.

(2) Linguistic inclusion is not just an aspiration for 'endangered' languages, but for all globalised societies where multilingual constellations have emerged through migration.

Language must be considered a significant factor in SD in the current era of globalisation where most societies are dramatically diversifying, yet do not always recognise or accommodate the variety of languages that are spoken or understood. Language is an individual resource—cultural, educational and social—and failure to cultivate it can result in inequality. In the form of multilingualism, language is a societal resource that also poses significant challenges for social cohesion. A focus on inclusion and participation would also serve the so-called endangered languages within a broader discourse.

Linguistics and its sub and cognate disciplines can thus make significant contributions to sustainability discourses that are concerned with the future development of society. A shift from 'the sustainability of languages' to 'languages for sustainability' would present a more holistic approach to issues of human and educational rights. However, as yet, the discourse of SD has no great presence in language studies research. As such, it is hardly considered to be a 'typical' topic in discussions on sustainability, both within and outside academic research.

3 Imbalance in the Discussion on SD

Although different research disciplines are engaged with and address various challenges for SD, there is a bias in the role of SD across these disciplines. While SD is virtually omnipresent in the natural sciences, engineering and economics, it is not explicitly defined in other fields of research (Jetzkowitz 2012). The lack of sociological involvement in discussions on sustainability becomes apparent by looking at the backgrounds of those engaged in such discussions: many of the contributors are natural scientists or engineers, or sociologists working in institutions concerned with environmental issues. In comparison, sociologists seldom participate in the same discussions. When they do, as stated above, their engagement tends to relate to lifestyle choices that threaten stocks of natural resources and cause environmental pollution. Sustainability in the field of sociology is thus largely bound to sustainability in terms of the natural environment and is mainly discussed in the sub discipline of environmental sociology (Lange 2011). The role of sociologists here is to monitor and collect data, as well as to analyse, identify and explain the impact of society on environmental changes (Renn 1996). In some sustainability projects dominated by natural sciences, the role of sociologists is seen to be restricted to developing communication or acceptance strategies that push society towards more sustainable development.

At the same time, this interpretation does not mean that research on SD only considers environmental challenges. Up to the mid-1990s social concerns were generally brought into the environmental context, for example the role of poverty in exacerbating ecological degradation (Lozano et al. 2013). Following this period there were calls for a broad sense of social responsibility in various sustainability initiatives, ranging from the 2005 Graz Declaration to the 2009 Turin Declaration that emphasised new models of social and economic development consistent with sustainability principles. Newer research has highlighted the role of cultural influences on the implementation of sustainability and has integrated social concerns alongside environmental and technical orientations of sustainable development (Stephens et al. 2008).

But this widening of theoretical discussions on sustainability (Graedel 2002) has not yet led to a broad usage of the term in all kinds of research. The Intergovernmental Panel on Climate Change, which may itself symbolise the dominance of environmental subjects in discussions of SD, acknowledges that discourses of SD have historically focused on the environmental and economic dimensions and have overlooked the need for more holistic change. The importance of issues such as poverty, social equity or governance, has only recently gained recognition in climate change research that is dominated by the natural sciences (Sathaye et al. 2007). Much research in the social sciences engages with the development of society and the distribution of (social) resources, yet there is hardly an explicit link between such research and debates on SD.

In many disciplines, either the sustainability concept still appears as externally 'bolted on' to the core disciplinary pursuit or, conversely, researchers in sustainability appear to 'bolt on' findings from other disciplines. The lack of translation or, say, transformation of the concept of SD across disciplines is clear when we look at research on linguistic diversity (as referred to in Box 2). Interdisciplinary thinking has investigated the links between biological, cultural and linguistic diversity "as manifestations of the diversity of life" (Maffi 2005 p 599). Threats to linguistic and biological diversity have been linked regarding the importance of local knowledge —which is contained in and transmitted through language—to sustaining the natural environment. A language and the physical environment in which it is spoken most certainly have links. However, the inclusion of language and linguistic disciplines in SD discourses has a much greater potential than merely adding to the environmental issues that already dominate such discourses, as discussed in Box 2.

The imbalance in the discussion on SD also becomes obvious when we look at how higher education institutions deal with the concept. Here, environmental and ecological aspects are stressed and are often seen as the most essential component of any SD initiative. Even recent initiatives on sustainability that take a 'whole university' approach may weight 'greening' and (natural) resources more heavily than, or separately from, social justice concerns (McMillin and Dyball 2009).

4 The Imbalance Impedes a Full Understanding of the (Diverse) Challenges for SD

We have argued that there is a lack of engagement from SSH in debates on sustainability. This imbalance impedes a full understanding of the diverse academic and societal tasks related to SD and inhibits a mutually beneficial cross-fertilisation of different research disciplines.

Although a number of researchers from SSH already actively engage with SD topics, their inclusion in the field is clearly not as 'taken for granted' as that of natural scientists and engineers. Indeed, there is often a sense of having to justify or explain the role of SSH research here. The question has been raised within the humanities, "how might those of us who do something broadly defined as 'literary and cultural studies' contribute to the creation of a more sustainable world?" (Phillipon 2012 p 163). This question seems to be fair regarding an area that, after all, does have environmental roots. Yet Philippon notes that "the humanities certainly need to make a place for ourselves at the sustainability table" (ibid p 169), thus confirming that certain disciplines remain on the relative outskirts of this supposedly interdisciplinary field.

In response to these inequities in research, we propose a model (see Fig. 1) to illustrate our argument that the goals of SD can touch upon very different issues and involves a range of perspectives. The division between the circles represents our interpretation that the diversity of perspectives on the goals of SD is not always recognised. The inner circle of this figure illustrates the 'mainstream understanding' of SD that is characterised by topics such as climate change and corporate responsibility. The additional circle highlights the research on social inequality as

Fig. 1 Conceptual model of the proposed broadening of sustainability research subject areas

part of the research on SD as it seeks to enable disadvantaged or marginalised groups with skills for societal participation and social mobility. Language and linguistic diversity are considered as a particular cultural resource in this area.

A Meta research level is implemented whereby the exploration and realisation of sustainable development can be assisted by the analysis of sustainability goals, interests, actions and processes. Sociologists and economists can develop practice-oriented decision support systems in close theoretical and methodical collaboration. This integrated approach introduces a wide and explicit understanding of sustainable development. It enables researchers from more disciplines to engage with sustainability and the way to a future balance in sustainability research remains open.

A recent research controversy highlights the shifts in disciplinary mind-set when a more holistic approach to SD research is generated. The following example (Box 3) from the natural sciences helps demonstrate that while traditional research goals are justified within a narrow SD framework, the means and methods used to generate expected research outcomes may be at odds with a broader sustainability perspective. A multi-tiered approach is then seen to bring a broader, more nuanced set of benefits.

Box 3: Broadening the scope of sustainability research shifts the focus, means, and intent of that research—an example regarding the ethics of genetic disease mutation research

A recent research goal in the field of disease genetics was to expand understanding of the pivotal factors that favour the occurrence of a mutation of the virus H5N1 (avian influenza) towards resembling strain H1N1 (swine flu). Such a mutation has the potential to generate millions of casualties through the faster spread of these lethal viruses. The ability to study the evolution and properties of such a virus in a confined research laboratory was seen within the natural science discipline to fully satisfy the SD definition from a natural science perspective in that it helps to secure the existence of humans on earth. Indeed, in a series of ground-breaking studies, two research teams could demonstrate that conditions exist under which an influenza virus can naturally emerge which combines the large mortality rate of H5N1 with the high transmission rate of H1N1 (Herfst et al. 2012; Schrauwen et al. 2013; Zhang et al. 2013).

However, this research and its announcement generated considerable public outrage and controversy and produced considerable doubt as to whether *practical* conduct of an *idea* aiming at SD is actually sustainable at all. To allow time to develop a more holistic perspective on SD in this research subject, the natural scientist research teams then decided on a 60-day moratorium to pause research on this subject (Fouchier et al. 2012). The pause was later expanded to 12 months in response to the growing critique that had evolved to include ethicists and policy makers, some of them arguing that these studies should not even have been approved for funding in the first place (Fauci and Collins 2012; Williams-Jones et al. 2014).

The pause resulting from the H5N1 mutant case stimulated a more holistic SD approach to assess the risks and benefits of such research projects. It became evident that rating a study simply by the anticipated outcome and knowledge gain can be short-sighted and does not meet the criteria of a real and integrated SD objective. During the 12-month pause, the possible benefits of the study were extensively communicated to the public and discussed openly and with room for debate. New biosafety recommendations were then announced as mandatory for laboratories interested in engaging in such research. The remaining risks were then evaluated by independent, public organizations and measures for their mitigation have been detailed. A majority of scientists and policy-makers have since decided that the research on H5N1 is to be continued in the light of a now much more positive benefit/risk relationship (Fouchier et al. 2013).

This case study represents an example where a one-sided approach to SD in the natural sciences has been replaced by a more general interpretation including not only scientific aspects but also risk-management, communication strategies, policy-making, etc. As Fauci and Collins (2012) have phrased it, "a social contract among the scientific community, policy-makers, and the general public that builds trust is essential for success of [such research projects]." In many ways, this multi-disciplinary and publically engaged approach applies to SD as well.

A complete integration of sustainability as a trans-disciplinary approach is necessary in both the questions asked by sustainability researchers and the methods used to resolve those questions. Lehtonen (2004) argues that the essence of SD can be found exactly at the interfaces and trade-offs between the often opposed objectives of economic and social development and environmental protection. Attention to these issues is highlighted by a recent focus in hydrological sciences towards 'socio-hydrology' that recognises the need to rebuild and expand hydrology as a discipline that is much more balanced with social actors, interests and disciplines (Lane 2014). This redesigned field would recognise more explicitly that any environment examined in hydrology is essentially socially constructed and determined—for example, the effects of land use change and climate change on hydrological systems are often analysed and both are clearly anthropologically influenced. Similar trans-disciplinary approaches have also been advocated in biodiversity research (Alves et al. 2013) where increased citizen engagement and inclusion of social perspectives have been recognized to expand both research and conservation outcomes.

5 Conclusion

The fact that SD is (mostly implicitly) considered in terms of environmental challenges leaves us with the impressions that (1) the concept has not yet been transferred to SSH as a relevant research topic and (2) the social questions of SD have not been fully transferred into the natural sciences and engineering disciplines. We argue that the incompleteness of this project may be the result of a fragmentary understanding of what SD aims to address. So long as the research community—either explicitly or implicitly—continues to conceive SD in terms of natural environmental challenges that have social and economic aspects, it seems reasonable that SSH researchers will fail to adequately connect with the concept. A wider and more explicit debate on the aims of SD could see more researchers engage with the topic from their respective disciplinary perspectives.

We conclude with some practical ideas toward the more complete integration of sustainability into research. First, the Brundtland report, which served to widen the understanding of aims in SD, does not determine how SD is to be implemented or negotiated. For this reason, further development and clarification of the target goals of SD are necessary. One critical point is that the report does not assess how to deal with the fact that developing certain courses of action may imply the narrowing of others. There is some tension in the concept of sustainability as both (1) improving the lives of individuals and (2) working toward the common good and the 'just' distribution of collective goods. There is moreover the question of the effects of actions taken in the name of sustainable development and their reach; for what might be considered sustainable at a local level may not be the case at the global level—and vice versa. It is thus clear that interpretations of what sustainability is and how it might be implemented provide ample material for conflict and debate.

Second, when SD is presented, an operational definition of SD should be provided to make clear its inclusion of many disciplines and cross-cutting goals (environmental, social, economic, cultural, etc.).

Third, educational and awareness efforts within each discipline by disciplinary practitioners themselves may encourage a realisation that the research already performed in these fields can fit within a framework of sustainability. If all fields are more aware of their potential and experience in SD research and can articulate their contributions to these goals, then an inclusive approach may be more quickly reached. This suggestion extends previous arguments (Fien 2002) that have encouraged all kinds of research approaches as part of the advance of sustainability in higher education.

Fourth, it should be more strongly acknowledged that studying the natural world is not sufficient to generate a holistic SD research program, even within the natural and applied sciences. Including non-environmentally oriented sustainability concerns from those perspectives is just as important for SD as well.

Fifth, truly cross-disciplinary research should be further stimulated and encouraged by funding agencies and research institutions to enable practical collaborative experiences that would help internalise the integrated SD approach.

Distinctions between different types of interactions within, between, and beyond disciplines, and a road-map towards their support in project development are proposed in Box 4.

Box 4: Multi-, inter-, trans-disciplinary research for sustainability?

Whether multi-, inter-, or trans-disciplinary research is best suited for sustainability is of course a matter of target (research question) and context (case study). Nevertheless, we have the impression that there is an implicit assumption that only inter- and trans-disciplinary research is sustainable research (see e.g. Brandt et al. 2013).

In **multi-disciplinary research** diverse disciplines work separately in the same research field. The knowledge users can rely on discipline specific quality standards. But the knowledge user has to determine to what extent the chosen theory, method and related assumptions are suitable for the problem that has to be solved.

In **inter-disciplinary research** diverse disciplines work together to jointly generate knowledge that addresses a specific problem. The knowledge user trusts that the theory, methods and related assumptions reflect the scientific perspective and needs of the problem. When the scientific problem perspective does not match the perspective of the knowledge user it may be difficult to transfer the knowledge generated from this research into the user domain.

In **trans-disciplinary research** non-scientists (e.g. stakeholders, problem-owners, decision-makers) are involved systematically in the whole process of inter-disciplinary research (see e.g. Lang et al. 2012). The involved knowledge user is able to frame the research question in ways that her/his problem is addressed. But some scholars question the scientific quality of such practical and action-orientated research because of the normative orientation (for a good approach see Stokols 2006). One of the challenges of this approach is therefore to create a cooperative process while keeping the research independent and insulated against the dominance of interest groups (Gottschick 2014b). We suggest that the trans-disciplinary research project is distinguished into phases where trans-, inter-, multi- and mono-disciplinary tasks alternate (see Fig. 2).

And sixth, there should be more focus on the issue of evaluation to examine what exactly is the impact of a given research project on SD. There should be a more critical debate on how the concrete research projects handle (a) the tension in the concept of sustainability between the common good and the 'just' distribution of collective goods on the one hand, and the interest in improving the lives of individuals on the other and (b) the effects and their reach on a local level and at the global level—which do not necessarily correspond. Finally, further work is necessary to expand the scholarly concept of sustainability research questions, to

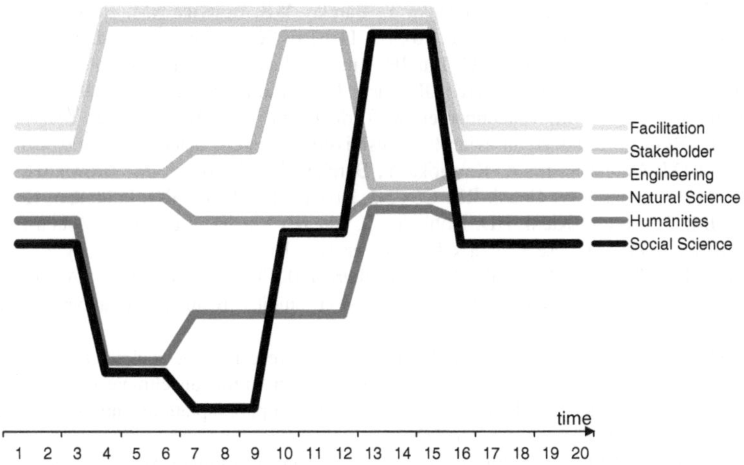

Fig. 2 Schematic time-phases of trans-, inter-, multi- and mono-disciplinary tasks, the distance between lines symbolises the level of collaboration in this time phase (Gottschick 2014a)

incorporate new methods and techniques of trans-disciplinary interaction, and to push for greater recognition of social and cultural resources within the SD discourse.

The frameworks presented are broad and are meant to nudge community discussion towards a more inclusive framework without being explicitly proscriptive. The implementation of extra-disciplinary notions of sustainability in research will necessarily take on a different flavour at different institutions and for different targeted questions. The authors feel strongly that all disciplinary researchers should expand their notion of sustainability. The consequences for this paradigm shift may greatly hasten research outcomes in all the diverse sectors of sustainability research.

References

Alves F, Leal Filho W, Araújo MJ, Azeiteiro UM (2013) Crossing borders and linking plural knowledge: biodiversity conservation, ecosystem services and human well-being. Int J Innov Sustain Develop 7(2):111–125

Anastas PT, Zimmerman JB (2003) Design through the 12 principles of green engineering. Environ Sci Technol 37(5):94A–101A

Beyer J (2006) Pfadabhängigkeit. Über institutionelle Kontinuität, anfällige Stabilität und fundamentalen Wandel. Campus, Frankfurt, New York, 296 p

Brandt P, Ernst A, Gralla F, Luederitz C, Lang DJ, Newig J, Reinert F, Abson DJ, von Wehrden H (2013) A review of transdisciplinary research in sustainability science. Ecol Econ 92:1–15

Burns TR (2012) The sustainability revolution: a societal paradigm shift. Sustainability 2012 (4):1118–1134

Fien J (2002) Advancing sustainability in higher education: issues and opportunities for research. High Educ Policy 15(2):143–152

Fauci AS, Collins FS (2012) Benefits and risks of influenza research: lessons learned. Science 336:1522–1523

Fouchier RAM, García-Sastre A, Kawaoka Y, Barclay WS, Bouvier NM, Brown IH, Capua I, Chen H, Compans RW, Couch RB, Cox NJ, Doherty PC, Donis RO, Feldmann H, Guan Y, Katz JM, Klenk HD, Kobinger G, Liu JH, Liu XF, Lowen A, Metten-Leiter TC, Osterhaus ADME, Palese P, Peiris JSM, Perez DR, Richt JA, Schultz-Cherry S, Steel JSK, Swayne DE, Takimoto T, Tashiro M, Taubenberger JK, Thomas PG, Tripp RA, Tumpey TM, Webby RJ, Webster RG (2012) Pause on avian flu transmission research. Science 335:400–401

Fouchier RAM, García-Sastre A, Kawaoka Y, Barclay WS, Bouvier NM, Brown IH, Capua I, Chen H, Compans RW, Couch RB, Cox NJ, Doherty PC, Donis RO, Feldmann H, Guan Y, Katz JM., Kiselev OIKlenk HD, Kobinger G, Liu JH, Liu XF, Lowen A, Metten-Leiter TC, Osterhaus ADME, Palese P, Peiris JSM, Perez DR, Richt JA, Schultz-Cherry S, Steel JSK, Swayne DE, Takimoto T, Tashiro M, Taubenberger JK, Thomas PG, Tripp RA, Tumpey TM, Webby RJ, Webster RG (2013) Transmission studies resume for avian flu. Science 339:520–521

Gottschick M (2014a) Investigative and transformative research for societal transformation to sustainability. In: Adaptation frontiers—conference on European climate change adaptation research and practice, Lisbon, Mar 10–12. http://adaptationfrontiers.eu/pdf/B-T-04-MGottschick2.pdf (last accessed 4/17/2014)

Gottschick M (2014b) Plädoyer für eine investigative transdisziplinäre Anpassungsforschung. In: Beese K, Fekkak M, Katz C, Körner C, Molitor H (eds) Anpassung kommunizieren?! Konzepte, Fallstricke und Perspektiven von Bildung, Kommunikation und Wissenstransfer für eine regionale Anpassung an den Klimawandel. oekom verlag, München, 35 p

Graedel TE (2002) Quantitative sustainability in a college or university setting. Int J Sustain High Educ 3(4):346–358

Hall P, Pfeiffer U (2000) Urban future 21. A global agenda for 21st century cities. E&FN Spon, London, p 41

Herfst S, Schrauwen EJA, Linster M, Chutinimitkul S, de Wit E, Munster VJ, Sorrell EM, Bestebroer TM, Burke DF, Smith DJ, Rimmelzwaan GF, Osterhaus ADME, Fouchier RAM (2012) Airborne transmission of influenza A/H5N1 virus between ferrets. Science 336:1534–1541

Hess PN (2013) Economic growth and sustainable development. Routledge, London (536 p)

Jetzkowitz J (2012) Verantwortung für die Zukunft. Soziologie und das Problem der Nachhaltigkeit. In: Unzicker K, Hessler G (eds) Öffentliche Sozialforschung und Verantwortung für die Praxis. Zum Verhältnis von Sozialforschung, Praxis und Öffentlichkeit. Springer Fachmedien, Wiesbaden, pp 67–83

Kroneberg C (2011) Die Erklärung sozialen Handelns. VS, Wiesbaden, 362 p

Lane SN (2014) Acting, predicting and intervening in a socio-hydrological world. Hydrol Earth Syst Sci 18(3):927–952

Lang DJ, Wiek A, Bergmann M, Stauffacher M, Martens P, Moll P, Swilling M, Thomas CJ (2012) Transdisciplinary research in sustainability science: practice, principles, and challenges. Sustain Sci 7:25–43

Lange H (2011) Umweltsoziologie in Deutschland und Europa. In: Groß M (ed) Handbuch Umweltsoziologie. VS, Wiesbaden, p 19

Lehtonen M (2004) The environmental-social interface of sustainable development: capabilities, social capital, institutions. Ecol Econ 49(2):199–214

Lottermoser F (2014) Der reflexive Konsument. Gesellschaftsinteresse im 21. Jahrhundert. Nomos, Baden-Baden, 270 p (forthcoming)

Lozano R, Lukman R, Lozano FJ, Huisingh D, Lambrechts W (2013) Declarations for sustainability in higher education: becoming better leaders, through addressing the university system. J Clean Prod 48:10–19

Maffi L (2005) Linguistic, cultural and biological diversity. Annu Rev Anthropol 34:599–617

Mankiw GN (2012) Principles of economics. South-Western, Cengage Learning, Mason, Ohio, 888 p

Maurer A, Schmid M (2010) Erklärende Soziologie. VS, Wiesbaden, 466 p

McMillin J, Dyball R (2009) Developing a whole-of-university approach to educating for sustainability linking curriculum, research and sustainable campus operations. J Educ Sustain Develop 3(1):55–64

Phillipon DJ (2012) Sustainability and the humanities: an extensive pleasure. Am Literary Hist 24 (1):163–179

Renn O (1996) Ökologisch denken—sozial handeln: Die Realisierbarkeit einer nachhaltigen Entwicklung und die Rolle der Kultur—und Sozialwissenschaften. In: Kastenholz H, Erdmann K-H, Wolff M (eds) Veröffentlichungen der Akademie für Technikfolgenabschätzung in Baden-Württemberg. Nachhaltige Entwicklung. Springer, Berlin, pp 79–117

Rockström J, Steffen W, Noone K, Persson Å, Chapin FS, Lambin E, Lenton TM, Scheffer M, Folke C, Schellnhuber H, Nykvist B, De Wit CA, Hughes T, van der Leeuw S, Rodhe H, Sörlin S, Snyder PK, Costanza R, Svedin U, Falkenmark M, Karlberg L, Corell RW, Fabry VJ, Hansen J, Walker B, Liverman D, Richardson K, Crutzen P, Foley J (2009) Planetary boundaries: exploring the safe operating space for humanity. Ecol Soc 14(2):1–33

Sathaye J, Najam A, Cocklin C, Heller T, Lecocq F, Llanes-Regueiro J, Pan J, Petschel-Held G, Rayner S, Robinson J, Schaeffer R, Sokona Y, Swart R, Winkler H (2007) Sustainable development and mitigation. In: Metz B, Davidson OR, Bosch PR, Dave R, Meyer LA (eds) Climate change 2007: mitigation. Contribution of Working Group III to the Fourth Assessment Report of the Intergovernmental Panel on Climate Change. Cambridge University Press, Cambridge, New York. https://www.ipcc.ch/pdf/assessment-report/ar4/wg3/ar4_wg3_full_report.pdf (last accessed 4/17/2014)

Schrauwen EJA, Bestebroer TM, Rimmelzwaan GF, Osterhaus ADME, Fouchier RAM, Herfst S (2013) Reassortment between avian H5N1 and human influenza viruses is mainly restricted to the matrix and neuraminidase gene segments. PLoS ONE 8:3

Stanford JN, Whaley J (2010) The sustainability of languages. Int J Environ Cultural Econ Soc Sustain 6(3):111–121

Stephens JC, Hernandez ME, Román M, Graham AC, Scholz RW (2008) Higher education as a change agent for sustainability in different cultures and contexts. Int J Sustain High Educ 9 (3):317–338

Stern N (2008) The economics of climate change. Am Econ Rev 98(2):1–37

Stokols D (2006) Toward a science of transdisciplinary action research. Am J Commun Psychol 38 (1–2):63–77

Theis T, Tomkin J (eds) (2012) Sustainability: a comprehensive foundation. Connexions. http://cnx.org/content/col11325/latest (last accessed 3/25/2014)

WCED (1987). Our common future. World commission on environment and development. Oxford University Press, Oxford, 400 p

Williams-Jones B, Olivier C, Smith E (2014) Governing 'dual-use' research in Canada: a policy review. Sci Public Policy 41:76–93

Wright TS (2002) Definitions and frameworks for environmental sustainability in higher education. High Educ Policy 15(2):105–120

Zhang Y, Zhang Q, Kong H, Jiang Y, Gao Y, Deng G, Shi J, Tian G, Liu L, Liu J et al (2013) H5N1 hybrid viruses bearing 2009/H1N1 virus genes transmit in guinea pigs by respiratory droplet. Science 340:1459–1463

Authors Biography

Biographical Note The authors are members of the *Postdoc-Kolleg Sustainable Future*, a post-doctoral research group at the *Centre for a Sustainable University* (*KNU*) at Universität Hamburg. The remit of the KNU is to define and develop the Universität Hamburg towards a University for a Sustainable Future and to explore sustainability in the areas of research, critical reflection upon research and teaching, new learning and teaching approaches and university governance (http://www.nachhaltige.uni-hamburg.de/en.html).

Astrid Ouahyb Sundsbø's research is located at the interface between social science and educational science theory and seeks towards a complex understanding of the (re-) production of educational and social inequality. She holds a Ph.D. in Social Science from the Humboldt University of Berlin.

Benjamin R.K. Runkle's research in environmental science examines the intersections of the carbon, water, and energy cycles on a landscape scale. He studies hydrological controls on CO_2 and CH_4 production in northern peatland and permafrost landscapes. He holds a Ph.D. from the University of California—Berkeley in Civil and Environmental Engineering.

B.R.K. Runkle is additionally supported by the Cluster of Excellence "CliSAP" (EXC177; Integrated Research Activity 08/2- 034) of the Universität Hamburg, as funded by the German Research Foundation (DFG).

Sarah McMonagle's research in language education and applied linguistics focuses on urban and virtual multilingualism. She is especially interested in minority languages in globalised contexts of big cities and the internet. She has a PhD in Language Policy and Planning from the University of Ulster.

Kerstin Jantke's research focuses on the interfaces and interdependencies of conservation biology, land economics and mitigation of climate change. Her study objects are ecosystems and protected areas in Europe. She holds a PhD in geosciences from Universität Hamburg and a Diploma in environmental sciences from Lüneburg University, Germany.

Florian Lottermoser's main interest lies in the sustainable transformation of the modern age and in the sociological analysis of renewable energy consumption from an action theory view. He holds a Ph.D. in Social Science from Universität Hamburg.

Manuel Gottschick's research is based on transition theory in which he integrates their Conflict-oriented Understand approach and his experiences on participatory research, reflexive governance and investigative transformative research to sustainable transformation. He holds a PhD in Technology Assessment from the Technical University Clausthal and an MA in Environmental Engineering.

Sönke Häseler's research focuses on market and non-market mechanisms to ensure the provision of flexibility to the electricity markets to accommodate increasing amounts of fluctuating renewable energy in the course of Germany's energy transition. He holds a PhD in Economics from Universität Hamburg and an MA in Financial Economics from St. Andrews University.

Juan Miguel Rodriguez Lopez' areas of research include financial market institutions, sustainability, and quantitative methods. He is currently researching cross-national comparisons of institutional factors, and market mechanisms for sustainability.

Marcus Scheele obtained a PhD in Chemistry from the Universität Hamburg and has been working in the field of material sciences and nanotechnology for many years. His current research interests focus on the development of new concepts to apply semiconductor nanoparticles for optoelectronic devices. An important aspect of his work is the size-property relationship of nanoscalic matter and an improved understanding of the impact of this novel material class on the environment. M. Scheele is currently at the Eberhard-Karls-Universität Tuebingen, Faculty of Science, Institute of Physical and Theoretical Chemistry.

Campus Ecological Footprint Versus the Global Influence of Sustainability Related Campus Activities: How Do They Compare?

Tim Lang and Christopher Kennedy

Abstract

Higher education institutions impact global environmental sustainability in two primary ways: through the ecological footprint of campus activities, and through the impacts of campus activities on global systems. Much attention has been given to the former in sustainability in higher education literature, while few attempts have been made to quantify the latter. The mechanisms by which higher education influences global sustainability are discussed, and the World3 integrated global model is modified to quantitatively explore their dynamics from the present day to the year 2100. Results show that the positive indirect influence of higher education activities on global sustainability vastly outweigh the direct resource consumption and pollution generation attributable to higher education institutions.

Keywords

Higher education · Sustainability · Modelling · Technological change · Social policy

T. Lang (✉) · C. Kennedy
Department of Civil Engineering, University of Toronto,
35 St. George St., Toronto, ON M5S 1A4, Canada
e-mail: tim.lang@utoronto.ca

C. Kennedy
e-mail: christopher.kennedy@utoronto.ca

W. Leal Filho et al. (eds.), *Integrating Sustainability Thinking in Science and Engineering Curricula*, World Sustainability Series, DOI 10.1007/978-3-319-09474-8_16

1 Introduction

While many higher education institutions recognize the importance of environmental stewardship through campus greening initiatives, their contributions towards addressing the world's sustainability challenges are also manifested in the students and knowledge they produce. Many institutions are making efforts to reduce their direct environmental impacts, assessed through measures such as the campus ecological footprint (Venetoulis 2001; Conway et al. 2008), but the role of higher education in sustainability extends beyond the eco-efficiency of university operations (Shriberg 2002; Cortese 2003). This is reflected in such assessment frameworks as the Association for the Advancement of Sustainability in Higher Education's (AASHE) Sustainability Tracking and Rating System (STARS), the predominant framework for sustainability in higher education in North America. While STARS does not entail a complete ecological footprint assessment, it does consider many measures of campus metabolism—e.g. energy and water use, waste generation and diversion—alongside metrics for sustainability related teaching and research, as well as student and community engagement, and governance.

While systems such as STARS do recognize that universities are involved in sustainability via mechanisms other than eco-efficiency, they fail to quantify them in terms that ultimately relate to global scale impacts. In this paper we develop a methodology for conducting such assessments in a way that is not recognized in current frameworks and literature. What we propose is a means of quantitatively assessing the influence of higher education on the eco-efficiency and environmental impacts of other anthropogenic systems—and ultimately the global system as a whole—through systems dynamics.

There is a class of systems dynamics models known as *integrated global models (IGMs)*, which seek to quantitatively model human-environment interactions at a global scale (Costanza et al. 2007). It is submitted here that to assess the role of higher education in global sustainability is to assess the influence of higher education on human-environment interactions at a global scale. Subsequently, augmenting an IGM to include the influence of higher education is a means of quantitatively assessing its role in global sustainability. Many of the models reviewed by Costanza et al. (2007) would perform satisfactorily in this function, but the remainder of this analysis was specifically conducted with World3, developed by Meadows et al. (1972) in the early 1970s for their seminal work, *The Limits to Growth.*

2 World3

World3 is a formal mathematical model of the world's major ecological and anthropological systems. Its model files and modelling platform, STELLA, are both publicly available and can be run on any personal computer. *Dynamics of Growth in a Finite World* (Meadows et al. 1974) provides a comprehensive description of

World3's development, including complete model code, schematic drawings, explication of model structure and parameter selection, and in depth discussion of incremental changes made for the each of the different model scenarios presented.

Meadows et al. (1974) designed World3 "on the assumption that mental models, rather than black-box computer models are and should continue to be the primary basis for social policy" (pp 24), and to make "conditional, imprecise projections of dynamic behaviour modes (pp 7)." They expressly state that World3 was not constructed for quantitative accuracy, and that "no assumption or parameter without real-world meaning should be added merely to improve mathematical convenience or historical fit (pp 24)."

Following from this modeling philosophy, we sought to augment World3 to include mechanisms that represent our mental models of how higher education affects various aspects of the global system. Parameter selections for these mechanisms were intended to reflect 'real-world meaning,' and we continued to interpreted model outputs as 'conditional, imprecise projections of dynamic behaviour modes.' We inspected the structure of World3, and identified modules that were logical connection points for higher education. This approach yielded three main mechanisms: technological change, policy, and culture.

3 Technological Change

The mechanism that might come to mind first for many is the role of higher education in technological change. In the case of human-environment interactions, one might think of Enrico Fermi's development of nuclear reactor technology, and the promise of "energy too cheap to meter," (speech by Lewis L. Strauss, September 16, 1954) or Norman Borlaug's revolutionary work in agricultural sciences, increasing land yield more than fourfold from the end of the second world war to the 1970s. Higher education has the potential to discover new technologies and improve existing ones, so that resources can be used more efficiently or substituted out, and deleterious environmental releases can be minimized, rendered more inert, or eliminated altogether.

There is a technological change module in each of World3's three main environmental sectors: non-renewable resources, persistent pollution, and agriculture. The structure of each module is the same, and at its heart, featuring a stock of technology, which could be thought of as technological potential defined by the state of the art. On the upstream side of this stock is the process of technology development—the advancement of the state of the art. On the downstream side is the process of technology adoption; the transition of state of the art technologies into commonly used technologies.

In World3, the process of technology development begins with a system performance multiplier, which is a function of a desired level of system performance compared to an actual level of system performance. As system performance becomes less desirable, more resources are directed towards technological development, and the multiplier accelerates the rate of change, within preset bounds —Meadows et al. (1974) capped this rate at 4 % per year in their scenarios.

The other key parameter of the technology development process is the policy year, which is the year in which technology development begins to take effect. The policy year is a simple binary function, which switches on technological development in the policy year. Technology development prior to the policy year is taken into account as an inherent component of the industrial development function.

In World3, technology adoption is modelled plainly as a third order exponential smoothing function. The governing (and sole) parameter of this function is an averaging time, reflecting how long it takes for technological advancements to become used after they have entered the stock of technological development. Meadows et al. (1974) use a value of 20 years for their scenarios. Contrary to the technology development, technology adoption does not respond to any other levels of system performance; averaging time is an exogenous parameter.

Higher education can be considered to have a (significant) role in both technology development and technology adoption. As alluded to at the beginning of this section, higher education is most typically thought of as influencing technology development. Basic scientific research and applied scientific research play a strong role in driving technological potential and advancing the state of the art. It should be noted however, that higher education is only one element of the technology development process, with private industry, other public research, and market factors playing significant roles as well.

What is less apparent, but equally true, is that higher education also plays a role in technology adoption. Only a fraction of post-secondary students and faculty produce original research; knowledge transfer is as much or more the aim of higher education as knowledge creation. The green shift away from a carbon intensive economy to one more based on renewable energy, for example, requires as much training of new technicians for new technologies as it does development of those new technologies.

Higher education also plays a role in both technology development and technology adoption through policy research and development. While basic or laboratory research are the activities that literally produce new technologies and technological advancements, the amount of basic and laboratory research that happens and how it is directed is influenced in part by policy research on the topic of technology development. Such research may concern itself with topics such as the allocation of research funds, financial support for graduate students, or the role of basic research in economic growth.

Policy arguably has a more significant role in technology adoption, through the development of policy tools to achieve goals that incidentally bring new technologies to market. Returning to the case of the transition to a lower carbon economy, it is widely acknowledged that even though further technology development is welcomed, the current problem is getting existing technologies into broad use. As mentioned earlier, training is one part of this, but the more active and contentious area of research is the development of the political case for policy action and the development of policy options, such as cap and trade systems, carbon taxes, feed-in tariffs, and other incentives and regulations for technology adoption.

4 Social Policy and Culture

Policy research need not be directed narrowly at technological development in order to improve system performance. The general analysis of the world system and identification of leverage points and strategies that promote desirable outcomes can be as constructive as technology research, or even more so. In fact, one of the main conclusions of Meadows et al. is that technological development alone will not produce stability in the world system.

In *Limits to Growth: The 30-year Update*, (Chap. 7) Meadows et al. (2004) describes a truly sustainable scenario, Scenario 9. This scenario is based on the implementation of population policies shifting the desired family size to two children and introducing perfect contraception, as well as policies that set a fixed goal for industrial output and extends the life of capital, in addition to the technological improvements of previous scenarios. It is noted that in the absence of social policies all technological advances do is facilitate overshoot and collapse at higher levels of population and consumption; changes to system objectives effect stability.

To return to the focus of this paper, it is not so much being suggested here that intense social engineering is the recommended path forward, but that when assessing the role of higher education in sustainability, it extends well beyond basic STEM research and technical training. Higher education activities surrounding social policy and dynamics can plausibly exert some influence, as can the simple act of education itself. Rates of not only higher education but education more broadly, especially in developing countries, can significantly influence birth control and family size (Shirahase 2000; UN 2011), two factors identified as crucial to long term stability by Meadows et al.

One final note on the role of higher education in global sustainability is that it plays an absolutely pivotal role in our understanding of the world system, generally speaking. Antecedent to the discussion of knowledge creation and transfer around technology and policy development is an understanding of how the world works and what problems are emerging. Higher education is playing an enormous role not just in addressing climate change, but knowing that climate change is even a phenomenon, and a phenomenon of concern. It may eventually show up a mathematical model like World3 as a policy year or climate sensitivity, but more importantly it leads to the creation of the model in the first place.

5 Quantitative Analysis

To quantitatively assess the influences of the above higher education mechanisms on global sustainability, the corresponding parameters in World3 were identified and modified. The most substantial modification to the model was the addition of a parameter called higher education intensity (HEI). HEI is a continuous exogenous parameter, manually entered, having a value between 0 and 1. A value of 1 represents the complete observed level of resource use and pollution generation by higher education, and also the full effect of the mechanisms outlined above. A value of zero removes all of the direct environmental impacts of higher education, but also removes its contributions to the mechanisms outline above. The equations for how parameters respond to a level of HEI between 0 and 1 are outline below.

5.1 Technological Change

As discussed above, there are three modules corresponding to technological change: one each for non-renewable resource use, persistent pollution, and land yield. Each module has a technology development process and a technology adoption process. The structure of these modules, describe above, is presented below in Fig. 1. The black blocks and their connectors are World3's original structure, as constructed by Meadows et al. and the grey blocks and connectors are those added to model the influence of higher education.

The technology development mechanism begins with system performance multiplier, which is a graphical function of actual level of system performance and desired level of system performance than ranges from 0 to 4 % improvement per year. Actual level of system performance is the state of an endogenous system parameter, such as non-renewable resource use rate, and desired level of system performance is an exogenous parameter representing the ideal level of the corresponding endogenous parameter.

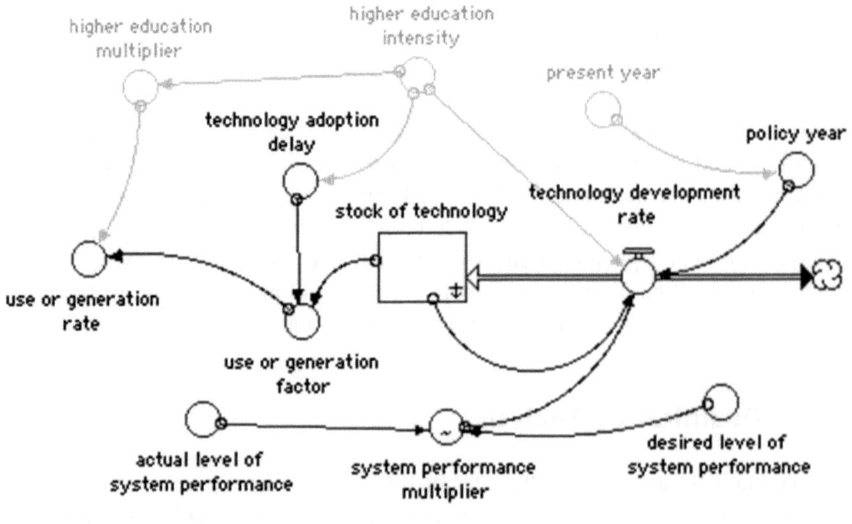

Fig. 1 World3 technological change module structure

Originally, the relationship was:

$$[\text{Technology development rate}] = [\text{stock of technology}] * [\text{system performance multiplier}]$$
$$\text{when } [\text{time}] > [\text{policy year}], \text{ and } 0 \text{ otherwise}$$

To model the influence of higher education, this was modified to,

$$[\text{Tech. dev. rate}] = [\text{stock of tech.}] * [\text{sys. perf. mult.}] * 0.5 * (1 + \text{HEI})$$
$$\text{when } [\text{time}] > [\text{policy year}], \text{ and } 0 \text{ otherwise},$$

The influence of higher education on technology development was modelled thusly to reflect the fact that some technology development would happen, even in the complete absence of higher education. As modelled, the technology development rate will cap out at 2 %, or half of its maximum with the full contribution of higher education. Aside from the magnitude of the technology development rate, when technology development begins is influenced by policy year, which is influence by higher education intensity, which will be described in more detail below with its role in social policies.

The technology adoption mechanism is driven by the technology adoption delay, which is the averaging time for the 3rd order exponential smoothing function in of the use or generation factor, which regulates the rate at which the stock of technology can reduce the use or generation rate; use for non-renewable resources, generation for persistent pollution and land yield.

Originally,

$$[\text{technology adoption delay}] = 20\,\text{years}$$

As with technology development, the influence of higher education is modelled as a deviation from its full effect, so this was modified to,

$$[\text{tech. adoption del.}] = 20 + 20 * (1 - \text{HEI}) \text{ years}$$

At its maximum value, the averaging time will remain at 20 years, and increase to 40 years in the complete absence of higher education.

The technological change module culminates with the use or generation rate, which is the rate at which the stock of non-renewable resources is depleted, at which persistent pollution is generated, and the rate at which a unit of land generates agricultural output. (These rates are not all the same, but they all occupy the use or generation rate position in their respective modules)

Originally,

$$[\text{use or generation rate}] = [\text{system parameter A}] * \ldots * [\text{system parameter N}] * [\text{use/gen. factor}]$$

where, [system parameter x] varies from resources to pollution to land yield

To include *the effect of throttling higher education*, this was modified to,

$$[\text{use/gen. rate}] = [\text{syst. param. A}] * \ldots * [\text{syst. param. N}] * [\text{use/gen. fact.}] * \text{HEI}$$
$$\text{where, } [\text{higher ed. multiplier}] = 0.985 + 0.015 * \text{HEI}$$
$$\text{when, } [\text{time}] > [\text{present day}]$$

for non-renewable resources and persistent pollution only.

The purpose of the higher education multiplier is to model reduced resource use and pollution generation *by higher education institutions* as a result of throttling higher education activity. When HEI = 1, there is no reduction in resource use and pollution generation, and when HEI = 0, there is a 1.5 % reduction. The model does not include any modification to land yield, because it is not assumed that the product of land yield, i.e. food, is directly attributable to higher education the same way that resource consumption and pollutant emissions are.

The value of 1.5 % as an estimation of direct resource use and pollution generation attributable to higher education was derived from two sources. First, using the 2002 US producer price EIO-LCA model (Weber et al. 2009), and 2002 data for total US higher education expenditures (Weber et al. 2009), total US GDP (World Bank 2014), and total GHG and SO_2 emissions (US EPA 2013), colleges, universities, and junior colleges sector were calculated to be responsible for 1.4 and 1.6 % of total US GHG and SO2 emissions, respectively in 2002. As an LCA, it is important to note that these percentages are not just the emissions released by higher education institutions, but fully attributable to them.

Second, using 2011 data from the Natural Resources Canada Office of Energy Efficiency (2014), it was determined that the education services sector consumed 1.6 % of the total secondary energy use in Canada. These data only represent the fraction directly consumed by the education services sector, not attributable upstream sources, but the education services sector includes primary and secondary education in addition to higher education. Furthermore, secondary energy use includes all end uses of energy at the consumer, including industrial consumers and energy for transportation, but excludes such things as line losses for power distribution and pipeline fuel.

Considering the above results, it was decided that 1.5 % was an acceptable estimate for the fraction of non-renewable resources and persistent pollution attributable to higher education. For the purposes here, a precise estimate is not required, nor even particularly feasible. First, in World3, non-renewable resources and persistent pollution are measured by the composite indicators of *resources units* (p 389) *index units* (p 423), respectively (Meadows et al. 1974). The process of reproducing these indicators for higher education would be extraordinarily laborious for higher education, if at all possible. Furthermore, the purpose of this analysis is to explore the interplay between the direct environmental impacts of higher education and its influence on the impacts of elements of the global system, not make precise predictions about the relative magnitudes of the two.

5.2 Social Policy and Culture

Contrary to technological change, where higher education influenced the rate of technology development, this analysis is not modifying World3 in such a way that higher education would affect the magnitude of policy goals relating to fertility control efficiency, desired family size, desired industrial output per capita, and the average lifetime of capital. However, as with technology adoption, this analysis modifies World3 to include the influence of higher education on the time of implementation for these policies. As mentioned above, the time at which technology development begins is treated similarly.

In *Limits to Growth: the 30-year Update*, Meadows et al. present scenarios for social policies being implemented in 2002 and 1982. The work was published in 2004, with the analysis being completed before publication, so the 2002 scenario can be interpreted as reflecting immediate policy action. Meanwhile, the 1982 scenario represents what would have happened if the world had acted 20 years earlier. In Meadows et al.'s version of World3, a number of exogenous parameters are changed manually to reflect the time of policy implementation, these parameters are:

[t_fert_cont_eff_time_45], time at which policy goal for fertility control is achieved
[t_zero_pop_grow_time_38], time at which policy goal for desired family size is achieved
[t_ind_equil_time_57], time at which desired industrial output per capita feedback is initiated
[t_policy_year_150], time at which technology development begins
[t_land_life_time], time at which erosion control policies are implemented
[t_fcaor_time], time at which non-renewable resource acquisition costs decrease.

As all of these time parameters are exogenous, and changed to the same value simultaneously by Meadows et al. to assess the influence of time of policy action, in this analysis, they were all set equal to [t_policy_year_150]—named [policy year] in our analysis. Policy year was then defined as below:

$$[\text{Policy year}] = [\text{present day}] + 20 * (1 - \text{HEI})$$

where [present day] is the time considered to be the present dayof the model.

Under this formulation, policy action is immediate in the presence of full higher education intensity, and linearly increases to 20 years in the future with no influence of higher education. In combination with the structure of the technological change modules outlined above, present day and higher education intensity are the only two exogenous parameters that need to be manually varied in order to explore the dynamics of all the mechanisms discussed in this analysis.

The only mechanisms omitted from the quantitative analysis described here are the broader cultural mechanisms described above, i.e. the role of higher education in general understanding of the world system, and the cultural influence of education as a whole on parameters such as desired family size aside from guided policy research. This analysis only considered the influence of higher education on mechanisms that were varied in Meadows et al.'s original scenarios, and only considered its influence as linear multipliers of those mechanisms.

This analysis did not even consider the influence of heightened levels of higher education; it used Meadows et al's Scenario 9 as the maximum level of policy action, and assessed model response to diminished levels of policy action as a function of diminished levels of higher education. The consideration of other mechanisms would require much more research into their structure and magnitude to even explore "conditional, imprecise projections of dynamic behaviour modes."

6 Results

Using the modified World3 model described above, a series of model runs were conducted for a variety of HEI, and 2014 as the present day. HEI = 1 corresponds to Meadows et al's Scenario 9, but with [policy year] = 2014 instead of 2002. The other seven levels of HEI run were, 0.99, 0.95, 0.9, 0.75, 0.5, 0.25, and 0. These values were chosen to see how the model would respond as HEI gradually left 1, and then larger steps were taken to see the model response all the way down to no contributions from higher education whatsoever. Both the direct effects of resource use and pollution generation by higher education, and the indirect effects of the higher education mechanisms discussed above respond to the varying of HEI. The time scale for the figures runs from the year 2010, four years before present day, to 2100, and the time horizon of the original World3 scenarios. One general observation to note, is that many of the figures do not appear to be reaching a steady state at 2100.

Ecological footprint—a function of arable land, urban land, and land required for pollution mitigation in World3—increases for all model runs before eventually declining to 2100. The higher the HEI, the less pronounced the increase before eventual decline, the sooner the peak, and lower overall ecological footprint at 2100. Figure 3 shows that persistent pollution exhibits similar trends. A noticeable difference between Figs. 2 and 3 is that ecological footprint peaks sooner than persistent pollution, which may be attributable to the use of persistent pollution generation rate in the ecological footprint calculation instead of the level of persistent pollution, a stock which takes time to dissipate.

Figures 2 and 3 show that higher HEI leads to better environmental performance of the world system when accounting for both direct environmental impacts of higher education, and its influence on other system elements. Note that the direct and indirect impacts of HEI have opposite signs, with the positive indirect effect of a higher HEI outweighing the negative direct effect.

Figure 4, however, shows the resource return on investment for choosing a higher level of HEI. For HEI = n, Fig. 4 is showing the resources that would be saved if HEI = 1 instead of n, divided by the immediate resources savings of only

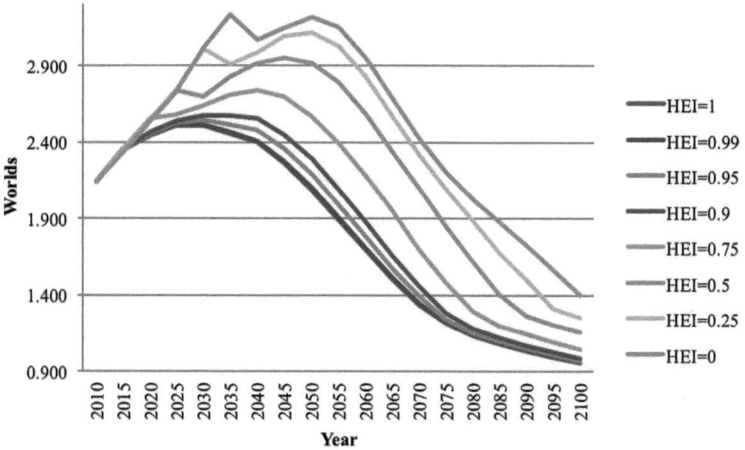

Fig. 2 Impacts of higher education intensity on global ecological footprint using adapted scenario 9 of the World3 model

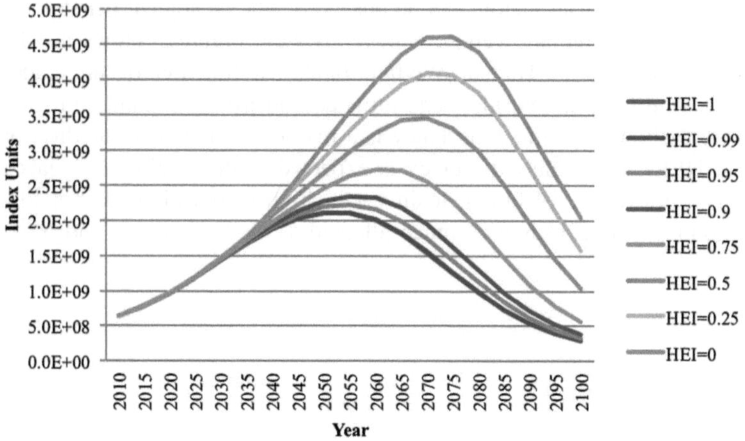

Fig. 3 Impacts of higher education intensity on persistent pollution using adapted scenario 9 of the World3 model

investing in HEI = n instead of 1. It is interesting to note that for HEI = 0.99 (almost 1) there is a completely different behaviour mode than for 0.75 < HEI < 0.95. For all HEI, there is a very brief period time horizon (1 year) where not investing in higher education is optimal, i.e. ROI < 0, but this quickly reverses and generally continues to increase as the time horizon is extended. The prevalence of focusing on quick ROIs in economic decision making, combined with this result may partly explain why we focus on quick operational resources savings instead of investing in the large longer turn returns of higher education.

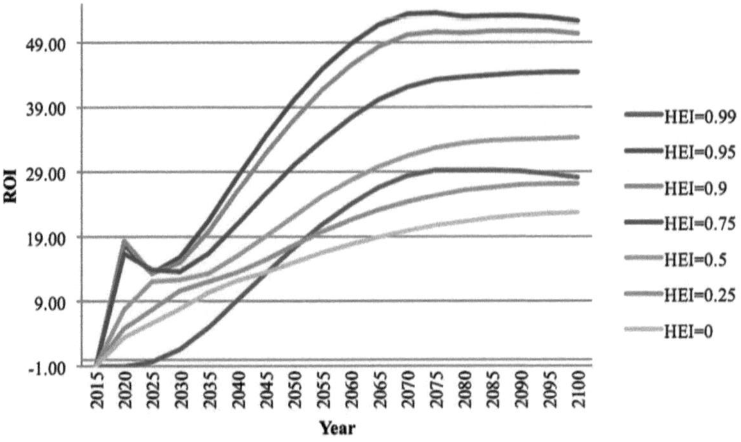

Fig. 4 Non-renewable resource return on investment in higher education

One of the many benefits of using a systems dynamics model is that the interplay of ecological and social factors can be observed. The ecological benefits of a high HEI correlate with human welfare benefits as well. Figure 5 shows that HEI = 1 minimizes the drop in Human Welfare Index—a function of life expectancy, education, and GDP in World3—and turns around from the minimum sooner.

Figure 6 shows that life expectancy is a major contributor to the drop in HWI. For 0 < HEI < 0.25, life expectancy is expected to drop below 35 years around 2065. The drawbacks of failing to invest in higher education are not limited to consumption; wellbeing is very much affected as well. Figures 5 and 6 also both

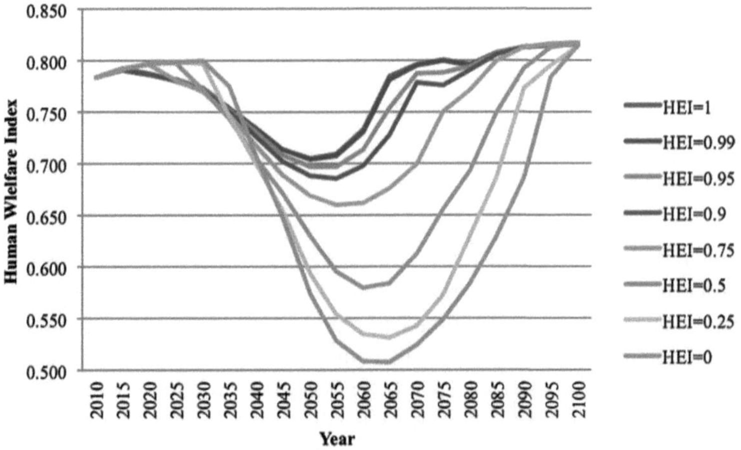

Fig. 5 Impacts of higher education intensity on human welfare index using adapted scenario 9 of the World3 model

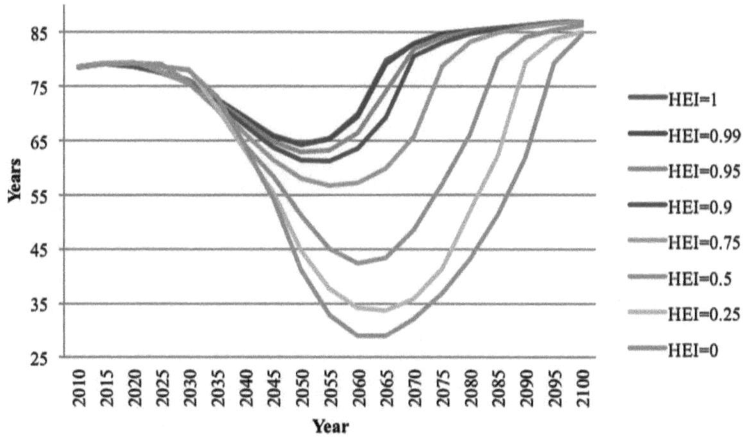

Fig. 6 Impacts of higher education intensity on life expectancy using adapted scenario 9 of the World3 model

demonstrate overshoot and collapse; HWI and life expectancy are both highest for the lowest levels of HEI until 2030, after which they plummet catastrophically. As with Fig. 4, restricted investment in higher education may seem attractive in the short run, but in the long run it has disastrous consequences.

Figures 5, 6, and 7 may seem to indicate that after a mid century rough patch, all levels of HEI will result in roughly the same outcome. Two other factors temper this conclusion. First, Fig. 8 shows that while things may appear fine on a per capita basis, the population in 2100 for HEI = 0 is only slight higher than one half the population in 2100 for HEI = 1. Furthermore, as mentioned earlier, very few metrics of system performance appear to be stable at 2100, and the gradients for the

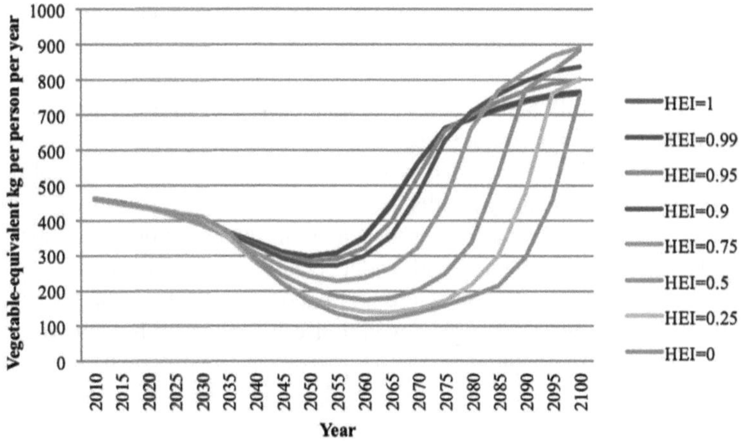

Fig. 7 Impacts of higher education intensity on food per capita

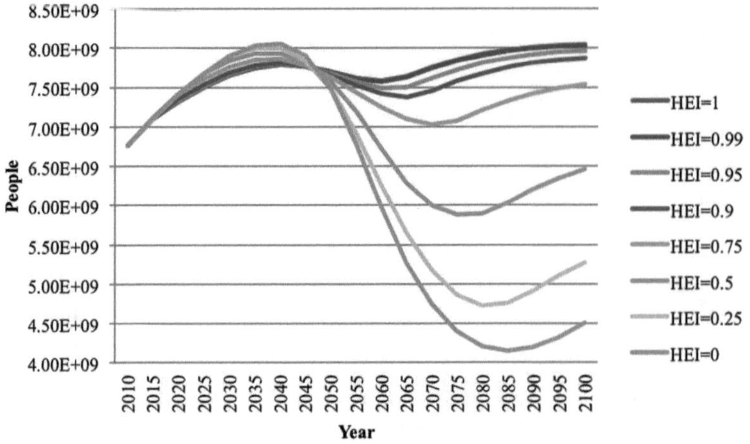

Fig. 8 Impacts of higher education intensity on population using adapted scenario 9 of the World 3 model

lower HEIs are unanimously higher than the gradients for the higher HEIs. This means that, magnitude of system performance metrics aside, higher HEIs lead to more stable system performance.

7 Conclusion and Discussion

By modifying the World3 integrated global model to include higher education, this analysis has been able to show that, as modelled, the positive influence of higher education activities on global sustainability vastly outweigh the direct resource consumption and pollution generation attributable to higher education institutions. This result suggests that reducing a campus' ecological footprint is only advisable if it will not sacrifice an institutions contribution to global sustainability via the mechanisms of technology change, social policy, and knowledge transfer to students. Maintaining full higher education intensity also appears to increase the stability of the global system, in addition to improving its long term sustainability.

While this analysis has provided proof of concept for conducting such a quantitative analysis and promising results, both the method and results are preliminary. The structure is justified and built around a sound mental model, but speculative and largely unconfirmed by empirical evidence all the same. The method should be further developed to include a structure and parameters than can be tested against real world data. The method's rigour would also be enhanced by applying it to an integrated assessment model more actively used in policy, such as IMAGE or International Futures.

Finally, whatever model it is applied to, there is the question of whether the influence of higher education on global sustainability is being looked upon too favourably. This analysis assumes higher education can only be a force for good; what if it's not? Research also generates new, harmful materials—CFCs, PCBs, DDT—and finds ways to exploit resources faster. Some policy research also influences faster economic expansion and resources utilization. Higher rates of education reduce desired family sizes and birth rates, but they might also promote more destructive Western ideals, such as unsustainable consumerism.

References

Conway T, Dalton C, Loo J, Benakoun L (2008) Developing ecological footprint scenarios on university campuses: a case study of the University of Toronto at Mississauga. Int J Sustain High Educ 9(1):4–20

Cortese A (2003) The critical role of higher education in creating a sustainable future. Plan High Educ 31(3):15–22

Costanza R, Leemans R, Boumans R, Gaddis E (2007) Integrated global models. In: Constanza R, Graumlich L, Steffen W (eds) Sustainability or collapse: an integrated history and future of people on earth. MIT Press, Cambridge

Meadows DH, Meadows DL, Randers J, Behrens W (1972) The limits to growth. Universe Books, New York

Meadows DL, Behrens W, Meadows DH, Naill R, Randers J, Zahn E (1974) Dynamics of growth in a finite world. Wright-Allen Press, Cambridge

Meadows DH, Randers J, Meadows DL (2004) Limits to growth: the 30-year update. Chelsea Green Publishing Company, White River Junction

Natural Resources Canada Office of Energy Efficiency (2014) Comprehensive energy use database, 1990–2011. http://oee.nrcan.gc.ca/corporate/statistics/neud/dpa/comprehensive_tables/list.cfm?attr=0. Last accessed 30 Mar 2014

Shirahase S (2000) Women's increased higher education and the declining fertility rate in Japan. Rev Popul Soc Policy 9:47–63

Shriberg M (2002) Institutional assessment tools for sustainability in higher education: strengths, weaknesses, and implications for practice and theory. High Educ Policy 15:153–167

Strauss L (1954) Speech to the national association of science writers. New York (September 16)

United Nations Economic and Social Council (2011) Speakers link higher education among girls to declining fertility rates as commission on population and development continues session. Press release. 44th session of the Commission on Population and Development. http://www.un.org/News/Press/docs/2011/pop994.doc.htm. Last accessed on 30 Mar 2014

United States Environmental Protection Agency (2013) Inventory of US Greenhouse Gas Emissions and Sinks: 1990–2011 Government publication. http://www.epa.gov/climatechange/Downloads/ghgemissions/US-GHG-Inventory-2013-Main-Text.pdf. Last accessed 26 Mar 2014

Venetoulis J (2001) Assessing the ecological impact of a university: the ecological footprint for the University of Redlands. Int J Sustain High Educ 2(2):180–196

Weber C, Matthews D, Venkatesh A, Costello C, Matthews H (2009) The 2002 US benchmark version of the economic input-ouput life cycle assessment (EIO-LCA) model. http://www.eiolca.net/docs/full-document-2002-042310.pdf. Last accessed on 30 Mar 2014

World Bank (2014) Data: GDP (current US$). Web page. http://data.worldbank.org/indicator/NY.GDP.MKTP.CD?page=2 Last accessed on 30 Mar 2014

Authors Biographies

Tim Lang is a graduate student in the Department of Civil Engineering at the University of Toronto, where he is currently completing his doctoral work on interpreting and quantifying the role of higher education in global sustainability through systems dynamics. He has also been managing the Sustainability Office at University of Toronto Scarborough since 2007. Tim was one of the founding organizers of the Ontario College & University Sustainability Professionals, and is currently their co-chair. Through this organization, he is working with the Council of Ontario Universities to revise provincial sustainability reporting for universities. Tim has recently presented research at AASHE's annual conference in Nashville, and has previously spent time as a Fulbright Scholar at Politechnika Krakowska to research Poland's environmental management infrastructure.

Christopher Kennedy is a Professor in the Department of Civil Engineering at the University of Toronto, where he teaches courses in Infrastructure Economics, Engineering Ecology, and the Design of Infrastructure for Sustainable Cities. He has recently returned from secondment to the OECD in Paris, where he worked on Cities, Green Growth and Policies for Encouraging Investment in Low Carbon Infrastructure. He has also served as a consultant or advisor to the Ontario Ministry of Finance, Infrastructure Canada, Clinton Climate Initiative, C40, US National Science Foundation, UNEP, UN-HABITAT and the World Bank. Amongst Chris's publications are studies of urban metabolism, greenhouse gas emissions from global cities and processes for developing sustainable urban transportation systems. His wider work includes contributions to probability theory, regional economics and engineering education. He has been a visiting scholar at Oxford University, ETH Zürich and UFZ Leipzig. His book The Evolution of Great World Cities: Urban Wealth and Economic Growth was published by University of Toronto Press in August 2011.

Developing Action Strategies for Sustainable Living Amongst Employees

Alexandra Mifsud

Abstract

Employees are frequently not considered as a specific target group in education for sustainable development campaigns. Adults at their place of work are the decision makers and problem solvers of the present generation. It is these decisions and actions that will impact the quality of life of present and future generations as well as the sustainability of our planet. This paper presents the work in progress of research taking place with members of staff at a University in the UK. It aims to provide a valid contribution to education for sustainable development in organisations by exploring routes to a needs-based training programme for employees that would bring about (i) behavioural change; (ii) improved sustainability performance and (iii) financial savings across the organisation. The paper will put forward an example of the 'how' element, or rather the process required by practitioners to design education for sustainable training programmes that are relevant to different audiences. The emphasis in the process will therefore need to be its adaptability in that it possesses elements that would render it possible to be used for employee training in other work contexts and flexibility to make it work and fit around the day to day demands and challenges of any given organisation. Finally the paper will demonstrate that such action strategies need to be attractive and user-friendly for practitioners to believe that they can adopt the model and make it their own (or rather make it that of their target groups).

A. Mifsud (✉)
Centre for Environmental Education and Research, University of Malta,
Msida MSD 2080, Malta
e-mail: alexandra.mifsud@um.edu.mt

W. Leal Filho et al. (eds.), *Integrating Sustainability Thinking in Science and Engineering Curricula*, World Sustainability Series, DOI 10.1007/978-3-319-09474-8_17

Keywords
Education · Sustainable development · Employee/staff training · Thinking skills

1 Introduction

The key questions at the outset of this research study were: (i) how can the use of de Bono thinking skills lead to employee participation during the course design in a manner that facilitates their own training programme?; (ii) how can the application of thinking skills to the design of employee programmes in education for sustainable development bring about employee empowerment in order to induce positive behavioural changes necessary in achieving an improved quality of life?; and (iii) what form should employee programmes in education for sustainable development take if it is to assist employees lead a sustainable lifestyle? Herman B. "Dutch" Leonard in Epstein (2008), refers to two forms of corporate social responsibility programmes. Those that talk a lot but "don't actually do very much or generate much impact" and those that are engaged in "socially responsible activities being carried out on a material scale and significant results are actually being achieved". Like Leonard, it appears to the author that there is "still far too much of the former and not nearly enough of the latter" (ibid.). This ongoing research study explores the level of contribution education for sustainable development (ESD) training programmes for employees within an organisation could possibly have to create a shift for more corporations forming part of the latter of the two described above. The sustainable development factors that organisations are expected to take into account when drafting their corporate plans are many. Whether these factors are regarded as constraints or challenges depends on the adaptability, flexibility and commitment of the organisation and its work force.

1.1 Education for Sustainable Development Within Organisations

Businesses and organisations operate on models that, at best, acknowledge the need to improve their environmental performance as part of their compliance to regulatory and legal measures. However, the spirit of sustainable development in the Brundtland report suggests that they ought to be more "flexible and responsive, more diverse and devolved organisational forms than those dominant at the present stage in industrial society." Roome and Oates, in Huckle and Sterling (1996). Hence there is requirement for a learning framework that is (a) able to be sensitive to the ever changing needs of society and the environment and; (b) equipped with the skills to respond appropriately to these needs. Here is the challenge for education in business management and employee training and development. All members within an organisation must share the vision for a sustainable future alongside working on improving environmental performance in the present.

1.2 Thinking Skills

This study includes experimental work on the implementation of de Bono thinking programmes both in its methodology as well as in the process that may be adopted in the design of ESD programmes for employees. The author is of the opinion that there is significant relevance between thinking and sustainable development because as Morris et al. (2009), state "Our contention is that learners cannot deal with the wicked problems of sustainability without learning to think and act systematically." However, whilst there is ample discourse on systems thinking and ESD, there appears to be none about the specific implementation of the worldwide acclaimed thinking programmes by de Bono in the field of ESD. It is for this reason that this study investigates the concept further. The emphasis on thinking processes has in fact been highlighted by Cloud when discussing the UN Decade for Sustainable Development (2005–2014) in Cloud et al. (2009), whereby he suggests that "one of the greatest opportunities that EFS can offer to EE that will strengthen its capacity over the next 10 years is the contribution of the tools, concepts, archetypes, and "habits of mind" of systems thinking and system dynamics education—a core content area of EFS."

Hence, the overall aim of this research study is to assist employees acquire cognitive and practical skills that will help them to improve their quality of life in relation to sustainable development. The subsidiary aims are: (i) to provide the author with an opportunity to participate in dialogue with the research participants that would bring to the fore their expectations and needs regarding their participation in a course in ESD; (ii) to assist research participants move from dialogue to action by devising a formula for a course design that would serve as the basis for the process involved in the development of employee training on ESD; and (iii) to improve the author's own skills as a change agent through the experience of engaging in an interactive and dialogical research.

2 Employee Training and Sustainable Development at the Workplace: Constraints and Opportunities

This paper discusses the enquiry currently underway in order to explore whether it would be beneficial for organisations to invest in training and development of their own employees on matters relating to sustainable development and if so, what would the best approach be.

2.1 Legal Obligations and Reporting

Most organisations use performance based tools such as the ISO 14,000 series and the EU Ecolabel which are based on environmental management principles. Improving the environmental performance of an organisation does not necessarily bring about an increase in the level of environmental literacy of its employees.

Neither does it empower the employees to become proactive in working towards sustainable living. There is little or no room for social transformation. Furthermore, organisations are bound by national and international environmental legislation. For example, the European environmental policy (Article 174 of the Treaty on European Union and of the Treaty establishing the European Community, 2006 C321 E/9) provides the basis for a legal framework for

- "preserving, protecting and improving the quality of the environment,
- protecting human health,
- prudent and rational utilisation of natural resources,
- promoting measures at international level to deal with regional or worldwide environmental problems."

One of the instruments within this general framework is a piece of specific legislation on sustainable development with a first draft document in May 2001 'A Sustainable Europe for a Better World: A European Union Strategy for Sustainable Development' (Commission proposal to the Gothenburg European Council) (COM (2001) 264 final—not published in the Official Journal). In December 2005 the European Commission presented the review of the Sustainable Development Strategy—A platform for action [COM (2005) 658 final—not published in the Official Journal] (http://europa.eu/legislation_summaries/environment/sustainable_development/l28117_en.htm, accessed on 27/03/2014).

Environmental legislation may be regarded as a minefield for organisations and many of them have addressed the matter by outsourcing the task for legal compliance requirements to specialist service providers or have engaged staff in house. Whilst legislation is an essential tool that assists countries on their journey towards meeting sustainable development goals, it is often implemented using a top-down approach in organisational structures with staff merely being instructed to change their workflows or work practices with very little environmental knowledge being passed on in the process. The author is of the opinion that this is potentially seen as something that employees have to do because they have been ordered to do it. If that is the case, then whilst legislation is helping organisations improve their sustainable development performance, little is being achieved in raising the environmental literacy of the employees or creating a true institutional change. There appears to be a gap in creating a sense of ownership and relevance of initiatives taking place within organisations that would sufficiently bring about behavioural and institutional change leading to social transformation.

2.2 Maintaining Financial Sustainability

Compliance to environmental legislation poses potential financial challenges to organisations as many fear significant capital outlay or productivity losses. Addressing financial challenges and opportunities arising from working towards sustainable development is key. The conventional argument that environmental protection comes as an extra financial and productivity burden for companies has been challenged in recent years. Indeed, some analysts argue that "improving a company's environmental performance can be associated with better economic performance for a variety of reasons." Ambec and Lanoie in Jackson et al. (2012). These reasons include:

1. Environmental products may allow a firm to reach new customers;
2. Environmental innovation may reduce operational costs;
3. Adopting a sustainable development plan may stimulate a change towards increased performance in the firm;
4. Improved sustainability performance increases the likelihood of accessing financial capital and grants;
5. Better sustainability performance may improve a firm's access to human capital.

(adapted from Ambec and Lanoie in Jackson et al.2012)

It is case studies of successful implementation of sustainability initiatives as documented in Jackson in Jackson et al. (2012) that could offer the confidence in other industries to shift their organisation towards sustainability. Investing in staff development programmes on sustainable development may well be instrumental in attaining success.

2.3 Social Responsibility of Organisations

Businesses are becoming increasingly aware of their impact on the environment and society leading them to take responsibility of managing sustainability. This is sometimes referred to as the 'triple bottom line' or 'people, planet, and profits'. Trends that provide a good impetus for businesses to shift their mindset towards the 'triple bottom line' include green consumerism, socially responsible investing and employees wishing to work for green companies. Whilst these trends do exist they are still not strong enough to encourage some sectors within industry to make the change. However, they are trends that are likely to become important in the future as the level of society's awareness and knowledge increases. One may argue that the key to creating a shift towards sustainability amongst organisations is to embark on education campaigns aimed at raising the awareness and knowledge of citizens. Simply put, if citizens' environmental literacy increases then they will strengthen the green trends thus providing industry the confidence to change their own sustainability performance.

2.4 Education for Sustainable Development for Employees

Literature reviewed in this study highlights that significant research has been carried out on ESD in the formal education sector. Fien (1995, 1996, 2002), Huckle and Sterling (1996), Haigh (2009), and Blewitt and Cullingford (2004), amongst others. Not only is research in the informal and non-formal educational settings lacking, similarly so are ESD programmes not taking place sufficiently in these settings. They could serve as an ideal platform to target key players in the attainment of sustainable development. These include adults at their place of work where they are the decision makers and problem solvers of the present generation. It is these decisions and actions that will impact both the quality of life of present and future generations as well as the sustainability of our planet. The fact that employees are present in an organised structure at their place of work, providing ESD programmes in a bottom-up approach would not pose insurmountable difficulties (or one would think!). Nonetheless, the main challenge in the provision of employee training remains a logistical one: managing employees' time for training in a manner that would not disrupt the work output of the organisation. This study looks at the barriers that might exist which would prevent an organisation from engaging with their work force in offering training programmes in ESD. Moreover, it is hoped that based on the findings, a model for the manner in which employee empowerment can effectively take place through the design and implementation of ESD programmes can be presented. This model would include sufficient room for flexibility and adaptability so that it would serve as a framework for other organisations to use with their own employees in their endeavours to assist in attaining a sustainable lifestyle for all. The focus will therefore be on the *process* and not the *product* since as rightly put in the UNESCO Concept statement, 2004 as cited by Higgitt (2009):

> There is no universal model of education for sustainable development. While there will be overall agreement on the concept, there will be nuanced differences according to local contexts, priorities and approaches....[these] must therefore be locally defined to meet the local environmental, social and economic conditions in culturally appropriate ways.

The author believes that this central characteristic of ESD is paradoxically its strength and its weakness. The fact that no set model has been put forward in any prescriptive manner has left many practitioners in the field and the general public grappling with the 'how' element of ESD. Yet, the absence of a pre-set universal model is what enables practitioners to become intrinsically involved with their respective target groups making ESD relevant to them. This study aims to put forward an example of the process required by practitioners when designing ESD training programmes relevant for employees. The emphasis in the process therefore needs to be its adaptability and flexibility. It needs to be attractive and user-friendly for practitioners to feel confident in adopting the model and making it their own (or rather that of their target groups).

3 Methodology and Study Design

3.1 Research Paradigm and Methodology

It was determined that this research best fits within the ontological position of constructionism as it "asserts that social phenomena and their meanings are continually being accomplished by social factors…in constant state of revision."

Bryman (2004). As a social researcher, the author is aware that personal values and beliefs can "intrude at any or all of a number of points in the process of social research: choice of research area; formulation of research question; choice of method; formulation of research design and data collection techniques…" ibid. (2004). In fact the author's own experiences of working at grass root level with environmental NGOs as well as the involvement in de Bono thinking programmes bore a strong influence on the choice of the research both in the content and methodology adopted. Furthermore, social constructionist grounded theory provides a framework for researchers to be reflexive about how "their prior interpretative frames, biographies and interests, as well as the research context, their relationship with research participants, concrete field experiences, and modes of generating and recording empirical materials" Charmaz (2006), influence their analysis. The author is influenced by the paradigm of critical educational research since "its intentions are transformative: to transform society and individuals to social democracy" Cohen et al. (2007). Upon examining the definition for action research by McNiff and Whitehead (2010), it became clear that the aims of this study would best be met through educational improvement particularly the nature of processes of improvement where the researcher has "the capacity to influence the future through acting intentionally in the now….[the researcher has] the responsibility of ensuring that we act well, to create the kind of future we wish to live in." Through the application of de Bono's thinking tools the author attempts to devise a process for the design of educational programmes amongst employees that would bring about employee empowerment. This is being done through action research whereby the author explores the process for social and institutional transformation within the organisation because "the community and researcher together produce critical knowledge aimed at social transformation" and "theory and practice are integrated", Anderson and Herr (2005).

3.2 Research Context and Design

The study is with a set of employees at DeMontfort University (DMU) in Leicester, UK. Despite having the full support of its Environmental and Sustainability Officer who introduced the study to DMU's Environmental Champions Network, the response has been limited. To date it is only employees holding a post that allows them to potentially manage their time in a flexible manner have come forward to be interviewed. This highlights the intrinsic problem that exists in institutions and organisations: accessing employees especially those within the lower ranking posts. It could also suggest that ESD and sustainability in general is not met with much enthusiasm by individuals in a work place setting even when the organisation, as is the case with DMU boasts of becoming "a leader in the field" in its sustainability strategy https://www.dmu.ac.uk/about-dmu/sustainability/sustainability-strategy.aspx (accessed on 19/03/2014).

3.3 Triangulation and Sampling

For the purposes of this study *data triangulation* was adopted since it uses a variety of data sources such as "interviewing people in different status positions or with different points of view", Patton (2002). Indeed data is being collected through interviews and casual conversations with employees occupying different roles and from various departments at DMU to obtain a multi-faceted perspective on the area being investigated. *Methodological triangulation* using multiple data collection methods to study a single problem, namely documents, interviews, casual conversations and observations has also been adopted. Since accessing research participants for data collection was a major difficulty it was decided to opt for *opportunistic sampling* which "takes advantage of opportunities that open up", Miles and Huberman (1994). This was supported by the *snowball sampling* technique whereby other research participants were contacted through suggestions made by the first interviewees.

3.4 Interviews

A number of the interview questions were designed by using de Bono's thinking tools. The Direct Attention Thinking Tools recently renamed Power of Perception™ Tools were selected as being most appropriate for the design and analysis of the interview questions since the tools enable both the employees and the author to have a broad and inclusive viewpoint; assist in creating a framework for defining a situation with an improved ability to consider consequences before any action is taken, de Bono Consulting, 2009. The interview questions fed into the five main themes of the study as can be seen in Table 1.

Furthermore, the data collected falls under the categories: (i) level of awareness and understanding of sustainable development issues; (ii) relevance between the context of the participants' workplace and sustainable development; (iii) needs of the workplace as a commercial entity; and (iv) individual needs of participants in terms of thinking processes that would lead to empowerment and sustainable living.

4 Preliminary Results

The members of staff at DMU that have participated in the data collection for this study are not representative of the organisational structure of DMU. This is because it has proven challenging to find members of staff occupying lower level posts able to participate in the research. Nonetheless it is hoped that with some further perseverance the study will obtain the views of staff within such categories. That said this observation raises concerns about accessibility to all members of staff. If it is so difficult for the organisation to release them for a thirty minute research interview, it puts into doubt the organisation's commitment to engage with lower level members of staff and offer training programmes on thinking skills and ESD to them.

Table 1 Themes

Themes of the study	
Type of organisation	Commitment to ESD
	Staff development programmes
	Staff initiatives
Characteristics of ESD	The strands in ESD
	Constraints and opportunities
	• Legislation
	• Financial, social, environmental considerations
DeBono's thinking skills programmes	Thinking skills
	deBono's thinking programmes
	Applicability of deBono's thinking programmes to
	• The design of employee training
	• ESD
	• A sustainable lifestyle
Employee training	Staff development programmes for SD
	Skills for sustainable lifestyle
	Design process of needs based employee training
Behavioural change, social transformation, institutional change	Empowerment skills
	The role of a change agent
	Social transformation and institutional change
	• Employees
	• Transfer of behaviour changes beyond the workplace

4.1 Commitment to ESD; Staff Development Programmes; and Staff Initiatives

The responses from the respondents so far indicate that they are aware of DMU's commitment to sustainable development with some referring to the university's sustainability strategy. However, apart from the work done at the Institute of

Energy and Sustainable Development, it appears that the respondents are not aware of or see little need for DMU's education and learning role in sustainable development. None of the respondents were offered, and therefore have not participated in any ESD training by DMU. The initiatives predominantly led by the Environmental and Sustainability Officer at DMU such as the Green Impact and Smart Spaces have good visibility as all respondents know that they exist with some being heavily involved in them.

4.2 Strands in ESD; Constraints and Opportunities; Financial, Social and Environmental Considerations

In order to explore the process ESD practitioners may need to follow when designing staff training programmes, it was necessary to obtain DMU staff views on what they perceive as key factors to be considered together with any constraints (Tables 2, 3).

On the other hand, the respondents were of the opinion that by introducing ESD DMU would benefit from several opportunities (Table 4).

Table 2 Key factors to be considered before DMU commits to ESD

Clear aims on what is to be achieved
Understanding and commitment from the senior management
An assessment of the existing knowledge, skills and attitudes of staff
An assessment of the process, costs and resources required for delivery of ESD training
High level of staff engagement and motivation

Table 3 Constraints

Resistance from members of staff
Lack of commitment by departments and faculties across DMU
Time and financial limitations
Difficulty in selling the concept that ESD ought to be a priority for all across the organisation

Table 4 Opportunities

Giving DMU a better public profile
Contributing to creating a more sustainable society
Improving its position in the university league tables
Increasing awareness on sustainable development issues
Achieving a more efficient and healthier work environment for DMU staff and students

4.3 Applicability of DeBono's Thinking Programmes to the Design of Employee Training; and to a Sustainable Lifestyle

If ESD training programmes are to be relevant to employees then the latter need to be consulted upon as part of a needs assessment exercise. Informal conversations between the researcher and DMU members of staff suggest that employees do not always feel sufficiently empowered to recognise their role in this stage of their staff development training. Neither do they feel equipped with the competencies to make the necessary behavioural changes towards a more sustainable lifestyle. This study has therefore carried out some exploratory work on the applicability of thinking programmes to ESD for employees. Respondents gave mixed responses about training in thinking skills or other 'soft skills' with most stating that they do not know of and have not been offered any such training at DMU. The respondents are of the view that equipping employees with 'soft skills' has the potential to assisting them to lead a more sustainable lifestyle with one respondent suggesting that this is because the 'soft skills' lead to more responsible decisions. Whilst another believes that it would help staff to carry out their role at DMU better and more in line with sustainable development goals of the organisation. Most were able to see the relevance between 'soft skills' training, employee engagement and empowerment yet one respondent believes that DMU staff are already equipped with the skills to design training.

4.4 ESD for Employees; Skills Needed for Sustainable Lifestyle; and the Design Process of Needs-Based ESD Employee Training

It was necessary to obtain the views of DMU staff on the process they perceive as key when embarking on ESD for employees. Interestingly not all respondents are of the opinion that ESD training would involve employees across the entire organisation. Respondents believe that the senior management team, Human Resources (HR) department, Institute for Energy and Sustainable Development (IESD), Estates department and the Environmental Champions Network would play a major role. The respondents did comment that some of the people involved may offer a degree of resistance and/or feel aggravated, whilst others may receive the initiative positively or simply view it as their role (particularly the HR and IESD teams). Upon being asked to identify the factors that should be kept in mind when designing ESD for DMU staff, the respondents listed (Table 5).

The findings highlight that DMU staff consider the following aims should drive ESD programmes for employees across DMU (Table 6).

The respondents identified that for them to lead a more sustainable lifestyle they would benefit from (i) stronger ability to make and maintain behavioural changes by exploring alternative ways of doing things; (ii) improved thinking; (iii) more

Table 5 Key design factors for ESD programmes for employees

Ensuring that the training leads to tangible changes in behaviour
Relevance to the role members of staff play within the organisation
Time, funds and expertise available for ESD initiatives

Table 6 Aims of ESD programmes at DMU

Assessing existing knowledge of DMU staff and building upon it
Assisting staff in identifying the need to lead a sustainable lifestyle
Establishing and maintaining employee motivation and engagement
Creating an impact
Obtaining support from senior management
Ensuring the training time is kept to a minimum and is enjoyable

support both in knowledge, funds and time; and (iv) better skills on how to engage effectively with other colleagues in order to communicate sustainable development issues.

4.5 Social Transformation and Institutional Change; Transfer of Behaviour Changes Beyond the Workplace

Since this study explores ways of how ESD supported by thinking skills can potentially bring about behavioural and institutional changes necessary for the attainment of sustainable development, it was important to seek the opinion of DMU staff on the skills that would assist in this process. Awareness; knowledge and understanding; motivation; change of outlook and being given the authority to make the changes were identified as being key skills that staff should acquire in order for them to become sufficiently engaged in a process of change to sustainable development. The study also found that providing incentives and allocating time would increase staff motivation and engagement. It was encouraging to note respondents believe that having staff involved in the design of their own ESD programmes would have a positive impact as it would create an effective process for behavioural and institutional change (Fig. 1).

The study revealed that for the changes in behaviour for sustainable development to be extended beyond the workplace setting it is necessary that the ESD programmes for staff infuse ethical responsibility and commitment by assisting them to internalise the principles of sustainable development. It was suggested that this could be done by exploring the differences between the work and home setting and through examples from outside the work environment as tasks or case studies forming part of the staff training module. Sharing the organisation's vision on sustainable development was another key element that would assist in the process of institutional change and social transformation.

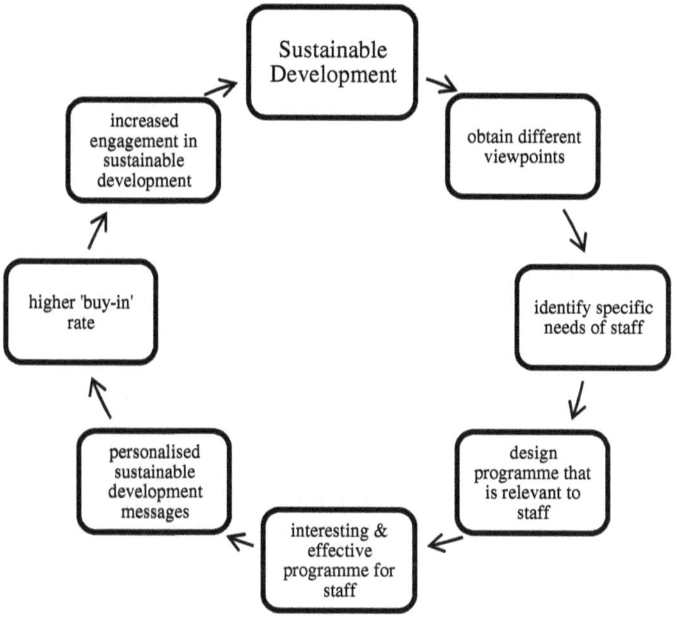

Fig. 1 Suggested process offering adaptability and flexibility in the design of ESD employee programmes

5 Conclusions

The results obtained point towards sufficient scope to integrate thinking skills in the process of designing ESD programmes for employees. There is strong evidence that a needs-based approach is regarded as bringing about higher success rate both in attaining sustainable development goals for the organisation and instigating behavioural and institutional change. Embarking on a process for the design of ESD programmes for employees does indeed present constraints and challenges. The study reveals that lack of commitment from senior management, resistance to change by employees coupled with financial and time constraints are the major challenges that could hinder ESD programmes for employees. Yet organisations may already have a network such as the Environmental Champions Network at DMU which could be used to infuse the needs-based approach in order to seek viewpoints of employees to tailor-make the ESD programmes. This is not far from the Sustainability Strategy at DMU:

> *Teaching - Inspiring Sustainability*: To help staff and students become responsible 'global' citizens in the face of the environmental challenges ahead of them.
>
> *Research - Thinking Sustainability*: To develop ground-breaking and interdisciplinary research that advances knowledge on sustainability and establishes DMU as a leader in this sector.

Built Environment - Being Sustainability: To reduce our environmental impact and be an exemplar of best practice.

Health and Well-Being - Living Sustainability: To ensure DMU is a healthy, creative and inspiring 'space' to study and work.

Community Engagement - Leading Sustainability: To establish DMU as a beacon of best practice, a centre of research excellence for sustainability and an ambassador locally, regionally and beyond." Adapted from

https://www.dmu.ac.uk/about-dmu/sustainability/sustainability-strategy.aspx. (accessed on 15/03/2014).

Unfortunately, the preliminary results in this study reveal some of the above aims are not being met. This appears to be particularly the case when equipping staff with the necessary skills, competencies and motivation to make behavioural changes for a more sustainable lifestyle. From the various informal conversations held at DMU it emerged that much focus is placed on reducing the organisation's environmental impact through rigorous monitoring of energy use and carbon footprint. This is commendable and ought to be given further support and recognition. Yet there is clearly a lack of vision on ESD for employees. If the organisation is to create an impact on sustainable development then it requires a workforce that is driven by ethical responsibility towards the planet and its people. The study indicates that investment in employee training would be one of the leading factors contributing to an effective social transformation process.

5.1 Limitations of This Study

This paper forms part of a doctoral research being undertaken by the author. It is therefore limited in scope as not all the data from interviews and casual conversations with employees has been collated and analysed. Whilst significant content analysis of literature has been reviewed, the need to review further literature as the doctoral research develops may arise at a later stage after the writing of this paper. Nonetheless, from the content analysis and data collected to date, it is encouraging to see that the niche initially identified by the author at the outset of the research does indeed exist. It is evident from the work carried out thus far that there is a gap in work on ESD through the application of thinking skills amongst employees. The author is of the opinion that on full completion of the doctoral research, there will still be potential areas for future research work in the area. These are highlighted in the following section.

5.2 Future Work

This paper is based on research work currently taking place with a small set of employees at DeMontfort University in Leicester, UK. It is therefore not representative of employees as a whole however it should be sufficient to shed light on the process required by practitioners when designing ESD programmes for employees. It would be interesting to trial out the needs-based process in different organisations through comparative research on how public and private sector organisations adopt this process to give a deeper insight to organisations wishing to embrace the concept of ESD for their employees. Another area of future research would be an analysis of the financial, social and environmental impacts the implementation of a needs-based programme has within the organisation and its immediate community.

References

Anderson GL, Herr K (2005) The action research dissertation. California: Sage Publications Inc

Blewitt J, Cullingford C (eds) (2004) The sustainability curriculum: the challenge for higher education. Earthscan Publications Limited, London

Bryman A (2004) Social research methods, 2nd edn. Oxford, Oxford University Press

Charmaz K (2006) Constructing grounded theory: a practical guide through qualitative analysis. Sage Publications, London

Cloud JP (2009) In: Chalkley B, Haigh M, Higgit D (eds) Education for sustainable development: papers in honour of the united nations decade of education for sustainable development (2005–2014). Routledge, London

Cohen L, Manion L, Morrison K (eds) (2007) Research methods in education. 6th edn.Routledge, London

Epstein MJ (2008) Making sustainability work. Greenleaf Publishing Limited, Sheffield

http://europa.eu/legislation_summaries/environment/sustainable_development/l28117_en.htm Last accessed on 19 Mar 2014

https://www.dmu.ac.uk/about-dmu/sustainability/sustainability-strategy.aspx. Last accessed on 15 Mar 2014

Huckle J, Sterling S (eds) (1996) Education for sustainability. Earthscan Publications Limited, London

Jackson SE, Jackson SE, Ones DE, Dilchert S (2012) Managing human resources for environmental sustainability. Jossey-Bass, San Francisco

McNiff J, Whitehead J (2010) You and your action research project, 3rd edn. Oxford: Oxford University Press

Miles MB, Huberman AM (1994) Qualitative data analysis: an expanded sourcebook, 2nd edn. Sage Publications, Thousand Oaks

Morris D, Martin S, Stibbe A (2009) The handbook of sustainability literacy: skills for a changing world. Green Books Limited, Totnes, UK

Patton MQ (2002) Qualitative research and evaluation methods, 3rd edn. Sage Publications, Thousand Oaks

An Instructor's Experience: Implementing Sustainable Development in the Curriculum

Sima I. Mpofu

Abstract

The end of the International Decade of Education for Sustainability warrants an analysis of developments in Canadian Community Colleges especially how sustainable education is incorporated into the curriculum. External forces are likely to derail the agenda for sustainable education. Educational institutions are dealing with reduced funding and enrolment issues. Innovative strategies are required to integrate sustainable development into the curriculum. Most publications are focused on developments in Universities. Best practices are still debatable and it is not very clear how faculty is incorporating sustainable practices into the curriculum. The paper describes an instructor's experience incorporating sustainable education into the curriculum and presents a conceptual model to address challenges. The experience gained through curriculum development and delivery in the Bachelor of Applied Science Degree in Horticulture at Olds College and a review of literature was used to develop a strategy for a conceptual project-based approach to integrating sustainable education into Applied Degree Programs offered in Alberta. The strategy involves a stand-alone course, Research Methods, in the Applied Degree Program. The program has three majors in Golf Course Management, Landscape Management and Production Horticulture. Student groups worked on solving an industry problem. The students had an industry and faculty mentor. The process was guided by the instructor through different stages by a diversity of assignments starting with problem identification, developing a research proposal and culminating in project presentations organized in a professional symposium

S.I. Mpofu (✉)
School of Animal Science and Horticulture, Olds College, 4500-50 Street, Olds, AB T4H 1R6, Canada
e-mail: sima.mpofu@gmail.com

W. Leal Filho et al. (eds.), *Integrating Sustainability Thinking in Science and Engineering Curricula*, World Sustainability Series, DOI 10.1007/978-3-319-09474-8_18

at the end of the third year. The symposium was open to all stakeholders. These included all students, faculty, staff and management at Olds College, and industry and community members. The main benefit to this format was providing a forum to discuss sustainability issues among all stakeholders thus raising awareness of sustainability issues and how these can be addressed. The concept addresses challenges without compromising academic integrity and freedom. Psychological theory suggests that sustainable education and problem based learning have the ability to shape the intellectual, moral and identity development of college-age students. Both a top-down and bottom-up strategy that address college policy frameworks and more wide spread implementation are recommended.

Keywords
Sustainable education · Applied sustainable development · Curriculum · Strategy · Applied degree programs · Research methods

1 Introduction: Applied Sustainable Education in Practice

Sustainable education is defined as lifelong learning designed to equip learners with skills to solve world problems in a responsible way. Sherren (2008) describes this type of education as an education that develops critical thinking skills, broad and integrated contextual knowledge and the desire and capacity to apply that knowledge.

The integration of sustainable education into the curriculum started in 2001 through teaching an Introduction to Horticultural Scientific Methods course to cohorts in a Bachelor of Applied Science in Horticulture at Olds College. There are three majors in the Bachelor of Applied Science in Horticulture Program. They are Golf Course Management, Landscape Management and Production Horticulture. A curriculum reform changed the course name from Introduction to Horticultural Scientific Methods to Research Methods in 2009. Students engaged in group work to solve a real world problem which involved economic, social and environmental sustainability issues in horticulture. Research Methods is a common course in most educational programs. Integrating sustainable development into the Research Methods course presents a great opportunity to expose a large majority of students to sustainable development concepts directly applicable to their areas of specialization. Integrating the two areas offers an opportunity to break the problem of program silos within educational institutions.

1.1 Applied Degree Programs

Canadian Colleges started offering Applied Degree programs in the 1990s. This was a result of increased demand for degree studies in combination with employers that were demanding job-ready graduates with advanced specific skill sets (Hamilton et. al. 2008; Sargsyan 2008). The training includes 6 semesters of instruction and 2

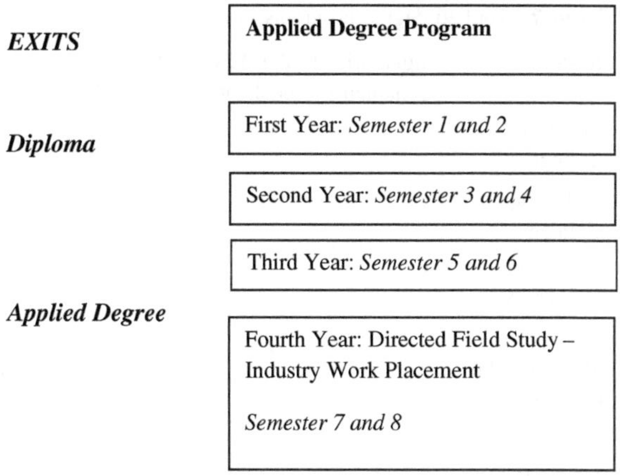

Fig. 1 Current structure of applied degree programs

semesters of industry placement geared towards gaining clearly identified and employment-focused competencies (Sargsyan 2008). The applied degrees at Olds College are four year programs (Fig. 1). Students graduate with a diploma after the first two years (first 4 semesters). The third year (2 semesters) had 10 courses that included Introduction to Horticultural Scientific Methods/Research Methods, Ethics and Integrated Pest Management common to all majors. The students engaged in Directed Field Studies in the fourth year (2 semesters). Work place competencies clearly identified in a learning plan are implemented under the supervision of an employer and the program coordinator. This is what differentiates Directed Field Study from a general work place experience at times referred to as co-op. The students spend the 4th year doing paid work structured according to the learning plan. More competencies can be accommodated in the 4th year using online learning tools. The Applied Degree Programs currently offered in Alberta are presented in Table 1.

1.2 Problem Solving Approach

The problem solving approach used in the Introduction to Horticultural Scientific Methods/Research Methods course, the focus on applied competencies in the Applied Degree Program, and the applied research mandate for Canadian Community Colleges are complementary to the concept of applied sustainable development proposed by Leal Filho (2011). Leal Filho (2011) defined applied sustainable development as: "An action-oriented and project-based approach, which uses principles of sustainable development and applies them to real contexts and to real situations, yielding the benefits which can be expected when methods, approaches, processes and principles of sustainable development are put into practice". The aspect of action or active learning discussed by Cook and Khare (2012) is related to Leal Filho (2011)'s applied sustainable development. Active learning is defined as any strategy that "involves students in doing things and thinking about the things they are doing" (Cook and Khare 2012). Dalhousie

Table 1 Applied degree programs in Alberta

Institutions	Signatories to the ACCC's Pan-Canadian Protocol for sustainability	Applied degrees
Lakeland College		Environmental management
		Emergency services (suspended)
Lethbridge College	√	Conservation enforcement
MacEwan University	√	Business administration—accounting
		Human service administration
Medicine Hat College	√	Health science
		Visual communications
		Ecotourism and outdoor leadership
Mount Royal University		Business and entrepreneurship—sport and recreation (suspended)
		Child studies (suspended)
		Ecotourism and outdoor leadership (suspended)
		Interior design
Olds College	√	Horticulture
		AgriBusiness
NAIT		Business administration—accounting
		Business administration—finance
		Information systems technology
Red Deer College	√	Motion picture arts
SAIT		Business administration
		Geographic information systems technology
		Petroleum engineering technology

University's Environmental, Sustainability and Society undergraduate program emphasizes teamwork, problem-based, and experiential learning to effect change in behavior (Wright 2013).

1.3 Sustainable Education Developments in Canadian Community Colleges

The United Nations declared 2005–2014 to be the International Decade of Education for Sustainable Development (IDESD) (UNESCO 2005). During this time period the Association of Canadian Community Colleges (ACCC) held a Symposium on Environmental Sustainability (2007) (Pan-Canadian Protocol for

Sustainability n.d.). The symposium enabled sharing best practices on incorporating sustainability principles in vision statements, missions, core values, strategic plans, priorities, policies, operations, procurement strategies, ethical investments, and student and staff engagement. Olds College and approximately 60 other educational institutions signed an ACCC's Pan-Canadian Protocol for Sustainability. This signified serious commitment to providing sustainability leadership to internal and external communities, and to playing an important role in ensuring an environmentally, economically and socially sustainable future. This led to significant developments in adoption of sustainable practices in areas of strategic planning at Olds College that was followed by the development of sustainability policy and the formation of a sustainability committee in 2013 (Changing the perspective: Unlocking the potential, students, supporters and success n.d.). The signatory institutions agreed to integrate the principles of sustainability within curriculum to enable students and communities to develop competencies and commitment to contribute to a sustainable future (Pan-Canadian Protocol for Sustainability n.d.).

Five of the nine Alberta educational institutions with Applied Degrees are signatories to ACCC's Pan-Canadian Protocol for Sustainability (Table 1). It is not clear how these institutions are incorporating sustainability into the curriculum. Most publications are focused on developments in Universities. More community college contributions to the debate are needed. Wright (2013) and Raufflet (2013) are examples of contributions from universities and field specific contributions to the debate on best practices. Education was amongst the lowest reporting sectors (0.75 %) in sustainability reporting (Fonseca et al. 2011). Fonseca et al. (2011) reported few disclosures of how sustainability is being incorporated into the curriculum. Lozano (2011) recommended 4 indicators for the Global Reporting Initiative (GRI) (Table 2). Fonseca compiled 7 curriculum and teaching indicators from several campus sustainability assessment tools that included the Association for the Advancement of Sustainability in Higher Education (AASHE)'s Sustainability Tracking, Assessment & Rating System (STARS) (Table 3). Educational institutions are adopting sustainability reporting systems in response to growing expectations of accountability (Fonseca et al. 2011). Olds College plans to utilize the Association for the STARS in developing goals, metrics and bench marks for sustainability reporting (Changing the perspective: Unlocking the potential, students, supporters and success, n.d.).

Olds College's strategy is focused more on infrastructure improvements such as landscape and buildings that have seen major developments of the Olds College Botanic Gardens and Wetlands and the construction of a sustainable student

Table 2 Amendments to the GRI on curriculum (Lozano 2011)

Indicators	Description
1	Sustainable development in the curriculum
2	Sustainable development capacity building
3	Sustainable development monitoring in curriculum
4	Administrative support

Table 3 Curriculum and teaching indicators compiled by Fonseca et al. (2011) from campus sustainability assessment tools

Indicators	Description
1	Policies related to sustainability in curriculum
2	Courses related to sustainability
3	Students taking sustainability-related courses
4	Sustainability literacy assessment
5	Degree programs related to sustainability
6	Non-curricular teaching initiatives related to sustainability
7	Scholarships offered to sustainability-related education

residence on campus as well as sustainable research in biofuels, compost, wetlands and land reclamation (Changing the perspective: Unlocking the potential, students, supporters and success, n.d.). Landscapes and buildings reflect a hidden curriculum that powerfully influences the learning process (Orr 2002). Fonseca et al. (2011) points out that the landscape; buildings and research have a significant impact on the world. Creating an ecologically sensitive academic environment and establishing a result oriented culture of sustainable action provides examples of sustainable effort and provides a vision of innovation and technological advances that can be reflected in a curriculum (Savelyeva and McKenna 2011). Conceited effort is required to complement an ecologically sensitive environment with coordinated curriculum development efforts specifically designed to incorporate sustainable education into the curriculum. Traditional lecturing methods are not likely to change behavior. A problem solving approach used in the Research Methods course is more likely to change behavior.

The 2013 budget cuts to post-secondary education in Alberta and declining enrolment may derail the sustainable education agenda. Significant financial resources are required for problem-based learning and incorporating sustainability into the curriculum. An emerging trend of Educational Enterprises may partially address the financial challenges. Critics raise questions about the ethics of publicly funded institutions competing with the private sector. Cook and Khare (2012) presented a viable construct on exploring academia-corporate collaboration to address university sustainability issues. The Federal Minister of Employment and Social Development who is also the Regional Minister for Southern Alberta called for increased Canadian Businesses expenditure on educational training and emphasized the importance of collaboration between the education system and employers ("Kenney says Canada can learn from Germany on skills training." GlobalPost. N.p., 13 Mar. 2014. Web. 22 Apr. 2014.). The minister led a delegation of politicians, business and labour union representatives on a tour of Germany and Great Britain to learn about apprenticeship programs. There is a growing demand for a pathway that allows tradespeople the opportunity to move from a trade certificate to the applied degree program. This pathway would satisfy the workforce demand for tradespeople with managerial, leadership and entrepreneurial skills

(NAIT launches trades to degrees initiative to meet Alberta's needs 2012). This pathway is likely to address declining enrolment issues and incorporating sustainable education into the curriculum is likely to increase student satisfaction. Innovative strategies are required to both engage the private sector in educational training and increase student satisfaction without compromising academic integrity and freedom. The interplay of political, economic and social forces dictates the urgency to address sustainability issues in educational institutions, and the end of the IDESD calls for a review of developments in Canadian Community Colleges.

The paper describes an instructor's experience in incorporating sustainability into the curriculum and presents a conceptual approach to align strategy with practice in order to keep pace with the evolving challenges of sustainable education.

2 Curriculum Reform Procedures at Olds College

Program Curriculum Committees (PCCs) were introduced in 2007 to implement a college wide curriculum reform. The main primary objective was to facilitate credit transfer. A secondary objective was to solve timetabling issues associated with highly variable course credits and formats of delivery not correlated to hours of delivery. The PCCs were made up of faculty and instructional assistants who teach in each of the programs and was led by the program coordinator. The purpose of the PCCs was to review the curriculum by sorting through the existing competency profiles organized into 35–45 courses and repackage them into 21–25 courses per program. This involved deletion of some old competencies and addition of new competencies based on the feedback from industry advisory committees. The competences in each program were to include general employability skills such as oral and written communication, use of numbers (computation), management of information, managing self (personal), interacting appropriately with others (interpersonal), solving problems/managing projects, and pursuing self-directed learning. The basics of these competences are introduced in two communications courses that included professionalism and ethics. The PCCs also ensured these competences are threaded through other courses in the curriculum to allow for appropriate application to the areas of specialization. Sustainable development competencies could be introduced using the same model for communication, professionalism and ethics. After the implementation of the college wide curriculum reform in September 2009 the PCCs were tasked with the responsibility for conducting annual reviews based on evolving job market research and industry needs (Olds College Policies and Procedures n.d.). Program outcomes for horticulture diploma programs included articulation of the ecological, economic and social implications of industry decisions and processes. Instructors can also present any major changes to their courses for PCC approval in order to avoid content duplication. This is a plausible method of introducing sustainable development competences into the curriculum. When people focus on sustainable development, institutional governance becomes less relevant whether an institution is considered efficient in stimulating sustainable development or not (Platje 2008). However, leadership motivation, attitude and planned behavior are still desirable.

Comprehensive program evaluations every few years are the responsibility of Comprehensive Program Review Teams appointed by the Dean (Olds College Policies and Procedures n.d.). The process is led by the Dean. The schedule for Comprehensive Program Reviews is determined by the Vice President for Academics and Research (VPAR) based on the Director for Educational Technology and Curriculum's recommendations. When deemed necessary by the Dean and VPAR, the process will include a qualified external reviewer who will conduct interviews with all stakeholders and present a report for consideration by the Comprehensive Program Review Team. The stakeholders include industry, students, faculty and staff.

3 Integrating Sustainability into Introduction to Horticultural Scientific Methods/Research Methods

More than 300 students enrolled in the Bachelor of Applied Science in Horticulture program have taken the Introduction to Horticultural Scientific Methods/Research Methods course since 2001 (Applied Degree Symposium: Students apply themselves to research and solutions 2011). Then the course had 7 credits and was completed in two semesters in the third year of the Applied Degree Program. The students enrolled in the program are mostly from Western Canada. A few students come from Ontario. The first semester involved class instruction in research methods. The content was focused more on statistical analysis and critical analysis of scientific publications. Four scientists were invited to make class presentations. The presenters were mostly from the Olds College Centre for Innovation (OCCI) and the Prairie Turfgrass Research Centre. This linkage of the research methods course to OCCI, the college research branch, is a strategic advantage.

The second semester involved project based problem solving. The students identified an industry problem and developed a research proposal in the first month of the second semester. The students worked in groups of three. The first stage of the project was an individual assignment. Each group member identified a problem based on their personal interest, work experience or talked to their previous employers or industry networks. The second stage was an assignment designed to facilitate discussing the feasibility of the three projects. At the end of the discussions the group selected one of the projects based on feasibility, budgetary and time constraints. The selections were facilitated by the instructor and in most cases the instructor's involvement was in refining the scope of the project ideas.

The majority of project ideas in the first year were based on student interest. As the years progressed students worked more on ideas originating from industry partners. In the first few years the ideas were mainly from industry partners based on campus. Ideas for projects in the Golf Course Management Major originated from the Prairie Turfgrass Research Centre based at Olds College. Many of the projects originated from the OCCI. After approximately 3 years private organizations not associated with the college made inquiries about possibilities of students working on industry projects. These ideas were presented to the students but the students were not obligated to select these projects. In most cases students made the choice to work on industry projects. The projects ranged from biodiversity, waste, fertility and water management. An example of a biodiversity project was a project proposed by the Alberta Native Plants Council. There was concern about seed packages from outside Alberta being sold as mixes of native plants thus threatening the biodiversity of native plants in Alberta. The project investigated the presence of alien invasive species in commercially available wildflower seed mixes. Related projects also involved research and creation of landscape designs that incorporated native plant species and required minimum maintenance, that is, less mowing, fertilizer and watering. Another noteworthy project was a waste audit of garbage and creating a waste management strategy for The Crossing, a cafeteria on campus run by the Olds College Student Association. The project proposed ways for the college to divert waste from the landfill. Some interesting projects originating from the OCCI looked at ways to utilize compost and Biochar in horticulture, production or managing algae in aquatic environments, and using plant material for land and water reclamation. Some projects involved testing patented plant growth regulators from the National Research Council Plant Biotechnology Institute. The products increased drought tolerance in horticultural plants. Crop protection projects included evaluating pest monitoring methods and effectiveness of biorational products in pest management. Every year there was a project dealing with some aspect of water salinity. Some of the projects were implemented as per student recommendations. These included landscape designs for schools, brownfields, municipal parks, Ermineskin Cree Community and a community garden in the Town of Olds. Some students published their results in trade magazine journals and local newspapers, and had radio interviews.

The research proposals were presented to the class and industry partners for feedback. The projects were carried out under the supervision of an industry mentor and a member of faculty. The students were required to complete assignments that included planning, technical and statistical analysis meetings. This involved developing meeting agendas, presentations and minutes. The student's learning outcomes included oral and written communications, problem solving, project management and critical analysis.

The students presented the results of their project in a one day symposium planned for the end of the third year. The students were actively involved in organizing a professional symposium. The symposium had invited keynote speakers external to the college. In multiple years keynote speakers were representatives from Agrium Inc. and the National Research Council Plant Biotechnology Institute. One keynote address presented by Westhoff Engineering Resources was on developing sustainable systems within our landscapes. The event was open to the college community and industry partners. The VPAR was in attendance and presented cash awards to the student groups with the best oral and poster presentations. The event and awards were sponsored by OCCI. The judges were members of faculty, scientists from the OCCI, the Community Learning Campus, the Alberta Association of Colleges and Technical Institutes (AACTI) and private industry partners.

All projects ideas in the first year involved some aspect of sustainability. Thus the theme for the Applied Degree Symposiums was about sustainable horticulture. The background research on the projects presented an opportunity for the students to learn about sustainability issues. The VPAR commented that he had learnt more about Biochar from the Applied Degree student symposium than he had from any other conference or meeting he had attended (Bob Wilson, personal communication).

4 Benefits to Stakeholders

Curriculum, research, operations and relations with local communities are often considered separate entities but they are not (Cortese 2003). This fact became evident in the diversity of projects selected by students enrolled in the Bachelor of Applied Science in Horticulture at Olds College. The students became constructive critics of college operations, common practices in the industries associated with their areas of specialization and common practices in the communities they worked with. Their confidence in questioning these practices and suggesting alternative solutions increased with each stage of the project. The students were learning more from their physical learning environment and this helped them make connections among the research and curriculum at the college and community and industry practices. The intent of the course had been to teach problem solving skills. The organic growth in the comprehension of sustainability concepts evolved naturally from the nature of industry and community issues they were addressing. Organic growth is defined as the potential for facilitated independent learning and drawing linkages to learning outcomes in other courses. The sustainability issues tended to be the same every year although the community and industry clients were different. Different solutions to the same problems also became evident depending on the objectives and the nature and vision of the community and industry clients. The symposium presented a unique opportunity for administrators, faculty, research scientists, staff, students, community and industry to have an open and honest dialogue about how to solve real world problems in a responsible way.

The Applied Degree Symposium provided a forum to illustrate teamwork among faculty who voluntarily participated with enthusiasm in the one day symposium. The teamwork was also evident among faculty, industry and community clients thus helping the college to build stronger industry and community linkages. The industry was enthusiastic about contributing to student learning. Many donated materials and a few gave financial contributions. This increased the integrity of the Olds College credential or the Olds College brand as employers were involved in the learning process and witnessed the growth in employability skills. Industry advisory committee members have raised concerns about the inability of new graduates to work independently to solve problems. Industry participation in this manner would serve to help alleviate some of these concerns. The response of community and industry to the student presentations showed that the industry and community were also learning about sustainable development. Sustainable education is an important element of competitive advantage (Maloni and Paul 2011).

Psychological theory suggests that sustainability education can shape the intellectual, moral and identity development of college-age students (Myers and Beringer 2010). Data suggests that active, problem-based learning can be a very effective form of sustainability education (Myers and Beringer 2010). Some projects conducted by Applied Degree students utilized college facilities and networks. Project-based courses that use the campus as a learning laboratory can strongly support the dimensions of learning implied by psychological theory (Myers and Beringer 2010). Myers and Beringer (2010) say "By contextualizing these processes of psychological growth within an identity movement from "sustainability learner" to an empowered, confident "sustainability change agent," students themselves as well as academic staff, university management, and off-campus stakeholders can clearly witness the sustainability learning outcomes many higher education institutions now seek for their graduates".

5 Why Is Integrating Project-Based Sustainable Development into the Curriculum Challenging?

Students in the Applied Degree program struggled with adapting to the active project-based learning. Instructors should be prepared for resistance. There is a high risk of getting poor course and instructor evaluations based on cohort dynamics. The culture of an organization affects the willingness of instructors to take that risk. Instructors must accommodate different styles of learning such as including formal lectures and quizzes from time to time. This has implications for both student and instructor workload. Myers and Beringer (2010) reported similar observations and state that sustainability calls for new levels of intellectual challenge appropriate to ill-structured problems also referred to as "wicked" problems. Wicked problems are problems that are difficult to solve because of incomplete, contradictory and changing requirements. Because of complex interdependencies, the effort to solve one aspect of a wicked problem may create other problems. Learners find it challenging to deal with questions that do not have a clear cut solution that is applicable in all situations.

Evaluating individual competency in group based projects is also a big challenge for instructors. Peer evaluations are commonly utilized. The effectiveness of this evaluation strategy is questionable. Negative group dynamics can come in the way of performance evaluations and active learning. Students were evidently uncomfortable with the idea of evaluating each other. There is also no way of evaluating a change in behaviour or transformative learning. The expectation of shaping the intellectual, moral and identity development through sustainable education is considered a lofty ideal. Not all educational institutions are ready to embrace teaching a values based education (Sherren 2008).

Project timelines were difficult for students to manage for several reasons. The skills for time management were in the process of being developed. The industry clients had variable degrees of cooperation and at times their demands were outside the scope of curriculum requirements. The instructors have to be flexible in order to accommodate these complex variables. The college policies on student evaluations can limit the degree of flexibility.

6 Recommendations for Integrating Sustainable Development into the Applied Degree Curriculum

The limited financial resources, college policies and organizational structure present significant challenges in implementing sustainability strategy and purpose into curriculum. These challenges give credence to a bottom-up approach to integrating sustainability into the curriculum in addition to a top-down strategy. Organizational structure, culture, policies and procedures should be reviewed in order to facilitate incorporating sustainable education into the curriculum. The role of sustainability director could be combined with that of educational technology and curriculum director. Alberta Enterprise and Advanced Education should put together a working Group to develop a code of conduct on the nature of Directed Field Study collaboration between corporations and colleges with Applied Degree programs.

The organic growth in student comprehension of sustainability concepts witnessed in the Introduction to Scientific methods/Research Methods course can be more intentional and enhanced by more coordinated efforts across campus and the province. Alberta Enterprise and Advanced Education are focused on growing the eCampusAlberta brand ("Letter of expectation between The Minister of Alberta Enterprise and Advanced education (as representative of the Government of Alberta) and the Board of Governors of Olds College (as representative of Olds College)" n.d.). A focus on sustainable education is one way to grow the eCampus Alberta brand. It is recommended that eCampus Alberta offer online courses, for the faculty and students in this area (Fig. 2).

The first course will be designed to equip faculty with skills to incorporate sustainable development into the curriculum (Fig. 2). The Alberta Government should provide the resources to support an online model similar to the Pacific Spirit Project at the British Columbia Institute of Technology (Rockall 2012). The Pacific Spirit Project is modelled after the Piedmont/Ponderosa Model of faculty development (Barlett and Chase 2012). Most educational institutions have a budget for professional development activities. The course could also be offered globally to generate financial resources for incorporating sustainable development into the curriculum. The model will help replicate the effectiveness of the Piedmont/Penderosa Model of faculty development in an online environment that is more accessible.

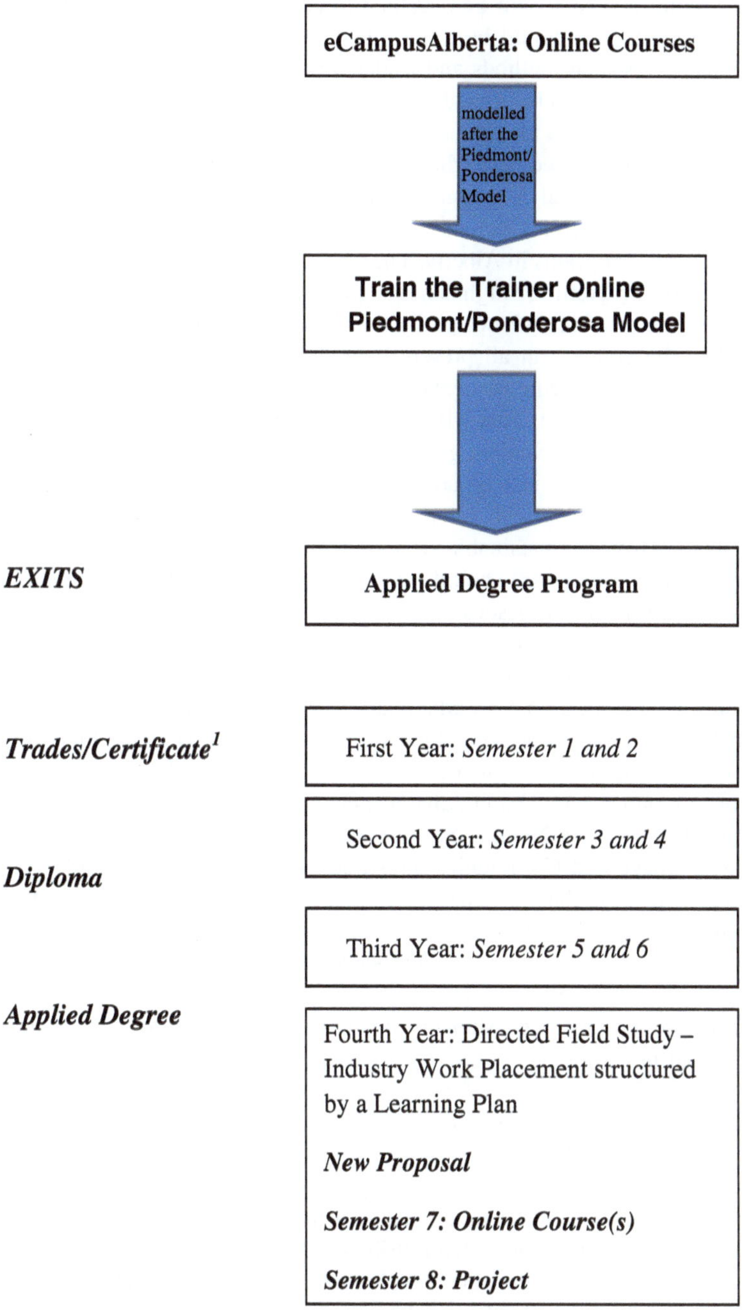

Fig. 2 Conceptual model for integrating sustainable development in the curriculum. ([1] Emerging market trend)

The second course(s) will be designed to equip students with problem solving skills. The course (s) would be more intentional designed to provide sustainable development, research methods and leadership training in the fourth year (Fig. 2). The online course would be offered in Semester 7 of the applied degree program. The students will be required to conduct a project on a sustainable development issue in Semester 8. The project could be directly applied to the student's directed field study placement. The proposed conceptual model could be piloted with all undergraduate Applied Degree programs across a campus and eventually implemented in all Applied Degree Programs in Alberta. The course would also be a good candidate for the dual credit program for high schools, post-secondary institutions, business and industry in the province advocated by the Alberta Dual Credit Strategy. Students enrolled in the course can all present their results at an Annual Applied Degree Symposium that will bring students together at least once in the final year of the Applied Degree Program. This will break the program silo problem evident within educational institutions. Student groups in different programs can work on different aspects of the same project broadening the learning outcomes. eCampusAlberta could form partnerships with international educational institutes thus infusing a global perspective on sustainable development issues. Sustainability Committees and Educational Enterprises should provide financial resources for student projects that address sustainability issues in campus operations.

The Alberta Association of Colleges and Technical Institutes in collaboration with Alberta Innovates should develop a sustainable development program agenda that connects the eCampusAlberta courses to private industry and community sustainable development projects. Industry placements for Directed Field Study can be encouraged by a tax credit for companies that hire students enrolled in these types of programs. AACTI and Alberta Innovates should also tap into the national resources through the Natural Sciences and Engineering Research Council of Canada—College and Community Innovation Program. A well-developed framework would meet global sustainable development goals, the needs of educational programs, industry and community needs, and provide access to resources for conducting applied projects. Funding models would tap into national and provincial government funding sources, and industry sources. The desired outcome would be to turn educational institutes, students, businesses and communities into sustainable development practitioners.

7 Conclusions

Canadian colleges are in the initial stages of implementing sustainability strategy into the curriculum. The end of the IDESD (2005–2014) is approaching (UNESCO 2005). Implementing sustainable development into the curriculum in Canadian Community Colleges should be accelerated. There is a need for innovative strategies to address the limited resources and to encourage widespread adoption of sustainable education and business practices. These present significant challenges in incorporating sustainable development into the curriculum. It is time to deconstruct the silos separating corporations and higher education based on mutually beneficial opportunities (Cook and Khare 2012). This paper describes how an instructor incorporated sustainable development into the curriculum using a problem based approach that involved college administrators, community and industry partners. The main benefit of this approach is presenting an opportunity for dialogue among stakeholders thus increasing the commitment to and understanding of sustainable development among all parties. The paper proposes moving some competencies from the 3rd year to the 4th year of the Applied Degree Programs. This shifts responsibility for resource allocation from the educational institute to the employer thus addressing the resource limitations. A unified front in developing a more structured code of conduct for consistent implementation across Applied Degree programs and industries is required to preserve academic integrity and freedom. Faculty training is required to address the significant challenges experienced by instructors using the problem-based applied sustainable education approach. A review of college policy would be beneficial in order to address some of the pedagogical challenges. A conceptual model that addresses these challenges in Canadian Community Colleges is proposed. It is designed to incorporate an online learning environment to increase accessibility to sustainable education for both faculty and students. It involves faculty professional development activity based on the Pacific Spirit Project at the British Columbia Institute of Technology modelled after the Piedmont/Ponderosa Model (Rockall 2012). It also involves redesigning the structure of Applied Degree programs. It proposes an online course(s) incorporating research methods, leadership and sustainability for students. The interplay of political, economic and social forces supports the exploration of this concept. The concept expands applied sustainable education beyond campus operations to industry operations without compromising academic integrity and freedom.

References

Applied Degree Symposium: Students apply themselves to research and solutions. Horizons Magazine. Version Fall/Winter. Olds College (n.d.) Web. 20 Mar. 2014. http://www.oldscollege.ca/Assets/OldsCollege/shared/communications/Horizons/Past-Issues/2011-horizons-5-4_001.pdf

Barlett PF, Chase GW (2012) Curricular innovation for sustainability: the piedmont/ponderosa model of Faculty Development. Liberal Educ 98(4):14–21

Changing the perspective: unlocking the potential, students, supporters and success. Olds College. N.p. (n.d.) Web. 18 Mar. 2014. http://www.oldscollege.ca/Assets/OldsCollege/shared/BottomNav/Administration/reports/Comprehesive-Institutional-PlansBusiness-Plans/CIP%202013-2016.pdf

Cook P, Khare A (2012) Academia–Corporate on campus sustainability collaboration: an exploration of the construct: handbook of sustainability management. In: Madu CN, Chu-Hua Kuei (eds) World Scientific Publishing, Singapore, pp 487–512. ISBN: 9789814354820 (March 2012/Chapter 23)

Cortese A (2003) The critical role of higher education in creating a sustainable future. Plan High Educ 31(3):15–22

Fonseca A, Macdonald A, Dandy E, Valenti P (2011) The state of sustainability reporting at Canadian Universities. Int J Sustain High Educ 12(1):22–40
Hamilton K, Charlton S, Elmes R (2008) Developing a four-year baccalaureate degree in applied psychology. Plan High Educ 36(4):23–32
Kenney says Canada can learn from Germany on skills training. GlobalPost. N.p., 13 Mar. 2014. Web. 22 Apr. 2014. http://www.globalpost.com/dispatch/news/the-canadian-press/140313/kenney-points-study-late-us-democrat-income-splitting-promis
Leal Filho W (2011) Applied sustainable development: a way forward in promoting sustainable development in higher education institutions. In W. Leal Filho (Ed.), World Trends in Education for Sustainable Development. Vol. 32, pp. 11−29. Frankfurt: Peter Lang.
Letter of expectation between The Minister of Alberta Enterprise and Advanced education (as representative of the Government of Alberta) and the Board of Governors of Olds College (as representative of Olds College). Alberta Innovation and Advanced Education. Alberta Government. (n.d.) Web. 10 Apr. 2014. http://eae.alberta.ca/media/letters/Olds-College.pdf
Lozano R (2011) The state of sustainability reporting in universities. Int J Sustain High Educ 12 (1):67–78
Maloni MJ, Paul RC (2011) A service learning campus sustainability project. Decis Sci J Innovative Educ 9(1):101–106
Myers J, Beringer A (2010) Sustainability in higher education: psychological research for effective pedagogy. Can J High Educ 40(2):51–77
NAIT launches trades to degrees initiative to meet Alberta's needs (2012) NAIT Newsroom. NAIT. N.p. Web. 20 Apr. 2014. http://www.nait.ca/44779_89417.htm
Olds College Policies and Procedures. Olds College. N.p., (n.d.) Web. 18 Mar. 2014. http://www.oldscollege.ca/Assets/OldsCollege/shared/BottomNav/Administration/policies/D/D22%20Program%20Review.pdf
Orr DW (2002) The nature of design: ecology, culture, and human intention. Oxford University Press, Oxford
Pan-Canadian Protocol for Sustainability. Pan-Canadian Protocol for Sustainability. N.p., (n.d.) Web. 20 Mar. 2014. http://www.accc.ca/xp/index.php/en/members
Platje J (2008) Institutional capital as a factor of sustainable development—the importance of an institutional equilibrium. Technol Econ Dev Econ 14(2):144–150
Raufflet E (2013) Integrating sustainability in management education. Humanities 2:439–448
Rockall M (2012) Incorporating sustainability in curricula while meeting entrenched traditional learning outcomes: sustainable development at Universities: New Horizons. In: Leal Filho W (ed) Environmental education, communication and sustainability, vol 34. Peter Lang, Bern, pp 319–331 (Chapter 27)
Sargsyan H (2008) Canadian applied degrees: struggling for recognition. WENR, May 2008: Feature. N.p. (n.d.) Web. 20 Mar. 2014. http://www.wes.org/ewenr/08may/feature.htm
Savelyeva T, McKenna JR (2011) Campus sustainability: emerging curricula models in higher education. Int J Sustain High Educ 12(1):55–66
Sherren K (2008) A history of the future of higher education for sustainable development. Environ Educ Res 14(3):238–256
UNESCO (2005) UN decade of education for sustainable development (2005–2014): international implementation scheme. United Nations Educational, Scientific and Cultural Organization Education Sector, Paris (ED/DESD/2005/PI/01)
Wright T (2013) Stepping up the challenge-the Dalhousie experience. In: Johnston LF (ed) Higher education for sustainability: cases, challenges, and opportunities from across the curriculum. Routledge, New York

Author Biography

Dr. Sima Mpofu spent the last 12 years teaching at a Canadian Community College in Alberta. She has a Ph.D. in Environmental Biology from the University of Guelph specializing in crop protection. She has been involved in plant pathology research as a Canadian Research Fellow with Agriculture and Agri-Food Canada. She is in the process of completing an Executive MBA with Athabasca University. This paper was completed in partial fulfillment of the requirements for the Sustainable Development and Business Course with Athabasca University. She is currently working on an Applied Project in Sustainable Development. She is most grateful for Dr. Anshuman Khare's support and encouragement. She would like to take this opportunity to thank all the students who have taken the Introduction to Horticultural Scientific Methods or Research Methods Course at Olds College over the years for providing the inspiration for this paper.

Sustainability into the University of Sao Paulo (USP), São Carlos Engineering School (EESC)—Brazil

Yovana M.B. Saavedra, Fabio Neves Puglieri, Vitor Ranieri, Rosane A. Aranda, Patrícia Silva Leme and Aldo Roberto Ometto

Abstract

Nowadays, in the educational context, environmental issues demand a range of skills that professors and technicians are supposed to have in order to prepare undergraduate students for sustainability challenge. In this sense, this chapter

Y.M.B. Saavedra (✉)
Environmental Engineering Sciences, Engineering School of São Carlos, University of São Paulo, 400 Trabalhador São-Carlense Blvd, Arnold Schimidt, São Carlos-SP, Brazil
e-mail: ybarrera@sc.usp.br

F.N. Puglieri · A.R. Ometto
Department of Production Engineering, Engineering School of São Carlos, University of São Paulo, 400 Trabalhador São-Carlense Blvd, Arnold Schimidt, São Carlos-SP, Brazil
e-mail: puglieri@usp.br

A.R. Ometto
e-mail: aometto@sc.usp.br

V. Ranieri
Department of Hydraulic and Sanitary Engineering, Engineering School of São Carlos, University of São Paulo, 400 Trabalhador São-Carlense Blvd, Arnold Schimidt, São Carlos-SP, Brazil
e-mail: vranieri@sc.usp.br

R.A. Aranda
Engineering School of São Carlos, University of São Paulo, 400 Trabalhador São-Carlense Blvd, Arnold Schimidt, São Carlos-SP, Brazil
e-mail: rosane.aranda@usp.br

P.S. Leme
USP Recicla Engineering School of São Carlos, University of São Paulo, 400 Trabalhador São-Carlense Blvd, Arnold Schimidt, São Carlos-SP, Brazil
e-mail: pazu@sc.usp.br

W. Leal Filho et al. (eds.), *Integrating Sustainability Thinking in Science and Engineering Curricula*, World Sustainability Series, DOI 10.1007/978-3-319-09474-8_19

aims to present the results obtained from an exploratory review about the activities developed at Engineering School of Sao Carlos (EESC) regarding sustainability integration in the campus. Thereby, the program entitled "Sustainable EESC", created in 2011, was analyzed through three methodological procedures: an exploratory review about EESC initiatives, four phases to assess campus responses to sustainability and a SWOT analysis. The results show that EESC is developing five initiatives towards sustainability: Teaching and Training of Human Resources; Research and Innovation; Environmental Management; Community Extension; and Information and Communication. Although many EESC initiatives are found in the high leadership levels, especially in phases 3 and 4, some barriers still exist and need to be overcome in order to achieve better sustainability integration. Finally, it is expected that EESC experiences can support other universities that intend to become more sustainable.

Keywords
Higher education · Sustainability solutions · Sustainable EESC program · Brazil

1 Introduction

The World Sustainable Development Forum that took place in Johannesburg in 2002 was where the United Nations Decade of Education for Sustainable Development (DESD) (2005–2014) was presented, stating that universities must act as institutions for sustainable development research and teaching. The United Nations Educational, Scientific and Cultural Organization (UNESCO) adds that education for sustainable development allows every human being to acquire knowledge, abilities, skills and values to create a more sustainable future in terms of environmental integrity, economic viability and a fair society (Desha and Hargroves 2010; UNESCO 2014).

In this sense, Higher Education Institutions (HEIs) play an important role to disseminate sustainability by concepts, methodologies, procedures and relevant technologies to prepare their students (Almeida et al. 2013).

The last decade presented an increment in the number of universities which initiated the sustainable issues inclusion in their institutions. In this context, there are several researches applying methodologies and alternatives that can be used to identify and guide universities onto this integration. An example is a research developed by Krizek et al. (2012) presenting four phases that may typify or predict campus responses to sustainability: grassroots, executive acceptance of the business case for sustainability, the visionary campus leader and fully self-actualized and integrated campus community.

Almeida et al. (2013) mention four alternatives to integrate sustainability within universities, identified through a literature review. These alternatives are general initiatives, environmental management systems implementation, inclusion of sustainability content into curriculum courses, and diagnosis and evaluation of those sustainable initiatives implemented.

However, it is important to say that, inside engineering courses, sustainability has the purpose to let future professionals aware about the concept and how it should be considered in a practical way (Boyle 2004). In this sense, the World Federation of Engineering Organizations (WFEO/FMOI) uses the following expression to encourage engineers: "Become knowledgeable of sustainable development principles and be continuously trained about the current sustainable development technologies applicable to their work" (WFEO 1997).

According to Kamp (2006), Sammalisto and Lindhqvist (2008), Lozano and Lozano (2013) and (Ramirez 2006), examples of this sustainability integration process in engineering courses come mainly from Australia, Netherlands, Sweden and United States. However, there are just a few studies regarding this topic in Brazil. Performing a review in international databases such as ISI/Web of Knowledge and Scopus with the keywords "sustainability", "environmental", "Brazilian", "Brazil", "courses" and "engineering courses", just five results were found: Almeida et al. (2013), Jabbour et al. (2013), Pereira et al. (2014), Palma et al. (2011), Jabbour (2010). From these five results, only one referred to engineering courses about an energy synthesis application at Universidade Paulista (Almeida et al. 2013). In this sense, it is necessary more incentives and disclosure about sustainability at Brazilian universities.

This article aims to present an exploratory review about Engineering School of Sao Carlos (EESC) initiatives, and its barriers and opportunities towards a more sustainable campus. The importance of studying EESC is because the campus is part of University of Sao Paulo (USP), a Brazilian university known by its academic and scientific relevance, with 249 courses in several areas of knowledge, 42 research and teaching units, and more than 58,000 students. EESC is also known for its nine engineering courses: Aeronautical Engineering; Civil Engineering; Computer Engineering; Manufacturing and Materials Engineering; Production Engineering; Electrical Engineering; Mechanical Engineering; Mechatronic Engineering; and Environmental Engineering.

To guide the current exploratory study, the following research question was used: *What are the main activities, barriers and opportunities about sustainability integration at EESC*?

2 Methodology

The methodology adopted in this paper in order to answer the research question already presented was based in an exploratory review about the main initiatives implemented at EESC through the "*EESC Sustentável*" (Sustainable EESC) program. Web pages were reviewed to identify the main characteristics and activities developed at EESC. Then, information elaborated by working groups was compiled in five initiatives (Information and Communication, Education, Research and

Innovation, Environmental Management and Community Outreach) and classified according to the four phases (grassroots, executive acceptance for sustainability, the visionary campus leader and fully integrated sustainability approach) presented by Krizek et al. (2012).

Lastly, a SWOT (Strengths, Weaknesses, Opportunities and Threats) analysis was presented in order to identify barriers and potential opportunities for EESC.

3 Exploratory Case Study

3.1 General Context

USP was created in 1934 and it is the most important university in Brazil. Nowadays, USP is responsible for 22 % of scientific production in Brazil (USP 2014). Related to sustainability at USP, one of the first initiatives was the USP Recycles Program, created in 1994 at Sao Paulo campus. The program aims to generate a model focused on responsible consumption and an appropriate waste destination. USP Recycles Program also involves around 500 people including students, professors, researchers, employees and visitors (SGA 2014a).

In 2012, USP Recycles was bound into the Superintendence of Environmental Management (SGA) that promotes sustainability in the environmental area at all different USP campuses. The main principles of the SGA are based on actions development for natural resources conservation, promotion of the environmental welfare and environmental security, rational use of resources, and sustainable education. All actions at USP intend to have community participation to build a sustainability model for society (SGA 2014b).

In the context of EESC, USP Recycles Program developed some initiatives. Some examples include the furthering of sustainable events with the guidelines developed by USP Recycles (Leme and Mortean 2010), waste management and waste reduction in the campus (Leme et al. 2012a), educational projects to reduce waste in the restaurant, promoting sustainable mobility, human resources training on environment, composting, and sustainable student housing (SGA 2014a).

Recently, in 2012, Sustainable EESC program was developed in partnership with USP Recycles Program to extend EESC action on sustainability, integrating sustainable concepts in the university activities, such as teaching, research, extension and management. Moreover, the program intends to involve local community, projects and actions to be implemented at EESC (EESC Sustentável 2014).

According to Jabbour et al. (2013), initiatives related to sustainability tend to be incorporated by teaching, research and activities involving the community. In the Sustainable EESC program case, the initiatives will be promoted in five groups, which will be described in the next topics: (1) Education and training in human resources; (2). Research and Innovation; (3) Environmental Management; (4) Community Outreach and (5) Information and Communication.

4 Current Scenario for Sustainable EESC Program

In this section, the main initiatives developed during the Sustainable EESC program are presented. Working groups (WGs) created at the university developed these initiatives, and they are composed by professors, graduate students and undergraduate students, and other university employees. Thereby, a group manager was suggest to assist the WGs, having as a goal the coordination and integration of activities, projects, groups and opportunities to strengthen the actions developed within Sustainable EESC program.

The following items describe the main initiatives developed by the WGs at EESC.

4.1 Education and Training of Human Resources

This WG aims to elaborate policies to integrate sustainability into EESC engineering courses and training for university community to meet the challenges towards sustainability. The main purpose of this WG is curriculum review of all engineering courses, in order to identify its content, the sustainability integration level and to suggest additional alternatives to include sustainability in EESC undergraduate courses. Moreover, the results intend to encourage other USP campuses towards a more sustainable education (Ometto et al. 2013).

Currently, the WG analyzed what is the potential for sustainability integration in in nine engineering curricula, considering all mandatories, free optional and electives disciplines. The results show that Environmental and Production Engineering have major sustainability integration with 57 and 13 % respectively, and Electrical and Computer Engineering present a lower sustainability-orientation, around only 3 % (Lopes et al. 2014). The WG is also working in a proposal to create a new discipline for all engineering courses at EESC and it will be focused more on proactive and preventive attitudes, with a conceptual and methodological base. Finally, in the next step, questionnaires will be applied to all engineering professors at EESC, a study with experts will be carried out, and the result will be disclosed in a workshop to identify opportunities and contributions for EESC initiative (Ometto et al. 2013).

Research and Innovation

This WG maps the competences, initiatives and researches that involve sustainability inclusion, with the goal to disclose and to introduce specific knowledge and solutions for the environmental concerns in the campus. This WG is composed by professors and graduate students, and it has the aim to identify projects and researches developed at EESC, and the creation of a database to promote contact between researchers from the same area.

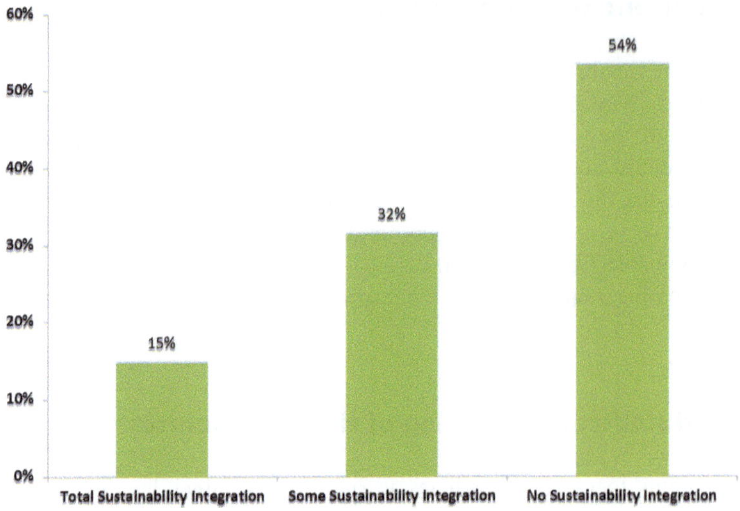

Fig. 1 Results about sustainability integration level in research projects

Currently, this initiative is organized in three steps, and the results will be integrated in an online platform entitled "Sustainability Project". The steps are: (1) identify the professor working in sustainability area, (2) curriculum analysis in categories and (3) evaluate the professor according to their own opinion. The Fig. 1 presents the results about sustainability integration level in professors' research projects.

The results demonstrate that sustainability integration level in research projects is still low. Only 15 % (34 researches) presented a high level of integration, 32 % (72 researches) with some integration, while 54 % (122 researches) did not include sustainability content.

Finally, in the third step, 113 professors from different engineering courses were contacted to evaluate, according to their own opinion, whether their researches include sustainability in some level. As a result, only 12 researchers were included in an online platform for having their projects with integration in sustainability, and the main areas identified were sustainability indicators, renewable energy, sustainable mobility, Life Cycle Assessment (LCA), water treatment, and industrial processes and machine tools. Another important finding is related to the high number of researchers without a clear idea whether their projects had or not integration with sustainability, including those researches that did not demonstrated interest for this topic.

As future steps, the WG aims to identify all research projects developed at EESC and to propose new tools to analyze sustainability integration within professors' academic researches (Ranieri et al. 2013a).

Environmental Management

This WG aims to intensify the current actions related to the institutional programs at EESC. Two of these programs are the Water Rational Use Program (PURA) and the Energy Rational Use Program (PURE). This WG is also working with the following activities: Sustainable Indicators, Sustainable Construction, Sustainable Mobility, Waste Management, Sustainable Procurement and Electronic Management.

Sustainability Indicators

Currently, the university has some initiatives to develop sustainability indicators at EESC related to education, institutional policies and environmental management. In this sense, two research projects are being developed about environmental management indicators in order to evaluate sustainability, using Ecological Footprint and Global Reporting Initiative tools. For educational indicators and institutional policies, AISHE—Assessment Instrument for Sustainability in Higher Education (AISHE 2014) has been used. Finally, interviews and workshops were performed with experts in order to develop sustainability indicators (Ranieri et al. 2013a). Figure 2 presents these indicators for education, institutional policies and environmental management.

The next steps in this initiative are directed to the identification of limitations related to the indicators, methodologies selection, interviews with engineering courses coordinators and other collaborators at EESC, questionnaire elaboration for data collection, and database systematization (courses that integrate sustainability,

Indicators for Education

1. Number of mandatory disciplines introducing sustainability concepts;
2. Number of optional disciplines introducing sustainability concepts;
3. The presence of a standard institutional tool to identify sustainable-oriented disciplines;
4. Whether there are initiatives to encourage professors to include sustainability content in their disciplines;
5. Number of student groups discussing or sensitizing the local community regarding sustainability;
6. Number of undergraduate final research related to sustainability;
7. If sustainability education is included in university policies;
8. Support materials to promote sustainability inclusion in undergraduate curriculums;
9. Disciplines making use of university campuses to teach about environment and sustainability;
10. Tools to assess student learning evolution in sustainability;
11. Number of academic conferences related to environment and sustainability;
12. If there is an equal integration of sustainability within disciplines traditional content;
13. If there is cooperation among professors from different areas, aiming an interdisciplinary approach.

Indicators for Institutional Policies

1. If there is a strategic planning aiming the inclusion of sustainability within university activities;
2. A group in charge for a sustainable strategic plan development;
3. Professors, employees and students involved in institutional policies development;
4. If the university supports professors, employees and students' initiatives for sustainability;
5. If the university is opened for criticism from external groups regarding its policies and strategies;
6. Indicators to assess and follow goals achievement defined by the university;
7. If there is a news communication plan about sustainability;
8. Sustainability and social responsibility reports elaborated and disclosed regularly.

Indicators for Environmental Management

1. Water consumption (total and by person);
2. Energy use (total, by square meter and by person);
3. Paper consumption (total and by person);
4. Non-motorized transportation (%);
5. Perceived comfort;
6. Waste (total and by person);
7. Total CO2 emission (kg CO2 eq.);
8. Environmental criteria for building.

Fig. 2 Indicators for education, institutional policies and environmental management

final course paper and Pedagogic and Policies Projects). Finally, the construction of indicators is a continuous process and is necessary to review them, and to propose alternatives to solve their limitations (Ranieri et al. 2013a).

Sustainable Purchases

In this WG, the goal is to promote studies and actions to integrate sustainability in materials and services purchasing. In this context, the main activities are related to encourage the inclusion of sustainability principles into purchasing activities, orientation on materials procurement, and contracts made under social and environmental ethics. Furthermore, the WG is elaborating a guide the sustainable purchase at the EESC. This guide will have information to choose products and actions to promote the environment protection and sustainable development (Malachias et al. 2013).

Sustainable Construction

There is an initiative at USP to propose aspects related to sustainability inclusion in buildings projects since 2013, and this document is still under Superintendence of Environmental Management (SGA) analysis. Moreover, there is a WG elaborating guidelines to incorporate sustainability concepts at USP buildings.

As a pilot study, EESC developed a WG in sustainable construction, which is incorporated into Sustainable EESC program. Meetings were done from October 2012 to July 2013 with the WG members, employees, professors and people from Architecture and Urbanism department (IAU-USP) to identify improvement opportunities. As result, a checklist was proposed to assist the analysis of future buildings, to be reviewed by other institutes at USP Sao Carlos campus, such as the Institute for Mathematics and Computer Sciences (ICMC) and the Physics Institute of Sao Carlos (IFSC) (Ranieri et al. 2013b).

In the same way, three meetings involving 33 people from other USP campuses were done. The results from these meetings include that sustainability incorporation to USP buildings should be view as a strategic issue, not as a secondary plan or just for specific buildings. Furthermore, the final document obtained from the meetings must be seen as a standard to be followed at USP, and to reinforce the principles and values to be regarded, not limited only to specific aspects of buildings (SGA 2014b; Ranieri et al. 2013b). The adoption of an environmental certification system may be a feasible alternative, although it cannot replace a general standard from the university (SGA 2014b).

Finally, EESC campus is planning to build a building based on sustainability concepts to serve as headquarters for the Superintendence of Environmental Management and to all the administrative and educational activities developed by USP Recycles program (SGA 2014b).

Sustainable Mobility

The main result in Sustainability Mobility was based on a research developed by graduate and undergraduate students. Initially, it is possible to quote the research developed by Stein (2013), researcher from Transport Department, entitled "Barriers, motivators, and strategies for sustainable mobility in the campus of USP". Through the results was possible to identify that the users inquired would change their transportation modality to walking (27.1 %), bicycle (46.2 %) and automobile (13.6 %). As main barriers, the study mentioned motor vehicles. Related to alternatives, Stein's research identified two strategies to promote sustainable mobility: the construction of cycling paths connecting the campus with the urban street network, and parking lots for bicycles in the campus. Other strategy named is the construction of more pedestrian walkways in the campus (Silva et al. 2013).

Another research, still in development, has the aim to characterize non-motorized users' behavior that attend the campus. Finally, a study from Civil Engineering undergraduate can be mentioned. This research is developing preliminary indexes to evaluate the sustainability mobility conditions in the campus.

Solid Waste

In this WG, between 2012 and 2013, it was performed a quantitative and qualitative study to propose alternatives to improve waste management in the campus. This research was conducted by undergraduate students during a discipline and under professor's supervision. Besides, in this research was considered the Brazilian Solid Waste National Policy. In 2014, the WG established the following goals: reassessment and restructuring of waste selection, and further quantitative diagnosis to determine waste generation per student, for example, to support better sustainability indicators adoption (Leme et al. 2013b).

Other initiatives in this WG can be found in the "Practical Guide of Minimization and Waste Management at USP Sao Carlos", where some kind of wastes, such as printer cartridges, electronics, civil construction waste, conference waste, fluorescent lamps, chemical waste, non-recycling waste, etc. are presented, including tools and management practices to minimize their generation and final disposal alternatives (Leme et al. 2012a).

Community Outreach

This WG promote the interaction and collaboration with different societal instances to contribute with the generation of solutions for environmental issues. The main goal here is the integration of EESC community in different topics related to sustainability and environmental education. For instance, in 2013, this WG developed activities such as a short course entitled "Sustainability and Environmental Education" with other WGs participation, a diagnosis EESC wastes through photos exhibition and interviews with EESC employees to identify their doubts about waste separation (Nassour et al. 2013).

Information and Communication

This WG discloses and documents all program's steps in different sorts of media to spread the knowledge and to promote discussion with community participation. Finally, it is important to elaborate reports about sustainability implementation at EESC (Aranda and Luzia 2013).

5 EESC Initiatives Analysis and Discussion

As presented in the methodology session, the following Table 1 summarizes the main EESC's initiatives in comparison with Krizek's et al. phases.

In the first phase known as Grassroots, USP Recycles Program revealed to be one of the main drivers for sustainability at EESC. Created in 1993, USP Recycles had in mind since the beginning the purpose to consumption and waste management, including recycling. USP Recycles Program became an integral part of USP and EESC as well, extending its actions solid waste management, environmental education to local communities, such as lectures, and partnership with municipal governments, non-governmental organizations (NGOs) and other institutions. Thus, considering all USP Recycles' assignments, this program embraces other two phases: Executive Acceptance and The Visionary Campus Leader. The first one is justified by actions headed to cost reduction through solid waste management,

Table 1 Analysis of EESC initiatives under Krizek's et al. phases

Four phases	EESC initiatives
1. Grassroots	Re cycling programs leaded by USP recycles program
2. Executive acceptance	Some initiatives here are related to USP recycles program, for example, solid waste management; other WGs are in charge for actions such as water consumption reduction and energy use (environmental management)
3. The visionary campus leader	Environmental ethics and education within schools, and education for selective collection with local communities by USP recycles program (community outreach); Developing alternatives to assist administrative and operational activities, and university transparency (information and communication); CO_2 emission, sustainability indicators development for education and institutional policies (environmental management)
4. Fully self-actualized and integrated campus community	Engineering courses curriculum review, analysis of current sustainability integration level in engineering disciplines and new disciplines creation for all EESC engineering courses (Education and training in human resources); interdisciplinary cooperation and sustainability integration within professors' academic researches (research and innovation); sustainable construction and sustainable mobility (environmental management)

Table 2 SWOT analysis

	Positives points	Negatives points
Internal environment	*Strength (S)*	*Weaknesses (W)*
	USP recycles program Sustainable EESC Program supported by its directors Sustainable EESC WGs responsible to reach each initiatives goals Actions focused on rational use of water and energy Guidance for waste minimization at the campus (such as printer cartridges, electronics, civil construction waste, conference waste, fluorescent lamps, chemical waste, and non-recycling materials) Guidance for sustainable conferences organization Development of a guidance for sustainable buildings construction	Low-level of sustainability inclusion in curriculums Low concern level about topics related to sustainability in researches Low-level of partnerships between professor and sustainability projects Lack of updated databases about researches involving sustainability Low-level of sustainable mobility at EESC Low-level of results disclosure about rational use of water and energy inside EESC members
External environment	*Opportunities (O)*	*Threats (T)*
	Development of partnerships between EESC and other USF campuses for sustainability projects EESC can become a national and international reference for sustainable curriculum in engineering Indexes developed at EESC can assist other universities which are interested in sustainability	Lack of integration of sustainable actions can be a factor to decrease university reputation in a medium and long-term

while the second phase is explained by stakeholders' involvement and transparency of EESC environmental initiatives.

Regarding Krizek's et al. phase 2 (Executive Acceptance), other actions from Environmental Management initiative deserve attention. In this phase, EESC has WGs aiming to water consumption reduction and energy use, typical cost-reduction practices.

In the third phase (The Visionary Campus Leader), because of the EESC directors' commitment with sustainability, many actions were identified: CO_2 emissions, sustainability indicators development for education, institutional sustainable policies, local community and municipal government involvement (stakeholders participation), and transparency. One more time, USP Recycles Program plays an important role in this phase, specially extending EESC's sustainable actions outside the university.

Finally, some initiatives related to education and training, research and innovation, and environmental management suggest that EESC is achieving important elements present in phase 4. Some examples include sustainability rules for buildings construction, more environmental-friendly access and transportation across the campus, the integration of environmental content in all EESC's undergraduate courses, and also interdisciplinary cooperation for new academic researches about sustainability involving professors from different departments. Table 2 shows a SWOT matrix about sustainability integration at EESC.

A SWOT analysis demonstrated that several initiatives are being developed and other can be done to include sustainability topics in the university. However, it is clear that some barriers can negatively affect this integration process, for example, by lack of professors' motivation regarding sustainability topics. This situation was experienced by Pereira et al. (2014) at another USP campus, where there were just a moderate involvement with professors from other courses besides engineering, such as medicine, business administration, nursing and dentistry. In this sense, in the EESC specific case, new actions must be used to involve and motivate professors, for example, through workshops. Another barrier is a lack of an updated database containing information about sustainability projects at the university. This situation can affect the researches performance involving a large number of researchers.

Related to the opportunities, the actions under development at EESC can assist other universities, in Brazil or even in other countries, to integrate sustainable concepts in their activities. A special case is a curricular reformulation, which can encourage other universities to follow the same EESC steps.

6 Conclusions

This chapter brought many information about the initiatives developed at University of Sao Paulo (USP) regarding sustainability, and USP Recycles program has been an important contributor in this path to continue the development of sustainable initiatives at the university. However, the project still presents challenges that need to be analyzed and developed in order to identify and take opportunities.

Regarding the future actions, the authors realize how important incentives for professors are to integrate sustainability in their projects, create new partnerships and continue with the process to include sustainability in undergraduate curricula.

Lastly, this article brought relevant information for scientific community about how sustainability is being incorporated into Latin-American and other international universities. It is expected that the results here presented can support other universities that are just beginning their path towards a more sustainable performance.

Acknowledgments We acknowledge the Superintendence of Environmental Management (SGA), USP Recycles program (Sao Carlos) and all working groups supporting Sustainable EESC program for the valuable information provided.

References

Almeida CMVB, Santos PAZ, Bonilla SH, Giannetti BF, Housing D (2013) The roles, perspectives and limitations of environmental accounting in higher educational institutions: an energy synthesis study of the engineering programme at the Paulista University in Brazil. J Clean Prod 52:380–391

Aranda R, Luzia LF (2013)"Atividades do Grupo de Trabalho de Gestão Eletrônica". 6 slides: color Slides gerados a partir do software PowerPoint. (In Portuguese)

Assessment Instrument for Sustainability in Higher Education AISHE (2014) Assessment Instrument for Sustainability in Higher Education AISHE-Netherlands http://www.eauc.org.uk/theplatform/assessment_instrument_for_sustainability_in_hig. Last accessed 20 Mar 2014

Boyle C (2004) Considerations on educating engineers in sustainability. Int J Sustain High Educ 5 (2):147–155

Desha CJ, Hargroves KC (2010) Surveying the state of higher education in energy efficiency, in Australian engineering curriculum. J Clean Prod 18:652–658

ESSC Sustentável (2014) EESC Sustentável. http://www.eesc.usp.br/portaleesc/index.php?option=com_content&view=article&id=188&Itemid=349. Last accessed 16 Mar 2014 (in Portuguese)

Jabbour CJC (2010) Greening of business schools: a systemic view. Int J Sustain in Higher Edu 11 (1):49–60

Jabbour CJC, Sarkis J, Jabbour ABLS, Govindan K (2013) Understanding the process of greening of Brazilian business schools. J Clean Prod 61:25–35

Kamp L (2006) Engineering education in sustainable development at Delft University of technology. J Clean Prod 14:928–931

Krizek KJ, Newport D, White J, Townsend AR (2012) Higher educations's sustainability imperative: how to practically respond? Int J Sustain High Educ 13(1):19–33

Leme PCS, Mortean AF (2010) Guia prático para organização de eventos mais sustentáveis Campus USP de São Carlos. http://www.sga.usp.br/wp-content/uploads/Guia_eventos_sustentaveis.pdf. Last accessed 14 Mar 2014 (in Portuguese)

Leme PCS, Martins JLG, Brandão D (2012a) Guia Prático para Minimização e Gerenciamento de Resíduos-USP São Carlos. http://www.sga.usp.br/wp-content/uploads/Guiapr%C3%A1tico-para-minimiza%C3%A7%C3%A3o-e-gerenciamento-de-res%C3%ADduos-USP-Recicla-Digital.pdf. Last accessed 14 Mar 2014 (in Portuguese)

Leme PS, Nassour AC, Silva CAS, Romanelli GB, Valin SAA (2013b) Atividades do Grupo de Trabalho de Resíduos Sólidos. 6slides: color slides gerados a partir do software PowerPoint (in Portuguese)

Lopes BP, Assumpção L, Saavedra YMB, Puglieri F, Azanha A, Leme PS, Ometto AR (2014) Analysis of sustainability content into undergraduate engineering curriculum at the Engineering School of Sao Carlos, University of São Paulo, Brazil In: 2nd world symposium on sustainable development at universities (WSSD-U-2014) in press

Lozano F, Lozano R (2013) Developing the curriculum for a new bachelor's degree in engineering for sustainable development. J Clean Prod 64:136–146

Malachias CE, Giampá JGC, Mariano JP, Rui PA (2013) Compras Sustentáveis. 9 slides: color. Slides gerados a partir do software PowerPoint (in Portuguese)

Nassour AC, Silva CAP, Romanelli GIB, Maragno CA, Leme PS, Valin SA, Silveira WM, Mattos WAO (2013) "Atividades do Grupo de Trabalho de Formação ambiental da Comunidade e Coleta Seletiva". 11 slides: color Slides gerados a partir do software powerpoint (in Portuguese)

Ometto AR, Belini E, Montaño M, Musetti M, Ranieri V, Leme PS, Kuri NP, Aranda R, Puglieri F, Saavedra YMB, Azanha A, Shiguenaga TM, Lopes BP, Assumpção L (2013) Ambientalização Curricular nos Cursos de Graduação da EESC/USP. Diagnóstico Preliminar e proposta em construção" 64 slides: color Slides gerados a partir do software powerpoint (in Portuguese)

Palma LC, Oliveira LM, Viacava KR (2011) Sustainability in Brazilian federal universities. Int J Sustain 12(3):250–258

Pereira GSM, Jabbour CJC, Oliveira SVWB, Teixeira AA (2014) Greening the campus of a Brazilian university: cultural challenges. Int J Sustain High Educ 15(1):34–47
Ramirez M (2006) Sustainability in the education of industrial designers: the case for Australia. Int J Sustain High Educ 7(2):189–202
Ranieri V, Malheiros T, Leme PS, Kanbay AS, Guijarro MO, Brandão MS, Calixto KG, Kiyuna LSM, Marques J (2013a) Mapeamento de Pesquisas em Sustentabilidade na EES e Indicadores de Sustentabilidade para a EESC-USP 32 slides: color slides gerados a partir do software powerpoint (in Portuguese)
Ranieri V, Garcia CC, Sichieri EP, Salado GC, Schenk LBM, Daniel LA, Corrêa MRS, Brito SD, Pessotta RD, Pessotta RC, Nassour AC, Silva CAP (2013b) Edificações Sustentáveis e Áreas Verdes". 8 slides: color Slides gerados a partir do software powerpoint (in Portuguese)
Sammalisto K, Lindhqvist T (2008) Integration of sustainability in higher education: a study with international perspectives. Innov Higher Educ 32:221–233
Silva ANR, Stin PP, Brandão D, Ragonesi R (2013) Mobilidade Sustentável 2013.19 slides. color slides gerados a partir do software powerpoint (in Portuguese)
Stein PP (2013) Barreiras, motivações e estratégias para mobilidade sustentável no campus São Carlos da USP" Dissertação de Mestrado, Planejamento e Operações de Sistemas de Transporte. Escola de Engenharia de São Carlos (in Portuguese)
Superintendência de Gestão Ambiental—University of São Paulo USP (SGA) (2014a) USP Recicla-Historico. http://www.sga.usp.br/?page_id=1011. Last accessed 14 Mar 2014 (in Portuguese)
Superintendência de Gestão Ambiental—Universityof São Paulo USP (SGA) (2014b) Relatório de Atividadesfevereiro/2012 a Janeiro/2014. http://www.sga.usp.br/wp-content/uploads/Relat%C3%B3rio-SGA-janeiro-2014-Final-Reitor.pdf. Last accessed 14 Mar 2014 (in Portuguese)
UNESCO—United Nations Educational Scientific and Cultural Organization (2014) Education for sustainability—from Rio to Johannesburg: lessons learnt from a decade of commitment 2002. http://www.unesco.org/new/en/jakarta/education/education-for-peace-and-sustainable-development-psd/education-esd/. Last accessed 20 Feb 2014
Universidade de São Paulo (USP) (2014) 80 Anos de Excelência. http://www5.usp.br/institucional/a-usp/historia/. Last accessed 10 Mar 2014. (in Portuguese)
WFEO (1997) Commitment to sustainable development, WFEO General Assembly. Available at: www.wfeo-comtech.org. Last accessed 20 Feb 2014

Authors Biography

Yovana M.B. Saavedra has a degree in Environmental Engineering from Universidad Manuela Beltrán in Colombia and a Master degree in Production Engineering. Currently, she is a Ph.D. Candidate in Environmental Engineering Sciences at EESC, USP. She has experience in Environmental Management, Strategies for End of Life Products, Life Cycle Assessment (LCA), Material Flow Analysis (MFA), Industrial Ecology (IE) and Management Life Cycle Product.

Fabio Neves Puglieri is a production engineer from Federal University of São Carlos (UFSCar). He got his Master's degree in Production Engineering from EESC, USP, and currently he is a Ph. D. Candidate in Production Engineering at EESC, USP. Puglieri's research comprises Sustainable Business Strategies, Strategic Planning, Life Cycle Management (LCM), Life Cycle Assessment (LCA) and Design for Environment.

Vitor Eduardo Lima Ranieri is an agronomic engineer from USP, with Master's degree in Environmental Engineering Sciences from USP, and Ph.D. in Hydraulic Engineering from USP.

Currently, he is a professor at USP, working with protected natural areas, environmental impact assessment, environmental zoning and environmental licensing.

Rosane Aparecida Aranda has graduation in Pedagogy from Federal University of São Carlos (UFSCar). She is an expert in social-environmental sustainability from USP and MBA in Public Management from GPublic-USP. Currently, she is analyst in administrative affairs and administrative advisor at EESC-USP.

Patrícia Cristina Silva Leme is biologist from Federal University of Sao Carlos (UFSCar). She has Master's degree and a Ph.D. in education by UFSCar. She is an invited researcher at College of Education, da University of South Africa (2013–2015), and educator at USP working with the Environmental Management Superintendent (EMS).

Aldo Roberto Ometto is a chemical production engineer from Federal University of São Carlos (UFSCar). He has Master's degree in Environmental Engineering Sciences and a Ph.D. in Hydraulic Engineering, both from EESC, USP. Currently, he is a professor at the Production Engineering at EESC, USP, working with researches related to Engineering and Life Cycle Management of Products.

Quality Air Monitoring in a Brazilian University: The Use of Passive Samplers

Eduardo Pavan Korf, Denise Daris, Luciana Londero Brandli, Marcos Antonio Leite Frandoloso and Giulius Sebastian Soares Schwanz

Abstract

The aim of this paper is to monitor the air pollution at Campus I of Passo Fundo University, south Brazil, focusing on atmospheric pollutants SO_x, NO_2 and O_3. To monitor the air quality we used passive samplers placed in several locations in the university campus, namely: energy generator, pool heating boiler, entrance porch, an exhaustion chamber and a place with less anthropic influence to represent the blank. The monitoring period went monthly from July 2012 to March 2014. After monitoring the sample's membranes with absorbing solutions were analysed in the laboratory, by analytical methods specific for each gas. The results obtained for NO_2, O_3 e SO_x from all monitoring sites varied between 0.11–8.20, 0.66–9.50 and 0–1594 μg/m^3, respectively. The results of the various monitoring locations on campus were, with the exception of SO_x, below those recommended by the legislation. It was also observed that the energy generator was the site where

E.P. Korf (✉) · M.A.L. Frandoloso · G.S.S. Schwanz
Faculty of Engineering and Architecture, University of Passo Fundo, Campus I, Passo Fundo, RS 99052-900, Brazil
e-mail: eduardokorf@upf.br

M.A.L. Frandoloso
e-mail: frandoloso@upf.br

G.S.S. Schwanz
e-mail: 120298@upf.br

D. Daris · L.L. Brandli
Environmental and Civil Engineering, University of Passo Fundo, PPGeng, Campus I, Passo Fundo, RS 99052-900, Brazil
e-mail: dedaris@hormail.com

L.L. Brandli
e-mail: brandli@upf.br

W. Leal Filho et al. (eds.), *Integrating Sustainability Thinking in Science and Engineering Curricula*, World Sustainability Series, DOI 10.1007/978-3-319-09474-8_20

pollution was highest. Thus, it is recommended that combustion gases are treated before they are discharged into the atmosphere. Finally, by showing a case study, this paper emphasizes the importance of monitoring the environmental aspects from the different activities in the university campus.

Keywords
Air pollution · Fick's law · Higher education institutions · Nitrogen dioxide · Ozone

1 Introduction

Many studies have produced evidence that air pollution has adverse effects on human health and that it is responsible for the increase of cases of respiratory diseases, which reflect in an increase in public health costs (Li et al. 2012). There are established standards within legislation for air quality; however, to ensure that these legal requirements are carried out, a monitoring process is necessary.

Passo Fundo University (UPF), situated in the north of the state of Rio Grande do Sul, Brazil, is involved in activities that potentially cause air pollution. Currently, it has approximately 20,000 students, who transform this campus into a small town (Brandli et al. 2011). Therefore, besides the movement of vehicles, other varied activities occur, such as sewage treatment, energy generation, chemical laboratories, and others. These all produce atmospheric pollutants that require monitoring for their control. Among the main ones, we have methane (CH_4), volatile hydrocarbons (VOC), nitrogen oxide (NO_x), sulfur oxide (SO_x), carbon monoxide (CO), carbon dioxide (CO_2), particulate matter (PM) and acid vapors.

The difficulty in monitoring air quality occurs due to the fact that pollutants are generally found in low concentrations in the atmosphere as a result of the process of dispersion and dilution (Cruz and Campos 2008). Besides, active monitoring, using automatic portable equipment or fixed installations is almost always burdensome, requiring considerable investment, trained personnel to operate it and the supply of constant electrical energy. This makes this monitoring model ineffective, giving rise to the necessity of alternatives in regions where this technique is not viable (Cruz and Campos 2002).

Because of the emissions from Passo Fundo University, air pollution monitoring has become a necessity and the idea is to make a diagnosis of air quality in this location and also to study a viable technology for the reality of the institution. One of the viable alternatives, from a practical and economic perspective, is the use of passive samplers to measure air pollutants. It is also important to emphasize that no official (e.g. by environmental agencies) air monitoring is done in Passo Fundo or in nearby cities. Fixed air monitoring stations are not a common reality in Brazil, especially because of the costs associated with them; therefore, passive samplers can be a good qualitative way to verify air quality at sites outside the university campus as well.

There are various types of passive samplers with various applications. One of them is the use of plants, mosses, and lichens, known as biomonitoring (Wolterbeek 2002). Another one is the construction of badge or tube type samplers with low cost materials which are based on principles such as diffusion or absorption with subsequent permeation of gas or vapor molecules through a membrane (Cruz and Campos 2002).

Many studies have been carried out using passive samplers for monitoring the more diverse types of pollutants, such as SO_2 (Cruz et al. 2004), NO_2 (Sekine et al. 2008; Melchert and Cardoso 2006), O_3, H_2S (Campos et al. 2010), polycyclic aromatic hydrocarbons (Tao et al. 2008; Cornelissen et al. 2008; Zhang et al. 2011), persistent organic pollutants (Melymuk et al. 2011; Chaemfa et al. 2009), polychlorinated biphenyl (Cornelissen et al. 2008), metals, and particulate matter (Brait and Antoniosi Filho 2010). The majority of these studies were conducted in emerging countries, which illustrates that passive samplers represent an accessible alternative for monitoring air quality. Among the advantages is the ease of installation at the site, the low construction and installation costs, the limited need for maintenance after installation, low cost laboratory analyses, no electrical energy required, and little variation in price for their use in different regions (Krupa and Legge 2000).

Air quality standards in Brazil are established by CONAMA 003/90 Resolution (Brasil 1990), which determines the primary standards—maximum concentrations of a pollutant which can be found in the atmosphere, and secondary standards—concentration levels that produce minimum adverse effects in the population. Ideally, pollutants should be below the secondary standard to ensure a healthy and favorable environment.

With the aim of evaluating the applicability of passive samplers for the monitoring and diagnosis of air pollution at Passo Fundo University, the general focus of this study is on evaluating and verifying the viability of using a passive sampler for atmospheric pollutants SO_X, NO_2 and O_3, using the methodologies available in the literature.

2 Methodology

2.1 The Study Area and Pollutants

The Passo Fundo University campus is located near Passo Fundo town in South of Brazil. The samplers were installed on Campus 1 of Passo Fundo University in locations which have different sources of pollution. The monitored pollutants were SO_X, NO_2 and O_3. These pollutants were chosen because they are common in urban regions. The sulfur dioxide (SO_2) is a primary pollutant mainly resulting from the combustion process of fuels that containing sulfur such as heavy fuel oil. This combustion is also the main source of NO_2, and ozone is a secondary pollutant which can be formed from NO_2; its presence is also responsible for the increase in cardiovascular and respiratory diseases (Hoek et al. 2002).

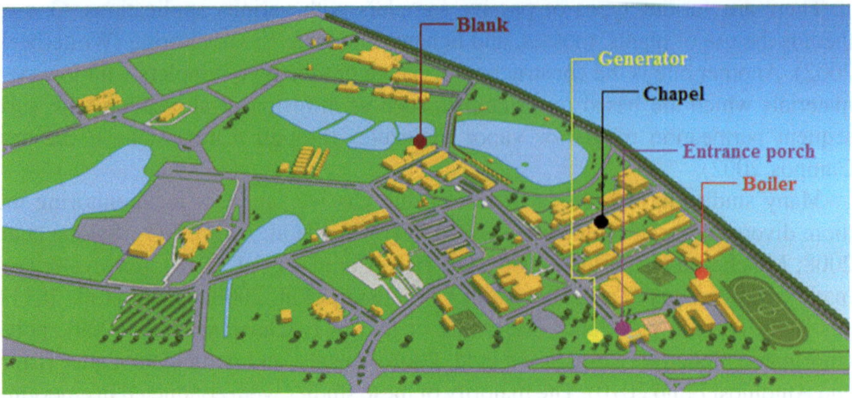

Fig. 1 Sampling sites at UPF

2.2 Monitoring Sites

The choice of the points was based on pollutants sources existing in the university, as illustrated in Fig. 1. P1: Generator: located near the exit of the gas power; P2: Entrance Porch: located in the portico of vehicular access; P3: Boiler: near the boiler heating the pool; P4: Chapel: fume cupboard chemistry lab P5: isolated places, without contaminants, used as "blank".

2.3 The Experiment

A passive sampler was constructed, based on the configurations proposed by Campos et al. (2010) and showed in the Vieira et al. (2013). The basic process of the sampler is the molecular diffusion of gas through a static layer of air, and its structure is made up of PVC connections, which are closed at the top to prevent convective transport. A Teflon membrane (Millipore, PTFE, 0.5 μm, 25 mm in diameter, hydrophobic), found at the air inlet, is used to minimize the interference of particles and turbulent diffusion. Below that, there is a stainless steel screen (0.08 × 0.125 mm), with the purpose of protecting the membrane from mechanical damage. After the diffusion space has a filter, which is immersed in an absorbing solution, both characteristic for each pollutant. For the NO_2 and SO_x pollutants, the solutions were produced according to Campos et al. (2010), being used cellulose filter (Whatman 40). The O_3 pollutant solutions were produced in the laboratory according to Bucco (2010) and used the filter paper Fiberglass (50GF/A).

Air samples were collected from the samplers, installed during 5 days in a week for 1 week for month for all the 18 months monitored (July 2012–January 2014).

2.4 Analysis of Pollutants

To analyse the concentration of pollutants in the filters, UV/VIS color spectrophotometry methods were used. For the analysis of NO_2, the modified Griess-Saltzman (Saltzman 1954) method, described by Campos et al. (2010), was used, and Bucco's method (2010) was used for O_3.

The concentration of pollutants in the external environment was determined, following the principle of Fick's first law of diffusion, as demonstrated in Eq. 1, described by Campos et al. (2010).

$$C = \frac{m \cdot L}{D \cdot A \cdot t} \quad (1)$$

where:

C—External concentration (environment) of gas ($\mu g\ m^{-3}$);
m—Total collected mass (μg);
L—Length of line of diffusion (m);
D—Coefficient of diffusion ($m^2\ h^{-1}$), obtained from the literature, according to Massman (1998). 0.3769 $m^2\ h^{-1}$ for NO_2 and 0.3999 $m^2\ h^{-1}$ for O_3
A—Cross-sectional area of the line of diffusion (m^2);
t—Sampling time (h).

The total collected mass (m) is the one obtained by the previously mentioned analytical methods and determined by Eq. 2.

$$m = C_{extract} \times V_{extract} \quad (2)$$

where:

$C_{extract}$—is the result obtained in the analyses which have a unit of mass/volume;
$V_{extract}$—is the quantity of extract used to obtain the value of $C_{extract}$. $V_{extract}$ is 0.005 L for NO_2 and 0.025 L for O_3.

3 Results and Discussion

3.1 NO_X Concentration

The NOx concentrations determined on the basis of the experiment performed for the period from July 2012 to January 2014 are shown in Fig. 2 as well as the rainfall. Results indicate that concentrations are higher in the Generator point. The Generator point showed the highest value of nitrogen dioxide concentration throughout the monitoring period, with peaks in the months of November 2012,

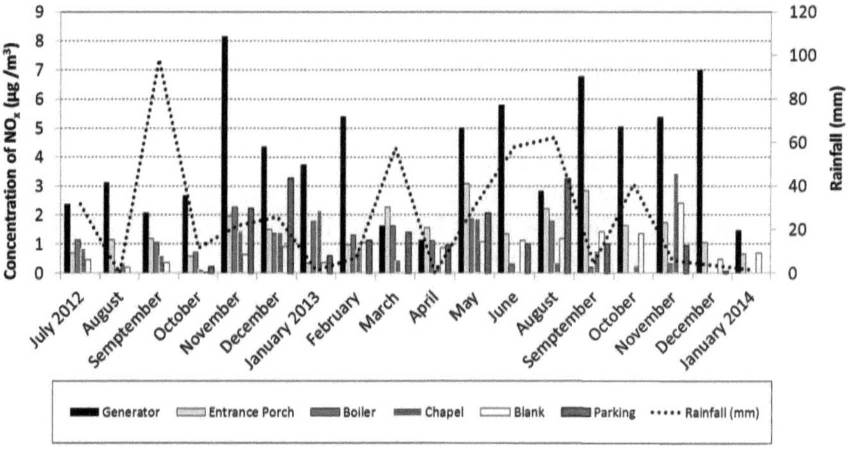

Fig. 2 NO_x concentration

September and December 2013 with values of 8.13, 6.75 and 6.97 µg/m^3 respectively, instead of April 2012, which showed the lowest value 1.12 µg/m^3.

The highest values can be justified by the burning of diesel. This occurs because gas is discharged by the exhaust system without any previous treatment. Although these are the highest values found this pollutant is not of great concern because it has not reached the limits established.

After the generator, parking presents the second highest value, which can be seen in the months of December 2012 to 3.25 µg/m^3 and August 2013 with 3.23 µg/m^3. Thirdly, the entrance porch on May 2013 with 3.06 µg/m^3 and on September 2013 with 2.83 µg/m^3.

These values are mainly related to mobile sources coming from the vehicle, because in this university on average 5,000 cars, 300 buses and 230 intercity buses and vans circulate daily (UPF 2012).

According to Perkauskas and Mikelinskiene (1998), who conducted their research with passive samplers in Lithuania, the vehicular pollution is the main source of emission of NO_2. this confirms the results observed in this study for highs levels of NO_2 and NO_x in the parking.

Analysing the data has not found correlations with the NO_X concentration and the rainfall.

The maximum concentration obtained considering the variations of place and period, was 8.13 µg/m^3, the value below which determines the CONAMA Resolution 03/90 (Brasil 1990) that establish as Primary Standard average annual concentration of 100 µg/m^3 and average concentration of 1 h of 320 µg/m^3.

3.2 O_3 Concentration

In Fig. 3 it can be observed that the sampler near the generator showed the highest value of ozone concentration, with a peak in May 2013 about 9.5 μg/m^3. This month it has presented the highest O_3 concentration of all the points.

Vieira et al. (2013) observed an evident relationship between the presence of NO_x and O_3 compounds, since ozone is a secondary pollutant, and their formation depends on the presence of compounds such as NO_x and VOCs. The same result was observed in this study considering the monitored data of both pollutants, within the reporting period.

The main source of O_3 concentration in UPF campus is the burning of fossil fuel for the buses, which can be an inducer of formation of such secondary pollutant, due to the emission of Volatile Organic Compounds (VOCs) and NO_x, as observed in May 2013, for the latter.

In a study conducted by Watson et al. (2002), the exhaust gases from the burning of diesel fuel (as in the case of the power generator of the UPF and some vehicles) and the exhaust gases from the burning of gasoline (as in the case of vehicles) were presented as the main inducing gases of O_3 formation.

Regarding rainfall, this can show a small influence on the ozone concentrations. During the wetter months it is noted that the concentrations of O_3, for all points tend to reduce as the overcast or rainy weather protecting the surface of the radiation falling on the ground, it may be too wet deposition and inhibiting the formation of O_3 from NO_X and VCO_S by photosynthetic reactions.

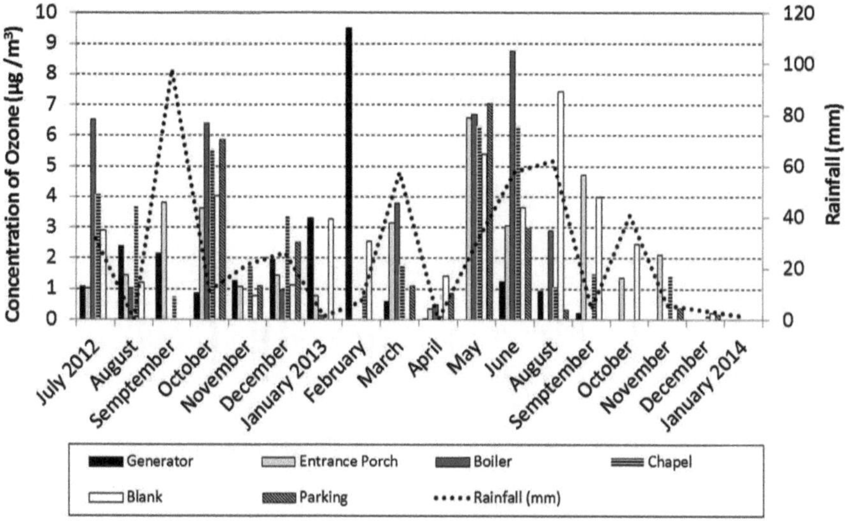

Fig. 3 O_3 Concentration

Comparing the concentrations obtained with the Brazilian law, it is observed that the maximum ozone concentration was 9.50 μg/m^3, much lower than the limit set by CONAMA Resolution 03/90, that define do not exceeded more once a year 160 μg/m^3 (Brasil 1990).

3.3 SO_x Concentration

Figure 4 presents the levels of concentrations for SO_X. The concentration data regarding Blank point were not considered due to the values found for some represented peak levels, possibly by the influence of an unknown external source.

The point that showed the highest level of pollution was in the boiler with values in July 2012 and August 2013 from 1,376.7 to 1,164 μg/m^3, respectively. Followed by the boiler, the porch had values of 1,300 μg/m^{3}in August 2012 and 1,191 μg/m^3 in August 2013 and the chapel had values of 1,216.91 μg/m^3 in July 2012 and 1,230 μg/m^3 in August 2013.

These values in the boiler and porch are justified by the fact of SO_x gases such as SO_2 were being emitted into the atmosphere mainly by burning fossil fuels, since each of these points has more intense traffic. (Cullis and Hirschler 1980).

Analysing the results of Fig. 4, it is observed that the highest values of SO_x were in August 2013 on most collected points, followed by the month of July 2012, which showed similar results. The lowest values occur in the months of January and February for most points, probably due to the holiday period of the institution, where the circulation of the vehicles decreases.

Comparing the concentrations obtained with the Brazilian law, it is observed that the maximum SOx concentration was 1,230 μg/m^3, above the limit set by CONAMA Resolution 03/90, that define do not exceeded more once a year 80 μg/m^3 (Brasil 1990). However, during the monitoring period, only two values were above this limit (1,216.91 μg/m3 in July 2012 and 1,230 μg/m^3 in August 2013).

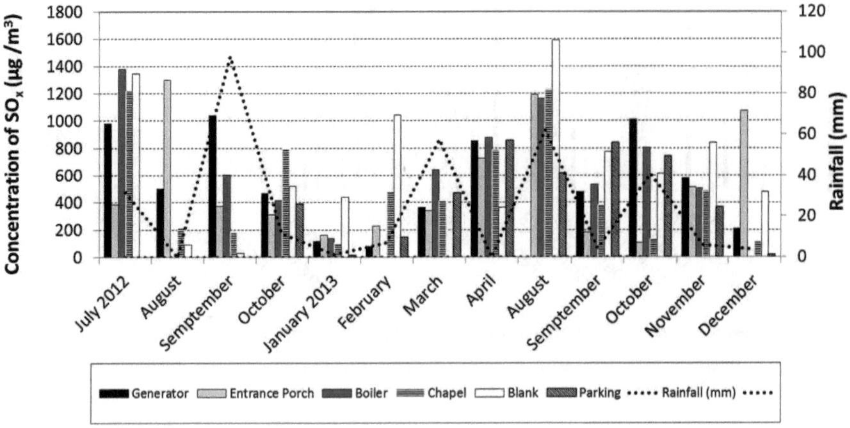

Fig. 4 SO_x Concentration

The explanation of the variation of SO_x concentration can be explained by the emission of sulfur dioxide to be related to the use of fossil fuels containing sulfur, both vehicles as for power generation.

Analysing the data has not found correlations with the SO_x concentration and the rainfall.

3.4 Perspectives

Data from air quality monitoring are environmental aspects so the university can reflect on environmental quality improvement involving students and academic staff. The use of passive samplers is justified by its economic viability and facility to use. The traffic of vehicles and the generator are major sources of air pollution. With regards to the vehicles this is the result of a lack of a mobility plan by the municipality and also by the University. In this case, actions should be taken to reduce the volume of traffic, with associated environmental benefits.

For this, the UPF can use some educative initiative:

Public Transport: publish information about the bus at the bus stop (time, route); negotiate better conditions of bus (frequency) and a substantial discount in the tickets for the staff and students.

Car use: promote a car sharing scheme (car pool); introduce the use of paid/controlling parking in the university campus.

Cycling: promote bicycle use in the campus; construct lanes for cyclists, parking and make available bicycles for use inside the campus; campaigns and orientations about directions, security and others.

Information about travel on the website: the university could update all information about travel on the website, for example; groups for car sharing, time of buses, route of bus, campaigns.

4 Conclusion

The results presented in UPF campus, the burning of fossil fuels and vehicles are the main source of SO_x and NO_2, the latter, along with the VOCs, may have been an inducer of O_3.

It should be emphasized that these results are treated in a qualitative analysis, because there is no monitoring station or active monitoring able to prove that these values reflect the reality of the equipment. However, the findings are accurate in connection with the actual concentrations, since the same sampler was used and validated by Campos et al. (2010) in this study and obtained an accuracy of 10–19 %.

The passive samplers were presented as an alternative easily applied to academia regarding air monitoring, the structure adopted and the methodologies for analysing low cost. A perceived disadvantage is the time it takes to get a result from the experiment, which can take more than a week. The points with highest concentrations are the power generator followed by the boiler and portico. The maximum concentrations obtained in this research about the pollutants SO_X, NO_2 and O_3 in the UPF campus were below which determines the legislation.

This does not mean that the university doesn't need to worry about these pollutants, in contrast, the university must think about sustainable travel, reduce single occupancy vehicles, invest in and provide facilities for cyclists, encourage the use of public transport, and invest in awareness campaigns for staff and students to increase different, more efficient modes of transport.

Acknowledgments Acknowledgments the CAPES for the financial support for the research postdoctoral of Prof. Luciana L. Brandli under the supervision of Prof. Walter Leal Filho at Hamburg University of Applied Sciences (Process 9285-13-0).

References

Brandli LL, Frandoloso M, Tauchen J (2011) Improving the environmental work at University of Passo Fundo, Brazil—towards an environmental management system. Braz J Oper Prod Manag 8(1):1–24

Bucco MVS (2010) Construção e testes de validação de amostradores passivos para dióxido de nitrogênio e ozônio. Dissertação (Mestrado Profissional em Meio Ambiente Urbano e Industrial/Meio Ambiente Urbano e Industrial), Setor de Tecnologia, Universidade Federal do Paraná, Curitiba

Brait CHH, Antoniosi Filho NR (2010) Desenvolvimento e aplicação de sistema passivo de coleta de poluentes atmosféricos para monitoramento de Cd, Cr, Pb, Cu, Fe, Mn, Zn e particulados totais. Química Nova, São Paulo 33(1):7–13

Brasil (1990) Resolução CONAMA n°. 003, de 28 de junho de 1990. Diário Oficial da União, Brasília, de 22/08/90, 15937–15939, Seção I

Campos VP, Cruz LPS, Godoi RHM, Godoi AFL, Tavares TM (2010) Development and validation of passive samplers for atmospheric monitoring of SO_2. Microchem J 96:132–138

Chaemfa C, Barber JL, Kim K, Harner T, Jones KC (2009) Further studies on the uptake of persistent organic pollutants (POPs) by polyurethane foam disk passive air samplers. Atmos Environ 43:3843–3849

Cornelissen G, Arp HPH, Pettersen A, Hauge A, Breedveld GD (2008) Assessing PAH and PCB emissions from the relocation of harbor sediments using equilibrium passive samplers. Chemosphere 72:1581–1587

Cruz LPS, Campos VP (2002) Amostragem passiva de poluentes atmosféricos. Aplicação ao SO_2". Química Nova, São Paulo, 25, 3, 406–411

Cruz LPS, Campos VP (2008) Métodos de amostragem e análise para compostos reduzidos de enxofre atmosférico. Química Nova, São Paulo 31(5):1180–1189

Cruz LPS, Campos VP, Silva AMC, Tavares TM (2004) A Field evaluation of a SO_2 passive sampler in tropical industrial and urban air. Atmos Environ 38:6425–6429

Cullis CF, Hirschler MM (1980) Atmospheric Sulphur: natural and man-made sources. Atmosspheric Environment 14:1278

Hoek G, Brunekreef B, Goldbohm S, Fischer P, Brandt, PA, van den (2002) Association between mortality and indicators of traffic-related air pollution in the Netherlands: a cohort study. Lancet 360:1203–1209

Krupa SV, Legge AH (2000) Passive sampling of ambient, gaseous air pollutants: an assessment from an ecological perspective. Environ Pollut 107:31–45

Li J, Ewart G, Kraft M, Finn PW (2012) The public health benefits of air pollution control. J Allergy Clin Immunol 130(1):22–23

Massman WJ (1998) A review of the molecular diffusivities of H_2O, CO_2, CH_4, CO, O_3, SO_2, NH_3, N_2O, NO, and NO_2 in air, O_2 and N_2 near STP. Atmos Environ 32(2):1111–1127

Melchert WR, Cardoso AA (2006) Construção de amostrador passivo de baixo custo para determinação de dióxido de nitrogênio. Quim Nova 29(2):365–367

Melymuk L, Robson M, Helm PA, Diamond ML (2011) Evaluation of passive air sampler calibrations: selection of sampling rates and implications for the measurement of persistent organic pollutants in air. Atmos Environ 45:1867–1875

Perkauskas D, Mikelinskiene A (1998) Evaluation of SO_2 and NO_2 concentration levels in Vilnius (Lithuania) using passive diffusion samplers. Environ Pollut 102(S1):249–252

Saltzman BE (1954) Colorimetric microdetermination of nitrogen dioxide in the atmosphere. Anal Chem 26(12):1949–1955

Sekine Y, Watts SF, Rendell A, Butsugan M (2008) Development of highly sensitive passive sampler for nitrogen dioxide using porous polyethylene membrane filter as turbulence limiting diffuser. Atmos Environ 42:4079–4088

Tao S et al (2008) A directional passive air sampler for monitoring polycyclic aromatic hydrocarbons (PAHs) in air mass. Environ Pollut 156:435–441

UPF (2012) Duplicação da BR-285 é discutida. Disponível em: http://www.upf.br/vfjk. Acesso em: 15 Nov 2012

Vieira LC, Korf EP, Brandli LL (2013) Passive samplers for air quality monitoring in a Brazilian University. Int J Environ Pollut 53(1/2):148–158

Watson MA, Scott K, Griffith J, Dieter S, Jones CS, Nanda S (2002) The developmental ecology of mycorrhizal associations in mayapple, *Podophyllum peltatum*, Berberidaceae. Evol Ecol 15:425–442

Wolterbeek B (2002) Biomonitoring of trace element air pollution: principles, possibilities and perspectives. Environ Pollut 120:11–21

Zhang Y et al (2011) A passive air sampler for characterizing the vertical concentration profile of gaseous phase polycyclic aromatic hydrocarbons in near soil surface air. Environ Pollut 159:694–699

Authors Biography

Eduardo Pavan Korf received his Masters degree in Engineering from the Federal University of Rio Grande do Sul in 2011 and graduated in Environmental Engineering from the University of Passo Fundo in 2010. Since 2010, he has been a Professor in an Environmental Engineering Course at the Passo Fundo University and has been researching the topics: environmental monitoring such as air pollution, environmental modeling and environmental geotechnics. He is currently a PhD student in Engineering at the Federal University of Rio Grande do Sul, expected defense to March 2015.

Denise Daris holds a degree in Food Engineering from Regional Integrated University of High Uruguay and Missions (2000). Specialization in Food Technology (2004) and Word Safety (2011). She is currently an engineer of Work Safety, Professor at the Federal Institute of Education,

Science and Technology of Rio Grande do Sul, IFRS, Brazil. Student at Master in Environmental and Civil Engineering at Passo Fundo University since 2013 researching atmospheric pollution.

Luciana Londero Brandli graduated in Civil Engineering (1995), master's degree in Civil Engineering (1998) and Ph.D. in Production Engineering (2004). Pos Doctorial Research at Hamburg University of Applied Sciences (2014). She is currently an Associate Professor in the University of Passo Fundo, south of Brazil, working on the Masters Program in Engineering, Infrastructure and Environment. Her current research interests include sustainability in high education and green campus, environment management, management of urban infrastructure, sustainable cities and green building.

Marcos Antonio Leite Frandoloso has a degree in Architecture and Urbanism at Federal University of Pelotas (1986), Master in Architecture at Federal University of Rio Grande do Sul (2001), and a course in Urban Ecology at the Universitat Oberta de Catalunya (2008). Currently he is a PhD candidate at Universitat Politècnica de Catalunya, Barcelona, Spain, focusing on eco-efficiency and environmental management at universities. Besides his Architectural office, he is a Professor and researcher at the University of Passo Fundo—UPF—since 1995, nowadays he is Coordinator of Product Design course at the UPF. He has experience in Architecture and Urbanism with emphasis on building and environmental planning, acting on the following topics: energy efficiency, energy and environment, sustainable construction, bioclimatic architecture, architectural heritage, eco design and urban ecology.

Giulius Sebastian Soares Schwanz is student of Environment Engineering and has a CNPQ scholarship.

Part II
Best Practice

Embedding Environmental Sustainable Development Within Social Enterprise at the University of Northampton

Simon Mark Pole

Abstract

The two key themes from the United Nations Rio+20 conference of a green economy, and creating an institutional framework for sustainable development were developed at the University of Northampton Students' Union. The University of Northampton is currently the only accredited *Ashoka U Changemaker Campus* within the UK. The university has committed to embed the principles and values of Social Enterprise and Social Value within every course and to encourage entrepreneurial activity in a socially-minded manner in all graduates to influence the creation of a fairer and more sustainable society. Social enterprises aim to benefit society as their primary purpose before making profits which is a key component to overcome the challenges of sustainable development. However, the Students' Union felt that too much emphasis was put on social and financial elements, while the environmental aspects were being neglected. The answer was a successful bid for *Planet Too* a £250,000 two year project joining twenty-five other universities across the UK as part of the Students' Green Fund, government funded by HEFCE. *Planet Too* is a student-led program that delivers seven distinct interventions designed to transform thinking and behaviours around sustainable development at the University of Northampton. It combines aspects of volunteering and campaigning with social enterprise to increase employability and to engage students in leading change on their campus. Students, academics, landlords and the wider community are connected through the different elements. This essay highlights the ways environmental sustainability has been embedded into the social enterprise Changemaker ethos.

S.M. Pole (✉)
University of Northampton Students' Union, Boughton Green Rd,
Northampton NN2 7AL, UK
e-mail: simon.pole@northampton.ac.uk

W. Leal Filho et al. (eds.), *Integrating Sustainability Thinking in Science and Engineering Curricula*, World Sustainability Series, DOI 10.1007/978-3-319-09474-8_21

"The economy is a human construct. Economic "laws" are not like the law of gravity. They can be altered."
—(O'Neill, 2010)

Keywords
Education · Economics · Social enterprise · Environment · Sustainable development

1 Introduction

Forty years after the Stockholm Conference on the Human Environment, and twenty years after the first "Earth Summit" held in Rio de Janeiro, Brazil in June 1992, the world community met to outline the strategy and processes to move towards a global sustainable world. The historic UN Conference on Sustainable Development (UNCSD) in 2012, also known as "Rio+20", agreed on actions to make the world better and more sustainable in light of global climate change, and challenges we currently face such as resource use, food and energy security and striking global ecological changes. Priority attention was given to poverty eradication, food security and nutrition and sustainable agriculture, energy, sustainable transport, sustainable cities, health and population, and promoting full and productive employment due to their paramount importance (UN 2012). Poverty eradiation was given the highest priority with social protection systems being essential. Social protection can be defined as being concerned with preventing, managing, and overcoming situations that adversely affect people's lives (UNRISD 2010). Embedding sustainability into business ethics is a key part of *Planet Too* at Northampton University due to the interrelatedness with business, poverty, sustainability and environmental impact. The conference concentrated on two crucial themes a green economy in the context of sustainable development poverty eradication, and the institutional framework for sustainable development.

Leading from this and the significant role of universities as well as their multifaceted and collective knowledge, the first World Symposium on Sustainable Development at Universities was held in 2012. One of its strengths was the cross-sectoral range of participants including administrators, students, staff, operations staff, project officers and representatives. Universities acted as hubs for multidisciplinary knowledge on the complex topic of sustainable development as well as being an ideal environment for creating change as seen with the UK Student Green Fund movement (EAUC 2014). The main aims of the symposium were to provide opportunity to showcase university work, support exchange of information ideas and experiences, discuss methodological approaches and networking with the hope of powerful cooperation. The book 'Sustainable Development at Universities: New Horizons' was the highly comprehensive publication allowing the impact of the symposium to transcend beyond the event (Leal Filho 2012). Similar to Hill's (2007) concept of a 'mastermind' where the collective knowledge of a group is transcended by working together on the same topic, new ways of thinking were opened up.

There is a delicate interplay between economic systems, the environment of which they are part and cultural behaviors and beliefs which form them. On the one hand the pursuit of growth with the economy along with advancing technology is seen as the answer to current environmental problems (O'Neill and Ben-Ami 2013). However a global economic system driven towards endless growth conflicts with the finite resources which it is dependent upon. Beneath issues of poverty lie more complex issues of inequality, corruption and even how 'wealth' is measured (Dietz and O'Neill 2013).

The University of Northampton Students' Union felt too much emphasis was being put on social and financial aspects with the environmental side being overlooked. This created a successful bid for *Planet Too* a £250,000 twoyear project through the Students' Green Fund along with twenty-five other universities across the UK funded by Higher Education Funding Council for England. *Planet Too* is a student-led program that delivers seven distinct interventions designed to transform thinking and behaviors around sustainable development at the University of Northampton. These are Sustainability Changemakers, Student Switch Off+, Green House Award Scheme, Sustainable Business Ethics Loan Fund, Sustainable Employability Module, Sustainability Grants and Awards, and Changemaker Campus Delivery. The project has aimed to be innovative as student engagement has been historically low on campus. Dynamic methods combining the use of social media, competitiveness, social norms and the strong link between sustainability activities and engagement were used.

Behaviour change theories are discussed and applied to the *Planet Too* project (Miller and McFarland 1991; Toch and Klofas 1984; Berkowitz 2005). Ecological economic theories as alternatives to dominant paradigms are then critically analyzed (Ropke 2004; Daly 2013). This chapter explores different strategies for effective student involvement and engagement with sustainable development, highlighting the importance of developing student leadership.

2 Behaviour Change Theories

The *Planet Too* project was developed with a small group of seventeen students including the sabbatical officers with workshops where issues of sustainability were explored and what could be done to change the status quo. The group first agreed that general student preconceptions about sustainability need to be challenged. In social norm theory the 'norm' includes misconceptions regarding how other people behave, thus correcting "pluralistic ignorance" (Miller and McFarland 1991; Toch and Klofas 1984; Verplanken et al. 2008). Pluralistic ignorance means the majority think the norm is actually worse than it is. Social norm theory predicts that interventions which correct misperceptions by revealing the accurate healthy norm, will be beneficial to most individuals by reducing problematic behavior or encouraging participation in positive behavior (Berkowitz 2005). Social marketing techniques have been applied to present actual healthy norms in the example of alcohol consumption which reduced levels significantly with a 45 % reduction over ten years at Northern Illinois University (Haines and Barker 2003). Similar results were produced at the University of Arizona (Johannessen and Glider 2003). Using social media within this concept of social norms was also successful at reducing tobacco consumption at other universities (Hancock et al. 2002; Linkenbach and Perkins 2003). Furthermore, research conducted by the National Union of Students and the Higher Education Academy, has consistently demonstrated a serious interest in sustainability issues amongst students, with over 80 % believing sustainable development is something that universities should actively incorporate and promote, and 60 % wanting to learn more about (NUS 2013).

Within behavior change literature the 'habit discontinuity hypothesis' suggests that when people are going through a major lifestyle change it is easier to get people to adopt pro-environmental behaviors (Verplanken et al. 2008). Coming to university is a significant event for many people providing an opportunity to move out of home for the first time, become responsible for finances, focusing in depth on a chosen subject, creating a stronger sense of identity, develop passions through societies and be the catalyst to lifelong careers. Parallels can be drawn to anthropological rites of passage in non-industrialized indigenous cultures where children undergo a symbolic event and are transformed into adults (Gennep 2010). Turner (1969) states the three stages to a rite of passage are separation, transition and reincorporation. The transitional period of university is an ideal space and time to be used to transform and embody positive sustainable behaviors in young adults which may be continued as they are reincorporated into society throughout their lives.

Many students recognize climate change as a serious issue but research suggests that there is generally a poor understanding of climate change science and they are unaware of how their energy use contributes to the problem (Marcell et al. 2004). This divergence between environmental values and individual actions is well documented and referred to as the "value-action gap" (Blake 1999). Recent research demonstrated the significant influence of human psychological factors such as social norms, altruism, personal perceptions, beliefs, and attitudes in behavior decision making (Marcell et al. 2004). However, there are many complex and dynamic factors contributing to individual behaviors. Internal individual, psychological, and social barriers as well as external institutional, economic, social, or cultural barriers constrain many people from aligning there behaviours with their concern with the environment (Kollmuss and Agyeman 2002). Community based social marketing (CBSM) uses commitments, prompts, social norms, social diffusion, feedback and incentives to change behavior at a community level (McKenzie-Mohr 2000). It aims to produce behavior change via direct communication and community level initiatives which concentrate upon removing barriers to change (Kennedy 2010). Grounded in social psychology, CBSM is based on the principle that behavior change programmes are most effective when they are delivered at the community level using personal contact (McKenzie-Mohr Smith 1999). In a twofold process barriers are removed and benefits of behavior change promoted (Kennedy 2010). Barriers can include habits, financial limitations, society expectations or norms and time constraints, whilst incentives and personal motivations can include financial savings or social norms in taking positive action (Collier et al. 2010).

Public commitments have been shown to influence outcomes in a study on energy reduction in Iowa City (Pallak et al. 1980). Three groups were used for the study. The first was given information about reducing energy consumption and conservation strategies. A public commitment to the program was sought by telling residents that their names would be publicized with the results of the program. The second group were visited at home and asked to make a private commitment. The third group were not visited or asked to make any commitment. The first group where a public commitment was sought had a decrease of 10–20 % of energy usage with no significant decrease found in the other two groups.

The *Planet Too* project has utilized the theoretical frameworks discussed to encourage pro-environmental behavior change. Ranging from targeting students at university (habit-discontinuity hypothesis), communicating actual trends and positive reinforcement (social-norm theory), using marketing tools to communicate benefits and remove barriers (Community Based Social Marketing), and finally using public commitments to increase impact.

3 Conflicting Economic Paradigms in Sustainability

It can be argued that the economy itself is the direct cause of the imperative environmental global state of affairs (Dietz and O'Neill 2013). However, it is argued that a growing economy will provide technological ingenuity and escaping poverty will allow environmental issues to be transcended (O'Neil and Ben-Ami 2013). On the other hand Jackson (2009) states in *Prosperity Without Growth* "simplistic assumptions that capitalism's propensity for efficiency will allow us to stabilize the climate and protect against resource scarcity are nothing short of delusional". Furthermore, history has shown that when technological progress increases the efficiency with which a resource is used, the rate of consumption of that resource actually tends to rise (Czech 2000). Whilst global GDP has increased there also exists a drastic inequality with 2.7 billion living on less than $2 per day and two percent of adults owning half of all household wealth in the world (Dietz and O'Neill, 2013). Given the increase of technological advancements worldwide carbon dioxide emissions from fossil fuels have still increased by 80 % since 1970 (Jackson 2009).

The dominant economic paradigm of continuous growth and the culture of consumerism within global capitalism have been criticized as directly causing the current environmental problems (Daly 2013). Alternative theories to neoclassical economics are emerging after the global financial recession. "By conceptualizing the economy as a complex system, grounded in production, as well as exchange, full of contradictions and conflicts and incorporating political impasse, the heterodox political economy offers a far superior framework than does neoclassical economics with its isolated, acquisitive, rational and all-knowing consumer confronting equally powerless firms in isolated markets where conflicts of the real world, such as inadequate incomes, excess capacity, recessions, unemployment and poverty, never occur" (Klitgaard 2013). Ropke (2004) argues the human economy is inherently embedded within nature, and economic processes are actually biological, physical, and chemical processes, and transformations. He states 'Ecological Economics' requires trans-disciplinary work to describe economic processes in relation to physical reality. The worlds resources are finite and thus diametrically opposed to the pursuit of endless growth which fuels the dominant global economic paradigm (Dietz and O'Neill 2013). In contrast economic systems which strive for balance and equilibrate within the ecological constructs of which they are formed are being explored. A 'steady state economy' is an economy of a relatively stable size. It features stable population and stable consumption that remain at or below carrying capacity (Daly 2013). This system works within the ecological limits of energy and resource use and mantains a stable population, which replace the goal of maximizing economic output with maximizing the quality of life (O'Neill et al. 2010). Global capitalism which is often portrayed as the norm is a relatively new construct. The field of anthropology demonstrates countless examples of a whole array of economic systems ranging from subsistence farmers, to global trade, from non-industrialized indigenous cultures to globalized highly technological cultures (Chibnik 2011). If economies are formed from the environments of which they are part it is of paramount importance to maintain them and set limits based on ecological carrying capacities, which aim towards equilibrium as opposed to the endless pursuit of growth.

4 University of Northampton Students' Union and Planet Too

The University of Northampton is the only UK-based *AshokaU Changemaker Campus* of only two outside of the USA. *Ashoka* is the largest global network of social entrepreneurs with partners across the sectors that use entrepreneurial talent and new ideas to solve social problems. It is about empowering people with the skills, confidence, and belief that they can create solutions to the challenges faced in the 21st century, as entrepreneurs and interpreneurs alike (Drayton 2014). The university as part of the *Ashoka* accreditation has committed to embed the principles and values of social enterprise and social value within every course, and to encourage entrepreneurial activity in a socially-minded manner in all graduates to influence the creation of a fairer and more sustainable society. Social enterprises can be structured as profit or non-profit, and may take the form of a co-operative, mutual organization, a disregarded entity, social business, or a charity organization (Ridley-Duff and Bull 2011). The Students' Union however, felt that the focus was biased towards social benefits and the environmental aspects were neglected. The aim of *Planet Too* is to embed the principles of sustainability and the 'triple bottom line' across every aspect of enterprise, entrepreneurship, education, and social enterprise development within the University of Northampton and its partners, thus harnessing the energy and passion of its members to create a force for change.

The *Planet Too* project combines aspects of volunteering and campaigning with social enterprise and employability to engage students in leading change on their campus. Students, academics, landlords and the wider community are connected through the different elements. The strength of the project lies in bringing together the combined knowledge of university departments and the energy, creativity and ingenuity of students. The challenge has been to communicate sustainability in a way that not only engages, but empowers and gives ownership to students of the project allowing them to shape and direct its legacy into the future. A full time Mphil research student has been studying the ongoing project impact to provide feedback. This allows the effectiveness to be maximized by choosing to cease or continue strategies based on their effectiveness. As many strands of the project are pilots they are not intended to be flawless so long as lessons are learned contribute and support the evolution of the Students' Green Fund movement at universities.

The project amalgamates different schools and centers across the university to utilize their strengths. The project has a part funded marketer who focuses on communications and social media sites such as *Twitter*, *Facebook* and *YouTube*. Producing photos and videos for these sites helped to correct perceived social norms whilst giving volunteers a space to creatively market their own work. Social and visual media is of paramount importance when communicating to students and helped increase the impact for those who were not directly involved but could view the project. The University Centre for Employability and Engagement played a key role in developing the volunteers. Their networks have helped to promote the volunteering opportunities to other students. Boosting skills and employability has proven a successful recruitment strategy to a wide range of students new to sustainability and often motivated them to become involved. The Business School was commissioned to carry out the research with local businesses to form the employability module. The University Enterprise Club supported students with ideas to develop into robust business plans for the loan fund. Northampton is the only UK university with a Credit Union. This is run by students and is in partnership with the Northampton Credit Union who are responsible for managing the loan fund. The Department of Environmental and Geographical Science has played a key role in contributing research and experienced academics. They also oversaw the monitoring and evaluation of the project. The Accommodation Services are the first port of call to the network of student landlords for the strands which focus on off campus student housing. The Sustainability Team within Infrastructure and Estates has played a significant contribution in the project bid and the renovating and developing of the Students' Union buildings through the *Green Impact* sustainable accreditation scheme (GISUC 2014). The project board contained a representative from each of the groups and schools listed as well as students representatives to steer the project and to combine skillsets to help overcome the ongoing challenges and unforeseen circumstances the project encountered. The project joined into the wider work of the Northamptonshire Council and their Climate Change Strategy 2010–2014 (NCCS 2010).

The Sustainability Changemakers are the volunteer backbone of the project. The target is for one hundred to be recruited and trained over the two years. They are responsible for promoting and delivering the different elements of the project through leading campaigns for behavior change. Every volunteer receives a sustainability induction where they learn about transport, food, sourcing locally, saving energy, recycling and the well known definition of "development that meets the needs of the present without compromising the ability of future generations to meet their own needs" (WCED 1987). Efforts are made to understand their individual passions and skills often relating to their chosen degree to allow them to take part the in strands they are most interested in. Northampton is also part of the NET-positive initiative which highlights negative impacts as well as positive behaviors for volunteers to produce a 'net-positive' lifestyle. To boost their employability and effectiveness within the project specialized trainings are run for volunteers.

Student Switch Off+ is an energy saving competition building on the successful model carried out nationally in campus accommodation halls through the National Union of Students. The target is for ninety houses to be recruited over the two years. Key elements are competitiveness, prizes as incentives, and easy actions to carry out.

As part of the competition three hundred Green House Kits were distributed to the student houses. Half the households have received a kit with the other half acting as a control to measure their effectiveness in reducing energy consumption.

Building on the award winning pro-environmental university accreditation scheme Green Impact, a parallel scheme will be piloted with landlords in off campus student housing. Thc University Department of Environmental and Geographical Science established community and private sector landlord attitudes and activity regarding the impact of their housing upon the environment through questionnaires and focus groups. Landlords will use a bespoke version of the NETPositive tool as an action plan for each house with Bronze, Silver and Gold award levels.

The Sustainable Business Ethics Loan Fund offers a maximum of £3000 to student social enterprises where appropriate sustainable business ethics are evidenced in the plan and the enterprise has a focus upon enhancing and improving environmental sustainability. Loans are offered at a maximum low interest rate of 1 % (12 % APR), over terms of between 12 and 36 month payback.

The fifth strand Sustainable Employability Module with the Business School and School of Science and Technology carried out research with local businesses into key employer attitudes toward and requirements of Sustainable Business Ethics with particular emphasis upon what they would expect graduates to offer in this regard. The research also aimed to reveal their involvement with sustainability initiatives such as energy conservation, recycling, and cycling schemes. Using this research in collaboration with the Centre for Employability and Engagement an appropriate program will be developed and implemented to accredit practical experience of the required skills and knowledge that will form part of the Certificate in Employability. There is a significant opportunity for universities to embed sustainability within the curriculum. Plans to incorporate practical volunteering placements with the project are currently planned with the Department of Geography for Summer 2014.

The Grant and Award Scheme has been the most successful and diverse producing a multitude of projects across the student body. These deliver a range of activities all of which must support and embed behavior change on campus, thus challenging student pre-conceptions and building support for the other elements of the project. Ten grants of up to £300 each year are awarded to individual students to enable them to stage an event, activity or program that promotes behavior change or sustainability issues. Thirty Sports clubs and forty-six Societies are able to apply for one of up to ten each per year of up to £300 for the same purpose. Aims and objectives must be SMART, and focused clearly upon one of the project strands or another sustainability issue that the student, Society, or Club is passionate about. A sustainability panel of volunteer Sustainability Changemakers decides whether an application is accepted to go through to student council where it is voted for. The panel also suggests amendments or improvements which have proven to be highly beneficial in increasing impact amongst the student body and evolving ideas in transformative ways.

Lastly the final strand focuses on the new Waterside campus development set for 2018 to replace the existing campus. As part of *Ashoka* the university is committed to transforming the educational experience provided to students to one that is world-changing based on Changemaker and sustainable values. Part of that vision includes building a brand new campus in 2018 alongside a 'new university'—moving away from a teaching-led model of provision to a learning- and student-led model. A group of Sustainability Changemakers volunteers have the unique opportunity to influence both the campus and the curriculum to meet the sustainability demands of the future as members of a student-led Task and Finish Sustainability Sub Group which reports to the office of the Vice Chancellor.

5 Results

To date 42 volunteers have been recruited, inducted in sustainability and received additional training in communication, time management, leadership, project management, and effective meetings. 547 students, the sabbatical officer team, 41 university staff, and all six schools have been engaged with the project. Collaboration has been established with the *Wildlife Trust*, Northampton Council climate change team, and *Sustrans*.

All 300 'green house kits' were distributed to students and 50 student houses were recruited into the energy saving competition. The on campus energy saving competition has saved an average of 7 % of energy usage from the 2012 baseline. An online support group was formed between Bristol, Sheffield, Worcester, Falmouth and Exeter and Stafford universities to share best practice with similar pilot projects.

The research into landlords revealed physical improvements to infrastructure would be best suited for the scheme with criteria including draught proofing, loft insulation, double glazing, solar panels and rainwater capture. The main challenge will be to recruit landlords into the scheme without major financial rewards as the funding is for only two years. Research carried out and collaboration with the

Northamptonshire Landlord Association has recommended marketing the scheme together with the overall project to students to create brand identity and perceived value. If they recognize the name and logo of the project and understand the benefits of increased quality and welfare of the student houses taking part in the award scheme, this would increase marketing and demand for the student landlords.

Twenty-six sustainability grants were distributed with key projects being a student allotment, beer brewing kits, pedometer health campaign for primary school children and a *Fairtrade* campaign. Sports clubs focused grants on sustainable transport options to fixtures and local sourcing and manufacturing of equipment.

One student sustainability loan was distributed to start a bike renting business. Bikes are rented to students on a weekly, monthly or termly basis. The idea was developed through the enterprise club and the *Planet Too* support team.

The Students' Union was awarded a silver award for *Green Impact* through successfully achieving criteria relating to improving the behaviours of staff and environmental performance of its building.

Finally a website and film were produced to increase the impact and reach of the project. There are currently 269 follower on Twitter and 130 Facebook likes. The *Planet Too* project team has been invited to join the other Student Green Fund universities at the House of Lords parliamentary reception to celebrate the collective work towards sustainable development.

6 Discussion

The project made a successful start with interest gained from a wide range of students. The volunteers tailored ideas to effectively engage their student peers with the project strands helping to recruit new volunteers. However, the project went through a flat line period where momentum was slowed and efforts have been made to engage volunteers through giving them ownership and developing an executive committee. The emphasis on ownership and student-led is a key theme across the Students' Green Fund and was highlighted by Jamie Agombar the environmental manager for NUS. These attributes are paramount to creating a project legacy. The relationship between ownership and successful sustainability initiatives is apparent in comparing public attitudes towards a community-owned wind farm on the Isle of Gigha with attitudes towards several developer-owned wind farms on the adjacent Kintyre peninsula in South West Scotland (Warren and McFadyen 2010). This case study indicated that local attitudes could become more positive if future wind farms were owned by local communities. Increased positive psychological effects were indicative in the community owned turbines being affectionately named the 'Three Dancing Ladies'.

The Green House Kits have undergone an evolution in the first year of their piloting. They originally contained a sustainability checklist with coffee rewards for uploading photos on social media sites, local food posters, switch off stickers, bus and bike maps, recycling information and *Fairtrade* chocolate. However, after research and feedback it was felt they would not be successful due to the amount of one way information they contained which failed to engage. The kits did not significantly differentiate from other leaflets and takeaway flyers student houses receive. After reevaluating lettuce seeds, a pedometer and an eco-button for computers were included. The premise was that after students were interacting with the kits they would then be more likely to read the information within. Smart meters have been donated by the council to the project which will also be trialed. However, these meters are designed to be attached directly to the original electricity meter. It was believed that as students do not regularly check their meter readings and that they are often located in the basement it is unlikely they will have a substantial impact. A visual display with wireless technology which can be situated in communal areas such as the kitchen is likely to increase their impact. The Mphil researcher has considered carrying out a PHD study into the effectiveness of wireless smart meters which are synchronized to the internet so users can view them online.

The sustainability grants utilized society and sport team structures. These groups showed strong social structures, independence and often demonstrated confidence through group dynamics. They have also demonstrated strong channels of communication to the student body as the club president or committee applying for the grant is respected so sustainability initiatives through them are amplified. The best sports sustainability award was administered at the end of year awards ceremony by the CEO of the university adding prestige and value to this strand.

The business loan strand has been the most challenging to date. Finding students willing to take an additional loan on top of their compounding student debts whilst at university has been difficult. However, there is tremendous potential for significant environmental sustainable impact using a business model. A successful business will be self-sustaining so when the funding period is complete the impacts will be maintained. The one successful student bike renting business aimed at students shows potential for growth. A similar model has been successful in London with the *Barclays Borris Bikes* and also in Newcastle University.

Social media has been an effective way to communicate the project and reinforce the social norm that students are involved in pro-environmental behaviour. Tagging photos through Facebook marketed the project to people not directly involved. The sabbatical officer team were also very influential as they provided many contacts through campaigning for their positions so were useful allies.

7 Conclusion

Behaviour change theories were applied during the *Planet Too* project to create positive behavior change by communicating social norms, and Community Based Social Marketing methods. Utilizing different social networks and online media momentum was built with steady growth of students engagement.

The Sustainability Changemaker network of volunteers has gradually shown signs of self-leadership and a society structure. This is significant factor to support the continuation after the funding and needs to be supported and encouraged.

The energy saving competition Student Switch Off+ has potential for expansion. The effective use of social media could be used to recruit more houses and create a cultural norm or energy saving with incentives through competitions.

Research from the landlord awards is being used to create the Green House Award scheme to be piloted in September 2014. The research from the business school into local employer attitudes is also currently being used to create a module for students. On the one hand this will teach the students what employers value. It will also give students the skills to communicate sustainability in effective ways to create positive change within an organization.

There was scope and research into developing an online application for student smart phones for the project. This would firstly measure their carbon saving impact. Secondly it would be used to encourage and reinforce positive behaviours to build momentum to support long term positive behavior change.

Involving students through every stage of sustainability projects is crucial to making them relevant to and accepted by students as they are coming from the student body of which they are part. Collaboration between universities nationally and internationally will help create effective alliances, solutions, and momentum to overcome the global challenges of sustainable development.

References

Berkowitz AD (2005). An overview of the social norms approach. In: Lederman L, Stewart L, Goodhart F, Laitman L (eds) Changing the culture of college drinking: a socially situated prevention campaign, Hampton Press, New York

Blake J (1999) Overcoming the 'value-action gap' in environmental policy: tensions between national policy and local experience". Local Environ 4(3):257–78

Chibnik M (2011) Anthropology economics and choice. University of Texas Press, Austin, Texas

Collier A, Cotterill A, Everett T, Muckle R, Pike T, Vanstone A (2010) Understanding and influencing behaviours: a review of social research, economics and policy making in Defra. DEFRA, London, http://archive.defra.gov.uk/evidence/series/documents/understand-influence-behaviour-discuss.pdf . Accessed 14 Apr 2014

Czech B (2000) Shoveling fuel for a runaway train: errant economists, shameful spenders, and a plan to stop them all. University of California Press, Berkeley, California

Daly H (2013) From a failed growth economy to a steady-state economy. In: Cleveland CJ (ed) Encyclopedia of earth, Environmental Information Coalition, National Council for Science and the Environment, Washington DC

Dietz R, O'Neill D (2013) Enough is enough: building a sustainable economy in a world of finite resources. Berrett-Koehler Publishers, San Francisco, California

Drayton C (2014) Knowing history, serving it: Ashoka's theory of change. Ashoka. Arlington https://www.ashoka.org/knowinghistory . Accessed 5 Apr 2014

EAUC (2014). Environmental association of Universities and Colleges. http://www.eauc.org.uk/hefce_announces_plans_to_invest_5000000_in_high . Accessed 14 Apr 2014

GISUC (2014) Green Impact Students' Union and Colleges. NUS. London, UK. www.green-impact.org.uk . Accessed 10 Apr 2014

Haines M, Barker G (2003) The NIU experiment: a case study of the social norms approach. In: Wesley Perkins H (ed) The social norms approach to preventing school and college age substance abuse: a handbook for educators, counselors, and clinicians, Jossey-Bass, San Francisco

Hancock L, Abhold J, Gascoigne J, Altekruse M (2002) Applying social norms marketing to tobacco cessation and prevention: lessons learned from three campaigns. The report on social norms: working paper 6. Paper-clip communications. Little Falls, New Jersey

Hill N (2007) Think and grow rich. Wilder Publications, Radford, Virginia

Kennedy A (2010) Using community-based social marketing techniques to enhance environmental regulation. Sustainability 2(4):1138–1160. Basel, Switzerland

Jackson T (2009) Prosperity without growth? The transition to a sustainable economy. Sustainable Development Commission, London

Johannessen K, Glider P (2003) The University of Arizona's Campus health social norms media campaign Chapter 4. In: Perkins HW (ed) The social norms approach to preventing school and college age substance abuse: a handbook for educators, counselors, clinicians, Jossey-Bass. San Francisco

Klitgaard K (2013) Heterodox political economy and the degrowth perspective. Sustainability 5:276–297. Basel, Switzerland

Kollmuss A, Agyeman J (2002) Mind the gap: why do people act environmentally and what are the barriers to pro-environmental behavior? Environ Educ Res 8(3):239–60

Leal Filho W (ed) (2012) Sustainable development at universities: new horizons environmental education, communication and sustainability, vol 34. Peter Lang Publishers. Oxford, UK

Linkenbach J, Perkins HW (2003) Most of us are tobacco free: an eight-month social norms campaign reducing youth initiation of smoking in Montana Chapter 13. In: Perkins HW (ed) The social norms approach to preventing school and college age substance abuse: a handbook for educators, counselors, clinicians. Jossey-Bass, San Francisco

Marcell K, Agyeman J, Rappaport A (2004) Cooling the campus: experiences from a pilot study to reduce electricity use at Tufts University, USA, using social marketing methods. Int J Sustain High Educ 5(2):169–189

McKenzie-Mohr D (2000) Fostering sustainable behavior through community-based social marketing. Am Psychol 55(5):531–537

McKenzie-Mohr D, Smith W (1999) Fostering sustainable behavior: an introduction to community-based social marketing. New Society Publishers, Gabriola Island, British Columbia

Miller DT, McFarland C (1991); Toch. & Klofas. (1984) cited in Berkowitz AD (2005) An overview of the social norms approach. In: Lederman L, Stewart L, Goodhart F, Laitman L (eds) Changing the culture of college drinking: a socially situated prevention campaign, Hampton Press, New York

NCCS (2010) Northamptonshire climate change strategy 2010–2014. Northamptonshire Partnership, Northampton, UK

NUS National union of students (2013) Ethics and environmental round table: HEA/NUS research results. NUS, London

O'Neil D, Ben-Ami D (2013) Argument: is it time to ditch the pursuit of economic growth?. New Internationalist Editorial, Oxford, UK

O'Neill DW, Dietz R, Jones N (2010) Enough is enough: ideas for a sustainable economy in a world of finite resources. The report of the steady state economy conference. Center for the advancement of the steady state economy. Arlington, Virginia. Economic justice for all. Leeds, UK

Pallak MS, Cook DA, Sullivan JJ (1980) Commitment and energy conservation. In: Bickman L (ed) Applied social psychology annual, Sage, Beverley Hills, California, p 235–253

Ridley-Duff RJ, Bull M (2011) Understanding social enterprise: theory and practice. Sage Publications, London

Ropke I (2004) The early history of modern ecological economics. Ecol Econ 50(3–4):293–314

Turner V (1969) The ritual process. Transaction Publishers, New Jersey

UN United Nations (2012) The future we want. The general assembly. Sixty-sixth session. Agenda Item 19, United Nations. http://www.uncsd2012.org/content/documents/727The%20Future%20We%20Want%2019%20June%201230pm.pdf . Accessed 14 APr 2014

UNRISD United Nations Research Institute for Social Development (2010) Combating poverty and inequality: structural change, social policy and politics. UNRISD, Geneva, Switzerland

Van Gennep A (2010) The rites of passage. Routledge, London

Verplanken B, Walker I, Davis T, Jurasek M (2008) Context change and travel mode choice: Combining the habit discontinuity and self-activation hypotheses. J Environ Psychol 28(2): 121–127

WCED (1987) World commission on environment and development. Oxford University Press, Oxford

Warren C, McFadyen M (2010) Does community ownership affect public attitudes to wind energy? A case study from south-west Scotland. Land Use Policy 27:204–213

Author Biography

Simon Mark Pole has been involved in sustainability for six years and is particularly interested in sustainable food systems. He received his undergraduate degree in Social Anthropology from the University of Kent in 2011, where he focused on ecology, conservation and sustainable development. Through *Change Agents UK* he began working at University of Birmingham developing *Green Impact* a pro-environmental behavior accreditation scheme involving the staff. Whilst here he set up a staff and student allotment, and received a professional development award in project management in sustainable development. In 2013 he started at the University of Northampton Students' Union as the *Planet Too* coordinator, part of the Students' Green Fund. *Planet Too* is a student-led programme aiming to transform thinking and behaviours around sustainable development and sustainable business ethics. The project brings together students, staff, academics, estates and local landlords. Lastly he sits on the Sustainability Direction and Oversight Board for the National Union of Students which acts as a high-level think-tank on sustainability and social responsibility. It develops long term vision, strategy, effective strategic partnerships, and works to overcome barriers of sustainability within the wider students' union movement.

Improvements for a Sustainable Distance Education with the New UNED On-Site System for Virtualization of Exams: Malaga Region (Andalucía, Spain) as Case Study

M. Olga Guerrero-Pérez, Gustavo Román and Concepción Travesedo

Abstract

UNED University provides distance higher education to a high number of students, and, consequently, has to perform a large number of tests (ca. 780,000 exams in 2013). UNED has performed several improvements in the way these exams are made during the last year, with the implementation of a new protocol for evaluation procedures that optimizes transportation through on-site digitalization of exams being the most important. This new method is called virtual attaché return case. The virtual attaché return procedures optimize security and transportation costs through on-site digitalization of exams, bringing a new logistics paradigm. The great value is that exams are directly sent to evaluators in a digital manner. This speeds up evaluation procedures for our Bachelor and Master Degree students. The original documents remain thus safely stored at the Associated Centers. In the case of the Malaga region, two improvements have been performed in 2014 in order to save energy and money for the University and also to more than 6,000 students. Until this year, the exams of the entire region were performed in Malaga city, but in 2014, Marbella has been also habilitated as an examination center. By this manner, the students that are resident in the Marbella area (ca. 1,500) do not have to go to Malaga to complete their exams, saving time and money (Marbella is around 60 km far from Malaga). The other improvement is related to the use of the virtual attaché return, for the first time, in the Andalucia region. An economic study of the time, energy and money saved with these two innovations during the first exams celebrated in January–February 2014 is analyzed in the present paper.

M. Olga Guerrero-Pérez (✉) · G. Román · C. Travesedo
Associated Center of UNED in Malaga, 29006 Málaga, Spain
e-mail: oguerrero@uma.es

W. Leal Filho et al. (eds.), *Integrating Sustainability Thinking in Science and Engineering Curricula*, World Sustainability Series, DOI 10.1007/978-3-319-09474-8_22

Keywords
Education · Virtualization of exams · Sustainability · Distance education

1 Introduction and Objectives

UNED (Universidad Nacional de Educacion a Distancia) is a Spanish Public University. It is one of the largest European Universities, founded in 1972, with around 260,000 students and more than 1,400 Assistants, Associates and Full Professors. UNED has Associated Centers in many towns in Spain and also in some other countries. Malaga is a province of Andalucía, located on the southern Mediterranean coast. It has a total area of around 7,000 Km^2 with a population at an estimated 1,640,000. The province contains 101 municipalities, with Malaga city being its capital. In Malaga there is only one Associated Center of UNED, which serves all the citizens in Malaga province. The UNED Associated Center, with 120 tutors and around 6,000 students, is located in Malaga City, with delegated University classrooms in Marbella, Ronda and Mijas.

Distance higher education has been growing quickly during the last few years, due in part to new technologies and also to the fact that it seeks for students' autonomous learning adapting to their needs and schedules. Distance education for teaching/learning processes has been revised by different authors and researchers over the years (Litwin 2000; Barberà et al. 2001; Talbot 2004; Hope and Guiton 2006; Bernath 2009; Cleveland-Innes 2010; García Aretio 2011; Moore and Kearsely 2012; Moreno 2012). However, there are some basic common features: the separation of professor and learner which distinguishes it from face to face lecturing; the use of technical media; and the provision of two-way communication in ways that facilitate distance learning. The main way of controlling the academic performance of the students stands on the exams. The UNED evaluation system is based in simultaneously summons students at many locations in Spain and at selected venues across the World. Being the Faculties in Madrid, it is required to send all the exams to Madrid for their correction.

Nowadays our society is concerned about sustainability, in many different aspects such as the valorization of wastes (Guerrero-Pérez et al. 2011), the use of renewable energy sources (Schultz and Yoon 2014), alternative feed stocks (Guerrero-Pérez et al. 2009), or by saving water consume. As part of our society, Universities are also concerned about sustainability. College and university leaders recognize that progress toward finding solutions to sustainability problems requires engagement and input from all stakeholders within their campus and surrounding community through shared vision and face-to-face dialogue (Pollock et al. 2009). For instance, a number of Universities are engaged in educating the future leaders, decision makers, scientists, and engineers on how their decisions can help societies become more sustainable. Recently some Universities are offering Degrees in Sustainability related fields, such as the Bachelor's degree in Engineering for Sustainable Development at Tecnológico de Monterrery, in Mexico (Lozano and Lozano 2014).

It is clear that in different aspects, sustainability now plays a significant role in campus activities, governance and in the academic curriculum in colleges and universities around the world. The greening of higher educational institutions can be defined as the process of reducing the multitude of on- and off-site environmental impacts resulting from campus decisions and activities, as well as raising environmental awareness within the human communities of a college or university (Creighton 1999).

By greening our own campuses (Dahle and Neumayer 2001), buildings and practices, Universities can teach and demonstrate that it is possible to be closer to the main objective of sustainability: "development that meets the needs of the present without compromising the ability of future generations to meet their own needs" (WCED 1987). Since UNED Malaga Associated Center is also concern about sustainability, and also to give a better service to Malaga students, several improvements have been done in the way of making exams in the province during the present 2014 year:

a) Until this year, only one examination committee was sent from Madrid UNED Rectorate to Malaga for making the exams to the more than 6,000 students of Malaga province. These exams were performed in Malaga city, thus, the students from all over the province had to travel to Malaga. It should be considered that some important towns, like Estepona, are far from Malaga more than 100 km and traveling to Malaga takes more than one hour by car. The novelty this year is that two examination committees were sent to Malaga province, by this manner, the exams were performed in two different places: Malaga city and Marbella.
b) All the exams from all the students (more than 6,000, and obviously they use to take more than one exam each, corresponding to different matters) had to be sent to Madrid for correction. The novelty this year consisted in the use of a virtual attaché return procedure with no physical return of exams. With this system the transportation costs are almost negligible, since the exams are digitalized on site, and the files sent to the Professors, who correct them on their tablets.

The main objective of the present paper is to evaluate, in terms of money and time, the effects that these new procedures have had for both students and for our University. Also the environmental implications of these procedures are discussed, as an example of new practices, eco-friendly, at a Higher Education Institution.

2 Methods

In order to evaluate the concerns on sustainability of the new procedures, some estimations and calculations are necessary.

i) As an average, we have estimated that all the students have saved 120 km, since Marbella is 60 km far from Malaga.
ii) It has been also estimated that all the students have saved 2 h (estimated time for driving 120 km)
iii) An average CO_2 emissions rating fixed as 142 g/km (grams of carbon dioxide per kilometer driven).

3 Results: Marbella Examination Center

As has already been mentioned in the objectives section, during the last February 2014 exams, an additional examination committee was sent to Marbella, the second largest town in the province. In addition, it should be considered that population is mainly concentrated in the cost. With this improvement Marbella residents do not have to go to Malaga to take their exams, saving time and money. In addition, the residents in municipalities close to Marbella (Fig. 1) only have to go to Marbella and not to Málaga, saving more time and money. Like Estepona, which is quite a big town with ca. 66,000 inhabitants. Estepona is far from Malaga around 100 km, whereas it is only 30 km away from Marbella. Thus, it is clear than Marbella is a

Fig. 1 Map of Malaga province

more convenient place for many towns and villages like Ronda, Benahavís, Ojen, Casares and Istan, among others (Fig. 1).

Figure 2 shows the number of exams performed during the last February 2014 period in both Malaga and Marbella. As can be observed, more than 8,050 exams were performed in Malaga province, 7683 in Málaga and 889 in Marbella. The exams were performed during 2 weeks, with the students able to choose in which week they would prefer to take their exams. As expected, most of the students preferred to take the exams during the second week. If we consider that all the students are residents in Marbella, as an average, and Marbella is 60 km far from Malaga; we can assume that these students have saved 120 km (outward and return) and around 2 h for each trip. Since 889 exams have been performed in Marbella, these data indicate that Marbella region students have save around 21,000€ (considering 0.19€ per Km) and around 1800 h of their time. This data not only implies that time and money is saved for our students; the reduction in transport also implies less CO_2 and other atmospheric pollutants emissions and less oil consume.

Although it is difficult to calculate in an accurate way the amount of CO_2 that has been avoided to be emitted, it is possible to make an estimation (Nanaki 2009). The difficulty is because the CO_2 emission depends on the type of vehicle, the speed, and many other factors. With the lowest 0 g/km and the highest close to 400 g/km, a general value for the average carbon dioxide emissions rating is 142 g/km (grams of CO_2 per kilometer driven). Considering this value, each student has saved 17040 g/km (as an average, each student has saved 120 km to drive). Since 889 students have performed their exams in Marbella, this supposes more than 15 tons of CO_2 that have not been emitted.

Figure 3 shows the distribution of students from Malaga province that has chosen Marbella to take their exams. As expected, most students are Marbella residents, and the second bigger group are residents from Estepona, since it is a big town and very close to Marbella. Also residents from other municipalities have used the Marbella examination center: Ronda, Benahavis, Manilva, Mijas, Ojén, Monda, Istán, Arriate, Carratraca, Casares and Igualeja (Fig. 1, light blue color). This data

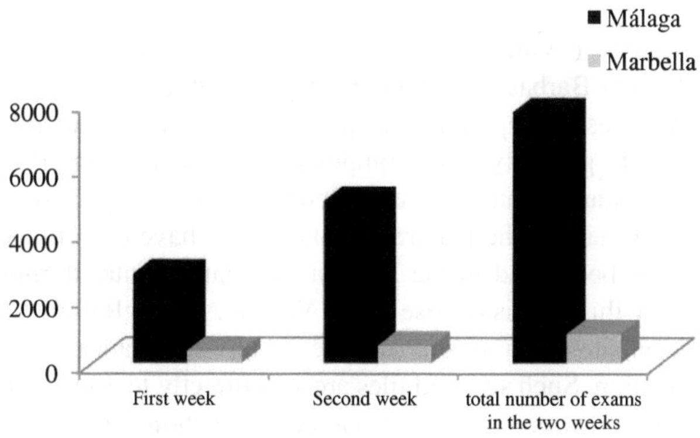

Fig. 2 Exams performed per week and per city

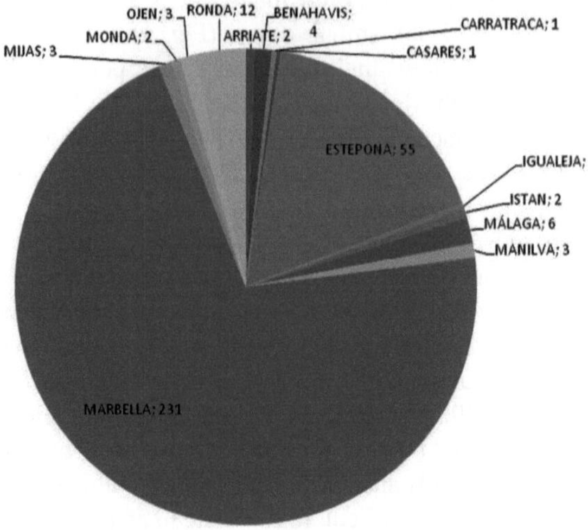

Fig. 3 Distribution of students from Malaga province that have use Marbella for making their exams (january–february 2014). *Note* residents in other provinces have not been included

clearly indicates that Marbella's new examination center not only benefits Marbella residents, it also benefits the citizens of a high number of municipalities that are close to Marbella. It is expected that the number of students from these towns and villages increase in the next academic years.

An analysis of the data also shows that students from other Spanish provinces have performed their exams in Marbella. Many of them are, for example, residents in Madrid. Since Marbella is a very nice location, and many people have a holiday residence in the Marbella region, this data could indicate that they have decided to spend some days in Marbella to perform their exams. Thus, Marbella is also a convenient examination center for some students who are not residents in the Malaga province.

4 Results: Use of the Virtual Attaché Return

The *Virtual Attaché* ("Valija Virtual" in spanish) application has been totally developed at UNED Barbastro Computer Department Center. The objective is to simplify the processes of preparing, transporting, distributing and collecting exams, maximizing security guaranty. This computer system is used in all UNED associated centers to conduct exams. A recent publication (Ortega-Navas et al. 2014) analyzes all the virtual attaché features and how they have been implemented. The virtual attaché has been used in the Malaga associated center during the last few years. The novelty this year is the use of the Virtual Attaché Return, which includes a digitalization on-site of the exams and the no physical return of paper-exams to Madrid for correction. Such scanned files are sent directly to each student and to the corresponding professor after 2 days. Upon authentication, students can access and view all their answers given on the exams, from home. Thus, Professors have ready access to all exams via a secured access network. UNED has also developed an evaluation program that allows the automatic correction of multiple-choice exams. Such a virtual attaché thus delivers not only efficient and safe timely delivery of exams to professors, but it also provides negligible paper transportation costs. To make the corrections, each Professor has two options: (a) Using the Digital Corrector application that allows on screen correction (PC, tablet or laptop) and export grades to a spreadsheet file for later reading by the Grades Management Application; or (b) Downloading the PDF exam files on their computer for later printing. The (b) option is not recommended since one of the main objectives of the system is saving paper.

The use of Virtual Attaché Return has been performed successfully in the Malaga Associate Center, both in Malaga and Marbella examination centers, during last February 2014 exams, and has been used in Andalucía for the first time this year. The system was implemented and no errors and/or incidents were detected. Since more than 8,500 exams have been performed (Fig. 1), this supposes that all the papers did not have to be sent to Madrid for correction, minimizing transportation costs and risk of documentation loss. As has been mentioned previously, saving on transport (of both people and paper) implies: less CO_2 and other atmospheric pollutants emissions and less oil consumed.

5 Conclusions

Two innovations in exam procedures have been performed during the present academic year 2013/14 in order to improve the sustainability of our center by saving transportation of both people and paper. Saving in transportation implies avoiding CO_2 and other atmospheric pollutants emissions and decreasing oil consummation. In addition to these initiatives, our students have saved time and money and have also obtained many other advantages.

On one hand, a second examination center has been habilitated in Marbella. The results have shown that our students have save around 21,000€ in transportation costs plus around 1800 h of their time. Many students from Marbella, from municipalities close to Marbella and also from other provinces, have benefited from habilitating Marbella as UNED examination center.

On the other hand, and for the first time in Andalucía, a Virtual Attaché Return has been used, avoiding sending 8,500 paper exams to Madrid for correction. In addition, the security of the system is improved, since loss of documentation is avoided, and the digital copy of the exam is accessible to both Professors and students for improving revision procedure. Also correction time is decreased, since the exam is available for correction just a few hours after the exam, whereas previously the exams had to be transported from Malaga to Madrid for correction.

Ackowledgments The authors wish to express their gratitude to all the persons that work at the evaluation procedures department of UNED, especially to Prof. Rosa M. Martín-Aranda and all her team. Thanks are extended to Barbastro Associate Center of UNED for the facilities and fruitful development of the new system for virtual attaché. We also thank Marbella Hall, especially to Education Delegation, for providing all the facilities which have been required to perform the exams in Marbella.

References

Barberà E; Badia A; Mominó JM (coords.) (2001) La incógnita de la educación a distancia. Barcelona. ICE-UB /Horsori

Bernath U (2009) Distance and e-learning in transition: learning innovation, technology and social challenges. John Wiley, New Jersey

Cleveland-Innes M (2010) An introduction to distance education: understanding teaching and learning in a new era. Routledge, New York

Creighton SH (1999) Greening the Ivory Tower. Improving the environmental track record of universities, colleges, and other institutions. MIT Press, Cambridge, MA

Dahle M, Neumayer E (2001) Overcoming barriers to campus greening: a survey among higher educational institutions in London, UK. Int J Sustain High Educ 2:139–160

García Aretio L (2011) La educación a distancia. De la teoría a la práctica. Ariel Educación, Barcelona(4ª ream.)

Guerrero-Pérez MO, Rosas JM, Bedia J, Rodríguez-Mirasol J, Cordero T (2009) Recent inventions in glicerol transformations and processing. Recent Pat Chem Eng 2:11–21

Guerrero-Pérez MO, Rosas JM, López-Medina R, Bañares MA, Rodríguez-Mirasol J, Cordero T (2011) Lignocellulosic-derived catalysts for the selective oxidation of propane. Catal Commun 12:989–992

Hope A, Guiton P (2006) Strategies for sustainable open and distance learning. Routledge, London

Litwin E (2000) La educación a distancia: temas para el debate en una nueva agenda educativa. Amorrortu, Buenos Aires

Lozano FJ, Lozano R (2014) Developing the curriculum for a new bachelor's degree in engineering for sustainable development. J Clean Prod 64:136–146

Moore M, Kearsley G (2012) Distance education: a systems view of online learning, 3rd edn. Wadsworth, Belmont, CA

Moreno M (2012) Veinte visiones de la educación a distancia. Edgvirtua, México

Nanaki EA (2009) Environmental comparison of the use of biodiesel and gasoline for transportation—a case study for Athens. Indoor Built Environ 18:416–423

Ortega-Navas MC, Muñoz-Mansilla R, Martín-Aranda RM, Latorre F (2014) Electronic logistics for a sustainable distance education: the new UNED on-site virtualization of evaluation procedure documents. Electronic logistics, chapter 8:344–349

Pollock N, Horn E, Costanza R, Sayre M (2009) Envisioning helps promote sustainability in academia: a case study at the University of Vermont. Int J Sustain High Educ 10(4):343–353

Schultz DM, Yoon TP (2014) Solar synthesis: prospects in visible light photocatalysis. Science 343:1239176

Talbot Ch (2004) Estudiar a distancia. Gedisa, Barcelona

WCED (1987) Our common future. Word commission on environment and development. Oxford University Press, Oxford, New York 400

Authors Biography

Prof. M. Olga Guerrero Pérez is Associate Professor of Chemical Engineering at Universidad de Málaga and Academic Secretary of Associated Center of UNED in Málaga. She received her B.Sc.

in Chemistry in 1997 from the Autonoma University of Madrid (Spain) and then she obtained her Ph.D. in 2003 at the same University, working at the Institute of Catalysis and Petrochemistry (CSIC) and at Repsol-Ypf. From 2003 to 2008 she was working as a postdoc fellow at Institute of Catalysis and Surface Chemistry (Krakow, Poland), Lehigh University (PA, USA), University of Cincinnati (OH-USA), Málaga University (Spain) and Institute of Catalysis and Petrochemistry (Spain). At the end of this period, she got a position as Assistant Professor at the Department of Chemical Engineering, in Málaga University, and in 2010 she was promoted to Associate. Her research interests are the development of new catalytic materials for application in Green Chemistry process. In this field, she is the author of more than 70 peer review papers and she has participated in more than 100 international conferences. She teaches classes on Separation Operations, Chemical Technology, and Water Pollution, among others.

Gustavo Román Ortiz is the IT Specialist at the "María Zambrano" Associated Center of UNED in Málaga since October 2010. He holds a degree in Computer Technical Engineering from the University of Málaga (Spain) in 2005. From 2005 to 2010 he worked as the Guadalinfo Project Coordinator for the Junta de Andalucia in the areas of Málaga and Cádiz (Spain). The main objective of Guadalinfo was the creation of projects that could have a direct impact in the areas where they are implemented, and in relation to TIC's. In 2010, he was awarded the "Premio Ratón Especial", granted by the Council of Economy and Innovation of the Junta de Andalucía government, in recognition of the project carried out in Hospital Costa del Sol of Marbella (Spain) based on the implementation of an education system using new technologies, so as to facilitate the recuperation of patients in the children's ward.

Prof. Concepcion Travesedo is the Director of Associated Center of UNED in Málaga and Associate Professor of Communication at Universidad de Málaga. She received her B.Sc. in Information Sciences by the Universidad Complutense de Madrid in 1994 and completed a pre doctoral program in the Jawaharlal Nehru University, India, in 1998. She worked at the Universidad Europea de Madrid from 1996 to 1998 and obtained her Ph.D. in Journalism by the Universidad Complutense de Madrid in 2000. In 2003 she got a position as Associate Professor at Universidad de Málaga, where she remains up to this day. University Expert in Virtual Environments for Education by the Universidad de Málaga in 2004, Travesedo conducted a post-doctoral research program in Leuven University in 2008. In 2013 she was appointed Director of Associated Center of UNED in Málaga. Her teaching and research topics are International Information, Communication for Development and Peace, Global Media System, and Methodologies and Techniques in Communication Research.

Overview of the Sustainability in Brazilian Information and Communications Technology Market

Bruna Grascyela Schaefer Paese and Márcia Cassitas Hino

Abstract

The share of ICT segment in Brazilian GDP grows every year and increased from 4.5 % in 2011 to 5.2 % in 2012. The expectation is that this growth rate will continue growing over the coming years and the share will reach 8 % in 2022 by when Brazil is projected to become one of the top three ICT centers in the world. On one hand, the number of companies adopting the sustainability theme in speeches and advertisements has been also increasing in an accelerated manner, but on the other hand, only a few of them are handling it in an effective way. In this context, this paper aims to present an overview of how ICT enterprises are handling the sustainable development issue and also compile major impacts of this industry on society. The measurement and assessment criteria adopted by the Global 100, issued by Corporate Knights. Lastly, the study suggests improvements to Brazilian ICT companies based on the model adopted by Vivendi, which was ranked as the 7th most sustainable company and the 1st in the ICT segment.

Keywords

Information and communications technology · Sustainability · Global 100 · Vivendi

B.G.S. Paese (✉)
Rua 2650, 190. Apto 72, Balneário Camboriú, Centro, SC 88330-380, Brazil
e-mail: bruna.paese@gmail.com

M. Cassitas Hino
Av. Visconde de Guarapuava 4517, Apto 172, Curitiba, Batel, PR 80240-010, Brazil
e-mail: marciahino@uol.com.br

W. Leal Filho et al. (eds.), *Integrating Sustainability Thinking in Science and Engineering Curricula*, World Sustainability Series, DOI 10.1007/978-3-319-09474-8_23

1 Introduction

Sustainability is no longer an unknown subject for companies. For some it is already part of the operational reality, being treated as a key element to ensure the business perennially and profitability.

Sustainability goes beyond preserving the environment. It encompasses people's awareness to significant changes in habits and socio-cultural. It also has foundation values such as ethics, respect, preservation, transparency and accountability of all actions.

In this context, there are organizational scenarios that bring, beyond the actions of individual employees, attitude and initiative from companies, through strategic positioning. The implementation of a sustainability framework in a company requires the involvement of several areas, people, concepts and actions.

The current Brazilian Information and Communication Technology (ICT) scenario shows that there are only a few real actions for sustainability. There is a large and strong social responsibility, considered one of the three pillars—economic, social and environmental—of sustainability framework, leaving consideration of the other pillars below expectations.

The aim is to identify ICT multinational company best practices classified as top 1 in an international sustainability ranking and analyze four significant Brazilian ICT companies.

2 Sustainability and ICT Companies

Companies must be economically viable, socially fair and environment-friendly, reaching several levels of human organization, from small villages up to the whole world (Srour 2008). Several factors make the ICT sector a great sustainability ally, mainly by reducing the impact caused without neglecting the contribution that technology can add to the development, optimizing the use of resources, promoting information through democratizing accessibility, developing products that improve the users' life quality and promoting social, ecological and economic development.

According to Nawi et al. (2013), ICT plays an important role in development, especially in ensuring data and communication. It helps to achieve development goals and spur progress in developing countries. There are also signs that the pervasive use of ICTs are globally leading to the transformation of some societies, the information or knowledge society more precise and recent.

Conforming to UN Economic and Social Council (2005), ICT revolution opens vast opportunities for economic growth and social development but also presents challenges and threats. While considering the impact of ICT on the creation of a global knowledge-based economy, the majority of the world population still lives in poverty and remains untouched by the ICT revolution.

Pursuant to Lewis (2013):

2–3 % is the ICT's contribution to global GHG emissions, similar to aviation industry;
4–6 % ICT's predicted contribution to global GHG emissions in 2020;
30 gigawatts is the current Internet power demand. This is around 1.5 % of the world's electricity and equal to the total supply from around 30 nuclear power stations;
50 billion is the predicted number of Internet-connected devices in 2020;
90 % is the estimated proportion of a typical data center's electricity consumption wasted;
90 % is the higher potential reductions in GHG emission due to ICT service migration to Cloud solution;
16.5 % is the potential reduction in global GHG emissions excluding ICT-related emissions that could be achieved through innovative ICT use.

Companies are responsible for the breakdown and changes happening in natural resources, since supplies are gathered from nature (Dias 2011). In recent years, actions in favor of the environment became secondary due to environmental problems caused by industry. However, fewer companies are caring and making more eco-efficient production processes. Even when they do, the initiative is taken more like an answer to government agencies demands than as a position of social and environmental responsibility.

According to Nawi et al. (2013), there is a need to integrate social, cultural, institutional, economic, political, and technological sustainability as vital elements in the planning and operation of ICT projects. Many studies refer to sustainability as being key to long term development outcomes for ICT projects. Aspects of institutional sustainability include well-defined ICT laws, participatory policy-making processes, and effective public and private sector organizations that develop a scenario in which the livelihoods of the community can be continuously improved.

ICT can help to create a versatile society in which changes in resource and consumption patterns lead to a drastic reduction in CO_2 emission, it could also have negative impact on the mental/emotional in case of utilizing technology without social and environmental consciousness (Fujimoto et al. 2009). ICT companies play a significant role in decreasing the resource intensity in society, aiming to have a positive impact on environment sustainability (Ahola et al. 2010; Scott Matthews et al. 2010).

Sustainability is still a complex subject and requires demystification; reaching the ones directly involved and the society in general. For this, it is important to highlight the major factors affecting the sustainable aspects of ICT area in order to pursue specific knowledge coming from this business model and how to use it to elaborate and execute specific actions to soothe the sector's impact.

3 Universities and Sustainability

According to Bekessy and Burgman (2003), leading society towards a sustainable future is part of universities' responsibility. Most of worlds' managers and decision makers are prepared by such universities and they have a major role in the economy, both global and national. Worldwide Universities are currently answering to this more sustainable ways of human activity needs. According to the author, "several key forces are driving this change. There is demand for expertise from students and employers". In the academic environment, there are social and economic advantages for including sustainability in the curriculum of research and operations. Also, University leaders have been recognized for their responsibility towards this. Altogether, such factor provides universities with a unique position to influence the society direction.

From Pidlisnyuk (2010) point of view, it is necessary a new ideology of life and modern innovative approaches for education in order to overcome the global and ecological crisis, resulting in a new cultural and ecological world outlook. When training students and specialists, mainly political scientists, economists, environmentalists and agriculturalists, sustainable developments' imperative is of upmost importance.

4 Methods

The basis for best practices comparison is Vivendi (2012a), a French ICT company, which received the prize for being the 7th most sustainable company in the world in 2012, according to the Global 100 ranking published annually by Corporate Knights. In ICT segment, Vivendi is the first in the ranking, for this reason was chosen to contemplate this review.

Moreover, it was chosen the fourth most important Brazilian companies in the ICT sector, based in relevance and impact in society. Sustainability reports from these companies were used for analysis. The ICT companies selected for this study are:

Globo (2013): the leading Brazilian TV Broadcaster. Globo is part of society, not only for content generation but mainly for cultural and social impact over it. Its influence is huge, being able to influence commemorative dates and affect economic and political decisions in country.

Vivo (2013): the largest telecommunications company in Brazil and one of the most sustainable according to Exame Magazine Sustainability Guide. It was recently acquired by Telefonica S.A. Since 2009 the company massively invests in sustainability under the "Vivo for the future" slogan with greater focus on environmental issues.

Positivo Informática (2013): the biggest computer manufacturer in Latin America. It produces much of the technology consumed in the country, serving primarily to governmental demands. Headquartered in Curitiba, it is the tenth largest computer manufacturer in the world.

GVT (2013): the fastest growing national company in the telecommunications industry. It has been in the market since 2000, attending both retail and business customers. It was the first telecommunication company to join Bovespa New Stock Market. Vivendi purchased 85 % of the company shares in 2009 and the remaining 15 % in 2010.

5 Results and Discussion

Vivendi (2012b) best practices support the comparison with Brazilian companies analysed and is considered a goal to be achieved.

6 Climatic Factors

The equipment power consumption that supports a telecommunication network represents 70–90 % from total consumption of an ICT company. To lessen the impact of this issue on society, service providers can work together with suppliers in order to increase the equipment energy efficiency. The increased efficiency also improves the ratio between data flow volume and power consumption.

Table 1 shows the performance comparison:

Brazil has serious problems concerning energy. The first factor is that demand is higher than supply, and although the structure has most of its supply composed by hydroelectric power plants, alternative sources are still poorly explored, leaving few alternatives for Brazilian companies, besides the fact that parts of the energy offered is from non-renewable energy, in this scenario the industries located on Brazil have no choice in type of energy they will use. Factors such as developing products with reduced environmental impact need much more focus from these companies, following efforts to increase energy efficiency. According to Hoover (2009), the electric power consumption of servers and other IT equipment doubled between 2000 and 2005. To reduce the impact of this increase, the industry should adopt Green IT measures that come from the selection of IT equipment with higher energy efficiency by adopting more efficient cooling systems.

Table 1 Climatic impact comparison

Global 100 factors	Vivendi	Globo	Vivo	Positivo Informática	GVT
Effort to increase energy efficiency	✓	×	✓	✓	×
Green IT concept	✓	×	✓	✓	×
Renewable energy sources use	✓	–	–	–	–
Search by developing or using products with reduced environmental impact	✓	✓	✓	✓	✓

Source Authors

Buildings, offices and shops usually compose the facilities of a company in the ICT sector. Thus, increasing the energy efficiency of these plants reduces the impact on the climate issue. A good measure would be using renewable energy that reduces emission of greenhouse effect gases. New services also contribute to the climate change issue. Video and teleconferencing, for instance, reduce the need for transportation of a company's employees and consequently CO_2 emissions. Other examples are cloud computing and e-commerce solutions.

7 Garbage and Disposal of Materials

Electronic equipment such as computer monitors, CPUs, modems and mobile phone batteries are harmful to humans and cause environmental damage due to the contamination of soil, air and water. According to the European Union (EU), the volume of electronic waste grows 3–5 % a year in its countries. In this scenario, ICT companies can develop actions to reduce the impact of issues related to this matter.

Table 2 summarizes the comparison:

Brazil has few companies that collect and recycle compared to the amount of waste produced, which makes the initiatives of companies that do recycle office supplies very hard, and for this matter, this item was disregarded for analysis. The reverse engineering instead of manufactured goods is another factor of low maturity in the country; outside government regulation, few industries offer the collection of unusable products.

The use of non-hazardous components is not part of the strategic planning of most researched companies. This is a particular worrying scenario for Positivo Informática that produces electronic equipment.

The development of longer life product by ICT companies contributes to the reduction of the electronic waste volume produced. The improper disposal of batteries for cell phones, modems and other equipment used by the industry's products is common. The adoption of measures such as reverse logistics (where the manufacture and the consumer are responsible for the product lifecycle, including disposal), is an alternative that may reduce the impact caused by them.

Table 2 Comparisons between garbage and disposal of materials

Global 100 factors	Vivendi	Globo	Vivo	Positivo Informática	GVT
Increased product life	✓	–	✓	✓	✓
Use of non-hazardous components	✓	×	×	×	×
New materials for packaging	✓	✓	✓	✓	✓
Decreased volume of electronic or physical waste produced	✓	✓	✓	✓	×

Source Authors

The selection of product components is another point that ICT companies may consider to action. The selection of materials that are not harmful and do not pollute the ecosystem is a measure that directly affects the environment where companies operate. Prospecting packaging compounds that pollute less and using renewable materials can also assist in reducing the impact of ICT companies over society.

Regarding electronic garbage, the inadequate management of products and networking equipment from service providers leads to an increase of electronic waste produced. Through the design of an effective garbage destination policy, the ICT companies can ensure the appropriate destination of hazardous materials and reuse or recycling of others. The separation and proper disposal of the waste produced by ICT company installations are their responsibilities and should be noted as an influencing factor on environmental impact of the regions served.

8 Communication and Information

Digital inclusion is an important tool for social and economic development in postmodern society. Therefore, the access dissemination to ICT products and services is a critical factor that contributes to sustainable development of localities and regions.

The main challenge in underdeveloped regions and emerging markets is providing access to information and communication technologies at a fair price to urban populations with lower social classes and provide technology to rural areas. Taking the Brazil example where access to broadband is higher or lower depending on the region, state, city and neighborhood, it is necessary to the State to act, through the National Broadband Plan (PNBL), to promote universal broadband access.

It is possible to verify in Table 3 if ICT Brazilian companies are engaged in fostering digital inclusion in less developed regions and fostering digital inclusion with disabilities is not part of the strategic planning of most researched companies.

Table 3 Comparison of communication and information approaches

Global 100 factors	Vivendi	Globo	Vivo	Positivo Informatica	GVT
Fostering digital inclusion in less developed regions	✓	✓	✓	✓	✓
Fostering digital inclusion of people with disabilities	✓	✓	×	×	×
Concern censorship group and generated content	✓	✓	✓	–	–
Access control tools for users	✓	×	✓	✓	✓
Inspection of the content provided	✓	✓	✓	×	×

Source Authors

This is a particular (disconcerting) scenario for GVT and Vivo, which provide telecommunication services for general people.

The ICT sector can promote social inclusion by creating products and services that enable accessibility for people of all ages, conditions and languages. Vodafone, for example, developed software that enables blind people to hear incoming text messages on their mobiles.

Service providers and software companies can provide services that allow users to control access or filter content. A common example is the tool that allows parents to set content filters avoiding their children from viewing specific information that they believe are inappropriate. These companies can also track content delivery such as encouragement to terrorism and sexual exploitation of young people. The difference in the access control is that the company, rather than the user, is controlling the access.

The ICT products dissemination significantly increased the power of freedom of expression by the world's population. It also increased government demands, laws and regulations that restrict access to specific contents. The number of states that actually restrict access to content reached 20 according to the OpenNet Initiative. Current dilemma is determining which requests are legitimate content filter (as terrorism encouraging, for example) or illegitimate (as censorship, for example).

9 Privacy and Security

The amount of data captured, stored and transmitted by ICT companies has grown significantly over the recent years, which also increases the responsibility to ensure privacy and security of such information.

Many of the actions in communication and access to information derive from laws and government determinations, with constant discussions regarding security, privacy and responsibility for information. This subject needs more attention from businesses that deal directly with the access to communication and information, (meeting the demand not only from government but the market itself.

The Table 4 shows that Brazilian companies in the sector are engaged in guaranteeing data privacy and information security for users. Users are concerned regarding the information security (that personal data will not be obtained by unauthorized persons, for example) and information privacy (personal information not being used for commercial purposes, for instance).

According to Sampaio (2011), the best age for a child to start using a smartphone is 12 years old. For children of this age the number of internet users and smartphones is decreasing notwithstanding the time it stays connected is increasing. This may increase the risk of inappropriate content to be accessed by the user. In this situation, service providers can act implementing appropriate content filters in order to create a safer environment interaction for minors.

Table 4 Privacy and security comparison

Global 100 factors	Vivendi	Globo	Vivo	Positivo Informatica	GVT
Information users security	✓	✓	✓	✓	✓
Users privacy	✓	✓	✓	✓	✓

Source Authors

10 Life at Work

ICT companies are potential employers in regions where they operate. Therefore, there are issues that can act to ensure employee welfare. The ICT sector should be concerned with matters relating to the welfare of its employees, as workplace safety ergonomics and impact of local pollution on health.

The gender diversity in the workplace is a difficulty encountered by the sector due to the predominance of male engineers, programmers and scientists. As companies become increasingly global as well, it's important to offer job opportunities to the local population and the relocation of the headquarters staff. The diversity on age is also important in the environment of ICT companies.

Refers to Table 5 Life at Work Comparison analysis:

Brazil went through a major evolution in relation to labor factors and, despite abuses still occurring and the low investment on quality of life at work, most companies already have the worker's life as one of their priorities to contribute in HR indicators. Some companies need to rethink how to retain their best talents.

The ICT industry is undergoing a process of expansion, besides the competition and the need of innovation (Guimaraes 2000). In this context, the acquiring and retention of talents becomes strategic for business success. Business productivity is affected by long working hours due to the increase in cases of physical and psychological damage and turnover of employees. In addition, companies in the sector must follow the law regarding working hours, payment rules and benefits.

11 Stakeholder Relations

There are several external factors that induce a response from companies to reduce environmental contamination, such as the state, local community, market and suppliers (Dias 2011). The convergence in the ICT industry means that the services

Table 5 Life at work comparison

Global 100 factors	Vivendi	Globo	Vivo	Positivo Informatica	GVT
Occupational Healthy and Security	✓	✓	✓	✓	✓
Gender diversity	✓	✓	✓	✓	✓
Age diversity	✓	✓	✓	✓	✓
Incentive and retention of talent	✓	✓	✓	×	×

Source Authors

Table 6 Comparison of stakeholders' relations

Global 100 factors	Vivendi	Globo	Vivo	Positivo Informatica	GVT
Transparency in the supply and billing	✓	✓	✓	✓	✓
Trusty marketing	✓	✓	✓	✓	✓
Concern about suppliers' sustainability	✓	×	✓	✓	×

Source Authors

provided by the companies are commodities resulting in an increased competition in the sector. In this scenario, the quality of customer service becomes a competitive item.

Table 6 summarizes the analysis undertaken:

Brazilian companies already have maturity in relation to stakeholders and believe in transparency and trust as a tool in managing good relationship. They promote transparency in products supply and billing and trust marketing. Globo and GVT still have a low return in relation to concerns about the sustainability of suppliers. Satisfactory customer experience such as social and economic benefits of digital inclusion are obtained only if the telecommunications network for data traffic is reliable.

Transparency in the supply and collection services and products are one of the keys to maintaining a good customer relationship. ICT companies should ensure that marketing offers are always honest, reliable, decent and fair. Inappropriate practices may negatively influence the company's reputation towards customers.

Over the past years, ICT business suppliers have expanded their production lines to developing countries and emerging economies aiming to have cost savings on raw materials and labor-intensive. This led the sector to have concerns as wage fair working hours and worker rights. In this context, the increase of supply chain sustainability is the major challenge for ICT companies.

Significant environmental impacts may exist along the supply chain, including the use of inadequate materials, pollution, waste of energy, use of harmful substances etc. ICT companies can determine policies and processes to require their suppliers to manage their environmental issues.

There are a growing number of companies serving as providers to others and needing to have a good environmental performance throughout its supply chain, requiring its own suppliers to have environmental certifications and respectful production units within the environment. Thus, even if a company does not suffer direct pressures from both the State and the Community, it is forced to take actions to avoid environmental contamination as clients companies, for which its consumer market requires integrating a supply chain, is environmentally correct (Dias 2011).

12 Product Use

A new concept of life cycle analysis of the product by the perception of the entire production chain brings the idea of cooperation and cyclical thinking around sustainability. According to Horne et al. (2009), the society went through a strong analysis of energy efficiency technologies and product services. The definition of LCA (Life Cycle Assessment) comprises the systematic evolution of environmental impacts resulting from the provision of products or services.

One point of sustainability is the integrated approach to managing the life cycle of products (Product Life Cycle Management—PLM), that covers business processes (Rozenfeld et al. 2012). According to Lewis (2013), most ICT lifecycle phases generate an environmental impact. The rising use of energy throughout the ICT lifecycle, with the emissions of GHG (Greenhouse Gas), could be the most evident impact. There is also a high usage of water, especially in the manufacturing process and in the data centers cooling process. Garbage is created throughout the product lifecycle.

Metallic matter and other materials utilized in the manufacturing process of equipment demand energy in order to be extracted and processed. This might damage the environment due production of toxic chemicals. There is also the possible depletion of rare minerals. Manufacturing hardware requires large amounts of energy, and components transportation often involves long distances from the factory to assembly, consuming fuel over the process.

More energy is used in the recycling process and the depositing of unusable materials to landfill at the end of useful life. Air pollution results from the recycling of hazardous and toxic materials as others that cannot be recycled are dumped in increasingly scarce lands, which cannot be recovered economically, e-waste and other types of waste, for example packaging materials, are generated at various stages of the ICT lifecycle. It is imperative to deal with them in an environmentally responsible way.

Based on Table 7 information, the scenario is what still needs to be improved and exploited. LCA is still a recent issue for ICT companies and have yet to be studied and adapted more precisely. Fact is that in addition to being recent, this information does not appear in reports and public information, complicating analysis.

A tool currently used to understand the effect of financial and environmental products and services is the Life Cycle Assessment. Since firms perform their activities to meet social needs, and receive authorization to act socially, they are ethically responsible to meet the wishes, beliefs and values of that society, incorporating in their environment better life quality and ensuring sustainable future (Fenker 1997). In this context, each part of the process, each action to be held, will bring great economic, social and environmental results.

Table 7 Product use comparison

Global 100 factors	Vivendi	Globo	Vivo	Positivo Informatica	GVT
Specific area of sustainable process	✓	×	✓	×	×
Managing product's life cycle	✓	×	×	×	×

Source Authors

Table 8 Responsibility for consumption comparison

Global 100 factors	Vivendi	Globo	Vivo	Positivo Informatica	GVT
Product social benefit	✓	✓	✓	✓	✓
Social programs	✓	✓	✓	✓	✓

Source Authors

13 Responsibility for Consumption

The impact of a company is not restricted to its operation, but also encompasses the way products and end user uses services. The information technology and communication products allow people to work continuously due facility and accessibility to equipment. Lack of balance between personal life and work routine generated by the continuous work can cause health problems.

On the other hand, some products and the ease of communication can help to balance the relationship between personal and professional life. ICT sector can bring social benefits to the community, such as distance learning and telemedicine.

Table 8 shows that Brazilian ICT companies invest massively in social actions, concerned with the evolution and development of society in general. Among all factors, this is the best in terms of investment. Actions such as "Criança Esperança", partnership between Globo and UNICEF, aimed at raising funds for social assistance in the country. Other examples are the awareness actions and appropriate use of the Internet leaded by GVT. There are several projects supported by Vivo, which aims at taking expanding technology infrastructure for educational purposes. Moreover, Positivo Informática encourages suppliers to seek sustainable practices hosting a sustainability program for suppliers.

14 Economic Development

The information technology and communication industry size affects the local economy through the number of jobs created and the amount of resources invested. According Stančík and Desruelle (2012), in 2009, over 6.1 million people worked in the EU ICT sector, representing 2.7 % of employment in the EU. ICT employment is highly concentrated in ICT services: 5.1 million people worked in ICT Services (85 %) whereas 0.9 million worked in ICT Manufacturing (15 %).

Table 9 Economic development comparison

Global 100 factors	Vivendi	Globo	Vivo	Positivo Informatica	GVT
Region's economic development	✓	✓	✓	✓	✓
Vision and economic growth	✓	✓	✓	✓	✓

Source Authors

Naturally, the companies' aim-profit and growth, however, these factors need to be sustainable. Based on Table 9, ICT companies are concerned with economic factors of the regions and the company itself. In some programs and projects, that concern is notable, for example, in Globo's reports and documentaries teaching economy and financial development. Through reductions in price and terms of payment for goods and services in the case of the telecoms and equipment companies, beyond the concern in bringing development to less privileged regions, ensuring quality products contributing to the economic development of all.

In addition, the products and services bring many indirect economic benefits, such as economic integration of rural and remote locations, mainly in developing countries through market information access and financial services.

15 Conclusion

Are Brazilian ICT companies making use of the best market practices to reduce the impact on the environment generated by them or by the use of their products and services? Are there opportunities to implement new sustainable practices?

A qualitative analysis was undertaken on the four most important Brazilian companies from ICT segment. Assessment criteria used came from Global 100, an institute responsible for ranking the most sustainable companies in the world. Vivendi best practices were analyzed as it was ranked in 2012 as the 7th most sustainable company and the 1st on ICT segment.

In Brazilian ICT scenario, companies have just started to realize the importance of investing in the factors that contribute to the sustainable development of this sector, having concerns regarding the environmental impact caused by the strong use of energy, the amount and disposal of waste generated, the responsibility of content management and consequently the business perpetuation.

In Brazil, the process of investing in sustainability criteria is still marketing imposition, external influence and legal obligations rather than companies' initiative or essence. There is not only lack of efforts, but it's necessary to expand the concept used by the companies in which sustainability is only related to the natural environment, leaving other aspects such as economic and social unexplored.

Despite the recent advancement, awareness and use of sustainability by Brazilian companies, when compared to Vivendi they still have to develop a lot in relation to social and economic actions, especially on environment. In order to reach a more mature stage, more investment is necessary, and also a significant change on the understanding of sustainability by the Brazilian society for the encouragement of universities is crucial.

References

Ahola J, Ahlqvist T, Ermes M, Myllyoja J, Savola J (2010) ICT for environmental sustainability: green ICT roadmap. VTT Tiedotteita, Finland. Research Notes, 2532. http://www.vtt.fi/inf/pdf/tiedotteet/2010/T2532.pdf. Accessed on 05 Apr 2014

Bekessy S, Burgman M (2003) Universities and sustainability. http://researchbank.rmit.edu.au/view/rmit:7691. Accessed on 30 Apr 2014

Dias R (2011) Gestão ambiental: Responsabilidade Social e Sustentabilidade. Atlas, São Paulo, pp 183–196

Fenker E (1997) Sustentabilidade ambiental: avaliação do ciclo de vida do produto. http://www.mma.gov.br/estruturas/a3p/_arquivos/sustentamb_ciclovida.pdf. Accessed on 30 Oct 2013

Fujimoto J, Poland D, Matsumoto M (2009) Low-Carbon Society scenario: ICT and ecodesign. Inf Soc Int J 25(2):139–151. doi:10.1080/01972240802701726

GLOBAL100 (2013) Methodology. http://global100.org/methodology/. Accessed 30 Apr 2014

GLOBO (2013) Relatório Social. http://redeglobo.globo.com/globocidadania/relatorio-social-2012/. Accessed on 07 Mar 2014

Global Village Telecom (2013) Institucional. http://www.gvt.com.br/PortalGVT/Institucional. Accessed on 05 Apr 2014

Guimaraes AS (2000) Competitive strategies adopted by business information technology. Dissertação de Mestrado, Curso de Pós-Graduação em Informática, Pontifícia Universidade Católica de Campinas, Campinas-SP. http://bd.camara.gov.br/bd/bitstream/handle/bdcamara/4085/estrategias_informacao_guimaraes.pdf?sequence=1. Accessed 30 Oct 2013

Hoover JN (2009) 10 ideias para impulsionar os projetos de TI verde. http://www.itforum365.com.br/noticias/detalhe/4880/10-ideias-para-impulsionar-os-projetos-de-ti-verde. Accessed 30 Oct 2013

Horne RE, Grant T, Verghese KL (2009) Life cycle assessment: principles, practice and prospects. Csiro Publishing, Australia, pp 1–7

Lewis A (2013) Sustainability and ICT: Exploring the relationship.http://www.sustainability-perspectives.com/perspective/sustainability-and-ict#article-more. Accessed on 05 Apr 2014

Nawi HSA, Shukor NSA, Basaruddin S, Omar F, Rahman AA (2013) Community ICT Hubs Sustainability: Result From a Field Study of ICT4d Project. Academic Conferences and Publishing International Ltd. pp 247–253

Pidlisnyuk, V (2010) Education in Sustainable Development: the role of universities. http://www.ees.uni.opole.pl/content/01_10/ees_10_1_fulltext_05.pdf. Accessed on 30 Apr 2014

Positivo Informática (2013) Institucional. http://www.positivoinformatica.com.br/www/institucional/instituto.aspx. Accessed 07 Mar 2014

Rozenfeld et al. (2012) Gestão do ciclo de vida de produtos sustentáveis. http://www.numa.sc.usp.br/grupoei/index.php/por/Projetos/Sustentabilidade-na-Gestao-do-Ciclo-de-Vida. Accessed on 05 Apr 2014

Sampaio T (2011) Crianças com smartphones: qual é a idade certa. http://www.techtudo.com.br/artigos/noticia/2011/11/criancas-com-smartphones-qual-e-idade-certa.html. Accessed on 30 Oct 2013

Scott Matthews H, Morawski TB, Nagengast AL, O'Reilly GP, Picklesimer DD, Sackett RA, Wu PP (2010) Planning energy-efficient and eco-sustainable telecommunications networks. Bell Labs Tech J 15(1):215–236

Srour RH (2008) Ética Empresarial. Elsevier, Rio de Janeiro, pp 252–254

Stančík J, Desruelle P (2012) The 2012 predict report: an Analysis of ICT R&D in the EU and Beyond. Seville: Institute for Prospective Technological Studies. Obtido em. ftp.jrc.es/EURdoc/JRC75513.pdf. Accessed on 05 Apr 2014

Vivendi (2012a) Breaking News. http://www.vivendi.com/press/news/vivendi-ranked-first-french-company-in-the-2012-global-100-list-of-the-most-sustainable-large-corporations/. Accessed 30 Oct 2013

Vivendi (2012b). Publications. www.vivendi.com/investment-analysts/publications-en/. Accessed on 30 Oct 2013

Vivo (2013) Relatório de Sustentabilidade. www.telefonica.com.br/servlet/Satellite?c=Page&cid=1386090999186&pagename=InstitucionalVivo%2FPage%2FTemplateRelatorioSustentabilidade. Accessed on 30 Oct 2013

Author Biographies

Bruna Grascyela Schaefer Paese has been researching sustainability in ICT since 2012, worked in GVT were contributed to the commercial area and with studies on sustainability for the data centre construction. She also worked at Unilever, and now work with her own business. Formed in business and MBA in Strategic Management.

Prof. Marcia Cassitas Hino has spent last 10 years teaching, researching and working on administration areas including sustainability, e-government and Information System. The PhD in Administration is in progress, and is supported by his background: Administration Master Degree, Marketing MBA and several post graduations comprising Administration, Finance, System Engineering and Didactic in Higher School. In 2008 participated in a multinational program that developed and implemented sustainability projects for poor community to in India.

Virtual Learning Environments for Transformation to Sustainability: A Case Study from the "South"

Rudi Pretorius, Melanie Nicolau, Jaco Immelman, Anna de Jager, Marié Nöthling and Anja du Plessis

Abstract

An appropriate pedagogy for supporting behavioural changes required to achieve more sustainable futures should include collaborative knowledge construction, within learning spaces allowing for critical discourse and reflection integrated with real-world applications. However, it is not easy to implement a challenge based pedagogy like this in institutions where the academic system utilises individual grading and related measures of success and with academic staff reluctant to use alternative approaches and to formulate course outcomes in terms of behavioural change. This chapter maps and critically reflects on the implementation of a sustainability focussed honours (junior post graduate, 4th level) degree, in which a transformative pedagogy is used. It is a fully online degree, offered by the Department of Geography, University of South Africa (an open and distance learning institution). Difficulties associated with collaborative learning for sustainability and the facilitation of project based research within the virtual learning environment are highlighted, and how these are addressed. These reflections include consideration of the Global South context within which Unisa functions, confronting students and staff with infrastructural issues such as access to information and communication technology, but not necessarily preventing transformational learning for sustainability.

R. Pretorius (✉) · M. Nicolau · J. Immelman · A. de Jager · A. du Plessis
Department of Geography, University of South Africa,
Private Bag X6, Florida 1710, South Africa
e-mail: pretorw@unisa.ac.za

M. Nöthling
Centre for Professional Development, University of South Africa,
PO Box 392, Unisa 0003, South Africa

W. Leal Filho et al. (eds.), *Integrating Sustainability Thinking in Science and Engineering Curricula*, World Sustainability Series, DOI 10.1007/978-3-319-09474-8_24

Keywords
Sustainability • Behavioural change • Transformative learning • Challenge based pedagogy • Open and distance learning • E-Learning

1 Introduction

1.1 Transforming to Sustainability in Higher Education

The theme addressed in this chapter concerns the type of teaching and learning conducive to and assisting students in adapting behaviours towards increased sustainability. The sustainability issues faced by the world involve so-called 'wicked problems' since they threaten the continuation of life, are urgent to attend to, embedded in uncertainty and associated with long-term impacts with no clear-cut solutions (Brundiers and Wiek 2013). Students cannot be prepared effectively to deal with these challenges in educational settings over-focussing on knowledge (Sarewitz et al. 2012) and/or prematurely consider solutions but without theoretical depth (Jerneck et al. 2011). As a result there is growing interest in transformative learning (TL) to achieve the required depth of learning to equip students to contribute meaningfully to address sustainability challenges (Blake et al. 2013). TL implies reflexive practice in action, i.e. learning involving deeper levels of knowing and meaning, thinking critically about what we do and why, and consequently implementing change and acting differently (Sterling 2011).

1.2 E-Learning for Transformation to Sustainability: The Global South Context

This chapter situates education for sustainable development (ESD) and the need for innovative practices such as TL, within the Global South and specifically Southern Africa. Despite the ESD agenda being embraced in many parts of the world, this is not necessarily the case in the Global South, including Africa, where initial enthusiasm dissipated with realisation that commitments need to be supported by multi-level actions (Manteaw 2012). In addition, the need for access to higher education is much greater in the Global South than elsewhere (McNaught 2005). This includes the need for access to ESD, described by Manteaw (2012) as lacking visibility in Africa. By implementing e-learning, open and distance learning (ODL) institutions have the ability to respond to this need. But they are limited in their ability to do so in the Global South due to amongst other things infrastructural constraints (Wright et al. 2009). This limits the access of students to information and communication technologies (ICT's), as indicated by Oyedemi (2012) for the South African context.

1.3 Context: ODL, Unisa and the Department of Geography

In terms of its size of more than 300,000 students in 2012 (Unisa 2012), experience and capacity as significant ODL provider (Tait 2008), Unisa is well positioned to contribute towards ESD (Unisa 2005). Although sustainability is embedded in the Unisa vision (Unisa 2007), a common perception is that ODL is better suited to a transmissive type of pedagogy (Barasa 2011), compared to TL and epistemic learning (encouraging critical questioning of assumptions, beliefs and values), and which are better suited to ESD (Sterling 2011). On a positive note, the constant review of pedagogical approaches at Unisa needs to be mentioned, with aim to address these criticisms. ESD at Unisa is hindered by the institutional structure of Departments, Schools and Colleges, which presents difficulties for collaboration and multi-inter-trans-disciplinary initiatives. Despite this, the Department of Geography managed to implement an undergraduate degree in environmental management in 2000, which has since developed into a flagship degree at Unisa.

1.4 Aim, Methodology and Value of Chapter

This chapter aims to take a reflexive position on the implementation of the re-curriculated Honours in Geography at Unisa during the 2013 academic year, in terms of contributing to the transformation towards sustainability. As a fully online degree, this includes consideration of the virtual learning environment (VLE) in facilitating a transformative sustainability learning experience. This chapter has been compiled from the reflective narratives by the development and teaching team of this degree and presents their pooled experience of the implementation and extent to which sustainability aims have been achieved. The value of this type of enquiry is to provide richer descriptions of contextual nuances than for traditional analytic research (McNaught 2005). The curriculum of the qualification and its facilitation through the VLE is firstly reviewed. This is followed by reflective narratives on the collaborative and project based research components of the qualification. The conclusion reflects on the extent to which the aim of transformation towards sustainability has been achieved, with reference to the Global South and teaching and learning through the online mode.

2 A Sustainability Focused Curriculum: The Honours in Geography at Unisa

2.1 The Sustainability Connection

Aligned to the discourse on ESD in higher education (Hopkinson et al. 2008), 'environment' and 'sustainable development' have been features of the curriculum of all modules offered by the Department of Geography (Unisa) for a number of years. Sustainable development is understood as the process enabling humanity to achieve an environmental, social, economic and cultural sustainable world. Recently, the Department of Geography moved from teaching 'about' sustainability to teaching 'for' sustainability (Pretorius et al. 2014) to produce graduates that could deal with the sustainability issues faced in the Global South. This shift aims to increase the graduateness, employability and citizenship of Geography graduates at Unisa.

2.2 Reasoning Underpinning the Curriculum

The Honours in Geography is a postgraduate qualification aiming to prepare students for research-based study in Geography. In South Africa this qualification follows on a Bachelor's Degree, and serves to deepen expertise in Geography and to develop capacity in the methodology and techniques of Geography. This qualification demands a high level of theoretical engagement and intellectual independence (Council of Higher Education 2013) and its development was premised on external and internal pressures to deliver graduates sensitive to a variety of environmental and societal issues in the Global South.

The consensus in the development team was that this qualification should focus on the ability of graduates to identify and solve geographical problems, with emphasis on problems with no specific 'right solutions' (Johnston 1997). Graduates have to be able to identify, analyse and deal with real-world problems through theory-driven arguments to suggest evidence-based solutions. Students therefore need to work through a range of research methods and techniques and develop ability to select those relevant to their specific research problems. This is aligned with the requirement of the Council of Higher Education (2013) that at least 25 % of honours degrees must include research.

2.3 Curriculum Components

This qualification consists of five compulsory modules (Table 1). Three of the modules focus on themes to address discourses underlying a variety of social and environmental challenges in the Global South. The fourth module allows students to identify a real-world research problem within the discipline of Geography, while the final module is about the actual research to explore this real-world problem and provide perspectives on possible solutions. Each of the modules relate to the perspective that the discipline of Geography is characterized by change and diversity, with a wide range of philosophical approaches blended around three interrelated concepts of space, place and environment.

Table 1 Curriculum for the Bachelor of Science Honours in Geography, University of South Africa

Module title	Purpose
Geography of people-resource interactions in the Global South	Guides students in developing a voices on issues related to interactions between people and resource use in the Global South
Geographic perspectives of environmental change	Develops capacity of students to deal with environmental change with reference to interrelationships of natural and anthropogenically-induced phenomena
Geography of everyday living in human settlements	Guides students to explore factors impacting on quality of life and everyday living in human settlements
Geography in action: from problem identification to methodology	Develops ability of students to identify a real-world geographical scenario justifying research and to prepare a research proposal
The geographer as a researcher	Guides students to undertake research on their chosen geographical scenario, and to report on research findings

2.4 Transformative Elements

Underlying the teaching and learning of the modules is provisioning of a variety of real-world problems in the Global South. Students are encouraged to make own interpretations on subject matter, rather than act on beliefs, judgments and perceptions of others. The overall purpose is for students to develop autonomous thinking on sustainability, thereby producing graduates who can act as moral citizens in situations of change. Online delivery supports the fundamental principles of ESD. In addition, this mode of teaching and learning provides students with current information and case studies, allowing them to become better citizens in the Global South. However, the VLE presents challenges for students and staff, especially in the context of the Global South.

3 The VLE for the Honours in Geography

3.1 Implications of Using a VLE

The Honours in Geography was re-curriculated with e-learning in mind. Based on information and communication technologies (ICT's), e-learning can take various forms (Lynch et al. 2008). This qualification was designed to be fully online, implying a shift from 'physical to virtual campuses' (OECD 2005). In the Unisa context, this is not a shift away from a physical campus, but rather from a distance learning to an online (virtual) environment and from paper based to electronic delivery. Online implementation was integrated with re-curriculation. The modules could be developed with consideration of the tools available in the VLE. This is in contrast with Unisa practice till recently, with students receiving all study material for a specific module upon registration.

3.2 Elements of the VLE for this Qualification

Online delivery and VLE's are diverse and often designed for particular courses, degrees or even disciplines. Thus it is essential to define the elements of the VLE for the Honours in Geography, which is based on a version of the Sakai collaboration and learning environment. It is an asynchronous platform, not requiring real-time interaction and includes a course content tool, a discussion tool, a blog tool, a wiki page and a file-sharing tool. While other applications of VLE's might apply creative and authentic scenarios, as Mundkur and Ellickson (2012), the tools for the Honours in Geography are straightforward. Students do not receive any printed study material, but have to access it via the course content and file-sharing tools. The course content tool guides students to explore general resources, however, most guidance to search for resources is provided in a module specific way.

3.3 The VLE and Transforming to Sustainability

The process of finding appropriate resources is crucial for student's transformative learning journey. This is something that very few students have done before, especially in terms of the legacy at Unisa of providing complete study packages up front. While some readings are recommended for individual modules, the qualification demands students to engage with various source materials to gain theoretical depth. Students therefore have to take initiative and responsibility to find relevant source material. This requires continual interaction and guidance to structure a process that students experience as arbitrary, something that would not be possible in the rigid structure associated with pre-prepared learning materials.

The VLE plays a significant role in guiding students and supporting transformative learning and enables creation of learning experiences to engage students in problems and paradigms rather than simply delivering content. For the Honours in Geography, guided discussion forums and blogs entice students not only to reflect on their frames of reference, but to learn about the viewpoints of others and be exposed to new frames of reference. This needs to be done in a collaborative environment where continual interaction supports transformation. VLE's are particularly well suited for this since they allow sharing of underlying theory, interactive collaboration and facilitation of problem-based learning. Problem-based learning (associated with Geography) is essential for transforming towards sustainability.

3.4 The VLE in the Global South

Learning experiences in this qualification are shaped by location specifics of students in the Global South. Even if disregarding disparities and digital inequalities, it is a fact that almost 63 % of students in South Africa are without Internet access at home (Oyedemi 2012). Although Oyedemi (2012) provides information on access to computers and Internet penetration, this is not necessarily a good indicator of ICT literacy. The experience gained with the Honours in Geography in 2013 showed that ICT literacy, as opposed to access, often deteriorated learning experiences. Problems with using the VLE (lack of access, poor ICT literacy and occasional downtime of the VLE) affect the online learning experience and hamper transformational learning, which relies heavily on collaboration and interaction through the VLE tools.

For this qualification, using appropriate tools, technologies and methods formed an integral part of course design. It was therefore critical to consider the needs of users during implementation (Wright et al. 2009). In the Global South, with issues concerning access and ICT literacy, the straightforward tools of the utilised VLE are well suited, as they are not data intensive and require minimal bandwidth. Furthermore, working asynchronously meets the needs of the majority users, with many relying on cellular or 3G mobile connections for Internet access (Oyedemi 2012). Implementation of this qualification illustrates that careful course design, cognizant of the capabilities and limitations of the rigid VLE, can support the collaborative and reflexive learning required for ESD.

4 The Role of Collaborative Work in the Honours in Geography

The reflection provided in this section focuses on the experience gained in the three thematically orientated modules (Table 1) of the Honours in Geography during the 2013 academic year.

4.1 Collaborative Work and Transforming to Sustainability in the Global South

Sustainability goals are reflected in the purpose statement and outcomes for each of the thematic modules in this qualification (Table 1). The sustainability discourse is the golden thread through these modules, while the Global South context is firmly established, especially in the module on people-resource interactions. Learning experiences are linked to real-world challenges in communities where students are from. The expectation is that local wisdom and experiences from different communities have to be incorporated in learning experiences, referred to by Manteaw (2012) as SCOLE (school-community learning exchange). In this way different cultural, social and ecological experiences feed into discussions as well as student's own ideas and value systems.

The tools provided in 2013 in the VLE facilitated various forms of lecturer-student and student-peer collaboration, thus supporting the transformation to sustainability (Lynch et al. 2008). This included "student lounge" forums, where students could introduce themselves, share information on professional/personal interests and discuss matters related to their studies. Despite the diverse and geographically dispersed student population, the VLE assisted in creation of a virtual student community. This is in line with the stages of student-peer interaction in e-learning (Siemens 2011), namely communication, collaboration, cooperation and community and supports the importance of socialization taking place before collaborative tasks for co-construction of knowledge.

4.2 Elements of Collaborative Work in the Three Thematic Modules

The people-resources module required students to apply theory to a region of their choice, identify issues in this region and suggest solutions. Each of the five focus areas of this module included participation in graded discussions or a wiki and reflective blogging. Several challenges manifested during this collaboration. Although participation in discussions was good, posts focused on the initial guiding quote, with little reaction to other posts. As different regions were selected, reflective blogs could have supported collaboration. But many students did not blog, with even fewer commenting on blogs by peers. The wiki was used to collaboratively compile a reference list, a sustainability glossary and guidelines for comparative studies. The challenge with the wiki seemed to be inability to use it. For various reasons, collaborative learning in this module was not totally embraced. However, students who indeed collaborated, obtained good final results.

In the environmental change module, changes in the physical environment were considered, together with risks and vulnerabilities for affected communities. Towards the goal of transformative learning, students were required to formulate their world view on environmental change and to reflect on it by means of comparisons with YouTube videos linked to the module site. This reflection had to be shared through a blog, with students commenting on each other's blogs. Various case studies were used to compare the impact of environmental change in different contexts such as the North and South, with an opportunity for collaborative learning in discussion forums. Credits were allocated to students for participating in some of these discussions, while other discussions were for formative purposes only. Some individual assignments consisted of a collaborative component, in which case the history of the wiki tool was used to allocate marks.

The design of the human settlements module involved contextual application of the geographical perspective, different discourses on quality of life in human settlements, performing authentic tasks on quality of life in human settlements and communicating feedback to interested/affected parties. The sustainability discourse and the Global South were considered for selecting reading material. Human settlements were explored in rural, urban and informal contexts using photographs, Google Earth and direct observations. Each of the learning units supplied opportunity for collaboration via discussion forums. For example, after observing and comparing neighbourhoods in terms of predetermined questions, students could share and discuss own experiences. Concluding each learning unit, reflective blogging was required. The wiki was used for creating a common product about using tools to evaluate quality of life, but unfortunately proved not to be very successful.

4.3 Pros and Cons of Collaborative Work with Reference to the Global South

Many of the challenges experienced with collaboration during 2013 with the three thematic modules are not unique to online learning, but generally characteristic of group work. Before commencing with their studies, students need to understand reasons for and benefits of collaborative learning and get to know the VLE. In 2013 different attitudes about collaborative learning were experienced. Some students were apathetic, while others were not keen to collaborate at all. Possible reasons could be individual unease of students after accidentally deleting work of peers or by posting in incorrect places, or having to repeat activities as they did not save their work. The lacking computer skills of the typical student could also be a de-motivation and barrier to e-learning. Collaborative work such as creating documents online was problematic as limitations experienced by some of the students, had an impact on the group as a whole.

Collaboration was more successful where clear structure and specific instructions were provided. An example from the people-resources module was using the wiki to create a sustainability glossary. The wiki listed the concepts, with each student assigned a concept to define. In the human settlements module, students had to identify tools to investigate quality of life as part of group work, but it might have been better to provide a list of tools on which can be commented in a structured way. A functional use of the wiki in the people-resources module was to provide a space for students to comment and ask questions while the lecturer reflects on marking of assignments. Further research is required on challenges of collaborative learning through the VLE at Unisa. This also refers to the required balance between individual initiative and collaborative learning to find solutions to Global South challenges and provide inputs for transforming to sustainability.

5 The Role of Project Based Research in the Honours in Geography

5.1 Project Based Research and the Sustainability Connection

Markham (2011) describes project-based learning (PBL) as the integration of knowing and doing. This type of learning pre-supposes that students will obtain knowledge within a core curriculum, and then be able to apply what they know from a combination of their present and previous formal training together with their life-worlds to attempt to solve authentic problems in a meaningful manner. PBL allows space for interdisciplinary, collaborative, experiential as well as transformative learning and within the VLE it can be used to facilitate inquiry, dialogue and reflection on various facets of sustainability (Thomas 2000).

5.2 Elements of Project Based Research in This Qualification

During the development stage of the Honours in Geography, the team shifted the curriculum focus towards the values of sustainability. The fact that students in this qualification come from a wide range of backgrounds with diverse interests, was used to incorporate collaborative learning within a system of PBL, thus allowing students to share the triple bottom line of sustainability (social, economic and environmental) from the perspective of the key interrelated concepts of Geography (space, place and environment).

The research proposal and project modules involve inquiry, knowledge building and progress towards resolution. This implies transformation and construction of knowledge, which is facilitated through student driven research activities. Students are guided to apply the perspectives and knowledge gained from the thematic modules in the context of real-world scenarios (Thomas 2000). Together with acquired knowledge and skills supportive to do research, students then proceed to identify a problem or issue for investigation and then embark in the research about it. The research proposal and project modules thus attempt to embrace transformative, experiential, and collaborative learning through the creation of a space in the VLE where various perspectives on geographical problems are presented in a way supportive of student driven research.

In the research project module, students are engaged in inquiry through application of the geographical perspective of space, place and environment. Within this perspective students are then required to identify problems/challenges. In order to address these problems/challenges, they have to develop a problem statement, seek further information on the topic, decide on a methodology to address the problem, collect relevant data, analyse the data and draw conclusions. Finally students are expected to communicate their findings, ideas and recommendations as an academic paper. Not only does the VLE allow for PBL in an authentic context, it also allows for collaborative learning and sharing between students and peers as well as between supervisors and students.

5.3 Project Based Research and Transforming to Sustainability

Reflecting on the experience with the research proposal module during the 2013 academic year, the problems initially selected by students were very generic, although related to sustainability issues in the Global South. The majority of identified problems required to be narrowed down to enable viable research. Most students participated in providing comments on the problems selected by their peers, which enabled them to narrow down the problems significantly. This activity was successful in terms of the number of comments submitted. In addition the comments indicated a high level of understanding of sustainability issues. As the process unfolded, it became clear that many students battled with contextualising their chosen problems within the discipline of Geography. It therefore became the task of the allocated supervisors to ensure the relevance of the selected research problem in this regard. To an extent this inhibited the student-driven research approach, as supervisors had to intervene and provide input at certain stages. To complete the research proposal module successfully, supervisor-student and student-peer collaboration through sharing and reflection via the electronic forums was critical. Eventually the majority of students were able to meet the minimum requirements and presented an acceptable research proposal.

The research project module followed a similar approach, but more individual supervisor-student and less student-peer interaction occurred. The student's research projects covered a range of real-world problems related to sustainability and improving quality of life in their selected local contexts, mostly in the Global South. The students were expected to present their research as a scientific article or report. In this module a variety of online activities via discussion forums and blogs were included to encourage supervisor-student and student-peer discourse to assist students in undertaking their research, but low participation rates occurred. Due to limited interaction between students, supervisors had to interact more intensely, thereby losing out on the value of student-peer interaction. The majority of students did not make efficient use of the collaborative or reflective opportunities throughout the study period. The low rate of participation in the reflective online activities as well as the lowered student-peer collaboration could be attributed to the fact that the research project is still based on individual grading. Despite the lack of collaboration between student and peers in the research project module, many students were able to submit acceptable work and achieved the outcomes of the module.

5.4 Pros Versus Cons of Project Based Research, with Reference to the Global South

The research proposal and project modules within the Honours in Geography were accompanied with challenges as well as opportunities for both students and supervisors during the 2013 academic year. The modules succeeded in contributing to the development of a VLE characterised by inquiry, dialogue and reflection on various facets of sustainability and placed PBL at the forefront. A student driven approach was encouraged, however, in the research project module lower participation rates reduced the advantages of PBL and collaborative learning. The collaborative facets were not fully embraced by all and therefore need to be placed at the forefront of all research proposals and projects in future. The technological realities related to access to the Internet and computer literacy need to be recognised and emphasised as this remains a huge challenge for both students and supervisors.

6 Concluding Remarks

Program review is an important process that should be well planned and executed with professional integrity in order to improve all educational offerings. It contributes to the overall quality of service to students, the reputation of Unisa as an ODL provider, and demonstrates the commitment of Unisa to professional accountability in offering quality educational programmes. An important first step of a review process is the opportunity provided to the academics involved in the design, development and teaching of the programme to share their personal experiences and reflections. These reflections can then contribute to the framing of problems and focus areas for further improvement.

In this chapter it may have become clear that designing and developing the Honours in Geography was no easy task. The development and teaching team had a well-articulated vision and aspired to introduce their students to the sustainability discourse in the context of Geography as a vibrant subject field. They also envisaged a learning scenario that would be dynamic and challenging rather than one associated with the conventional, transmissive learning model that fosters the expertise of the teacher only, merely stresses the importance of memory in assessment and offers learning experiences that assume the student to be a vessel waiting to be filled. Furthermore, they accepted the challenge to facilitate the learning in a fully online environment in order to follow the vision of Unisa to participate in the digital age.

This chapter presented the personal reflections of the academics that were involved in the three phases of the re-design of the Honours in Geography. Through these reflections certain key problems have surfaced, e.g. the resistance of students to participate in the interactive learning tasks and their problems related to the online environment. The academics believe that in terms of the geographical content, an excellent learning experience has been designed. However, to really realise their vision, the future focus needs to be on innovative learner support strategies to assist students to survive the shift into the digital age.

References

Barasa F (2011) Promoting open and distance learning in Africa: a critical reflection on rhetoric, real and ideal. http://events.aau.org/userfiles/file/corevip11/papers/fred_barasa_Promoting_ODL.pdf. Last accessed 1 Feb 2012

Blake J, Sterling S, Goodson I (2013) Transformative learning for a sustainable future: an exploration of pedagogies for change at an alternative college. Sustainability 5:5347–5372

Brundiers K, Wiek A (2013) Do we teach what we preach? An international comparison of problem- and project-based learning courses in sustainability. Sustainability 5:1725–1746

Council of Higher Education (2013) The higher education qualifications sub framework. Council of Higher Education, Pretoria 44p

Hopkinson P, Hughes P, Layer G (2008) Sustainable graduates: linking formal, informal and campus curricula to embed education for sustainable development in the student learning experience. Environ Educ Res 14(4):435–454

Jerneck A, Olsson L, Ness B, Anderberg S, Baier M, Clark E, Hickler T, Hornborg A, Kronsell A, Lövbrand E (2011) Structuring sustainability science. Sustain Sci 6:69–82

Johnston RJ (1997) Graduateness' and a core curriculum for geography? J Geogr High Educ 21 (2):245–252

Lynch K, Bednarz B, Boxall J, Chalmers L, France D, Kesby J (2008) E-Learning for geography's teaching and learning spaces. J Geogr High Educ 32(1):135–149

Manteaw OO (2012) Education for sustainable development in Africa: the search for pedagogical logic. Int J Educ Dev 32(3):376–383

Markham T (2011) Project based learning. Teach Libr 39(2):38–42

McNaught C (2005) Understanding the contexts in which we work. Open Distance Learn 20 (3):205–209

Mundkur A, Ellickson C (2012) Bringing the real word in: reflection on building a virtual learning environment. J Geogr High Educ 36(3):369–384

OECD (2005) E-learning in tertiary education: where do we stand? OECD, Paris 293p

Oyedemi TD (2012) Digital inequalities and implications for social inequalities: a study of Internet penetration amongst university students in South Africa. Telematics Inform 29:302–313

Pretorius RW, Hedding DW, Nicolau MD, Nkooe ES (2014) Campus greening in open and distance learning: curriculum initiatives in the Department of Geography, University of South Africa. In: Leal Filho W (ed) Implementing campus greening initiatives: approaches, methods and perspectives. Peter Lang, Frankfurt (Upcoming volume of book series "Environmental Education, Communication and Sustainability", in press)

Sarewitz D, Clapp R, Crumbley C, Kriebel D, Tickner J (2012) The sustainability solutions agenda. New Solut 22:139–151

Siemens G (2011) At the threshold: higher education, complexity and change. Keynote address: teaching and learning symposium during the University of South Africa teaching and learning festival. 1–2 Sept 2011

Sterling S (2011) Transformative learning and sustainability: sketching the conceptual ground. Learn Teach High Educ 5:17–33

Tait A (2008) What are open universities for? Open Distance e-Learn 23(2):85–93

Thomas JW (2000) A review of research on project-based learning. http://www.bobpearlman.org/BestPractices/PBL_Research.pdf. Last accessed 20 Mar 2014

Unisa (2005) 2015 Strategic plan: An agenda for transformation. http://unesdoc.unesco.org/images/0014/001486/148654e.pdf. Last accessed 31 Mar 2014

Unisa (2007) Unisa service charter. http://www.unisa.ac.za/cmys/staff/contents/docs/Final_Service Charter_120307.pdf. Last accessed 31 Mar 2014

Unisa (2012) An institutional profile of Unisa: Unisa Facts & Figures. http://heda.unisa.ac.za/filearchive/Facts%20&%20Figures/Briefing%20Report%20Unisa%20Facts%20&%20Figures%2020120215.pdf. Last accessed 31 Mar 2014

Wright CR, Dhanarajan G, Reju SA (2009) Recurring issues encountered by distance educators in developing and emerging nations. Int Rev Res Open Distance Learn 10(1):1–25

Authors Biography

Rudi Pretorius is senior lecturer at the Department of Geography, University of South Africa and academic coordinator of the undergraduate programme in environmental management. He holds masters degrees in Geography and in business leadership and is registered for a Ph.D in Geography

Melanie Nicolau is the Chair of the Department of Geography, University of South Africa and has taught a variety of modules related to improving quality of life in the Global South. She holds a Ph. D in Geography. Her passion is sustainable socio-economic change within rural communities

Jaco Immelman is a lecturer at the Department of Geography, University of South Africa and has participated in the online development and implementation of a number of modules. He holds a Master's of Science in Geography

Anna de Jager is a lecturer at the Department of Geography, University of South Africa and has developed and taught a variety of under- and post-graduate modules. She holds a Master's of Arts in Geography and is currently registered for a Ph.D in Geography.

Marié Nöthling is a specialist at the Centre for Professional Development, University of South Africa and has vast experience in curriculum and learning development. She is passionate about transforming social realities through higher education. She holds a Master's of Education.

Anja du Plessis is a lecturer at the Department of Geography, University of South Africa, and has lectured a variety of courses related to the environment and sustainability. She holds a Master's of Science in Geography and is currently registered for a Ph.D in Geography.

The Deployment of Sustainability in the Higher Education Business Studies Curriculum: Centrality, Pervasiveness and Practical Teaching and Learning

Clare Hagerup and Russell Woodward

Abstract

This piece examines the deployment of sustainability values in the business degree curriculum, in terms of its centrality, pervasiveness, staging and teaching and learning methods. Following an initial review of the meaning of sustainability with reference to business studies, the piece pursues discussion of sustainability's position in the business studies curriculum, using pedagogic and regulatory elements and in particular, application of the emerging curriculum design paradigm of the threshold concept criteria. This curriculum design conceptual application is taken forward through discussion of evidence based studies on the deployment of sustainability in the HE business curriculum, including specific experiences in the authors' own business school. Whilst sustainability within business fulfils the bulk of threshold concept criteria, the breadth of its integrativity in practice extends beyond the subject discipline such that the boundedness criterion of the paradigm does not apply. Furthermore, its discipline transformativity in the same criteria paradigm indicates that it should be deployed at the post-introductory level in business studies.

C. Hagerup (✉) · R. Woodward
HE School of Business Management and Computing, University Centre, Grimsby Institute Group, Nuns Corner, Grimsby, DN34 5BQ, UK
e-mail: hagerupc@grimsby.ac.uk

R. Woodward
e-mail: woodwardr@grimsby.ac.uk

W. Leal Filho et al. (eds.), *Integrating Sustainability Thinking in Science and Engineering Curricula*, World Sustainability Series, DOI 10.1007/978-3-319-09474-8_25

Keywords
Sustainability · Business · Threshold concepts · Curriculum · Transformative

1 Introduction: Meaning of Sustainability with Reference to Business Studies

Sustainability as a concept and topic has been on the rise in the higher education (HE) business curriculum for a number of years. The evidence from Christensen et al. (2007) on the increased inclusion of sustainability values in the courses of the leading UK business schools is just one representation of this powerful trend.

The bulk of texts looking at business and sustainability, for example Blowfield (2013) define the latter in context of the Brundtland Report (1987) depiction of sustainable development as being that which meets the needs of the present without compromising the ability of future generations to meet their needs. Though this is frequently associated with environmental aspects such as ecological resource depletion, it is very evident that sustainability need not be confined in that way. Pursuing activities in the current period without constraining equivalent activity in future periods can relate to all kinds of resources, including financial, and all sorts of outcomes, including income and wealth. Depicting sustainability thus corresponds with the Bridges and Wilhelm (2008) connection of sustainability to the triple bottom line of economic stewardship (prioritising financial continuity over current profit), alongside environmental stewardship (maintenance/renewal of natural capital) and social stewardship (maintenance of human and community long term well being).

Sustainability in this depiction can be connected, as shown by Nemetz (2013) and by Atfield and Kemp (2013) to issues ranging from renewables as energy sources, materials resource use, long term financial viability, social enterprises, treatment of workers and communities, and ethics. As such, sustainability is an approach concept that can apply to all kinds of decisions. All of them relate to organisations and households/consumers, and thus in this sense, the sustainability concept can be seen as relating in all its applications to the subject field of business studies. This breadth and connectivity of applicability within the business studies field, when looked at across the topic/module areas of the business studies curriculum, as evidenced in Weybrecht (2010), demonstrates the bearing of sustainability for Marketing, Human Resources, Finance, Operations and Supply chain, Strategy and Public and Third Sector Management.

In view of the above, it is interesting to see how the positioning and delivery of sustainability in business degrees can be informed by relevant pedagogic and regulatory themes—and in particular within the former—the contribution that curriculum design theory can make to the issue.

2 Pedagogic and Regulatory Issues: The Threshold Concepts Paradigm

For applied discussion of the position of the sustainability concept in the HE business curriculum, an important pedagogic framework suggested here is the Threshold Concept Criteria as first outlined by Meyer and Land (2003, 2005). Aiming at objective conditions for deciding core content within a course syllabus, it is held that a threshold concept should be Transformative, Irreversible, Integrative, Boundary-Setting and Counter Intuitive/Troublesome. The meanings of these criteria are as follows:

A threshold concept is depicted to be transformative. Meyer and Land (2003) portray this feature as being like a portal or gateway that opens up a fresh and previously unaccessed way of thinking about content within a subject.

The irreversibility criterion of threshold concepts is connected to their transformative nature. While it does not mean that a concept is not subject to challenge, it means that the transformed way of thinking facilitated by the concept ensures that in study of the subject concerned, the concept will not be dropped from use.

Threshold concepts are portrayed to possess the integrative criterion in that they serve to draw together content from within a subject field between which the connection was not previously appreciated.

Threshold concepts are viewed as probably boundary-setting, with a specific meaning particular to and defining of a distinct subject field. Meyer and Land (2005) link this to the ability of threshold concepts to define frontiers between conceptual areas with subjects seen as academic territories.

Drawing on the work of Perkins (1999), threshold concepts are depicted by Meyer and Land (2003, 2005) as being troublesome. This is connected to the student being required to move on from a customary way of seeing things to a perspective that may initially seem counter-intuitive.

Consistent with the above criteria is the further notion of threshold concept acquisition being associated with states of liminality. Meyer and Land (2006) depict this liminal state in terms of the crossing of a pedagogic threshold or transitional space. The fact that this is not a straightforward process is emphasised by Cousin (2006) who sees threshold concept acquisition as likely to involve messy journeys back and forth across conceptual terrain.

Fitting especially with the transformative and integrative criteria of threshold concepts, their identification is viewed as key to constructing courses that maximise relevant subject learning without overloading on descriptive content. This aspect is particularly recognised in Cousin (2006).

Given the transformative and boundary-setting criteria, it follows that threshold concepts are connected to development in students towards ways of thinking and practising that are particular to the subject field and the academic community of that subject. This aspect is explicitly noted by Davies (2013).

One should note that the threshold concepts paradigm has received some criticism, most notably by Rowbottom (2007), chiefly on the grounds that the threshold concept criteria are difficult to assess empirically because they are essentially matters of degree or extent rather than discrete properties. It is the view here that, while the criticism is valid in specific terms, it should not translate into the view that the threshold concept criteria are not a step forward from subjective viewpoints on what is a key concept in a subject curriculum. Albeit matters of degree, the threshold concept criteria are explicit specific criteria comprehensible to subject tutors, which can guide them towards the concepts most likely to aid in construction of an efficient curriculum from both a teaching and learning point of view.

Another aspect of potential relevance, one acknowledged by Land et al. (2005), is the challenge issues faced by the Threshold Concepts paradigm when a subject which stands as an intrinsic field in its own right, warrants inclusion in the curriculum of another subject course, for instance as a contextual element.

3 Pedagogic and Regulatory Issues: Applied Discussion

Much existing literature linking sustainability and threshold concept considerations is focussed in the field of engineering, though some of those pieces, e.g. Desha and Hargroves (2012) do include the theme of engineering business practice.

The Brooks and Ryan HE Academy report (2008) on Embedding Sustainable Development, reflects on the potential status of sustainability as a threshold concept though this is, significantly, in a strongly multidisciplinary context.

The main piece of academic work setting threshold concept status considerations of sustainability in the business curriculum context is the doctoral thesis of Garcia-Rosell (2013), in particular relating sustainability to the subject of marketing. The thesis depicts sustainability as a threshold concept for marketing though with partly a socially constructed meaning.

Moving into specifics of discussion here, sustainability can be seen as transformative in that it provides a fresh way of thinking about all decisions and actions in business. It can also therefore be seen as irreversible in that this fresh perspective is unlikely to be unlearned because of its pervasive value as discussed above. It should be noted though that much learning in business can and does proceed without consideration of sustainability. This has a bearing for the staging and sequencing of sustainability in terms of how early the concept, albeit transformative, should be brought into the business curriculum. One perspective on this issue can be reached from consideration of sustainability in context of the different levels of threshold concept within a subject as depicted by Davies and Mangan (2007)—namely basic, discipline and modelling. It may be that sustainability is transformative not in the sense of a basic threshold concept of business, but transformative in the sense of a discipline threshold concept whereby understanding of other subject field topics, e.g. marketing and finance, is transformed through acquisition of the concept.

Moving on, the noted fact that the Atfield and Kemp (2013) and Nemetz (2013) literature depicts sustainability as covering a diversity of business decisions, unifying them in a context of maintaining continuance of viability strongly resonates with the achievement of the integrative threshold concept criterion. Similar is indicated by Weybrecht's (2010) mentioned belief in the merits of demonstrating the cross-cutting bearing of sustainability for the main topic areas of business.

In addition to this, from a regulatory quality body point of view, the QAA (2007) subject benchmarks for undergraduate business and management cite sustainability as one of the pervasive issues, GMB 3.7. This again indicates an informed view that sustainability gives enhanced and connected meaning across many requisite curriculum topics in the HE business subject field, as would be the case with an integrative concept.

From a course design perspective, these points strongly suggest that sustainability should be applied at many topic/module points within business, though in terms of intrinsic explanatory coverage this may warrant locating in one place—with application payoffs being widespread—in line with Cousin's (2006) view that threshold concepts facilitate design of a less cluttered curriculum.

Looking at sustainability and the boundary-setting criterion, although it was noted that the decisions to which it applies can be set within the field of business studies, it is also evident that the term, in the same meaning, can fit into many other subject fields as well. This is shown not only by the threshold concept work linking sustainability to engineering, but also in the Brooks and Ryan (2008) multidisciplinary depiction of sustainable development embedding into subject areas once again across and beyond the field of business studies.

With regard to the counterintuitive/troublesome criterion and the liminality aspect, it is argued here that this is evidenced for sustainability partly by the varied way in which the concept is treated in academic debate in the business/economics field. A well known example here is Friedman (1970) who, although seeing a place for ethics, emphatically portrays sustainability specific concepts as having no merit unless they accord with the neo-classical depiction of firm profit maximisation. This can implicitly be seen as denoting a desire to depict sustainability as something artificial with regard to business practices, either to be ignored or to be deployed superficially as a tool, e.g. through marketing, to attract customers. The viewpoint here strongly contrasts with that of Porter and Kramer (2006), which effectively critiques the notion of dichotomy between environmental protection and business competitiveness. One might see this debate as representing standpoints from different positions across a portal or liminal space. The notion of the counterintuitive/troublesome aspect of sustainability is reinforced by consideration of the work of Stubbs and Cocklin (2008) who report efforts to engender a mindset shift, via sustainability conceptions, from reliance on the wholly neoclassical economics based paradigm across their studies, on the part of MBA students. The lofty level of the debate, as reflected in the opposing positions of Friedman and Porter, is instructive and indicates the counterintuitive threshold nature of the sustainability concept in business schools to be applicable to academics, and not just students. It would then be reasonable to suggest that many educational practitioners in the business studies field may not yet have ventured into this particular concept portal and liminal space. This has a bearing for business curriculum integration of sustainability with regard to the delivery team and staff development.

Another area of pedagogy which has a curricular bearing for sustainability in business studies is the cognitive domain taxonomy as set out by Bloom (1956). Sustainability, while not an easily measurable term, can be seen as a particular objective or criterion against which to assess business activity. In this respect then sustainability is essentially a critical analysis equipping concept. This consideration along with the discipline/basic aspect above does suggest that while centrality and pervasiveness of sustainability should not be compromised, there is merit in looking at deploying sustainability at a stage subsequent to introductory.

4 Relevant Evidence Reviewed in Context

The international multi-institution survey work by Naeem and Neal (2012) on business school deployment of sustainability in the Asia-Pacific region has significance, including with regard to threshold concept depictions. This is especially the case because as well as nature and extent of curriculum deployment, business faculty staff viewpoints on both the motives and obstacles to the integration of sustainability are key components of the survey evidence.

Looking at deployment, it is interesting that the two most dominant responses in the survey are (1) that sustainability is an integral component in one or more course modules, and (2) that there are no current plans to include sustainability in business programmes. Within lower scoring category responses on the deployment issue, it was evident that presence as an elective module clearly outscored presence as a core module, and that for course planning, development as an elective module strongly outscored development as a core module. It should be noted that these thematic results applied for undergraduate and postgraduate provision.

Initially, the single highest score going to integral presence across modules does indicate that the integrative nature of sustainability in the core modules of business is being carried forward by many institutions in the authors' survey. It should be noted however that this was not an overall majority score. The second highest score of non-deployment coincides with an earlier point that a great deal of business study and learning can take place without engagement of sustainability. These two highest 'extreme' scores taken together are suggestive that some liminality and a degree of troublesome engagement with sustainability, is being experienced across the business schools in the survey. The dominance of elective over core deployment in a sizeable minority of business schools both at the current and planned stages indicates centrality of incorporation is far from dominant, and that an appreciation of sustainability as transformative in business studies is not wholly prevalent.

Looking at the viewpoints and motives survey evidence from Naeem and Neal (2012), the main themes of interest include the disparities between strongly scoring respondent views on the core and fundamental nature of sustainability deployment within their business courses, and the weaker respondent scores on the perception that their institution or faculty views the matter in the same way. Regarding viewpoints on barriers to teaching sustainability in business schools, it is interesting that faculty being comfortable with their current teaching content scored highest while faculty scepticism about the relevance of sustainability scored very low indeed. These results alongside the high score for lack of sustainability case studies in the region suggest that some inertia, perhaps linking to a lack of tutor time and resources, rather than a lack of appreciation of sustainability on the part of business school faculty, is a major obstacle. One notes the externality/public good element here in that sustainability resource creation by a few tutors could increase sustainability resource availability across the field. This connects with the survey finding that development of regional networks on sustainability was seen as the strongest potential facilitator of progress in embedding the topic/concept into business studies.

The findings and reported experiences of incorporating sustainability into business by several of the chapter contributions to the Atfield and Kemp (2013) book on embedding sustainable development into business school subjects are significant here.

Findings presented by Emery (2013) regarding sustainability in marketing education reflect an appreciation of the transformativity of the concept in business, as shown by the self-evaluation evidence from the students. A key point to note is that this is a scenario of sustainability being introduced at the postgraduate (MBA) level, i.e. at a stage where students, through undergraduate studies and/or managerial experience, will have already engaged with the traditional core aspects of business.

The international survey evidence reported in Goderman et al. (2013) is interesting particularly in respect of a high proportion of business schools engaging sustainability through transdisciplinary means, i.e. delivery to business students alongside other subject cohorts. This thematic element coincides with the specific experience reported by Dobson and Tomlinson (2013) of successes in the approach of placing sustainability content as a university-wide open elective module for a wide variety of courses.

Looking across these selected contributions in Atfield and Kemp (2013), it would seem, firstly that the transformativity of the sustainability concept obtains even allowing for substantial business learning ahead of its introduction, and secondly, that the cross-curricular integrativity of sustainability means that it may not correspond to the boundary-setting threshold concept criterion.

Another area of research relevant to this issue of sustainability in the business degree curriculum is the work done by Benn and Martin (2010) regarding the use of boundary objects. Their study reports the use of boundary objects as having significantly enabled engagement with the sustainability concept on the part of business students in China.

The notion of boundary objects was first explored by Star and Griesemer (1989) who depicted them as points of intersection between different social worlds that can be highlighted to develop engagement towards pursuit of common objectives.

Similarly, Harvey and Chrisman (1998) have depicted boundary objects as common points of reference between different communities of practice, again with deployment viewed to enable engagement and cooperation across those communities.

In so far as, in line with the Benn and Martin (2010) study, boundary objects are key to effective and maximum engagement with sustainability for those focussed upon business studies, it does suggest that sustainability and business are different but intersecting fields of study and practice. This is consistent with the above threshold concept oriented interpretations of other studies in that sustainability may fulfil the bulk of threshold concept criteria, including integrativity, but does not correspond to the boundary-setting criterion in business studies. It is also consistent with the earlier point that much business learning can be achieved without engagement of sustainability. Further, in so far as these studies are correct, it would also seem that for the integrativity of sustainability within and across the business degree curriculum to be fully appreciated, boundary objects, as common reference points between sustainability and business, may need to be deployed. An example could be a stakeholder analysis focussed on an energy firm, where future employees, customers and communities are incorporated.

In view of these aspects, a number of points of curriculum design bearing can be drawn. First, in so far as sustainability is transformative and integrative in business studies, but is not a subject defining boundary-setting concept therein, its business curriculum deployment is very important, but does not require a wholesale change to the curriculum from start to finish. Secondly, if sustainability is to be delivered to business students via an open module, attended by other programme students, it is important that part of the delivery, e.g. seminars to complement open lectures, be contextualised to business students from the outset, so that the deployment of relevant boundary objects to maximise engagement, can be incorporated. Thirdly, it is advisable that some delivery of core traditional elements of business curriculum be undertaken, e.g. marketing, ahead of deployment of sustainability such that business oriented meaning of those boundary objects between sustainability and business, are strongly appreciated.

It is evident that the threshold concept depiction of sustainability within business studies encounters problems when it comes to the boundary-setting criterion. There are, however, reasons for seeing this criterion as the least meaningful in business studies curriculum design. Business studies as a partly vocational course must comprise a selective treatment of many subject fields thus the boundary-setting criterion may never apply when looking at the whole programme. An analogy to note is that while chemistry and biology are specific subject fields with genuine boundaries of scope, nursing as a professional/vocational course, incorporating elements of both of the above, has a more fluid curriculum to match developments in the job, including for example customer service.

Turning to the deployment of sustainability in the University of Hull validated business degree curriculum at University Centre Grimsby, this can be depicted as a mixed and developing picture.

Sustainability was formally introduced as a distinct module at the final level on the programme. Though this was consistent with being situated later than introductory in the course, it lacked the centrality to enable pervasive payoffs in other module areas. There was also the exacerbating issue that the module was optional—as with all final level modules on the programme.

The deployment has gone further in that in the intermediate level core module, Operations and Quality Management, sustainability is included via the latter's module tutor delivering a session as an internal guest speaker. This delivery aspect is supplemented in the sense that a module session in the following week is dedicated to reflection on the guest session in terms of its bearing and value. Setting this within the quality management scheme was considered effective in view of that subject's multiple perspective and criteria approach i.e. quality in terms of efficiency, customers, staff development, stakeholders, and alignment between activities and strategic mission.

External guest speakers, delivering to business degree students across all levels as a single audience, have tended to come from local industry, especially the food manufacturing sectors. In recent years pursuit and selection of such speakers has had, as an essential criteria, their ability to cover applied sustainability aspects within their talk.

As regards other modules, it is recognised that the teaching team, in line with the discussion point about the profession mentioned above, are at different places in terms of engagement of sustainability as a concept. It should be noted here that within cross curricular staff development festivals for the University Centre as a whole, sessions on sustainability and the higher education curriculum have featured consistently in recent years.

5 Conclusions and Recommendations

A number of points can be drawn here for business curriculum deployment of sustainability.

From application of the threshold concept curriculum design paradigm, sustainability is found to be both transformative and integrative in business studies. It is thus recommended that sustainability always be represented as a core rather than elective module in business degree courses, alongside pervasive engagement throughout other modules.

Sustainability is not a boundary-setting concept in business studies because of its multidisciplinary significance in HE. However, since business degrees are partly vocational, drawing content from many fields, this issue should not diminish the centrality of sustainability's deployment. It is however advocated that business tutors work on the use of boundary objects in their teaching and learning to maximise business student engagement of sustainability.

The transformative capacity of sustainability as a concept means that its inclusion does not require wholesale alteration to the business curriculum. Indeed, sustainability warrants location at a post-introductory level on business degrees as this is consistent with its criticality position in the cognitive domain and its discipline rather than basic position in threshold concepts context.

A final point relates to staff development of tutors in business. In view of the evident variation in levels of adoption of sustainability across business schools and inertia in progress in this regard, it is strongly advocated here that all curriculum area teaching staff have full engagement with sustainability, however liminal and counterintuitive the experience process may be. This development should take the form of contextualised sessions from sustainability tutors, and ideally in addition, global forums involving resource sharing and peer review thereof between business schools.

References

Atfield R, Kemp P (eds) (2013) Enhancing education for sustainable development in business and management, hospitality, leisure, marketing, tourism. Higher Education Academy, York

Benn S, Martin A (2010) Learning and change for sustainability reconsidered: a role for boundary objects. Acad Manage Learn Educ 9(3):397–412

Bloom B (1956) Taxonomy of educational objectives. Handbook I: the cognitive domain. David Mckay, New York

Blowfield M (2013) Business and sustainability. Oxford University Press, Oxford

Bridges CM, Wilhelm WB (2008) Going beyond green: the 'why and how' of integrating sustainability into the marketing curriculum. J Mark Educ 30(1):33–46

Brooks C, Ryan A (2008) Education for sustainable development interdisciplinary discussion series report. Higher Education Academy's Education for Sustainable Development (ESD) Project Report. Available at: http://www.heacademy.ac.uk/assets/York/documents/ourwork/sustainability/interdisc_discuss_series2008.pdf. Last accessed 8 March 2014

Brundtland Report (1987) Our common future. Oxford University Press, New York

Christensen LJ, Peirce E, Hartman LP, Hoffman WM, Carrier J (2007) Ethics, CSR and sustainability education in the financial times top 50 business schools: baseline data and future research directions. J Bus Ethics 73(4):347–368

Cousin G (2006) An introduction to threshold concepts. Planet 17:4–5

Davies P (2013) Threshold concepts: how can we recognise them? In: Meyer JHF, Land R (eds) Overcoming barriers to student learning: threshold concepts and troublesome knowledge. Routledge, London, New York

Davies P, Mangan J (2007) Threshold concepts and the integration of understanding in economics. Stud High Educ 32(6):711–726

Desha CJK, Hargroves C (2012) Applying threshold learning to teach sustainable business practice in post-graduate engineering education. In: The 119th ASEE annual conference, San Antonio, Texas, USA, June 10–13. Available at http://www.asee.org/file_server/papers/attachment/file/0002/2878/ASEE_2012_Paper_-_Final.pdf. Last accessed 8 March 2014

Dobson H, Tomlinson B (2013) Practical education for sustainable development through interdisciplinary problem-based learning. In: Atfield R, Kemp P (eds) Enhancing education for sustainable development in business and management, hospitality, leisure, marketing, tourism. Higher Education Academy, York

Emery B (2013) Sustainable marketing: embedding sustainability at the heart of business education. In: Atfield R, Kemp P (eds) Enhancing education for sustainable development in business and management, hospitality, leisure, marketing, tourism. Higher Education Academy, York

Friedman M (1970) The social responsibility of business is to increase its profits. New York Times, Magazine. 13 Sept 1970

Garcia-Rosell J-C (2013) A multi-stakeholder perspective on sustainable marketing: promoting sustainability through action and research. Doctoral dissertation, University of Lapland Faculty of Social Sciences, Acta Electronica Universitatis Lapponiensis 11. Available at http://doria17-kk.lib.helsinki.fi/handle/10024/88826. Last accessed 8 March 2014

Goderman J, Herzig C, Moon J (2013) Integrating sustainability into business schools: evidence from United Nations principles for responsible management education (UNPRME) sharing of information in progress (SIP) reports. In: Atfield R, Kemp P (eds) Enhancing education for sustainable development in business and management, hospitality, leisure, marketing, tourism. Higher Education Academy, York

Harvey F, Chrisman N (1998) Boundary objects and the social construction of GIS technology. Environ Plann A 30(9):1683–1694

Land R, Meyer JHF, Cousin G, Davies P (2005) Threshold concepts and troublesome knowledge (3): implications for course design and evaluation. In: Rust C (ed) Improving student learning: diversity and inclusivity. Oxford Centre for Staff and Learning Development, Oxford

Meyer JHF, Land R (2003) Threshold concepts and troublesome knowledge: linkages to ways of thinking and practising within the disciplines. ETL Project Occasional Report 4, May

Meyer JHF, Land R (2005) Threshold concepts and troublesome knowledge (2): epistemological considerations and a framework for teaching and learning. High Educ 49(3):373–388

Meyer JHF, Land R (2006) Threshold concepts and troublesome knowledge: issues of liminality. In: Meyer JHF, Land R (eds) Overcoming barriers to student understanding: threshold concepts and troublesome knowledge. Routledge, London, New York

Naeem M, Neal M (2012) Sustainability in Business Education in the Asia Pacific Region: a Snapshot of the Situation. Int J Sustain High Educ 13(1):60–71

Nemetz PN (2013) Business and the sustainability challenge: an integrated perspective. Routledge, Abingdon

Perkins D (1999) The many faces of constructivism. Educ Leadersh 57(3):6–11

Porter ME, Kramer MR (2006) Strategy and society: the link between competitive advantage and corporate social responsibility. Harvard Bus Rev 78–92

QAA (2007) Subject benchmark statement: general business and management. Quality Assurance Agency for Higher Education. Available at http://www.qaa.ac.uk/Publications/InformationAndGuidance/Pages/Subject-benchmark-statement-General-business-and-management.aspx. Last accessed 8 March 2014

Rowbottom DP (2007) Demystifying threshold concepts. J Philos Educ 41(2):263–270

Star SL, Griesemer JR (1989) Institutional ecology, 'translations' and boundary objects: amateurs and professionals in Berkeley's Museum of Vertebrate Zoology, 1907-39. Soc Stud Sci 19:287–420

Stubbs W, Cocklin C (2008) Teaching sustainability to business students: shifting mindsets. Int J Sustain High Educ 9(3):206–221

Weybrecht G (2010) The sustainable MBA: the managers' guide to green business. Wiley, Chichester

Authors Biography

Clare Hagerup teaches Sustainability modules across degree programmes in both the Business School and the Health and Social Care Sciences School at University Centre, Grimsby.

Russell Woodward teaches Economic Environment modules across the degree programmes of the Business School at University Centre, Grimsby.

A Sustainable Development Management System for the University of Coimbra

Maximilian Karl Votteler and Antonio Gomes Martins

Abstract

The University of Coimbra has been implementing several projects and activities towards sustainability, including advanced educational programmes in the framework of its Energy for Sustainability Initiative. This paper presents a Sustainable Development Management System tailored for the University of Coimbra (UC), to prepare the next institutional step towards a systematic planning and management approach to sustainable campus development. The term sustainable campus is introduced and defined for the UC. A framework to foster sustainability at the campus of UC is proposed and a management system is designed. The management system adapts the existing ISO14001 standard to manage campus environmental issues, in combination with a number of indicators of the STARS assessment system to also cover educational and social aspects of a sustainable campus. 27 STARS indicators of the STARS categories Academic and Engagement are analysed. Thereby, STARS indicators are applied to a "plan, do, check, act" management cycle. Applying the STARS indicators facilitates the recording to submit a STARS report for rating purposes.

M.K. Votteler (✉) · A.G. Martins
MIT—Portugal Program, University of Coimbra, Coimbra, Portugal
e-mail: maximilian_votteler@hotmail.com

M.K. Votteler · A.G. Martins
Energy for Sustainability Initiative, University of Coimbra, Pólo II—Pinhal de Marrocos, Coimbra, Portugal

A.G. Martins
Institute for Systems Engineering and Computers (INESC), Coimbra, Portugal

W. Leal Filho et al. (eds.), *Integrating Sustainability Thinking in Science and Engineering Curricula*, World Sustainability Series, DOI 10.1007/978-3-319-09474-8_26

Keywords
Sustainable campus • Sustainable development • Management system

1 Introduction: Sustainability on Campus

Growing environmental pollution and rising energy prices are common news in contemporary society. News about rising prices for a fossil resource or negative effect of human activities on nature (forest fires, floods, smog…) is broadcasted nearly at a weekly frequency. Public awareness for such events grew since the Club of Rome published the book "Limits to Growth" in 1972. Also the 30-year update of the book has raised awareness of environmental effects of human activities. The term "overshoot" is introduced to describe human behaviour that goes beyond natural limits, to use natural resources faster than they can be replaced by nature. Sustainability however, describes the situation wherein human activities are in equilibrium within natural limits (Meadows et al. 2004). During the same year as the book "Limits to Growth" was published, the Stockholm Declaration (1972) introduced the idea of education in environmental matters (United Nations 1972). With rising awareness of human impact on nature, the first measurable effects of the human footprint on the planet and the finite amount of natural resources, some nations started to care for environmental issues and decided to develop in a more sustainable way of interacting with nature. More than a decade after "Limits to Growth" was published, "Our common future", also known as Brundtland Report of the United Nations Organization (1987), describes the development towards a more sustainable relationship with nature and defines it as: "Sustainable development is the development that meets the needs of the present without compromising the ability of future generations to meet their own needs (United Nations 1987)." In this sense sustainable development (SD) addresses the responsibility for future generations. This was acknowledged by the Talloires Declaration in 1990, a 10 point action plan to implement sustainability at universities. With this declaration universities recognize their impact on SD and their special responsibility for future generations (UNESCO 1990). Also the Agenda 21 of the United Nations Conference on Environment and Development in Rio 1992 recognizes the importance and responsibility of education institutions towards SD (United Nations 1992). In 2005 the United Nations declared the UN Decade of Education for Sustainable Development (2005–2014) to mobilise educational resources towards SD.

Since then many developments occurred. Education for Sustainable Development has taken place in many countries and has, for example, led to situations in which children explain to their parents how to recycle in the correct manner. In this context, universities have a special responsibility in educating future leaders towards SD. In our contemporary society there are only 2 % of the world population attending higher education institutions and about 80 % of the decision makers in industry and politics have a higher education graduation (Salzburg Global Seminar 2013).

1.1 Education for Sustainability

Education For Sustainability (EFS) describes the education towards SD. It addresses all age groups and disciplines. EFS teaches how the world works and the effect of human activities on the environment in all its matters (Leal Filho 2006). It is designed to make people aware of the environmental effect in the choices they make and it provides tools to overcome existing problems in society and environment. Therefore it should be present in all educational institutions, from kindergarten to university.

Some universities already teach specialised courses regarding environment and SD matters, which is an incomplete response to the environmental problems as it is not the responsibility of a small number of experts to face and fight environmental problems (Cortese 1999). Universities can teach EFS in any course they offer to show students how they can act in their specific discipline towards SD and to clarify the following misassumptions defined by Cortese (2003):

- Humans are the dominant species and separate from the rest of nature.
- Resources are free and inexhaustible.
- Earth's ecosystems can assimilate all human impacts.
- Technology will solve most of society's problems.
- All human needs and wants can be met through material means.
- Individual success is independent of the health and well-being of communities, cultures, and the life support system.

It is therefore necessary to involve students in environmental projects during their time on campus, in that they learn applied and practical skills in dealing with environmental problems (Ferreira et al. 2006). Active participation in projects fosters deep learning in contrast to superficial learning. Thus, it facilitates overcoming of arising problems from environmental pollution and rising energy prices. If all students are equipped with knowledge of the environmental effects of their actions and answers to overcome problems arising from such actions, a big step within SD will be achieved. Furthermore, subject specific education in this field can also attract a future employer as such skills are often connected with savings in energy, water, waste and other environment related costs an organization has to pay (Disterheft et al. 2012).

Citing Cortese (2003): "If higher education does not lead the sustainability effort in society, who will?"

1.2 Sustainable Campus

Universities find themselves in a position of high responsibility in relation to society as they educate most future decision makers (Salzburg Global Seminar 2013). These decision makers face immense tasks in terms of climate change, environmental pollution and scarcity of fossil resources and should therefore be equipped with knowledge and answers to overcome arising problems. However, not only the curriculum is addressed, as a university teaches and usually hosts as many people as there are living in a small city, the energy consumption and pollutant emission being, likewise, quite high. Therefore, the university must apply energy efficiency measures to decrease its environmental footprint and lead SD transformation providing a positive example to society.

Since the mid 90s of the last century some universities joined together in networks and organizations to form agreements regarding energy efficiency (den Heijer et al. 2010), to act according to their special responsibility towards sustainability and promote the approach of a sustainable campus. Hereby campuses are motivated to encourage SD in order to represent a positive example for society and students.

The interpretation of sustainability at campus level varies among universities and has to be defined for this article. Some universities believe that sustainability is reached by signing an international declaration, others implement environmental guidelines and some get certifications for campus buildings from green building initiatives (Velazquez et al. 2006). The author of this article defines a sustainable campus as follows:

> A sustainable campus acknowledges its impact on environment and society, and engages itself in the promotion of sustainable development; to underline these ambitions, they can be anchored in the university policy. The sustainable campus recognizes all fields of interaction with society and environment and tries to improve in all fields. Therefore it has a Sustainable Development Working Group (SDWG) which develops a Sustainable Development Management System (SDMS) that contains actions to be carried out in order to meet the goals of sustainable development on a campus environment. Furthermore, this working group is responsible to measure and report ongoing processes on the campus regarding sustainable development.

To meet its responsibility towards nature and society, an organization has to reach a balance between its economic, environmental and social development (Alshuwaikhat and Abubakar 2008). These three pillars of sustainability are defined by the Triple Bottom Line (TBL) framework which can be seen as a tool to define and to report sustainability (Slaper and Hall 2011). In respect of the three P's of the TBL: people, planet and profits, the university has to turn into a living laboratory towards SD where students can be involved in programmes, activities, research and development regarding sustainable development.

To assess SD on a campus, a number of managing, rating and reporting tools were developed which differ from each other according to the developers' perception of sustainable campus. While some deal mainly with the physical structure of a campus (buildings, transportation and waste etc.) others focus only on the educational situation of a campus and evaluate only the level of Education For Sustainability (EFS). Only a few assessment tools provide a comprehensive evaluation which weights all fields of a campus activity. A campus must be evaluated in all its fields of action and not only according to its eco-efficiency (Shriberg 2002).

In order to assure a continuous improvement in all fields of importance of a sustainable campus, a comprehensive management system is needed, not just a simple assessment approach. Such systems not only evaluate indicators but set short and long-term goals and require continuous improvement.

2 University of Coimbra

The University of Coimbra is the oldest university in Portugal and one of the oldest in Europe. Its history goes back to the 13th century when the university was founded in 1290 and located in buildings in Coimbra-Sofia to be later moved uptown (Alta) in the Royal Palace of Alcáçova in 1537. On June 22 in 2013, UNESCO added the University of Coimbra Alta and Sofia to the list of world heritage.

Due to its long history, the university includes many old buildings and valuable treasures such as the Royal Palace of Alcáçova and the university library. UC is aware of its important role in both cultural and historical domains and puts effort in the protection of its heritage. This implies that the university also hosts several museums and engages in cultural activities as well as hosting a cinema and a theatre (University of Coimbra 2013).

Today UC has three campuses and hosts 12 faculties and similar units, 45 research units, two stadiums, 25 libraries, one botanic garden and two museums. All together 2,988 people (academic and non-academic) are directly employed and 24,403 students are registered at the university. More than 27,000 people are involved in the UC's activities, which makes up about 19 % of the population of the city of Coimbra (143,396), the university therefore has a major impact on the city (University of Coimbra 2013; Censos 2011).

2.1 Academic Association of Coimbra

The AAC is the oldest students' organization in Portugal. Established in 1887, it was founded by older student organizations. Nowadays it is the umbrella organization for all students' organizations at UC. It consists of numerous sport and culture clubs and has its own radio and TV station. The organization is led by a general direction which is made up entirely of students. Political activities against the government led to a close down in 1971, but AAC reopened in 1974.

2.2 Energy for Sustainability Initiative

The EfS initiative was launched in 2006 and engages in SD issues in research and education. It assembles researchers from 14 UC research units and works in close cooperation with industry and independent research and development units. The initiative works together with the Sustainable Energy Systems (SES) PhD programme of the MIT Portugal programme and offers a master programme and an advanced studies diploma in Energy for Sustainability (Energy for Sustainability Initiative 2014). For knowledge exchange the EfS initiative holds conferences and connects with partners worldwide.

3 Management Structure

Like the Triple Bottom Line framework, which focuses on commercial organizations such as companies, a complementary structure is created which evaluates the activities of a public university. Here, three core fields of activities regarding Sustainable Development on a university campus could be identified. These fields are built upon the major areas of campus influence on SD which are described as organization, education and social engagement (Fig. 1). These three fields can be seen as the three pillars of a sustainable campus. The block "Organization" manages the physical and financial matters of a campus including planning, buildings and all kinds of energy and waste flows. As UC is a public, non-profit oriented university, less attention is given to its economy, and this part is embedded within the "Organization" block. It seems to be logic that operations can only be carried out if finances are available. Block "Education" describes the educational initiatives regarding SD such as awareness programmes, specialization programmes (master, PhD), participation programmes and comprehensive EFS programmes. In the "Social" block the interaction within campus members and with the local community is addressed as well as internal justice and equity. It exists mainly from awareness programmes and participation programmes for the campus members and local community.

Within this three pillar system, an organizational structure can be worked out to describe which group participate and manage each field of the sustainable campus (Fig. 2). All the management regarding to SD is carried out by a working group that reports to the administration. The working group is responsible for awareness

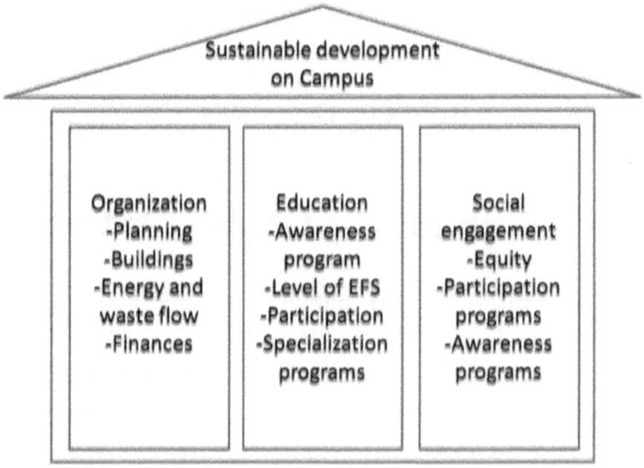

Fig. 1 Three fields of action at a sustainable campus

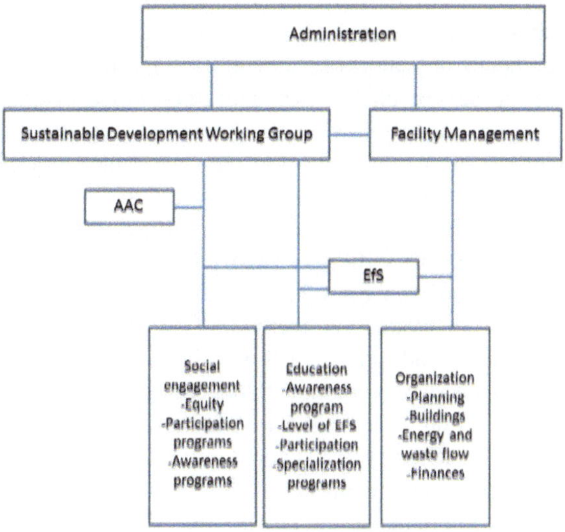

Fig. 2 Management structure for a sustainable campus Coimbra

programmes and the curriculum and is thereby supported by the EfS initiative, which consults. The AAC might support activities regarding social engagement and might promote the ambitions via university radio and TV to engage and reach more people. The facilities management department is responsible for the environmental management and is thereby supported by the EfS initiative which intervenes as a consultant and connects the facilities management department with interested students to engage them in environmentally oriented tasks on campus.

3.1 Management System

The proposed management system for the University of Coimbra is based on the ISO 14001 environmental management standard. This standard is chosen because the ISO 9001 quality management standard is already applied at the UC. Thus, experience with the ISO standard is earned, and the fact that both of the systems sometimes evaluate in the same fields facilitates the application of ISO 14001. To complement the ISO 14001 standard in order to manage SD on campus, performance indicators of the STARS assessment system for campus sustainability are applied to the environmental indicators of ISO 14001 (Appendix 1). Hereby these indicators are treated like any other indicator required by the ISO 14001 (energy use, raw materials, chemicals, air pollution etc.).

The implementation of the ISO 14001 requires six clauses which are further divided into 18 clauses and sub clauses (Whitelaw 2004). These clauses describe the routine to carry out to implement and remain with the ISO 14001. In order to function as a management system for SD, these 18 clauses are applied to all criteria of a sustainable campus. The commitment statement required in clause 4.2

"Environmental policy" can, for example, be within the sustainable campus commitment. STARS indicators are involved and managed the same way as the environmental indicators of clause 4.3.1 "Environmental aspects". Objectives and targets can be set for all STARS indicators which evaluate in percentage and absolute numbers (Academic Courses, Immersive Experience, Outreach Materials and Publications etc.), other indicators may be planned to implement as a target (Community Stakeholder Engagement, Participation in Public Policy). Some STARS indicators are even directly complemented by some of the ISO 14001 clauses.

The STARS indicators Staff Professional Development and Employee Orientation address clause 4.4.2 "Competence, training and awareness" directly and even suggest an evaluation method. Also clause 4.4.3 "Communication" is directly recognized and evaluated in the STARS indicators: Outreach Materials and Publications and Inter-Campus Collaboration, of which the former opens the way for target setting. Clause 4.5.1 "Monitoring and Measurement" addresses the Sustainability Literacy Assessment indicator of STARS which evaluates the success of the implemented courses.

The implementation process of STARS indicators within the ISO 14001 is facilitated through the fact that STARS indicators already suggest a measurement method. These methods can easily be applied to the objective and target setting tasks, required by the ISO 14001 as continual improvement. If these indicators are applied to evaluate a campus, the evaluation in the STARS system is hence facilitated as no extra exercise has to be carried out to collect data.

4 Conclusion

The ISO 14001 is a well-known and recognized framework which is widely used to manage the environmental issues of organizations. However, to manage sustainability it is an incomplete approach and was therefore complemented with additional indicators. Hereby indicators of the STARS assessment system for sustainable campuses were applied to satisfy the demand to manage a sustainable campus. It is believed that complementing environmental management systems with additional indicators in order to manage sustainable development has the potential for global reach, not only at universities but also in other types of organizations. Furthermore, applying STARS indicators facilitates STARS reporting. Thus, allows a university to participate in the STARS reporting and ranking system. Applying such systems underlines a university's engagement towards sustainability and fosters recognition as a leading member in the transition towards sustainability.

Acknowledgments This work has been framed under the Initiative Energy for Sustainability of the University of Coimbra and supported by the Energy and Mobility for Sustainable Regions Project (CENTRO-07-0224-FEDER-002004), co-funded by the European Regional Development Fund (ERDF) through the «Programa Operacional Regional do Centro 2007–2013 (PORC)», in the framework of the «Sistema de Apoio a Entidades do Sistema Científico e Tecnológico Nacional», and by the «Fundação para a Ciência e Tecnologia» and also partially supported by FCT under project grant PEst-C/EEI/UI0308/2011.

5 Appendix 1: Description of the 27 Additional Indicators from the STARS Assessment System and Their Complemented ISO 14001 Clauses (AASHE 2013)

Indicator	Description	Evaluation	ISO 14001 clause
ACADEMICS (Curriculum)			
Academic courses	Evaluates sustainability courses and courses which involve sustainability	Percentage of all courses	
Learning outcome	Evaluates the number of students that had at least one lesson regarding sustainability	Percentage of all students who graduate	
Undergrad programs	Evaluates the number of undergrad programmes (major, degree programme) regarding sustainability	Presence of the programme	
Graduate program	Evaluates the number of graduate programmes (major, degree programme) regarding sustainability	Presence of the programme	
Immersive experience	Practical experience with sustainability matters, on and off campus	Frequency and duration	
Sustainability literacy assessment	Evaluation of the success of education in sustainability through surveys or tests	Percentage of students assessed	4.5.1
Incentives for developing courses	Evaluates efforts towards sustainability in curriculum (professional training, funding)	Presence of action	
Campus as a living laboratory	On-campus engagement in form of thesis projects and research in: air and climate, buildings, dining services, energy, grounds, purchasing, transportation, waste, water, coordination (planning and governance), diversity and affordability, health (wellbeing and work), investment, public engagement, others…	Presence of incentives	

(continued)

(continued)

Indicator	Description	Evaluation	ISO 14001 clause
ACADEMICS (research)			
Academic research	Evaluates the number faculty staff engaged in research	Percentage of faculty staff engaged	
	Evaluates number of departments engaged in research	Percentage of departments engaged	
Support for research	Evaluates to what content students, faculty and research units are committed to do research in sustainability	Presence of effort	
Access to research	Evaluates the public access to university research	Percentage of public research results	
ENGAGEMENT (On campus)			
Student educators program	Evaluates the impact of eco-representatives or eco- ambassadors on the students	Percentage of students reached	
Student orientation	Evaluates the number of new students (freshmen) involved in orientation programmes dealing with sustainability	Percentage of new students	
Student life	Evaluates co-curricular programmes and initiatives to engage students in sustainability (conferences, student groups, outdoor programmes)	Total number of programmes	
Outreach materials and publications	Information media (website, newsletter, facebook) to inform about sustainability efforts	Total number of programmes	4.4.3
Outreach campaign	Evaluates programmes to encourage students and staff to save resources (energy, water)	Total number of outreach campaigns	
Employee educators program	Evaluate education programmes in sustainability for campus staff	Percentage of staff reache	
Employee orientation	Evaluates orientation programmes for new employees	Percentage of new employees reached	4.4.2
Staff professional development	Recognizes if staff has the opportunity to engage in professional courses regarding sustainability	Presence of initiatives	4.4.2

(continued)

(continued)

Indicator	Description	Evaluation	ISO 14001 clause
ENGAGEMENT (public)			
Community partnership	Evaluates partnership of the university with local community, including school districts, government agencies, non-profit organizations, business and/or other entities	Form of partnership (Supportive, Collaborative, Transformative)	
Inter-campus collaboration	Evaluates networking with other educational institutions to exchange knowledge regarding sustainability	Presence of network	4.4.3
Continuing education	Evaluates if courses (certificate) regarding sustainability or involving sustainability are offered	Percentage of continuing education courses (certificate)	
Community service	Evaluates the number of students and time involved in voluntary community service programmes	Percentage of students involved, and hours spent per year	
Community stakeholder engagement	Evaluates if a framework exists that identifies and involves community stakeholders in the campus	Presence of action	
Participation in public policy	Evaluates if the university promotes sustainability through public policy advocacy	Presence of action	
Trademark licensing	Evaluates the participation in social programmes (Fair trade, fair labour association, worker rights consortium)	Presence of action	
Hospital network	University hospital is engaged in global networks (Global Green and Healthy Hospital Network, Healthier Hospital Initiative, Practice Green health)	Presence of action	

References

AASHE (2013) *STARS* technical manual. Version: 2.0. Available: http://sustain.indiana.edu/overview/metrics-and-reporting/STARS_docs/stars_2.0.1_technical_manual.pdf. Accessed 22 Apr 2014

Alshuwaikhat HM, Abubakar I (2008) An integrated approach to achieving campus sustainability: assessment of the current campus environmental management practices. J Clean Prod 16 (16):1777–1785

Censos (2011) [Online] Available: http://www.ine.pt/scripts/flex_definitivos/Main.html. Accessed 19 Dec 2013

Cortese AD (1999) Education for sustainability: the need for a new human perspective, Second Nature, Inc. Available: http://www.mma.gov.br/port/sdi/ea/deds/arqs/antcort.pdf. Accessed 16 Apr 2014
Cortese AD (2003) The critical role of higher education in creating a sustainable future, Planning for Higher Education, pp 15–22. Available: http://www.aashe.org/resources/pdf/Cortese_PHE.pdf. Accessed 16 Apr 2014
Den Heijer A, Teeuw P, Aalbers K (2010) Visions for the future of higher education, In: ERSCP-EMSU 2010 Conference, pp 1–24. Available: file:///C:/Users/user/Downloads/034_DenHeijer.pdf. Accessed 16 Apr 2014
Disterheft A, Caeiro S, Ramos MR, Azeiteiro UM (2012) Environmental management systems (EMS) implementation processes and practices in European higher education institutions—top-down versus participatory approaches. J Clean Prod 31:80–90
Energy for Sustainability Initiative (2014) [Online] Available: http://www.uc.pt/en/efs. Accessed 10 Jan 2014
Ferreira AJD, Lopes MAR, Morais JPF (2006) Environmental management and audit schemes implementation as an educational tool for sustainability. J Clean Prod 14(9–11):973–982
Leal Filho W (2006) Towards the promotion of education for sustainability introduction: what is education for sustainability the education for sustainability movement, Available: http://www.revistaeducacion.mec.es/re2009/re2009_12eng.pdf. Accessed 16 Apr 2014
Meadows D, Renders J, Meadows D (2004) Limits to growth: the 30-year update. Chelsea Green Publishing Company
Salzburg Global Seminar (2013) SFLC, Academy, "Vision". [Online] Available: http://salzburgglobal.org/wp-sfa/?page_id=21. Accessed 06 Jan 2014
Shriberg M (2002) Institutional assessment tools for sustainability in higher education: strenghts, weaknesses, and implications for practice and theory. Int J Sustain High Educ 3(3):254–270
Slaper Timothy F, Hall Tanja J (2011) The triple bottom line: what is it and how does it work? Indiana Bus Rev 86(1):4–8
UNESCO (1990) The Talloires declaration, UNESCO, p 10, Available: http://www.ulsf.org/pdf/TD.pdf. Accessed 22 Ap 2014
United Nations (1972) Declaration of the united nations conference on the human environment, June 1–4. Available: http://www.unep.org/Documents.Multilingual/Default.asp?documentid=97&articleid=1503. Accessed 16 Apr 2014
United Nations (1987) Our common future-Brundtland report. Available: http://conspect.nl/pdf/Our_Common_Future-Brundtland_Report_1987.pdf. Accessed 16 Apr 2014
United Nations (1992) United nations conference on environment and development Rio de Janeiro, Brazil, 3–14 June 1992, Available: http://sustainabledevelopment.un.org/content/documents/Agenda21.pdf. Accessed 16 Apr 2014
University of Coimbra (Management) [Online] Available: http://www.uc.pt/governo/orgaos. Accessed 19 Dec 2013
University of Coimbra (2013) [Online] Available: http://www.uc.pt/dados. Accessed 19 Dec 2013
Velazquez L, Munguia N, Platt A, Taddei J (2006) Sustainable university: what can be the matter? J Clean Prod 14(9–11):810–819
Whitelaw K (2004) ISO 14001 environmental systems handbook. Second Edition, Elsevier Ltd

Authors Biography

Maximilian Karl Votteler has graduated as a B.Eng in Applied Chemistry at the Georg-Simon-Ohm University of Applied Sciences in Nürnberg, Germany. He wrote his bachelor thesis at RENA GmbH at the Research and Development centre in Freiburg, Germany. Following he joined the master course in Energy for Sustainability at the University of Coimbra where he graduated as a M.Sc. He is currently enrolled in the MIT Portugal PhD program on Sustainable Energy Systems at the University of Coimbra.

António Manuel Oliveira Gomes Martins has been a Professor at the University of Coimbra for more than 20 years. He obtained his PhD in Electrical Engineering in 1985 at the UC and joined the research institution INESC in Coimbra in 1991, of which he was president from 1999 to 2002. In his career at the UC he occupied also other positions such as Director of the Institute for Interdisciplinary Research of the University of Coimbra (2008–2011) and Vice-Rector at the University of Coimbra (2003–2011). He was co-founder (2007) and presently Coordinator of the Energy for Sustainability Initiative (EfS) of the University of Coimbra (since 2012). His research interests are Energy efficiency, Energy demand-side management, Energy policies for market transformation, Energy planning for sustainability, Smart networks and devices for intelligent use of energy resources and teaches several courses on energy management and market transformation at MSc and PhD levels.

Authors Biography

Encouraging Engineering Students to Question Technological Solutions for Complex Ecological and Social Problems

Sabine Pongratz and André Baier

Abstract
Sustainability in the curriculum of engineering students is either ignored or solely focuses on technological solutions. As the UN Rio+20 debate and its focus on Green Economy shows, there is a strong demand for technological innovations as a remedy for ecological destruction and as pathway to poverty eradication. Consequently, the predominant belief in technological progress is held up through technicians as well as a society wishing for easy technological solutions for complex ecological and social problems. Taking this into account, this paper outlines a course design that promotes socially and ecologically responsible engineering through a variety of alternative teaching methods. Engineering students acquire the competence to unveil the complex interdependency of their social, political, ecological and economic surroundings. This includes the consideration of different values, interests and needs within a global perspective as well as within one class(room). The course design encourages democratic decision-making not only to solve but also to define problems within the course itself and moreover outside of the classroom. This method is applied in order to adequately respond to the specific needs of users and to cooperatively develop technologies which are socially useful, locally adapted, durable, repairable and recyclable.

S. Pongratz · A. Baier (✉)
Faculty of Mechanical Engineering and Transport Systems, Technische Universität Berlin, Straße des 17. Juni 144, 10623 Berlin, Germany
e-mail: Andre.Baier@tu-berlin.de

S. Pongratz
e-mail: Sabine.Pongratz@tu-berlin.de

W. Leal Filho et al. (eds.), *Integrating Sustainability Thinking in Science and Engineering Curricula*, World Sustainability Series, DOI 10.1007/978-3-319-09474-8_27

Keywords
Engineering education · Sustainable development · Green economy · Democracy · Society nature relations

1 Introduction: Sustainability in Higher Engineering Education

Recognizing that engineers have played a key role in contributing to non-sustainable trends in our society and that some engineering innovations caused large scale disasters, there is now a call for a new kind of engineer who has the skills to address social, economic and environmental issues regarding technology (Conlon 2008; Zandvoort 2008). In the context of the UNESCO Decade on Education for Sustainable Development (2005–2014) many international conferences have drawn attention to the importance of including sustainable development issues in higher education. In addition, quite a few declarations on engineering education in sustainable development, such as the Barcelona Declaration, have been signed (Segalàs et al. 2008).

Meanwhile several technical universities accepted the challenge of embedding sustainability issues into their teaching and engineering programmes (Michel 2008; Boni and Pérez-Foguet 2008). While authors and organisations already defined relevant sets of learning outcomes and composed lists of competencies required to address sustainability issues, the results in implementing those in engineering curricula fall short of its objectives and pose still unresolved problems (Guerra 2012; Kastenhofer et al. 2010; Segalàs et al. 2008). Alongside, the belief in technological solutions as a cure for all problems still prevails (Brey 2009).

Taking this into account this paper explains our perception of the causes and approaches to social and ecological problems in a complex world as well as our perception of technology and engineering education. It differs from the role technology is assigned to in the context of the Rio+20 debate and its guiding concept Green Economy (UNEP 2011). We hold the view that economic growth is to be seen as a social relation and linked to the domination of the economy over people and nature. In order to address the existing social and ecological problems, the underlying causes and interdependent relations between economic, ecological and social factors have to be carefully analysed. After a sketch of our theoretical foundation consisting of the concept of society nature relations and an extensive understanding of democracy, it will subsequently be explained how they find their way into our course design at the Technische Universität Berlin (TUB).

2 Sustainable Development, Green Growth and Technical Solutions

At the United Nations Conference on Environment and Development in 1992, known as the Rio Summit, the notion of sustainable development was publically launched. It was supposed to stop climate change, resource depletion, biodiversity loss and lift the Global South out of poverty. Sustainable development was promoted as a concept to reconcile economy with ecology (Dietz 2012, 22). Sustainable development has largely failed as a magic formula, despite single successes (Brand 2012, 2). Overall, ecological, social and economic problems have not been solved; e.g. CO_2-emissions, environmental and forest degradation as well as social inequality are still increasing and reach a new worryingly level (Brunnengräber and Haas 2012, 15).

Twenty years later a new concept—Green Economy—was put on the agenda at the Rio+20 Summit as a remedy for the social-ecological crisis. The United Nations Environment Programme (UNEP) defines Green Economy "[...] as one which is low carbon, resource efficient and socially inclusive" (UNEP 2011, 16) and declares that "[...] the most prevalent myth is that there is an inescapable trade-off between environmental sustainability and economic progress" (Ibid., 16). The basic idea behind that new economic paradigm is that the greening of economies is seen as "[...] a new engine of growth, a net generator of decent jobs and a vital strategy to eliminate persistent poverty" (Ibid., 16).

According to the proponents of a Green Economy, the introduction of new energy and resource efficient technologies will decouple economic growth from resource use and therefore reduce ecological risks. This is seen as a way to enable economic growth, material prosperity and above social inclusion via the creation of green jobs (Dietz 2012, 23). In contrast to the debates on sustainable development, the discourse about The Limits to Growth (Meadows et al. 1972) has now simply been amended to green growth (Brunnengräber and Haas 2012, 16). This analysis falls too short for several reasons.

Limitless growth and rebound effect. The expectation of limitless growth due to the decoupling of economic growth from resource consumption can no longer simply be assumed. Although, more efficient production is already happening, there is at the same a rebound effect, i.e., gains in efficiency have been eaten up by increased consumption (Santarius 2012). As long as the pressure to accumulate capital is not called into question, economic growth tends to be stronger than resource and energy efficiency, resulting in an increasing resource consumption and environmental pollution (Brand 2012, 19). In addition, as green products and green technologies are resources-intensive, the possibility of absolute decoupling is highly questionable (Wissen 2012, 17).

Green jobs are an insufficient instrument for poverty eradication. Environmental technology in Germany, for example, is indeed creating many new jobs (Umweltbundesamt 2009). Despite the prevailing opinion, working conditions in this sector are often precarious, wages are frequently below average and the level of union organization is low (Schmiade and Becker 2008). Consequently, green jobs do not automatically enhance social equality and create good working conditions. Moreover, green products do not necessarily imply a green production. The required resources are often extracted in the Global South under ecologically and socially catastrophic conditions. The local population is confronted with massive environmental destruction and forced resettlements, while the workers are exposed to highly unsafe, health-threatening conditions on mining and extraction sites (Wissen 2012, 17). They often do gain little from the exploitation of the resources but have to bear many of the negative consequences. Accordingly, the generation of green jobs must go along with a strengthening of social rights and the promotion of global social justice.

Existing power structures. Global relations of domination in connection with the growth-oriented modes of production and consumption are not called into question (Brand 2014, 1). The praise for competition and the call for new "green" investments do not reflect that merely those who have access to power and capital can invest in green technology. Less powerful and economically competitive societies are not equally able to develop innovations. They are forced into the position of resource suppliers for the green tech industry and remain in a semi-colonial condition (Brand 2012, 30).

Technological solutions are not a panacea. Green Economy under conditions of world-market competition will push for centralized, capital-intensive technologies and large-scale projects e.g. offshore wind parks (Brand 2012, 3). These are rather controlled by corporations and free market forces than subject to a democratically organized negotiation process. Generating and maximizing profits play a more important role than social and environmental sustainability (Brand 2012, 4). In the context of the financial crisis, investments in green technologies seem as a strategy to create new markets as assets for released capital (Köhler and Wissen 2010, 217). Huge expectations rest on technology as a tool for environmental protection and a main source of economic development (UNEP 2011, 628). But astonishingly little attention is paid to the potential effects on nature and society; although history teaches us that risks and ambivalences of technologies cannot be avoided. Unintended side-effects can occur even more strongly the more far-reaching technological interventions into nature and society are (Grunwald 2011, 6). The development of certain environmental technologies is necessary, however we insistently point out that the huge expectations regarding technology within the Green Economy framework tend to rule out less profitable, locally adapted technologies as well as non-technological approaches such as sufficiency (Brand 2012, 30).

In short, the distribution of prosperity, the imperial, racist and gender-related forms of domination, the ownership structures and the domination of nature are all-together neglected. The prospect of a Green Economy which reconciles economy, ecology and society will fail, if it serves principally as a growth programme. For a substantial change towards an inclusive and just mode of production and living, democratic and sustainable societal nature relations are required. These include a fundamentally different organisation of key areas such as food, water, energy and housing. Additionally, a different understanding of technological and social innovation is needed (Brand 2014, 15).

3 Society Nature Relations

The concept of society nature relations is rooted in the tradition of Critical Theory (Köhler and Wissen 2010, 218). It opens a broader perspective to analyse social-ecological problems and provides a reference point for a critique of how society frames and deals with those problems (Brand 2014, 12; Becker and Jahn 2005, 4). The relationship between nature and society is neither theorised as a "naturalizing of society" nor as a "socializing of nature" (Ibid, 12). Nature and society are rather conceptualised as simultaneously different and mutually constituted (Jahn and Wehling 1998, 82). Their relationship is conceived as a "structure of mediation" and points to the reciprocal linkages between nature and society (Becker and Jahn 2005, 8). That way social-ecological problems can be considered as such and do not have to be treated as separate either social or ecological issues (Ibid, 12). Instead of perceiving nature in crisis, the dominant forms of the societal appropriation of nature come into perspective, i.e., the economic, political, cultural and technical forms of appropriating nature (Jahn and Wehling 1998, 81). In respect of this understanding it is the societal relations with nature that are crisis ridden (Becker and Jahn 2005, 4).

Max Horkheimer's and Theodor Adorno's quote of the "Dialectic of Enlightenment" illustrates that relation quite plainly: "Any attempt to break the compulsion of nature by breaking nature only succumbs more deeply to that compulsion" (Horkheimer and Adorno 2002, 9). The more people try to escape from their dependency on nature, the more they will get into that same dependency.

An example on how these dependencies are created, are the current political strategies of a Green Economy. To reduce CO_2-emissions, fossil fuels are increasingly substituted with agrofuels. The small gain of independence from fossil fuels leads to new powerful dependencies (Brand 2014, 12). The cultivation of palm oil plantations in Indonesia or Brasil deprives small-scale farmers of their land and livelihoods forcing them to sell their labour to the owner of these huge plantations. Moreover local food production has to compete with energy production for the world market.

This example illustrates, how social-ecological problems have to be discussed as social and political phenomena that cannot be isolated from issues of domination and power (Görg 2003, 129). Underlying economic and political conditions, i.e., matters of social inequality, power and domination, have to be taken into account as well as case-specific local circumstances (Köhler and Wissen 2010, 221). Therefore mere technological fixes and the belief in the possibility of an ever-more sophisticated domination of nature are rejected (Brand and Wissen 2013, 693). Technological progress can no longer be held up as precondition for social progress. In the context of the Green Economy, the crucial questions to be asked are: under what conditions is "green" growth taking place, who controls it, who has access to power and means of production, who profits and what are the socially, ecologically and locally different impacts (Brand 2012, 12)? Due to the inherent necessity to accumulate capital and to maximize profits, economic growth, whether green or not, will result in "the domination of man over man and society over nature" (Brand 2014, 13).

4 Democracy

The thorough analysis of society nature relations helps to perceive the ecological crisis not as a crisis of nature itself. Instead, conflicts among people and the domination over nature are identified as the primary cause for the interconnected social-ecological problems. Any sincere attempt to solve these problems may not only take the broader ecological surroundings into view. Instead, the societal relationships need to change at their core. Only a full shift towards a democratic organization warrants the chance to vanquish suppressive and exploitative societal relationships including the domination of men over nature (Demirović 2012, 70).

Democracy may not be reduced to its well-established form of parliamentary democracy in the Global North. It is claimed that democratic decisions are reached through the will of all people and legitimately bind the actions of every single one (Demirović 2007, 9). The process of democratic decision-making however, is mainly restricted to the governmental sector, i.e., the periodic election of representatives and majority votes. The execution of the decisions relies mostly in the hands of authoritarian bureaucracies which are subject to only a minimum of democratic control. This is done to ensure a continuity in state affairs (Demirović 2007, 13). Only organisations, notably international companies, are powerful enough to influence this process (Deppe 2009, 46). A direct participation of affected individuals is not seen as a viable option. In addition, there is a dilemma concerning nature in current politics. Either the rudimentary democratic process is emphasized, which leaves few chances for an ecological intervention, or governments take decisive actions, which will restrict parliamentary democracy even further (Demirović 1997, 195). Over and above, there is already a call of environmentalists for a total suspension of democratic procedures for the purpose of "saving" nature (Lovelock 2010).

Societal relationships need to be based on a broader implementation of democracy instead. People must have the same prospects to participate in the actual decision-making process comprising literally everyone who will be affected by the decisions taken. For this to happen, the decision-making process needs to be as inclusive and as transparent as possible. The execution of decisions must be controllable through the public, even up to the point of reverting any decision through a new decision-making process (Demirović 2007, 11). It may even be necessary to suspend decisions involving great uncertainty, where the impact on nature and society cannot be reversed. Bearing reversibility in mind may help to reduce the path dependencies for future generations which cannot participate in the decision-making process of today. In addition, equality of all people is crucial to any democratic process, thus an expansion beyond the equality before law is needed. Among other requirements, equal educational opportunities and just access to natural resources are central to ensure equal political participation. Based on that a more equal distribution of wealth is fundamental for a democratic participation (Demirović 2012, 66).

If democracy is good enough to run states, it cannot be illegitimate to manage companies (Krätke 2003, 58), as the various aspects of private property are granted by democratic decisions. Respectively there is no reason why democracy should be restricted to the political system and its limited influence on the economic system. An extensive economic democracy is possible, e.g. co-management through workers and cooperatives. Democratic participation provides the prospect of a design of technology which equally takes political, economic, social, ecological and technical aspects into consideration (Ropohl 2000, 40). Engineers and users may work together according to the demand of people on technologies, which need to be locally adapted, durable, repairable and recyclable.

Democratic participation is neither a panacea to avoid nor to resolve social-ecological problems. However, it seems that only a participatory, bottom-up democracy provides the chance to fundamentally change the current society nature relations which rely on a prevalent use of unsustainable technology and resources.

5 Design and Methodology of the Blue Engineering Course at the Technische Universität Berlin

The concept of society nature relations enables engineering students to acquire the competence to unveil the complex interdependencies of their social, political, ecological and economic surroundings. This includes the consideration of the various values, interests and needs of different actors within a global perspective as well as within one class(room). An extended use of democracy will help to resolve differences between individual actors and encourages them to participate as an individual or as a group in any decision-making and its implementation. The student-driven initiative Blue Engineering—Engineers with Social and Ecological Responsibility—successfully puts new topics on the agenda of one of the largest universities in Germany. It is a good example for the efficacy of self-organised groups as well as for the further development of any organizational structure through democratic participation (Baier 2013).

The underlying goal of Blue Engineering is to promote responsible engineering in universities and companies alike. Starting at their university in 2009, a group of engineering students outlined a course design for 25 students. They applied numerous alternative teaching methods to place the responsibility for a successful course on the participants. Thus, the role of instructors shifts from experts to moderators (Baier and Pongratz 2013). Accordingly, tutors were the only instructors for the successful test-run in 2011. It was immediately accepted as compulsory optional course in the curriculum of the TUB. In 2012 two lectures were assigned to the course and joined the Blue Engineering group as equal members. As of the summer semester 2014 three parallel courses with 30 participants each are offered.

Building blocks, i.e., self-contained study-elements, are at the core of the Blue Engineering course. They provide clear didactical instructions to facilitate a 90 min course as well as compact, yet multiple perspectives on a complex topic, e.g., ethical codes, recycling, pre-implementation diagnostics, social businesses and cooperatives (Blue Engineering 2011). Some of these study elements help to thoroughly analyse single technologies, e.g., energy saving light bulbs, in respect to the ecological, social, economical, political and gender related impacts of these technologies. In other study elements, engineering students learn to shift away from the general paradigm of engineers as problem solvers. They are encouraged to become problem definers in all areas of engineering, including their own proper working conditions. Along with the wide variety of topics, every single building block uses a specific set of wide-spread teaching formats such as case studies, storytelling and station learning. Most building-blocks, however, rely on a specific adaption and new combination of known methods, e.g. learning cascades, advocatus diaboli, triangular method, evaluation sculpture, crime scene investigations and court trials, educational games and challenges. On top, several building blocks make use of newly created methods or are build according to specific forms of pedagogy.

Since the building-blocks are well documented, groups of participants are regularly required to conduct them to their fellow students. In addition, any layperson may prepare and conduct a building block for any given group, as they are highly flexible and transferable. An initiative has been taken to make all thirty so far existing building blocks freely available, an online database has been set up (Blue Engineering 2014). Additionally, some building blocks are published in a freely available handbook which also provides further information about the genesis of Blue Engineering (Blue Engineering 2011).

The second core element of the Blue Engineering course is the preparation of a semester project in groups of about five students. According to their interests, they may choose any topic in the field of social and ecological engineering which they will reflect along various dimensions. Hence, the outcome of their reflection process is transferred into a format that is reusable in future courses or in other teach/learn settings. Consequently, the students have to combine their specific topic with various didactical methods and formats, e.g., newly constructed or refurbished building-blocks, exhibitions, card-/boardgames, encyclopedic treasure chests, picture books and e-learning-units. The prospect that their work will be used in future courses significantly motivates the students to prepare meaningful and useful semester projects. A continuous mentoring by the instructors, several rounds of peer-to-peer feedback and a public presentation and testing at the end of a semester ensure the overall quality.

After the course, students regularly engage in the further development of the course and the Blue Engineering initiative in general. Since its beginning, the work in this group has been democratically organized to a large extent, all procedures are carried out transparently and decisions are consensually made. The design of the various group processes is primarily based on a specific pedagogy of democracy, called betzavta (Hebrew for together). It turns conflicts of values and interests into personal dilemmas which helps to lead fruitful, result-oriented discussions where everyone can equally participate (Maroshek-Klarman and Vaddai 1993). Several exercises developed by Maroshek-Klarman and Ulrich (1997) have been specifically adapted to the context of engineering and are the foundation of some building-blocks which further strengthens the democratic claim of the Blue Engineering course.

Two similar student-driven initiatives fostered sustainability in engineering courses. In 1995 the University of Uppsala and the University of Agricultural Sciences in Uppsala founded the Centre for Environment and Development Studies (CEMUS) in 1995 due to a student initiative (Hald 2011). Up until today this student-run university centre offers each semester various courses to many different topics. In 2002 a research school with a clear interdisciplinary focus was added to CEMUS to provide a regular forum for PhD students. The second student-driven initiative was implemented within the democratic education programme at the University of California (DeCal). This programme gives students the opportunity to offer courses to their fellow students on subjects which are usually not found in a regular curriculum. In the spring term of 2013 students offered the first engineering ethics course in this programme (Sunderland 2013). The task in this course was to develop materials which would inspire others to reflect on the ethical implications of their actions as engineers and which might be used in educational contexts. The Blue Engineering initiative has drawn much inspiration from these two student driven approaches and may develop further in their respective directions, e.g. providing students with the chance of offering their own courses and taking PhD students in a stronger consideration.

6 Conclusion

The Blue Engineering course is an initiative to foster social and ecological engineering in the curricula at TUB. It is based on the observation that engineering is in urgent need of a critical transformation towards a strong democratic participation in the design of technology. In this process, the consideration of impacts on nature and society, working conditions and a commitment to serve the needs of all people as well as a focus on social justice is required.

The course concept is not designed to present students a proper definition of sustainability or to teach respective technical instruction. It rather provides a space where students from different disciplines with diverse perspectives come into contact to exchange and develop alternative ideas and get the chance to question common sense. With recourse to the concept of society nature relations they are encouraged to acknowledge that on the one hand everything engineers do, mutually affects nature and societies on a global level but with different local impacts. On the other hand they realise that the work of engineers itself is embedded in underlying structural conditions. The strong focus on alternative didactical methods, peer-learning and a democratic organisation of all processes helps to explore ways to collectively deal with those dilemmas. Engineering students learn to express their own points of view as well as to take interests and needs of other actors into account to find a common ground for democratic decision-making. As students have a strong say in the development and organisation of the course, they experience participatory democracy at first hand. The combination of these two theoretical approaches stresses the individual scope of action to induce social change, while at the same time reminding them that individual influence is restricted by structural factors on a societal level that also need to be changed. These structural constraints are reproduced within universities through a predominantly teacher-centered, hierarchically and undemocratic organization of processes. Therefore the transition towards participatory course structures is urgently needed. This phase can be time intensive and demands the willingness of both, teachers and students, to accept new ideas and ways of doing things.

Eventually, the objective of the course is to enable students to learn to ask critical questions as this is crucial for democratic participation. By doing so, they take their social and ecological responsibility seriously and will not fall easily for the promises of mere technological solutions. The great potential in the approach of the course design lies in the empowerment of students to abandon one-dimensional thinking and to open their mindsets. The ability of reflective and independent thinking may lead to the development of a more sustainable and considerate university and may be inspiring for other projects.

References

Baier A (2013) Student-driven courses on the social and ecological responsibilities of engineers. Sci Eng Ethics 19(4):1469–1472

Baier A, Pongratz S (2013) Collectively and critically reflecting on technology and society. In: Proceedings of the 41st SEFI annual conference, Leuven, Belgium, 16–20 Sept 2013

Becker E, Jahn T (2005) Societal relations to nature. Outline of a critical theory in the ecological crisis. In: Böhme G, Manzei A (eds.) (2003). Kritische Theorie der Technik und der Natur. Fink, München, pp 91–112. http://www.isoe.de/ftp/darmstadttext_engl.pdf. Last accessed 16 March 2014

Blue Engineering (2011) Blue Engineering Baukasten. Ein Handbuch. http://www.blue-engineering.org/baukasten.pdf. Last accessed 24 March 2014

Blue Engineering (2014) Blue engineering website. http://www.blue-engineering.org. Last accessed 24 March 2014

Boni A, Pérez-Foguet A (2008) Introducing development education in technical universities: successful experiences in Spain. Eur J Eng Educ 33(3):343–354

Brand U (2012) Beautiful green world. On the Myths of a Green Economy. Luxemburg Argumente 2, 3, Berlin

Brand U (2014) Growth and domination. Shortcomings of the (De-) Growth Debate. Forthcoming

Brand U, Wissen M (2013) Crisis and continuity of capitalist society-nature relationships: the imperial mode of living and the limits to environmental governance. Rev Int Polit Econ 20 (4):687–711

Brey P (2009) Converging technologies and ethics of the good life. In: Proceedings of the 16th international conference of the society for philosophy and technology, Enschede, Netherlands, 8–10 July 2009

Brunnengräber A, Haas T (2012) Rio+20: Die Grüne Beliebigkeit. Blätter für deutsche und internationale Politik 2(2012):15–18

Conlon E (2008) The new engineer: between employability and social responsibility. Eur J Eng Educ 33(2):151–159

Demirović A (1997) Demokratie und Herrschaft. Aspekte kritischer Gesellschaftstheorie. Westfälisches Dampfboot, Münster

Demirović A (2007) Demokratie in der Wirtschaft. Positionen, Probleme, Perspektiven. Westfälisches Dampfboot, Münster

Demirović A (2012) Marx Grün. Die gesellschaftlichen Naturverhältnisse demokratisieren. Luxemburg, 3, 2012, VSA, Hamburg, pp 60–70

Deppe F (2009) Kapitalismus und Demokratie? Reflexionen über ein problematisches Verhältnis. Luxemburg, 2, 2009, VSA, Hamburg, pp 39–50

Dietz K (2012) Trügerische Hoffnung. Green Economy und ökologische Modernisierung. Rosalux 2, 2012. http://www.rosalux.de/fileadmin/rls_uploads/pdfs/RosaLux/RosaLux_2-2012.pdf. Last accessed 11 March 2014

Görg C (2003) Nichtidentität und Kritik. Zum Problem der Gestaltung der Naturverhältnisse. In: Böhme G, Manzei A (eds) Kritische Theorie der Technik und der Natur. München, Fink, pp 113–133

Grunwald A (2011) Einführung in das Schwerpunktthema. Büro für Technikfolgen-Abschätzung beim Deutschen Bundestag (ed.), TAB-Brief Nr. 39, Berlin, p 6

Guerra A (2012) What are the common knowledge and competencies for education for sustainable development and for engineering education for sustainable development? In: Proceedings of the 40th SEFI annual conference, Thessaloniki, Greece, 23–26 Sept 2012

Hald M (ed) (2011) Transcending boundaries. How Cemus is changing how we teach, meet and learn. Cemus/CSD Uppsala, Uppsala University and Swedish University for Agricultural Sciences. http://www.cemus.uu.se/openuniversity/PUBLICATIONS/BOOK.transcending.bounderies.CEMUS.pdf. Last accessed 24 March 2014

Horkheimer M, Adorno TW (2002) Dialectic of enlightenment. Philosophical fragments. Stanford University Press, Standford

Jahn T, Wehling P (1998) Gesellschaftliche Naturverhältnisse—Konturen eines theoretischen Konzepts. In: Brand KW (ed) Soziologie und Natur. Theoretische Perspektiven. Leske +Budrich, Opladen, pp 75–93

Kastenhofer K, Lansu A, van Dam-Mieras R, Sotoudeh M (2010) The contribution of university curricula to engineering education for sustainable development. GAIA 19(1):44–51

Köhler B, Wissen M (2010) Gesellschaftliche Naturverhältnisse. Ein kritischer theoretischer Zugang zur ökologischen Krise. In: Lösch B (ed) Kritische politische Bildung. Ein Handbuch. Reihe Politik und Bildung, 54, Wochenschau-Verlag, Schwalbach/Ts, pp 217–227

Krätke M (2003) Wirtschaftsdemokratie und Marktsozialismus. Kritische Interventionen 8, Hannover, p 58

Lovelock J (2010) On the Value of Sceptics and Why Copenhagen Was Doomed. http://www.theguardian.com/environment/blog/2010/mar/29/james-lovelock. Last accessed in 24 March 2014

Maroshek-Klarman U, Ulrich S (1997) "Miteinander—Erfahrungen mit Betzavta. Ein Praxis-handbuch auf der Grundlage des Werks «Miteinander» von Uki Maroshek-Klarman, Adam-Institut, Jerusalem. Bertelsmann Stiftung, Gütersloh

Maroshek-Klarman U, Vaddai D (1993) There is no such thing as some democracy. On educating towards democracy and on democracy in the educational system. Kinneret Publishing House
Meadows DH, Meadows DL, Randers J (1972) The limits to growth: a report for the club of Rome's project on the predicament of mankind. Universe Books, New York
Michel J (2008) Editorial. A new issue devoted to sustainable development. Eur J Eng Educ 33 (3):245–246
Ropohl G (2000) Demokratische Technikgestaltung braucht konzeptionelle Konvergenz. Büro für Technikfolgen-Abschätzung beim Deutschen Bundestag (ed.), TAB-Brief Nr. 18, Berlin, p 40
Santarius T (2012) Green growth unraveled. How rebound effects baffle sustainability targets when the economy keeps growing. Heinrich Boell Foundation, Berlin
Schmiade B, Becker F (2008) Branchenreport Windenergiewirtschaft Europa. Arbeitsorientierte Fragestellungen und Handlungsmöglichkeiten. Hans-Böckler-Stiftung and IG Metall Vorstand (eds), Berlin
Segalàs D, Ferrer-Balas D, Mulder KF (2008) Conceputal maps: measuring learning processes of engineering students concerning sustainable development. Eur J Eng Educ 33(3):297–306
Sunderland M (2013) Using student engagement to relocate ethics to the core of the engineering curriculum. Sci Eng Ethics. doi:10.1007/s11948-013-9444-5
Umweltbundesamt (ed) (2009) Report on the environmental economy 2009. Facts and figures for Germany. http://www.umweltbundesamt.de/publikationen/report-on-environmental-economy-2009. Last accessed 21 March 2014
UNEP (2011) Towards a green economy: pathways to sustainable development and poverty eradication. United Nations Environment Programme. http://www.unep.org/greeneconomy/Portals/88/documents/ger/ger_final_dec_2011/Green%20EconomyReport_Final_Dec2011.pdf. Last accessed 11 March 2014
Wissen M (2012) Von Rio nach Rio. Sackgassen internationaler Umweltpolitik: Von 1992 bis 2012. Rosalux 2, 2012. http://www.rosalux.de/fileadmin/rls_uploads/pdfs/RosaLux/RosaLux_2-2012.pdf. Last accessed 11 March 2014
Zandvoort H (2008) Preparing engineers for social responsibility. Eur J Eng Educ 33(2):133–140

Authors Biography

Sabine Pongratz holds a bachelor degree in social sciences from Humboldt Universität zu Berlin, spent a semester at Sussex University pursuing a development studies programme and received her master degree in political science from Freie Universität Berlin. Since 2012 she is teaching the Blue Engineering course at Technische Universität Berlin. She worked for different non-governmental organisations in the area of social-ecological transformation, environmental politics and international development.

André Baier received a master 2 degree in philosophy of norms from Universíté Rennes 1, France, in 2010. Two years later, he completed his studies at Technische Universität Berlin receiving the magister artium degree in philosophy. He actively participates in the development and implementation of Blue Engineering—Engineers with Social and Ecological Responsibility—since its beginning in 2009. As of 2012 he is lecturer for this course.

Sustainability and Service-Learning: Creating a Rubric for Stewardship Across Courses

Amber Narro

Abstract

Since 2008, the researcher has advised a student-led initiative to promote sustainable living and learning on Southeastern Louisiana University's campus. A service-learning activity, the students have initiated research activities that have driven the following years' efforts to involve the university campus, other universities around the state, and the community at large in an Earth Day celebration. The researcher will present data derived from student projects each year the course has been taught and also will present best practices concerning involvement of administration and the student body in this valuable ongoing service-learning activity. Students rely on the ROPES (Hendrix in Public Relations Cases. Wadsworth, Belmont, CA, 1998; Kelly in Handbook of public relations. Sage, Thousand Oaks, CA, pp. 279–289, 2001) process for developing their campaigns and are assigned to develop Goals, Objectives, Strategies and Tactics to support their efforts. Utilizing the systems theory (von Bertalanffy in Science 111:23–29, 1950), students view the system as the entire community—citizens, students, staff and faculty, experts and business. Students collaborate with each arm of the system to develop a plan and utilize the "Stewardship" branch of ROPES in order to maintain community relationships that have developed over the years between the university and its various publics and to engage the next years' students who will take over the project and continue the service-learning activity. This paper will suggest a rubric by which instructors may devise a continuum of a single service-learning activity that may grow and develop over subsequent years to maintain a tradition of sustainability that

A. Narro (✉)
Department of Languages and Communication, Southeastern Louisiana University,
1220 SGA Drive, Hammond, LA 70402, USA
e-mail: amber.narro@selu.edu

W. Leal Filho et al. (eds.), *Integrating Sustainability Thinking in Science and Engineering Curricula*, World Sustainability Series, DOI 10.1007/978-3-319-09474-8_28

expands outside of the classroom and engages the various audiences of the service-learning system. This paper will be of interest to those who wish to develop service-learning projects that continue over time rather than relying on one course to develop, implement and end the project. Through a systems lens, the researcher focuses on branding the activity, initiating research processes in the course, and managing the system after the course is complete.

Keywords
Sustainability across courses · Sustainability · Rubric · Curriculum

1 Introduction

In 2001, Kathleen Kelly added stewardship to the ROPE process developed by Hendrix (1998). The ROPE process focuses on research, objectives, programming and evaluation in public relations (Hendrix 1998). Kelly (2001) added the important "S" to the end of ROPES to adopt stewardship into the acronym and recognize its importance as part of the public relations process. Stewardship includes activities such as reciprocity, responsibility, reporting and relationship nurturing (Kelly 2001).

This ROPES process has been an effective manner by which to organize the activities in this annual communication campaigns course. Each year, students build a campaign that focuses on the university's sustainability activities and those of the surrounding community. In the past couple of years, the students have begun including the activities of the entire university system to which Southeastern belongs. Students utilize the systems theory (von Bertalanffy 1950) to understand how the university is a mere part of a larger global effort and also to work within its limitations of being that one mere part.

It is important to understand that stewardship in this course, both with students to community as well as students to future students, is essential to the ongoing success and expansion of this service-learning project. This research will present a rubric by which a service-learning course in sustainability may be designed in order to sustain itself over time rather than ending the project abruptly at the conclusion of the course. The paper will present activities of the class in developing a plan of action for each year's Earth Day event, the implementation and the evaluation of the actual event and the stewardship activities the students initiate in order to appreciate their community partners and engage students in subsequent classes. It presents a manner by which instructors may facilitate relationships between classes to expand this important initiative rather than simply repeating processes year after year.

2 Literature Review

Laasch and Conaway (2011) state instructors can "demonstrate to students how creativity and communication messages integrate together for public awareness and motivation. Students who learn about these strategies can use similar ones with their future employers or generate their own ideas with a campaign" (p. 77). Texter and Smith (1999) found that the service-learning process may be one that can be of value to students and the community, and it also may present real world challenges to students. The researchers state, "Group projects, by their very nature, seem to generate an array of communication and relational issues and problems" (Texter and Smith 1999, p. 167).

Texter and Smith (1999) report that the "instructor should consider implementing methods of process and performance evaluation" (p. 167), suggesting that project grades should weigh heavily on the final grade for the course, students should provide feedback about the project, they should submit journals over the course of the class and should evaluate not only others in their group project but also themselves. In developing the rubric for a sustainable service-learning course that focuses on sustainability, instructors may also follow Cochran and Gibson (1979). Although the study is decades old, the Cochran and Gibson (1979) study represents a fundamental design of the communication course. The authors suggest instructors should utilize the systems theory, an integrative communication model and cases to fully develop understanding in applying communication models. The authors state that their model is "a representation of the real world" (p. 28) whereby the fundamental sender-receiver communication model is utilized from a systems perspective to study and analyze communication cases.

Utilizing these fundamentals, the researcher will develop a rubric by which students can utilize ROPES to develop an effective campaign that incorporates a systems approach to communicating and programming. Lindenfeld et al. (2012) report, "Environmental communication stands to benefit from intersecting with sustainability science in that it can be part of a larger, transdisciplinary endeavor focused on solving problems" (p. 37). The researcher argues that the addition of the service-learning activity to Cochran and Gibson's (1979) model may deepen understanding and meaning. Texter and Smith (1999) suggest that providing students a service-learning activity that allows them to utilize communication theory, research and skills in a practical application will teach them to assess organizational needs, establish goals and objectives, develop appropriate courses of action, implement a program and assess results. Additionally, Texter and Smith (1999) suggest activities will promote interpersonal skills with those they serve, and students will partner with organizations and assist with communication activities of those organizations.

3 A Rubric for Learning

Service-learning goals may include real-world goals that focus on teamwork, interaction and civic responsibility; the instructor of such a course must work with the students to design a program with evaluation in mind (Witmer et al. 2008). As Leahigh (1985) reports, evaluation is a simple process: "It is a matter of setting goals, designing programs to achieve those goals, then measuring the cause-and-effect results" (p. 23).

The researcher encourages students to begin designing their plans with the entire system in mind. Cochran and Gibson (1979) state that the systems concept can lead students to discover how communications fits into organizational success and that communication breakdowns can affect other areas of the process. This is why the researcher suggests students must begin with understanding the system in which they will develop the campaign. Therefore, the following rubric is suggested to focus course objectives.

Module 1: Students research the previous efforts of service-learning courses.

Outcomes:

- Students research previous plans to implement the Earth Day celebration and strategize on expanding the project to new audiences in the system.
- Students define potential audiences and narrow these to manageable audiences.

Of course, this module is not attainable until after the course has been taught at least once. Students rely on the evaluations of former service-learning students in order to focus their efforts on the current year.

Module 2: Students will define the system and create goals for the service-learning initiative.

Outcomes:

- Students focus their efforts and define the parts of the system.
- Students define the target audiences within the system.
- Students develop realistic goals to include members of the system in the Earth Day celebration.

Cochran and Gibson (1979) state that communication breakdowns can affect other areas of the process. Therefore, it is important that students define communication tactics early on and adopt a process by which contact is made and organized within the system. In Module 2, students begin collaborating with audiences in their defined parts of the system including such target audiences as administrators, those who are responsible for sustainability at the university, those in charge of organizations that focus on sustainability and business leaders who wish to collaborate with the university to help promote sustainable efforts. Students develop a mission and vision for their service-learning project and utilize the mission and vision to develop broad goals. Goals of students in the course have included those for researching the history of sustainability efforts at the university, understanding how the university fits within the rest of its system (made up of eight additional universities), how students view sustainability efforts at the university and what students actually know about "going green," and initiating a culture of respect for sustainability during the university admissions process. Students present well-defined goals that focus on community collaboration with businesses that focus part of their daily activities on sustainability efforts and support the systems (von Bertalanffy 1950) approach to this service-learning project.

Lindenfeld et al. (2012) report that interdisciplinary collaboration is difficult, but sustainability work is even more difficult because “it involves long-term collaborations with communities, cross-campus colleagues, and stakeholders” (p. 37). Students in this particular service-learning initiative have defined the system as the university as a whole, the university system, and the immediate community. Some groups of students in this course develop a micro-focus within the university and/or the community. For example, major parts of the university system have included campus organizations, the sustainability division (made up of university staff), the sustainability committee (made up of university staff and faculty), and individual departments.

Module 3: Students utilize ROPES (Kelly 2001) to develop a plan to reach their goals.

Outcomes:

- Students outline goals, objectives, strategies and tactics (GOSTs).
- Students create to-do lists and assign responsibilities within their groups.
- Students adjust their plans as necessary to accommodate the service-learning process.

This is the most time consuming portion of the course. The ROPE process (Hendrix 1998), along with the stewardship addition (Kelly 2001), guides students in developing goals, objectives, strategies and tactics. Students understand that the goals are broad, focusing on the mission and the vision of the project. They begin by researching their goals and the opportunities for implementing them. Secondly, students develop measurable objectives in order to begin the development of their programming and future evaluation process. The strategies are the broad activities that guide the tactics of achieving the objectives (see example in Appendix). Not required in the plan are the to-do items; however, students often find it easier to write those items into the GOSTs in order to better organize the process of evaluation and reporting.

The researcher ensures that the students understand that these items do not operate individually but depend upon each other. The goals guide the objectives, the objectives guide the strategies, and the strategies guide the tactics. The programming part includes the activities that will support the plan, and the evaluation depends on the measurable objectives.

Students are warned that measurable objectives can only be measured as successful if outliers also are considered, and students are expected to consider those outliers during the process of planning and implementing the campaign in order that these items may be reported as limitations in the evaluation. For example, Earth Day may not be considered a success simply because objectives for the event are met. Students must consider that people may be excited about Earth Day because of the tradition of the event and the media excitement surrounding it.

Teaching students to write effective goals, objectives, strategies and tactics is the most difficult portion of the course. Instructors should allocate ample time at the beginning of the semester and allow students to practice this craft until the GOSTs for the service-learning project will allow for a true measurement of successful activity.

Module 4: Students implement Earth Day activities.

Outcomes:

- Students stage the Earth Day event.
- Students implement the stewardship plan of their service-learning project.

This process varies from year-to-year. Students plan their activities in the programming portion of the ROPES process (Hendrix 1998; Kelly 2001) and then are expected to stage Earth Day on or as close to the official celebration as possible, which is April 22. Students are responsible for the activities, the implementation and the clean-up afterward.

Most importantly, students are responsible for the stewardship portion after the event. They are expected to debrief participants, properly thank them and then also to begin the evaluation of the event and the project to start the relationship with the next group of students, people they have not met.

Module 5: Students evaluate their service-learning project and their group members.

Outcomes:

- Students provide a complete plan that measures success of goals and objectives.
- Students evaluate the campaign and offer suggestions to future service-learners.

Pinkleton et al. (1999) note that the increasing pressure for public relations practitioners to prove their value of their efforts within the scope of organizational goals. The authors note, "If practitioners, supervisors and clients all value research, this suggests and opportunity for public relations professionals to begin to move the practice of research toward measurements having greater theoretical and practical value for the field" (p. 94).

It is important for students to understand that the importance of evaluation actually lies in the future and not necessarily the present. As Leahigh (1985) states, "Measurement not only helps refine current campaigns, but also enhances one's ability to make informed judgments about the effectiveness of public relations techniques in future marketing programs" (p. 17). Signitzer and Prexl (2008) agree that implementing sustainability measures "should guarantee long-term success and requires not only support of top management, but also networking of several other corporate systems" (p. 1).

Students report their efforts to the following classes not only to support the effort of the sustainability service-learning project, but also in order to explain what was successful and what was problematic in the campaign. Equipped with a well-organized evaluation that makes suggestion for the next class of service-learning students, the instructor for the course will have a clear plan for a sustainable service-learning project that will continue from year-to-year rather than starting over with every new group of students.

Lundy (2008) suggests that students demonstrate growth in understanding through journaling throughout the service-learning project. While students in this course did not journal their efforts, they are asked to reflect on their experience through group evaluations either in the form of a question on the final exam or through actual forms included in the syllabus for the class.

4 Additional Assignments

Students have assigned readings throughout the semester and analyze case studies for which they identify GOSTs for other campaigns. Oftentimes, forming the goals and objectives is the most difficult portion of the class because case studies often do not present measurable goals and objectives that were successful or unsuccessful; they simply state the problem and then list the steps the company took to solve the issue. Challenging students to determine goals and objectives allows them to broaden their knowledge about those GOSTs before they begin developing their own. Additionally, as Cochran and Gibson (1979) noted early on, analyzing case studies allows students to discover some of the underlying causes of organizational conflict and to make suggestions for appropriate solutions.

Students create a multi-media project focusing on an area of sustainability within the university branch of the system. This develops a history of the project that may be passed to future service-learning classes.

The final components of the grade for the communication course is made up of the following:

- The project and group evaluations
- The multi-media assignment
- The midterm and final examinations
- The case study assignment(s)

5 Conclusions

Student opinions of teaching have included praise for the real-world assignment, and students have often used experiences gained from the course to initiate conversation during job students interviews. Former students also report challenging future employers to develop formal mission statements in order that all goals and objectives are guided by the daily mission of the organization.

Kelly (2001) states the importance of stewardship in her research. Leeper and Leeper (2006) agree, stating that when developing relationships with the community, the university should focus on "quality, social responsibility, and stewardship" in order to ensure desired community relationships. Therefore, the researcher understands that stewardship is the single most important element of sustaining and growing the sustainability service-learning project.

Hall and Stahl (2012) state that "the University remains a symbol of those places where mass intellectuality can be consumed, produced and more importantly contributed to by all" (p. 197). The reports of previous students' experiences allows for early expansion of a project and growth from year-to-year. If students started over every year, members of the community may be overlooked with no record of contacts from the past. Students understand the value of maintaining relationships with some community members and that some relationships are not so beneficial to the process, as is in line with Witmer et al. (2008) who report that occasionally, community-based clients are unprofessional or unpredictable. Therefore, students understand that stewardship is not only for the present participants, but also for people who will be entering the course the following year, and part of the stewardship toward future students is informing them of the relationships that did not work out as expected.

Examination of previous plans with the goal of immediate growth allows sustainability of the service-learning project that actually concerns sustainability. With the growth in technology and opportunities for people to participate in "green efforts" expanding, it is important that students begin with growth in mind rather than hitting restart every year. Shell (2007) states that instructors must be committed to implementing the service-learning program as it is a physical, emotional and intellectual commitment to the course and its outcomes. However, it may be worth it, as Silverman (2008) found that students who participated in a service-learning project had a significant increase in their desire to volunteer.

Service-learning in the university setting is an opportunity for instructors to help students grow in their professional careers but also to strengthen the ties in the university's system. If there is an expandable initiative, there is opportunity for annual growth and improvement that opens opportunity for research across time. Instructors should embrace this growth of the project and encourage students' innovation and expansion. Students in this course have accepted its tradition and have come to expect and anticipate their participation in the service-learning project. Oftentimes, students come to the first day of class with ideas for the project.

This rubric has the opportunity for further exploration. It has only been tested with this particular course in communication, and it should be examined for opportunities for inclusion in other disciplines. Future research should further examine student reaction to the course and expectations as well as whether the university community understands the value of the service-learning project. Since this project is a successful initiative for the instructor, student reactions may provide more outcomes that should be measured within the scope of the modules.

Acknowledgments Special appreciation is expressed to graduate students Katlin Morse, Taylor Bergeron, Janel Page and Lori McAfee who allowed me to use their goals, objectives, strategies and tactics (GOSTs) as an appendix to this research.

6 Appendix: Setting up a Plan of Communication

Mission: Promote sustainability efforts in SLU Student Organizations
Vision: Encourage all current and future student organizations at SLU to be actively involved in the sustainability movement.

Goal 1: Build a relationship between SLU athletics and the sustainability movement

- Objective: Include every athletic team in SLU sustainability movement
  - Strategy: Incentivize student-athletes and coaches to participate

Tactic: Incorporate a point/reward system for highest participation
- To-do: Survey student athletes on use of sustainability practices.
- To-do: Compose rules and regulations for the contest
- To-do: Send out email to coaches to inform of the contest

Tactic: Host a kick-off event for all team members and coaches to explain contest
- To-do: Reserve time on football field
- To-do: Order pizza and drinks to serve
- To-do: Reserve DJ
- To-do: Promote event through Sports Marketing and Promotions Office.
  - Strategy: Provide opportunities for all student-athletes to mingle while learning eco-friendly practices

Tactic: Host "Go Green" slow-pitch softball game for student-athletes
- To-do: Choose date to host event
- To-do: Send emails to head coaches
- To-do: Reserve a field for game

Tactic: Host a trash clean-up "party" for student-athletes around athletic facilities with picnic lunch afterwards
- To-do: Choose a date
- To-do: Email head coaches
- To-do: Put in request for gloves and other tools from Sustainability or SGA
- To-do: Arrange for the food
- To-do: Assign athletes to teams.
- Objective: Reduce waste in Athletics by 30 %
  - Strategy: Increase eco-friendly practices for all athletic teams

Tactic: Encourage teams to conserve water
- To-do: Wash jerseys in cold water
- To-do: Suggest that athletes shower at home instead of in the locker room

Tactic: Encourage teams to conserve electricity

- To-do: Install energy efficient OR motion sensor light bulbs in facilities
- To-do: Avoid night practices that require lights to be in use
  - Strategy: Reduce plastic waste produced by athletic teams

Tactic: Promote the use of reusable water bottles

- To-do: Provide water bottles to each athlete
- To-do: Contact a vendor to purchase water bottles

Tactic: Encourage any plastic that is used to be recycled

- To-do: Provide recycling bins at each athletic facility
- To-do: Contact coaches to promote recycling efforts

Goal 2: Involve social student organizations (Greeks, BCM, Gamma, Beta, Phi) in SLU sustainability movement

- Objective: Have 50 % of waste from events and Homecoming activities be recycled.
  - Strategy: Promote student organizations to participate in recycling efforts during socials and events

Tactic: Implement that organization socials and events be more recycle friendly

- To-do: Provide recycle bins at all events and socials
- To-do: Contact organization heads to get them on board

Tactic: Set a plan to recycle wood from homecoming floats

- To-do: Pass along idea to organizational presidents.
- To-do: Contact sustainability about taking the recycled Homecoming wood.
- To-do: Designate a spot for wood drop-off.
  - Strategy: Promote student organization heads to endorse their recycling efforts

Tactic: Require organization heads to carry out sustainable efforts and influence their organization to do the same

- To-do: Contact leaders to introduce the plan
- To-do: Provide recycling resources

Tactic: Set up a plan to measure each organization's recycling

- To-do: Make a point scale as reward system
- To-do: Provide the necessary resources for recycling efforts
- Objective: Reduce internal (office, meetings, etc.) waste by 50 %.
  - Strategy: Increase the recycling efforts of student organization offices

Tactic: Implement a recycling plan for each organization

- To-do: Set up recycling bins in each office
- To-do: Establish a collection plan for item pick-up

Tactic: educate members on how recycling is good for the earth

- To-do: put together an educational power point
- To-do: posters and signage that promote recycling
  - Strategy: Increase the recycling efforts in the student organization meetings

Tactic: Implement the "Your Trash is Your Ticket" meetings, requiring members to bring a recyclable item to the meeting

- To-do: Set up a point system
- To-do: Set up recycling bin by the entrances

References

Cochran DS, Gibson CK (1979) Putting a square peg into a round hole: communication models and their application. Int J Bus 17(1):27–36

Hall R, Stahl B (2012) Against commodification: the university, cognitive capitalism and emergent technologies. TripleC 10(2):184–202

Hendrix JA (1998) Public relations cases, 4th edn. Wadsworth, Belmont, CA

Kelly KS (2001) Stewardship: the fifth step in the public relations process. In: Heath RL (ed) Handbook of public relations. Sage, Thousand Oaks, CA, pp 279–289

Laasch O, Conaway RN (2011) "Making it do" at the movie theatres: communicating sustainability in the workplace. Bus Commun Q 74(1):68–78

Leahigh AK (1985) Marketing communications: if you can't count it, does it count? Public Relat Q 30(4):23–27

Leeper KA, Leeper RV (2006) Crisis in the college/university relationship with the community: a case study. J Promot Manage 12(3/4):129–142

Lindenfeld LA, Hall DM, McGreavy LS, Hart D (2012) Creating a place for environmental communication research in sustainability science. Environ Commun 6(1):23–43

Lundy L (2008) The role of reflection in service-learning in the public relations campaigns class. Paper presented at the annual meeting of the NCA 94th annual convention, San Diego, CA

Pinkleton BE, Austin EW, Dixon A (1999) Orientations in public relations research and campaign evaluation. J Mark Commun 85–95

Shell A (2007) "More bang for the buck:" how an international public relations service-learning experience expanded pedagogical outcomes. Paper presented at the annual meeting of the NCA 93rd annual convention, Chicago, IL

Silverman DA (2008) Service-learning projects in the PR writing course. Paper presented at the annual meeting of the NCA 94th annual convention, San Diego, CA

Signitzer B, Prexl A (2008) Corporate sustainability communications: aspects of theory and professionalization. J Pub Relat Res 20(1):1–19

Texter LA, Smith MF (1999) Public relations and public service: integrating service-learning into the public relations seminar. Commun Stud 1999:165–175

von Bertalanffy L (1950) The theory of open systems in physics and biology. Science 111 (1950):23–29

Witmer DF, Silverman DA, Gaschen DJ (2008) Working to learn and learning to work: a profile of service-learning courses in university public relations programs. Pub Relat Rev 35(2):153–155

Author Biography

Dr. Amber Narro is an associate professor at Southeastern Louisiana University. She graduated in 2006 from the University of Southern Mississippi with a doctorate in mass communication. Her research interests include strategic communication tactics and Website design techniques. Campaigns, public relations writing and writing for the media are among her course load taught at Southeastern. In her annual communication campaigns course, she focuses the entire course on understanding how academia is an important piece of the sustainability effort.

University Networks in Italy and Their Intellectual Contribution to Implement Sustainable Development Issues in the CV

Carolina Collaro

Abstract

The paper analyzes two important Universities' network in Italy. The first network is the Venice International University, in partnership with some major universities in U.S. and the world; the second one is UNISCAPE, based in Florence, a consortium of 52 European Universities to implement the European Landscape Convention. Some case studies are proposed to discuss the education for sustainable development (ESD) in a globalized world. The aims are to evaluate not only the networks' contributions in Italy for ESD implementation in Universities' curricula, but also future directions to develop capacity building between these process' actors, considering particularly the priority of ecosystem services approach to Natural Capital theory. This approach recognizes an economic value to natural resources in opposition to the critique of nature commodification, regarding these common goods; that gives rise, instead, to alternative measures of progress and sustainable development goals.

Keywords

Education for sustainable development · Globalization · Ecosystem services · Corporate sustainability · Landscape

C. Collaro (✉)
Nova Gorica University, Campo della Chiesa, 3, 30132 Venice, Italy
e-mail: carolina.collaro@fastwebnet.it

W. Leal Filho et al. (eds.), *Integrating Sustainability Thinking in Science and Engineering Curricula*, World Sustainability Series, DOI 10.1007/978-3-319-09474-8_29

1 Introduction

Doing a retrospective analysis of the past few years after COP-15 in Copenhagen and after Rio+20 UN Conference, it is still difficult to assess how close we are to satisfy a development that utilizes natural resources in a sustainable way. We are far away from an efficient management, therefore, cost-effective, of our natural and cultural capital, as well as from an efficient educational system, equitable for all and the future generations. It is now widely questioned the traditional way of a country-wealth accounting, through the global development product (GDP) index, which no longer reflects the real welfare and countries' progress (Costanza et al. 2014). Given these difficulties, the education system has acquired a leading role to implement sustainable development (SD) principles (WCED 1987) enshrined in international agreements. 2004 has been the first year of the decade on ESD, promulgated by UNESCO, in which universities have acquired the catalytic role of agents of change, often in relation to regional socio-economic development as facilitators of SD (Sedlacek 2013) as well at the global level. Then the ties between universities have been strengthened, even among developing countries (Hansen and Lehmann 2006), and networks of universities have taken consistency and credibility, becoming in a few years the knowledge economy's actors. Much literature has been published about it, which enhances the role of networking, analyzing its essence. As reported by Boydell et al. (2008), the network's added value is an "intangible asset," in the form of knowledge, which will bring returns on investment in the future. Networking between society and research becomes more and more important (Mader et al. 2013) to understand a global-local economic context and to create knowledge and capacity building's synergies. Capacity building is the process by which individuals, groups, organisations and societies increase their ability to understand and solve problems in a sustainable manner. Then, it is a continuous process where the network's participants mutually share and benefit each other. Even in Italy some Universities' consortia are trying to pursue this direction. This paper explains the research and trainings offered by two networks: the Venice International University (VIU) and UNISCAPE, giving insights of potential effects on the territory of the city, extended on the networks' partners virtual offices. The examined context, then, is local and global together.

The paper describes, through the students or researchers close experience, what programs are affecting globalization and SD, and, as for UNISCAPE, affecting the European Landscape Convention (ELC) implementation (Council of Europe 2000) and commons goods' engagement.

According to Hesser and Siedersleben (2007), globalization is the ever increasing interdependence of economic, social, technological, cultural and political spheres in today's world. The word "Landscape" means area perceived by people, which character is an action and interaction's result of natural and human factors (Teofili and Clarino 2008). Through educational curricula's analysis, the aim is to illustrate how networks are helping to implement the international commitments about SD and ESD. In addition, we consider a particular perspective for capacity building, which although not expressly foreseen in the networks' Statutes, is essential for SD. We refer to the so-called Natural Capital (Costanza et al. 1997), a key element for ESD, to consider as a possible curricula evolution, with a major consideration. Our findings finally propend for an affirmative answer. However in the roadmap of the two networks analysed, these educational objectives will achieve curricula attention, insofar that network partners will share values also with humanities' sciences, that underlie them, thereby developing the networks' capacity building.

2 Venice and the VIU Genesis

The VIU was founded in 1995 by the Province of Venice, Ca' Foscari and IUAV University with other international universities. Over the years, partner universities have increased in number, and today the network has joined 15 partners offering courses on SD. A University such as the VIU is particularly significant for the city of Venice. In fact, for many years, important stakeholders have been discussing the future cultural offer of this unique city, addressed to a cosmopolitan audience of 22 million tourists a year, without finding viable alternatives than those established, often for lack of participatory management (Brochier et al. 2001). So we think just to name one, the well-known annual event of the Biennial, a film exhibition of worldwide resonance. This was founded by the industrialist Volpi in the '30s of the last century, and gave to the city a perennial trendy-face, which in truth does not always correspond to expectations and the reality in which inhabitants live.

Many of them are forced to leave their homes for the persistent moisture due to the high water's phenomenon, which puts a strain on daily life, together with the wave motion, due to the steamboats' passage. These serious environmental problems caused a population decrease from 165,000 inhabitants in 1960 to almost 58,000 in 2012, still persisting. Monuments themselves are threatened by climate change and pollution causing damage to the stones and ancient marbles.

Venice lagoon itself, that owes its survival to man's intervention on transitional waters is affected by sea-level rise and wetlands and habitat loss. Such endangered salt marshes and tidal flats should be restored in their ecological functions together with their ability of self-adaptation. The salt marshes' main characteristics are, in fact, to adapt to sea level growth, by increasing the vegetation and capture sediments. All this contributes, within set limits, to counteract the sea-level rise, at least one related to the soil's lowering, the subsidence, that is nowadays of 5 mm/year.

Ecological concerns in Venice have ancient causes, and not least is the debate on the future destination of Porto Marghera, built in a natural salt marsh area, to make way for a petrochemical center, that now is no longer functioning, leaving the area at high environmental risk (VIU 2011). Not even placated are the controversies over the Mose's construction, mobile barrier that will protect the town from flooding, but will distort, probably, the ecological balance of the lagoon. The debate in favor of more re-naturalisations inside the lagoon, to fight erosion and pollution, through wetlands' inclusion is becoming even more pressing. Wetlands have also the advantage of being a landscapes important element, also useful for aesthetic enjoyment and tourists and residents' recreation. Even though, this ecological projects' sustainability is clear, the ecosystem services approach, that would be useful to assess, for present and future economic and policy choices, has not yet been implemented, except in some projects on minor interventions. So the courses about SD, proposed at VIU, acquired a very high value, also to innovation's expectations in the territory. A very strong component of the VIU, together with SD are globalization's course and the courses of interchange with China. These last ones are with the Environment Ministry collaboration (VIU 2012). Since 2003, VIU through its research department, the Thematic Environmental Network (TEN) is a technical partner of the Italian Ministry for the Environment, for the Sino-Italian Advanced Training Program on Environmental Management and SD. The Ten center specialises in offering training on sustainability. Programs-cooperation with China are open to minister's functionaries and professors who need, for their work, to be specialized in this field. Through the VIU, a full program of exchanges was established with the Chinese University of Tongji and Tsinghua, then extended at the Japanese Waseda University. In the wake of the Italian economic and financial planning document, that considered education a cornerstone for employment policies, VIU established training courses at various levels. There are post-graduate until to graduate courses' level offered in addition to those specifically directed to officers already employed in government offices and third countries too (VIU 2011). All this happened in 1997, so VIU has been a forerunner in proposing a network with universities situated mainly in Europe, in U.S. and Asia. Every year, the partner universities send their teachers, always different, to teach courses related to SD and humanities' issues. Students follow, after the classes at VIU, stages of completion that typically are more tailored to the business and production's world. These stages are held in the country of the partner university. So students have the opportunity to meet each year, people from all parts of the world with whom to share 3 months of study, but also cultural programs open to the city's life with its monuments and current events. According to Russo and Arias Sans (2009), students in Venice become generators of "landscapes of creativity." Between the courses offered, some of them have annual recurrence and, therefore, have acquired a great importance for the presence of qualified teachers coordinators, who organize, aside to teach courses, interesting workshops with outside guests. Inside the Global Program, these relevant annual courses are titled: "Globalization, Environment and Sustainable Development," where the relationship between economy and environment is discussed; and "Cities, Global Change and Sustainable Development" where this relationship regards cities and megacities of the world. So far, it is possible to synthesize the ecosystem service benefits, with papers written by students in the form of case studies. As courses' achievement, the connection between economy-environment is the basis of SD discourse, whereas before these concepts were disconnected from each other.

This paper now highlights some case studies and relative conclusions, discussed from a Ph.D. student perspective, and reflecting the author's experience.

3 Case Study 1

In this first paper, required for accreditation of the course, the student reports critically what they learnt in the first module of the course. In summary, it explained the link between SD and adaptive capacity (Plummer et al. 2012), which is related to local culture, using the example of Venice.

Students explored answers to uncertainty, ambiguity and ignorance problems as well as looking at environmental decision making and the pursuit of sustainability. Very interesting was to understand the meaning of "irreversible damage's threshold," taken from the 1992 Rio Convention. After this first module, all students know that it may not be universally prescribed, but it is socially and politically determined. They realize that adaptation is a strategy that also includes the governance and institutions. Formulas and indicators, monitoring the relationship between economy and environment, find a dialogue between policy makers and the "post-normal" science, characterized by projections' uncertainty, both on environmental change both on models' imperfection (Roca 2002). However, it should also give answers to problems that affect environmental decisions, in deference to what is stipulated by the precautionary principle which, while not specifying what action to take, lets us understand that, unless you act, we will bear much higher costs. If facts are uncertain, decisions are to be supported by the extended community, through participatory processes. Porto Marghera, built in an area originally covered by saltmarshes, was largly debated, with regards to sustainable planning, because it is now, after the petrochemical dismantling, a very problematic site for the ambitious and legitimate government's objectives of contaminated soil reclamation (Critto and Agostini 2009). Students learned about the eco-park and industrial symbiosis thematics, that happen when industries share allotments and spaces, but especially environmental services, appointed to damages' mitigation usually implemented on the environment by companies.

At VIU, inside the Global Program's courses, many seminars are organized about SD issues in China, also in accordance with the Ministerial cooperation program. It is part of the network exchange, that Chinese functionaries learn about the Venice lagoon system and in turn students learn about Chinese culture toward sustainability. This is also in accordance with the Universities' role in influencing society in a two-fold manner: they train and educate people and they participate in governance at the national and regional level (Sedlacek 2013) as well as internationally. Now, a second paper for the course's attendance is discussed.

4 Case Study 2

China is a great country which we cannot look to without concern, because what it is decided in China, now effects the whole world. So, not only because it is the second largest economy in the world, but also because, it is facing environmental challenges, which are furthermore, the future possibilities executing SD theories.

Over the last 10 years China has experienced very steep economic growth and GDP increase, a lowering of poverty thresholds, a rise in life expectancy, a considerable population increase, and a frightening urbanization: namely, the emigration of an entire workers generation, from the countryside to the city (Knight Gunatilaka 2010). Today 40 % of the population resides in cities, and it is estimated that it will shortly become 54 % even though policymakers think that, from a planning issue perspective, rural-urban relationships are always current. The best way to approach the "planet China" knowledge, is trying to love it a bit more, precisely because of its inherent contradictions, and to go hereafter of statistical data, into urban stories and choices we know from policy makers of the remote and near past. How is the Chinese Government coping with such rapid development, that is causing irreversible environmental damage? What reforms, institutions, projects to implement or not? How can we help this giant? What is the extent to which we will be able to cooperate with China? These questions are of vital importance for future generations. We have been trying to find answers, at least briefly in an analysis framework of the problems. As an example, the case of hukou (household registration) system is very interesting, regulating the migration of Chinese from their own territories. (Wing Chan 2011). Namely, it is the transition from rural to urban, which must be accompanied by a control. In the time limit granted, it has been possible to read how the Chinese has developed some great mega regions, as the Pearl delta or Shengen region, which have merged with other notable projects stories that were presented in the course, such as China State Government's project of the South rivers diversion, to convey in the North rivers. In fact, China is a country where environmental problems converge, opposing the drought, to the water and to the great potential of hydropower. In this regard, China has built dams that are among the most impressive in the world, but although they provide so much energy, on the other hand they have displaced populations and created geomorphological context changes. In China there are, as in Europe, environmental impact assessment laws, but there is also awareness of how badly they are applied, for knowledge is lacking and corruption is still very strong, as in all developing countries. Within China, to a certain extent, there is still so much poverty and social conflicts among people. It has been said that to achieve a new SD model, China needs to strengthen the environmental policies' efficiency and better enhance environmental challenges' integration into economic decisions. So students have understood that environmental issues in China are becoming global, which is why there is today a strong interest to cooperate with China on an international level.

According to the Italian Ministry of environment, China as well as India are destined to be an important market with regards to hydroelectric power, with an untapped potential of 83,000 MW; economic growth and an increase in energy needs will stimulate hydropower's progress.

At this stage of strong expansion, it will still be important, for these countries, to introduce regulatory and construction standards, protecting the environment, minimizing the negative impacts, arising from large plant construction. European Governments should accompany, therefore, industries towards innovative and sustainable project preparation, such as the introduction of mini hydro and mini wind plants, which has minor impacts on the territory, compared to large plants. Moreover, these plants are interesting, especially if designed in the local development's context and co-benefits management with other related resources. With these remarks, there is reported another important aspect of VIU curriculum. Should universities maintain, in a sustainable way, this network collaboration with a giant like China? On what effort is worth pointing? At this regard, another vision for VIU students is the Corporate Sustainability, a module of the course "Globalization, Environment and Sustainable Development," whose students are asked to present a small paper for revisions. They learn how building potential to support SD more actively, and how increase capability building between development's stakeholders.

5 Case Study 3

The idea of Corporate Sustainable Responsibility (CSR), originates from SD, according to the report "Our Common Future" (WCED 1987). For organizations, it means the challenge to improve simultaneously social and human welfare, ensuring the effective achievement of organizational objectives (Karnani Aneel 2012). Corporate Sustainability is a multifaceted concept requiring a companies' organizational change and the adaptation on a different level. Numerous studies have been devoted to exploring the linkages between CSR and financial performance. In the existing literature, there is still no consensus on the question of whether CSR programs lead to a better corporate financial performance. In general, CSR means dealing proactively with the community base, engaging with stakeholders and providing a business venture that serves a social purpose (Commission European Communities 2001). CSR is not mandatory by law in many countries as in Italy, but instead it is on a voluntary basis. How to involve stakeholders? By our research's findings, the corporation's age, community involvement, and cultural diversity have a significant influence on CSR perception by stakeholders. Another important issue is the existence of a published CSR, the so called sustainability report (SR), which is crucial for stakeholders' understanding. Meanwhile, we distinguish companies, which choose corporate sustainability and the SR from companies with sustainable behavior. The SR involves an in-depth analysis by the firm, of so called environmental costs, as defined by the EU. Defining the 'sustainable business' is more than just an academic interest. In fact, environmental costs must prevent, reduce or repair environmental damages. If it is no possible to prevent, we could try to reduce and finally to repair. The "greening of accountancy" involves a reappraisal of how to identify and measure the relevant costs of processes and products and then the incentive mechanisms' redesign. Through these changes, managerial decisions and corporate behavior may be refocused towards the goal of achieving SD for the companies. Then environmental costs are not externalities but the companies responsibility. The company is aimed towards a continuous improvement of its organizational and technical capacity to reduce risks and protect the environment, effectively and at the same time with increasing efficiency. The goals related to these objectives are defined from time to time by plans called "Sustainable Reporting."

This openness to territory is the most fascinating side of CSR: to involve more stakeholders, trying to internalize the externalities, which are costs weighing on the community and are recognized by the companies as fundamental for SD (Dilling 2011). SR functions within the context of transparency about economic, environmental, and social impacts and is a "fundamental component in effective stakeholder relations, investment decisions, and other market relations" (GRI, Global Reporting Index 2008). Only when an organization is aware of its societal impacts (its social and environmental footprints), does it have a true and deep concern for these impacts, and consciously acts accordingly at both a strategic and operational level. Environmental reporting concerns a number of elements, including materials, energy, water, biodiversity, emissions, compliance and transport. SR covers four sub-themes: labour and work practices, human rights, society, and product responsibility. Economic reporting is also included and concerns such issues as financial performance, market presence, indirect economic impacts, and investment in the community. Each of these activities is reported against a performance indicators' series (GRI 2008). Such initiatives are to be understood as examples of 'soft' law (Zerilli 2010) based on aspirational voluntary declarations of intent. Nowadays, very few reports talk of impacts but they talk instead about performance of a company that means: the results of organization's activities or investment over a given period. SR has to show transparency, and information should be reliable and usable from others. SR must create financial value, even in the long term and enhance reputation. The GRI is the most widely known set of voluntary guidelines for SR; some examples of environmental indicators are: energy and water consumption, carbon dioxide emissions, and internal initiatives to improve energy efficiency. Now ISO 26,000 rules define all CSR key terminology, providing advice about the ways in which companies can identify their social responsibilities and how CSR can be integrated, not only into companies, but also into all types of organizations. Companies are increasingly striving to present a positive and reliable picture of their sustainability performance in their SR. The most interesting side is CSR extending beyond the company's doors, into the local community, that could be research within the field of territorial planners. Furthermore, Ca' Foscari University, one of VIU partner, has recently implemented its SR, published on the website for the university community and citizens. Finally, after these 3 case studies, we could not complete the picture of the courses taught in the context of SD, without mentioning the course "Coastal Wetlands Lagoons and Estuaries: Environmental Monitoring and Management" with Duke University's collaboration, concerning some of the most vulnerable ecosystems to climate change impacts (IPCC 2007). Specifically, the course is designed and taught by Padua University and Duke University, both VIU partners. Numerous stakeholders, also from outside universities and enterprises, have taken part, during the seminars. The foreign university's contribution regarded some wetlands' restoration in the U.S. like San Francisco Bay, the Louisiana and Florida coasts, as well as experiences in North and South Carolina where Duke University is very active. Why to restore the wetlands or just improve them or create new ones? Just the ecosystem approach has been deepened to consider a restoration of the lagoon, examining climate change impacts and environmental action for adaptation and mitigation measures, to prevent the further degradation of this unique and fragile landscape due to strong anthropogenic pressures. The wetland's restoration is also considered a strategy to mitigate climate change impacts, by virtue of the fact that wetlands absorb CO_2. The study showed that climate change is altering these ecosystem functions, so important to nature and society, by providing environmental but also cultural services. The conclusions are to prevent further impacts and biodiversity's loss, considering to encourage local adaptation strategies (Peake and Smith 2009). The cognitive analysis of the area, if it is true that is the subject of extensive studies, has benefited from the jointly established methods of Duke and Padua Universities, which essentially consider the impacts on salt marshes, subtidal platforms and tidal flats' ecosystems. The methods utilized were the modelling, literature review and the similar cases' experience, some field visits, and the remote sensing detections application. Students and researchers have understood that some problems remain, still difficult to predict, as the project's ability to achieve results and restoration targets. Then it is necessary to work, knowing the inevitable errors percentage. Students learnt that restoration projects differ in many factors, which help to predict the objectives potential achievements. It is fair to ask if ensuring these ecosystems' maintenance is easily possible; as a result, we must consider the whole resilience-system and the management experience acquired. In this regard, a second network operating in Italy is UNISCAPE, founded in 1998, composed by 52 universities, which are joined together, adhering to the European Landscape Convention (ELC) (Council of Europe 2000). As we know, this convention has been implemented and ratified by each EU member countries' law; the landscape is now a full-fledged element of planning (UNISCAPE 2010). In truth, it has always received consideration in Italy, because Article 9 of the Italian Constitution stipulates the landscape's importance, and even before, a Law of '39 protects the "natural beauty" (G.U. 1939). For UNISCAPE, the ES approach even considered a significant component for the landscape's protection, is always in danger of falling into "nature commodification "discourse. As an example, new ES, as carbon sequestration and watershed regulations have expanded the market of trading's resources, and that include new ecological functions into a price market. Then a commodification of the nature subsists, not having sharing with landscape values, as conceived by ELC. Then, it is still difficult to incorporate ES approach in the planning and ESD. These ES approaches are sanctioned, however, by major international conventions, such as the Convention on Biological Diversity ratified by Italian law.

Moreover, UNISCAPE, through workshops proposition and related publications, tries to involve local stakeholders especially situated in the region. With the Tuscany Region were organized the "Careggi Seminars," from the Medicea villa's name, in Florence. UNISCAPE is operating a researcher's network, whose members can propose initiatives, as long as it is related to ELC issues.

This Convention, which believes that perception is the landscape's essence, requires the active participation of people on which the landscape persists. In fact, according to Oles and Hammarlund (2011) what appear to be the common landscape values in local communities often conceals fundamental conflicts among individuals, groups, and institutions. Then this participation is to be understood as distinct from the consultation, which usually we use to define the plans, even landscape plans. Often there is confusion about these two concepts: the concentrations occur when the decisions have already been discussed and steps are taken for the plans final implementation.

The participation, instead, regards many complex and expensive processes. It requires stakeholder's identification and involvement, before implementing an engagement strategy. Lately, UNISCAPE organised a seminar on the "Common Goods" where it talked about natural resources and how to handle them. A new environmental governance's discourse raised, in which decisions complexity could be shared on properly designed platforms, extended to multiple stakeholders and knowledge systems. Also biodiversity is a common good, because it facilitate ecosystem functions, i.e. local climate and soil protection. However, to build capacity building around such complex issues, we need close ties with social sciences and public values construction too, which can only create trust and openness to dialogue between stakeholders. What could be done to pull the discussion's threads on more practice advancement in SD and ESD? We have a broad knowledge of biophysical phenomena affecting ecosystems, from which we drive economic and cultural benefits, but we lack proper integration with social sciences and human well-being's consideration. Instead, we should select, also in view of adaptation's scenarios to environmental changes, appropriate policies between different options. It follows that, from an ethical perspective, it is necessary to get interdisciplinary understandings memoranda, between ecologists and economists and planners too, and between natural and social sciences, to grow designated research. We, therefore, believe that, without losing the sight on local problems, we need to establish a "roadmap," where the sustainability path is finally not only drawn but critically evaluated and shared between stakeholders. We believe, therefore, that these two networks of Universities represent important realities for ESD in Italy, albeit with some differences and statutory characterizing objectives. The VIU can relate to so many issues in individual courses, due to worldwide students and professors' participation and has gained a significant place in the relational economy, context of the network.

Furthermore, joining the Sustainable Development Solutions Network (SDSN), a new UN network, VIU has strengthened the capacity to affect the social, through influential researchers accession too.

Concerning UNISCAPE, a greater opening to biodiversity and natural capital's thematics should be implemented in a more decisive manner besides participations theme. If we understand the landscape as the inhabitants perceive it, we must cultivate the ability to propose participatory methods and not just consultations. Even the research can benefit from these choices, nowadays widely systematized in methodologies. Then we await only the political will to build proactive participation's scenarios.

6 Conclusions

Both networks have established important paths on SD, in their CVs: in the globalization and environmental science perspective at the VIU and for landscape's study with the participatory methods at UNISCAPE. The results were reflected from global to the local level, as in the case of Venice, where highly specific courses are activated, pertaining the lagoon's ecosystems. The interdisciplinary contributions develop the climate change impact's thematic, and mitigations and adaptation's strategies, considering ecosystems services benefits for local populations and indirectly at the global scale. The UNISCAPE network wonders what are the landscapes commons goods, how we could safeguard natural resources, also designing possible new forms of environmental governance. Then in Italy, local and foreign partnerships and network creation have implemented the ESD, making also new contributions, generating awareness and sensitivity. At the same time, the natural capital and ecosystem services education still finds barriers, or a reduced consideration corresponding to its importance. To enhance capacity building in this area, that takes into account the interdisciplinary approaches to the environment, we suggest to work even along a humanistic perspective, reflecting on public values, to approach and better communicate these two worlds: the science, and the decision-makers and policymaker. So we believe that prospects for cooperation, which both networks gave themselves, are building a new environmental governance. They are giving trust to stakeholders, indicating those paths towards sustainability, which, already tracked by theory require finally its practical implementation in all countries.

References

Boydell L, Hoggett P, Rugkåsa J, Cummins AM (2008) Intersectoral partnerships, the knowledge economy and intangible assets. Policy Politics 36(2):209–224

Brochier F,Giupponi C, Sors J (2001) Integrated coastal zone management in the Venice area, potentials of the integrated participatory management approach. FEEM Venice. The Fondazione Eni Enrico Mattei, Note di Lavoro Series, 099. http://www.feem.it/getpage.aspx?id=935&sez=Publications&padre=73. Last accessed: 04/05/2014

Brundtland GH, WCED (1987) Our common future. World commission on environment and development. Oxford University Press, Oxford, New York, 383p

Commission of European Communities, Green paper (2001) Promoting a European framework for corporate social responsibility. /COM/2001/0366 final/, http://eur-lex.europa.eu/LexUriServ/LexUriServ.do?uri=COM:2001:0366:FIN:en:PDF. Last accessed 04/05/2014

Costanza R, D'arge R, De Groot R, Farber S, Grasso M, Hannon B, Limburg K, Naeem S, O'Neill RV, Paruelo J, Raskin RG, Sutton P, Van Den Belt M (1997) The value of the world's ecosystem services and natural capital. Nature 387(6630):253–260

Costanza R, Kubiszewski I, Giovannini E, Lovins H, McGlade J, Pickett KE, Ragnarsdottir KV, Roberts D, De Vogli R, Wilkinson R (2014) Time to leave GDP behind. Nature 505:283–285

Council of Europe (2000) European landscape convention. European treaty series, 176, Strasbourg. http://conventions.coe.int/Treaty/Commun/QueVoulezVous.asp?NT=176&CM=8&CL=ENG. Last accessed: 05/05/2014

Critto A, Agostini P (2009) Using multiple indices to evaluate scenarios for the remediation of contaminated land: the Porto Marghera (Venice, Italy) contaminated site. Environ Sci Pollut Res 16(6):649–662

Dilling PFA (2011) Stakeholder perception of corporate social responsibility. Int J Manage Mark Res 4(2):23–34

Gazzetta Ufficiale della Repubblica Italiana n. 241 of 14 Oct 1939, L. 29 June 1939, n. 1497 "Protezione delle bellezze naturali". http://www.gazzettaufficiale.it/. Last accessed: 05/05/2014

Hansen J, Lehmann M (2006) Agents of change: universities as development hubs. J Clean Prod 14(9–11):820–829

Hesser W, Siedersleben W (2007) Standardization goes East. The European-Asian academic network, international and multimedia based. Iso Focus, International Organization for Standardization. http://www.iso.org/iso/home/standards/standards-in-education/education_materials-list/educational_materials-detail.htm?emtype=paper&emid=18. Last accessed: 05/05/2014

IPCC (2007) Climate Change 2007: impacts, adaptation and vulnerability. Contribution of working group II to the fourth assessment report of the intergovernmental panel on climate change [Parry ML, Canziani OF, Palutikof JP, van der Linden PJ, Hanson CE (eds)] Cambridge University Press, p 976

Karnani Aneel G (2012) Corporate social responsibility does not avert the tragedy of the commons —case study: Coca-Cola India. Ross School of Business at the University of Michigan paper 1210. http://ssrn.com/abstract=2354022. Last accessed: 05/05/2014

Knight J, Gunatilaka R (2010) The Rural–Urban divide in China: income but not happiness? J Dev Stud 46(3):506–534

Mader M, Mader C, Zimmermann FM, Görsdorf-Lechevin E, Diethart M (2013) Monitoring networking between higher education institutions and regional actors. J Clean Prod 49:105–113

Oles T, Hammarlund K (2011) The European landscape convention, wind power, and the limits of the local: notes from Italy and Sweden. Landscape Res 36(4):471–485

Peake S, Smith J(2009) Climate change from science to sustainability. Oxford University Press, p 291

Plummer R, Crona B, Armitage DR, Olsson P, Tengö M, Yudina O (2012) Adaptive comanagement: a systematic review and analysis. Ecol Soc 17(3) art. 11

Roca J (2002) The IPAT formula and its limitations. Ecol Econ J 42(1):1–2

Russo AP, Arias Sans A (2009) Student communities and landscapes of creativity. Eur Urban Reg Stud 16(2):161–175

Sedlacek S (2013) The role of universities in fostering sustainable development at the regional level. J Clean Prod 48:74–84

Teofili C, Clarino R (2008) Riconquistare il paesaggio. La Convenzione Europea del Paesaggio e la Conservazione della Biodiversità in Italia. WWF Italia, MIUR Italia, p 368. http://www.aicare.it/documenti/cdd-documenti/riconquistare-il-paesaggio-wwf.pdf. Last accessed: 05/05/2014

Venice International University (VIU) (2011) Newsletter. Strategic Environmental Assessment and Risk Management, Venice, p 45

Venice International University (VIU) (2012) Newsletter. Sino-Italian Cooperation Program Environmental Training Community, Venice, p 70

Venice International University (VIU) (2012) Report on activities, Venice, p 51

UNISCAPE, Conference Materials (2010) Living landscape, The European landscape convention in research perspective [Bandecchi & Vivaldi (eds)], vol 2, 500p

Wing Chan K (2011) Internal migration in China: trends, geography and policies. In United Nations Department of Economic and Social Affairs Population Division, Population Distribution, Urbanization, Internal Migration and Development: An International Perspective. UN Publication, pp 81–108. http://www.un.org/esa/population/publications/PopDistribUrbanization/PopulationDistributionUrbanization.pdf. Last accessed: 04/26/2014

Zerilli F (2010) The rule of soft law: an introduction. Focaal. J Global Hist Anthropol 56:3–18

Websites

GRI, Global Reporting Index Guidelines: http://www.globalreporting.org. Last accessed: 05/04/2014
GDP, Global Development Product: http://www.beyond-gdp.eu/. Last accessed: 05/04/2014
Uniscape: http://www.uniscape.eu. Last accessed: 05/04/2014
SDSN, Sustainable Development Solutions Network: http://www.unsdsn.org. Last accessed: 05/04/2014
VIU, Venice International University: http://www.univiu.org. Last accessed: 05/04/2014

Author Biography

Carolina Collaro is an architect and teacher, Master in Territorial Planning and Real Estate at Turin Polytechnic, Italy, since more than 10 years involved in research and education's field for sustainable development at the global and local scale. She has a strong background in this field at the international level, acquired in prestigious universities: such as the EPFL and the University of Louvain-la-Neuve, and also in U.S. and France. Visiting scholar at the Institute for advanced studies on science technology and society in Graz Austria, she is Ph.D candidate at Nova Gorica University, in Economics and Techniques of Environmental and Architectural Heritage Conservation, a post graduated program with IUAV-Venice University, Italy. The core of her researches is the adaptation strategy to climate change and ecosystem services' conservation's relevance for cultural landscapes and protected areas. She collaborates with UNISCAPE, Universities' consortium for European Landscape Convention Implementation, and IUCN-CEM Commission.

What Are the European Legal Duties to Conserve Biodiversity in University Campuses?

Konstantinos Tzoulas, Philip James, Elizabeth Price and Phil Wheater

Abstract

Biodiversity is a key element of sustainable development in university campuses. However, integrating biodiversity in campuses requires strategic planning, beyond minimum compliance with protected species and habitats legislation. This leads to the questions: which university functions impact on biodiversity and what obligations are there under European environmental law for universities to consider biodiversity strategically? University functions and their consequent impacts were classified thematically into four and seven categories respectively. These categories were used to systematically search the Environmental Legislation Update Service and EUR-Lex for relevant legislation, which was also classified. Universities undertake capital projects, building alterations, grounds maintenance, and outdoor activities. These functions may cause loss, damage, disturbance, introductions, pollution to, and overuse of, biodiversity. Legislation applying to these impacts spans wildlife, plant health, planning, and pollution prevention disciplines. The interdisciplinary legal framework for biodiversity presents compliance and integration challenges, such

K. Tzoulas (✉) · E. Price
School of Science and the Environment, Manchester Metropolitan University, John Dalton Building, Chester Street, Manchester M1 5GD, UK
e-mail: k.tzoulas@mmu.ac.uk

P. James
School of Environment and Life Sciences, University of Salford, Peel Building, Salford M5 4WT, UK

P. Wheater
Faculty of Science and Engineering, Manchester Metropolitan University, John Dalton Building, Chester Street, Manchester M1 5GD, UK

W. Leal Filho et al. (eds.), *Integrating Sustainability Thinking in Science and Engineering Curricula*, World Sustainability Series, DOI 10.1007/978-3-319-09474-8_30

as overlooking legislation or duplicating efforts. This article will help those involved in university management, teaching and research to identify and integrate in their work the relevant legal obligations on biodiversity.

Keywords
Nature conservation • Urban ecosystems • Legal obligations • Legislation • Green infrastructure

1 Introduction: Biodiversity and Sustainable Development in University Campuses

Biodiversity conservation is central to sustainable development. However, biodiversity is often overlooked or not considered as a priority within the environmental management systems of universities (Wright and Wilton 2012; Dixon et al. 2007). Overlooking biodiversity may result in non-compliance with relevant legislation and in missed opportunities to promote a range of ecosystem contributions to sustainable development. This chapter systematically identifies the impacts that universities may have on biodiversity and the related European legal obligations. Moreover, this chapter outlines strategic opportunities to integrate nature conservation in universities' campus management holistically.

University and college campuses in the UK cover an area of 380 km^2 and include a variety of urban, peri-urban and rural habitats and species (Dixon et al. 2007). Therefore, universities have a role to play in nature conservation and in providing urban and peri-urban ecosystem services. Conservation of urban species and habitats could be enhanced through green infrastructure planning comprising greenways and green spaces as well as green walls, facades and roofs (Sadler et al. 2011; Tzoulas et al. 2007). The concept of greenways involves linear features lined with vegetation that form networks of paths (Walmsley 2005). Planning of university campuses could be based on creating greenways interlinking green spaces for walking and cycling between buildings (Balsas 2003); and on the principles of green infrastructure planning (i.e. integrating publicly and privately owned green spaces and promoting multifunctional ecosystem benefits; Tzoulas et al. 2007) and land use complementation (i.e. clustering together different green spaces to increase habitat and promote ecological processes in cities; Colding 2007). Importantly university campuses could be integrated in ecological networks of both local and European designated and non-designated sites and species for nature conservation.

While, additional research is needed to establish the ecological performance of green roofs (Berardi et al. 2014), green infrastructure planning (Tzoulas et al. 2007) and land use complementation (Colding 2007), as well as their socioeconomic and other environmental benefits, there is enough evidence to suggest that biodiversity on university campuses would bring about multifunctional benefits. These benefits are often referred to as ecosystem services and are grouped into four categories (i.e. supporting, provisioning, regulating and cultural; Millennium Ecosystem Assessment

2005). In order to support biodiversity conservation on campuses in the UK the Higher Education Funding Council for England (HEFCE) and the Environmental Association for Universities and Colleges (EAUC) have developed a practical guide for university estate managers (Dixon et al. 2007). Additionally the Learning and Skills Council, UK, in its sustainable development strategy identifies biodiversity conservation as a key area of actions and principles for higher education (Learning and Skills Council 2005).

Traditionally universities, along with businesses and corporate organisations, have seen biodiversity as a resource to be exploited and its protection as an additional legal constraint (Houdet et al. 2012). Corporate responses to biodiversity could take one of four forms or combinations: (a) securing contribution to production and/or sales; (b) avoiding costs of disservices; (c) mitigating impacts on biodiversity for other cultural reasons; and (d) assessing legal compliance (Houdet et al. 2012). Whilst the latter may be concerned with ensuring minimum compliance, the other responses indicate that organisations may consider biodiversity for strategic reasons. Do universities ensure minimum compliance? If so, can they go beyond minimum requirements and consider biodiversity in terms of wider nature conservation aims and the provision of ecosystem services?

2 Aim and Objective

Two problems are created when universities overlook biodiversity or do not consider it as a priority in their environmental management systems. First, universities may miss opportunities to promote a range of potential socioeconomic and environmental benefits. Second, they may not be compliant with legal obligations regarding biodiversity. Thus, it is important to identify what obligations are emerging for universities to consider biodiversity strategically from European environmental law. This research aimed to identify the European legal duties that universities have to protect biodiversity. The objective comprised a thematic classification of (a) university functions and consequent impacts on biodiversity, and (b) the European legislation that applies to these impacts.

3 Methods

This research comprised a critical review and thematic classification. The former provides a systematic way for interrogating (Hart 1998), and the latter for identifying patterns or themes in, written texts (Robson 1993). In combination these methods have the potential to reveal patterns, conflicts and synergies within and between policies and are thus often used in analysing legal documents. Critical reviews and thematic classifications of European legislation and policy have previously been applied to identify gaps in current academic research (O'Connell and Yallop 2002), adaptive capacity to climate change (Pettersson and Keskitalo 2011), and the compliance of protected areas with international legislation (Mauerhofer 2011).

University services that may impact on biodiversity were identified by a systematic review of the website of Manchester Metropolitan University (UK). This institution was chosen because it came first in the People and Planet Green League Table (2013), which is a UK ranking system for environmental management in universities. The contents of the University's website were searched and details of the services provided by different parts of the institution were noted. The search identified sixty two different services which were then classified in five broad themes—teaching, research, administrative, outreach and campus management—drawn from the institutional mission statement. The subsequent research focussed on campus management services because they could have direct and significant impacts on local biodiversity. The campus management services were then further classified using grounded theory (Lincoln and Guba 1985) until four functions emerged: capital projects, building alterations, grounds management, and outdoor activities.

Potential direct impacts on biodiversity were identified by establishing possible pathways between actions on the ground associated with each campus management function, and effects on species or habitats. For example, a new campus development (capital project), would change the land use, which in turn could impact on species and habitats (Tzoulas et al. 2007). Based on grounded theory (Lincoln and Guba 1985) such possible pathways were thematically classified in seven emergent categories of impacts: loss, damage, disturbance, introductions, pollution, use, and disease associated with species or habitats.

In the UK a not-for-profit organisation called new ground provides an Environmental Legislation Update Service (ELUS; www.legislationupdateservice.co.uk), which is a database of up to date law. ELUS was searched using the key word for each impact category (loss, damage, disturbance, introductions, pollution, use, and disease associated with species or habitats) and campus management function (capital projects, building alterations, grounds management, and outdoor activities). This ensured that legislation covering both functions and impacts were retrieved and collated. The second step focused on identifying the UK legislation that had a European origin by reviewing the preambles of UK legislation that was collated in the first stage. This way, all European Union legislation that had been translated in UK law was identified.

The final stage involved finding and reviewing the originating European legislation using the European Commission's service EUR-Lex (www.eur-lex.europa.eu), which provides access to legal documents including most recent updates and proposals for forthcoming legislation. All relevant European Union legislation that is legally binding across its member states was collected (i.e. Regulations and Directives). Decisions, Recommendations, Opinions, policy documents and international conventions were excluded because they may not apply to all member states or are not legally binding.

The collated Regulations and Directives were critically reviewed to collect legal obligations. Using grounded theory (Lincoln and Guba 1985) these obligations were then classified into eleven emergent categories relating to protected species; protected sites; controlled species (invasive, alien, traded and pests); managed species (wild, ornamental, exceptions); management methods; prohibitions (species, methods, substances, uses or actions); need for authorisation; controlled substances; waste disposal; environment (plants and animals generally); and planning considerations. These categories of obligations are the same for all European Union member states to which the relevant legislation has been translated into national law (Regulations are transposed as they are; but Directives can vary in the form and means of implementation amongst member states).

4 The Body of European Legislation Relating to Biodiversity

The European legal framework for biodiversity conservation comprises thirty four legally binding Regulations (six original, nine amending) and Directives (fifteen original, four amending); and proposals for two forthcoming Regulations and one Directive (Table 1). Thirty four legally binding documents, many of which have long and technically complicated sections and annexes, create a complex legal framework to comply with.

The European legislation for biodiversity can be grouped into four discipline areas: wildlife, spatial planning, plant health, and pollution (Table 1). The Birds [(COM) 2009/147/EC] and Habitats [(COM) 92/43/EEC] Directives are often quoted as the landmark legislation for European nature conservation, perhaps because they are the only two to have obligatory monitoring requirements (Henle et al. 2013), or because they inform much of the other environmental legislation. However, the obligations arising from these two Directives are only part of the legal duties that universities have towards biodiversity.

Pollution legislation that applies to biodiversity is the most fragmented and spread out. The Environmental Liability Directive [(COM) 2004/35/EC] and the Water [(COM) 2000/60/EC], Marine [(COM) 2008/56/EC] and Waste [(COM) 2008/98/EC] Framework Directives are most likely to be overlooked; because they are mainly associated with pollution compliance rather than biodiversity conservation. These findings indicate that the majority of legal obligations relating to biodiversity may be about preventing and mitigating risks and damage to it from pollution, disease or development; rather than incentivising proactive conservation.

5 The Legal Obligations for Biodiversity Conservation

The legal obligations arising from European legislation to protect and conserve biodiversity that apply to universities can be grouped into eleven categories (Table 2; top row). The majority of legal obligations relate to gaining authorisation for interfering with specific species, habitats or processes; to protecting the natural environment including non-protected plants and animals; to following specific permitted management methods; and to prohibitions (Table 2). The Habitats [(COM) 92/43/EEC] and Birds [(COM) 2009/147/EC] Directives introduce the

Table 1 EU biodiversity legislation

Wildlife legislation	***Pollution legislation***
R. *338/97: wildlife trade (basic)*	R. 850/2004: *POPs*
R. *865/2006: wildlife trade (implement)*	• R.1195/2006: amending R. *850/2004*
• R. 100/2008: amending R. *865/2006*	• R.172/2007: amending R. *850/2004*
• R. 791/2012: amending R. *865/2006*	• R.323/2007: amending R. *850/2004*
R. 792/2012: *wildlife trade (permit)*	• R.304/2009: amending R. *850/2004*
R. 587/2013: *wildlife trade (suspensions)*	• R.756/2010: amending R. *850/2004*
D. 92/43/EEC: *habitats*	• R.757/2010: amending R. *850/2004*
D. 2009/147/EC: *birds*	• R.519/2012: amending R. *850/2004*
COM (2013) 620: *alien and invasive species*[a]	R. 528/2012: *biocides*
	D. 2000/60/EC: *water framework*
	D. 2003/4/EC: *access to information*
	D. 2003/35/EC: *public participation*
	D. 2004/35/EC: *environmental liability*
	D. 2008/98/EC: *waste framework*
	D. 2008/56/EC: *marine framework*
Plant health legislation	D. 2009/128/EC: *pesticides*
D. 77/93/EEC: *plant pests* (first)	
• D. 86/546/EEC: amending D. *77/93/EEC*	
• D. 92/103/EEC: amending D. *77/93/EEC*	
D. 91/682/EEC: *ornamental plants*	
D. 93/49/EEC: *ornamental plants (pests)*	
D. 2000/29/EC: *plant pests (second)*	
• D. 2004/102/EC: amending D. *2000/29/EC*	
• D. 2009/118/EC amending D. *2000/29/EC*	
COM (2013) 141: *pests*[a]	***Planning legislation***
	D. 2001/42/EC: *SEA*
	D. 2011/92/EU: *EIA*
	COM (2012) 0297: *amend EIA*[b]

R Regulation; *D* Directive; *POPs* Persistent organic pollutants; *SEA* strategic environmental assessment; *EIA* environmental impact assessment

Notes [a]These are proposals for regulations; [b]this is a proposal for a Directive; all documents can be accessed on www.eur-lex.europa.eu using year and number of legislation

most legal obligations. This explains why these are often cited and at the top of the legal compliance list. However, it would be short sighted to assume that protected species and habitats are the centre of biodiversity legislation. This is because eight pieces of legislation refer to protected species and sites, but fourteen refer to environment including non-protected plants and animals (Table 2).

Table 2 Obligations arising from European biodiversity legislation

European Legislation	EN	SP	CS	MS	ST	M	BN	SB	AU	W	PL
Wildlife trade[a,d]		•		•			•		•		
Habitats[b]		•	•	•	•		•		•		•
Birds[b]		•		•	•		•		•		•
Alien species[a,c]	•		•			•	•		•		
First plant pests[b,e]	•		•			•	•		•		
Ornamental plants[b]				•		•			•		
Ornamental plant pests[b]	•		•	•		•	•		•		
Second plant pests[b,f]	•		•			•	•		•		
Pests[a,c]			•	•		•	•		•		
POPs[a,g,h]	•					•	•	•	•	•	
Biocides[a]	•					•		•	•		
Water Framework[b]	•	•			•	•					
Access to information[b]	•						•				•
Public participation[b]	•						•				•
Environmental liability[b]		•			•	•			•		
Waste Framework[b]	•					•				•	
Marine Framework[b]					•	•		•			•
Pesticides[b]	•					•		•	•		
SEA[b,j]	•	•			•				•	•	•
EIA[b,k]	•	•			•				•	•	•
Amend EIA[b,c]	•	•			•				•	•	•

Abbreviations *EN* environment, flora and fauna generally; *SP* protected species; *CS* controlled species (invasive, alien and pests); *MS* managed species (wild, ornamental, trade); *ST* protected sites; *M* permitted management methods; *BN* banned methods, uses or action and substances; *SB* controlled, restricted substances; *AU* authorisation, notification to/from regulator needed; *W* waste disposal; *PL* must be considered at planning stage; *POPs* Persistent organic pollutants; *SEA* strategic environmental assessment; *EIA* environmental impact assessment

Notes [a]Regulations; [b]Directives; [c]Proposal; [d]includes six Regulations; [e]includes three Directives; [f]includes three Directives; [g]includes eight Regulations; [h]persistent organic pollutants; [i]strategic environmental assessment; [j]applies to universities as major consultees and with regards to protected sites; [k]environmental impact assessment

6 University Functions, Impacts on Biodiversity and Legal Obligations

Capital projects in university campuses often involve master planning and large-scale land development and building work. Consequently, the whole range of legal obligations for biodiversity applies to capital projects (Table 3). To ensure legal

Table 3 University functions and corresponding potential legal obligations

Functions	EN	SP	CS	MS	ST	M	BN	SB	AU	W	PL
Capital projects	•	•	•	•	•	•	•	•	•	•	•
Building alterations			•			•	•				
Grounds maintenance	•	•	•	•	•	•	•	•	•	•	
Outdoor activities	•	•	•					•			

For abbreviations of obligations see Table 2 footnote

compliance it is common to undertake ecological impact assessments, and if necessary mitigation, during the master planning and development of new university campuses.

However, university estate managers may overlook that day-to-day grounds maintenance functions have to comply with almost as many legal obligations as capital projects (Table 3). The use of biocides, heavy machinery and intensive maintenance methods; and dealing with ornamental plants, pests and invasive species make grounds maintenance liable to a number of legal obligations, related to both protected and non-protected species. While the former may be taken into account during maintenance, the latter most often will not.

Universities are prohibited (BN column; Table 4) to impact in any way on protected species or habitats (SP and ST columns, Table 4). If any university functions are likely to cause a significant impact on protected species or habitats the relevant authority must be notified and authorisation gained (AU column; Table 4). Additionally, universities must ensure that day-to-day grounds maintenance follows permitted methods for dealing with introduced species, pollution discharges, handling biodiversity and pests (M column, Table 4). Also, it is important to be aware of the legal obligations relating to dealing with controlled substances and species (SB and CS column respectively, Table 4) as well as managed species (MS column, Table 4).

Table 4 University biodiversity impacts and corresponding potential legal obligations

Impacts	EN	SP	CS	MS	ST	M	BN	SB	AU	W	PL
Loss		•			•		•		•		•
Damage		•			•		•		•		•
Disturbance		•			•		•		•		•
Introductions	•	•	•	•	•	•	•		•		
Pollution	•	•		•	•	•	•	•	•	•	
Use		•	•	•	•	•	•	•	•		
Disease	•	•	•	•	•	•	•		•		

For abbreviations of obligations see Table 2 footnote

7 Discussion: Interdisciplinary Legal Framework

Wildlife legislation has been the main focus of European nature conservation law reviews (Henle et al. 2013; Mauerhofer 2011; Pettersson and Keskitalo 2011; O'Connell and Yallop 2002). Some limited aspects of pollution legislation (e.g. Water [(COM) 2000/60/EC] and Marine [(COM) 2008/56/EC] Framework Directives) and spatial planning policy [e.g. Strategic Environmental Assessment Directive; (COM) 2001/42/EC] appear in reviews (Henle et al. 2013; Pettersson and Keskitalo 2011). However, this research revealed a wider range of wildlife, planning and pollution legislation than previous studies. Also, it revealed that European legislation from the area of plant health has not been covered before. Organisations may assume that enforcement of biodiversity legislation depends on avoiding damage to the species and habitats covered in the Habitats [(COM) 92/43/EEC] and Birds [(COM) 2009/147/EC] Directives (Henle et al. 2013). However, requirements to protect non-listed as well as listed flora, fauna and landscapes exist throughout the plant health, planning and pollution related legislation.

8 Compliance Challenges

More legislation refers to managing impacts on non-listed flora and fauna and of the environment in general than to protected species and sites (Table 2). This suggests that organisations ought to consider both protected and non-protected elements of the natural environment. In the UK, several universities have established campus biodiversity audits and action plans (e.g. University of Brighton, Brunel University, and University of Chester). These plans often audit protected and non-protected species and/or sites, even though the focus tends to be on the former. Information and vertical integration gaps make it difficult to comply with European biodiversity legislation (Henle et al. 2013; Mauerhofer 2011; O'Connell and Yallop 2002). So, complying with duties to protect non-listed species and sites may be unlikely.

Although legislative measures to protect species and sites are necessary they tend to be reactive, static and fragmented (Pettersson and Keskitalo 2011; Kundis Craig 2010). This increases the complexity of biodiversity legislation and reduces its flexibility. The former present challenges to identifying and complying with all relevant obligations and the latter tends to concentrate efforts only on listed species or sites. Consequently, legislative measures alone may not be effective in addressing biodiversity conservation holistically (Pettersson and Keskitalo 2011; Kundis Craig 2010). Additional voluntary, financial and other instruments are needed to complement legislative measures (Pettersson and Keskitalo 2011; Kundis Craig 2010).

9 Integration Challenges

There are opportunities to integrate biodiversity considerations into existing mechanisms rather than duplicating policies (e.g. potential damage to plants, animals or sites integrated in health and safety, risk assessment and control of substances hazardous to health mechanisms). The main challenge in integration is the lack of clear communication between legal, research and operations departments (O'Connell and Yallop 2002).

Biodiversity related issues are mentioned passively but not discussed in detail in existing literature regarding environmental management of universities (Wright and Wilton 2012; Alshuwaikhat and Abubakar 2008). Poorly maintained, unmanaged or wild looking green spaces tend to be perceived negatively by people in urban areas and may be associated with ecosystem disservices (Sadler et al. 2011; Tzoulas and James 2010). However, there could also be strategic reasons for maintaining high quality green spaces on campus.

For example, the Green League Table ranks UK universities according to their environmental performance against eight impact areas: waste, transport, procurement, water, construction, emissions, community, and biodiversity (People and Planet 2013). Green League requires universities to have a policy on biodiversity with time bound targets to improve species and/or habitat diversity, which must be reported annually (People and Planet 2013). These requirements create an incentive for UK universities to integrate biodiversity at a strategic as well as operational level in order to gain a high ranking.

It has been possible to map university functions and impacts and their relevance to different legal obligations for biodiversity protection. The matrices developed can be used for quick referencing relevant duties to consider when different operations are undertaken. Mapping functions and impacts against obligations allows for easy navigation though a complex legal framework. The research presented here could be used in universities across Europe by decision-makers, environmental managers ensuring compliance with, and educators teaching and students studying biodiversity legislation.

10 Conclusion

There is a complex European legal framework for conserving biodiversity that applies to the higher education sector. Legal obligations cover protected habitats and species; non-protected plants, animals and landscapes; controlled plants, animals and substances; and pollution. By covering nature conservation, spatial planning, pollution and plant health legislation this research facilitates a holistic approach to auditing and ensuring compliance with European biodiversity legislation. Moreover, throughout this chapter opportunities for integrating strategically and operationally biodiversity conservation in campus management have been outlined.

Whilst impacts on biodiversity during capital projects are usually considered, and when necessary mitigated, this may not be the case with day-to-day grounds maintenance. Non-compliance with biodiversity legislation is more likely to come from the latter than the former. However, since there is no reporting mechanism for harming un-protected plants or animals such impacts may be overlooked or even knowingly ignored. Further research is needed to establish how biodiversity conservation is integrated in the management of university campuses and the reasons why it may or may not be seen as a priority.

Developing green infrastructure and land use complementation on campus present opportunities to integrate biodiversity conservation at a strategic and operational level across universities. For example, applying land use complementation principles from the planning stage and throughout the life time of a new university campus (Colding 2007). However, this would require the integration of biodiversity issues throughout key university policies. Additional research is needed to establish the role of universities have to play in local, regional and international biodiversity conservation though the management of their campuses.

The research presented here offers a starting point for ensuring compliance with European biodiversity legislation. However, like all publications this chapter can only provide a snapshot in time. Thus, continuous updating of legislation necessitates ongoing reviews of the legal obligations reported in this chapter. Even if universities were fully compliant with legislation relating to protected sites and species it is likely that they could overlook impacts to other non-listed plants and animals. University policies need to explicitly identify whether the aim is compliance with biodiversity legislation or contributing to nature conservation. Whilst the latter could ensure the former, the opposite may not be the case.

References

Alshuwaikhat HM, Abubakar I (2008) An integrated approach to achieving campus sustainability: assessment of the current campus environmental management practices. J Clean Prod 16:1777–1785

Balsas CJL (2003) Sustainable transportation planning on college campuses. Transp Policy 10:35–49

Berardi U, Ghaffarian Hoseini AH, Ghaffarian Hoseini A (2014) State-of-the-art analysis of the environmental benefits of green roofs. Appl Energy 115:411–428

Colding J (2007) Ecological land-use complementation' for building resilience in urban ecosystems. Landscape Urban Plan 81:46–55

Dixon M, Webb M, Cook G, Forster J (2007) Biodiversity on campus: an EAUC practical guide. The Environmental Association for Universities and Colleges, Cheltenham

Hart C (1998) Doing a literature review: releasing the social science research imagination. SAGE Publications, London

Henle K, Bauch B, Auliya M, Külvik M, Pe'er G, Framstad E, Schmeller D (2013) Priorities for biodiversity monitoring in Europe: a review of supranational policies and a novel scheme for integrative prioritization. Ecol Ind 33:5–18

Houdet J, Trommetter M, Weber J (2012) Understanding changes in business strategies regarding biodiversity and ecosystem services. Ecol Econ 73:37–46

Kundis Craig R (2010) 'Stationary is Dead'—long live transformation: five principles for climate change adaptation law. Harvard Environ Law Rev 34:9–74

Lincoln YS, Guba EG (1985) Naturalistic inquiry. SAGE Publication, Newbury Park

Learning and Skills Council (2005) From here to sustainability: the Learning and skills council's strategy for sustainable development. Learning and Skills Council, Coventry

Mauerhofer V (2011) A bottom-up 'convention-check' to improve top-down global protected area governance. Land Use Policy 28:877–886

Millennium Ecosystem Assessment (2005) Ecosystems and human well-being: synthesis. Island Press, Washington DC
O'Connell M, Yallop M (2002) Research needs in relation to the conservation of biodiversity in the UK. Biol Conserv 103:115–123
People and Planet (2013) People & planet green league guide 2013: driving UK universities' transition to a fair and sustainable future. People & Planet, Oxford
Pettersson M, Keskitalo ECH (2011) Adaptive capacity of legal and policy frameworks for biodiversity protection considering climate change. Land Use Policy 34:213–222
Robson C (1993) Real world research: a resource for social scientists and practitioner-researchers. Blackwell Publishers, Oxford
Sadler J, Bates A, Donovan R, Bodnar S (2011) Building for biodiversity: accommodating people and wildlife in cities. In: Niemelä J, Breuste JH, Elmqvist T, Guntenspergen G, James P, McIntyre NE (eds) Urban ecology. Patterns, processes and applications. Oxford University Press, Oxford, pp 286–97
Tzoulas K, James P (2010) Peoples' use of, and concerns about, green space networks: a case study of birchwood, Warrington New Town, UK. Urban Forest Urban Greening 9:121–128
Tzoulas K, Korpela K, Venn S, Yli-Pelkonen V, Kazmierczak A, Niemelä J, James P (2007) Promoting ecosystem and human health in urban areas using green infrastructure: a literature review. Landscape Urban Plan 81:167–178
Walmsley A (2005) Greenways: multiplying and diversifying In the 21st century. Landscape Urban Plan 76:252–290
Wright TSA, Wilton H (2012) Facilities management directors' conceptualizations of sustainability in higher education. J Clean Prod 31:118–125

Author Biographies

Konstantinos Tzoulas is Lecturer in Environmental Management at the School of Science and the Environment at Manchester Metropolitan University. His research focuses on urban green infrastructure planning and management and he is the Chairman of the University's Biodiversity Management Group.

Philip James Professor of Ecology in the School of Environment and Life Sciences at the University of Salford. He is a board member of the Local Nature Partnership for Greater Manchester and Chair of the Rivers Return Partnership (the River Irwell Catchment Partnership).

Dr Elizabeth Price is a Principal Lecturer at the School of Science and the Environment at Manchester Metropolitan University. Her research focuses on habitat restoration and she teaches ecology, ecosystem management and restoration. Liz is Head of Division of Geography and Environmental Management.

Phil Wheater is Professor of Environmental and Geographical Sciences and Dean of the Faculty of Science and Engineering at Manchester Metropolitan University. Much of his research focuses on human impacts on the environment (including on biodiversity), especially in the contexts of countryside management and urban ecology.

The Benefits and Challenges of Developing and Implementing an Environmental Management System Using a Participatory Approach: A Case Study of Manchester Metropolitan University, UK

Helena Tinker and Konstantinos Tzoulas

Abstract

Participatory approaches to developing and implementing environmental management in universities could be central in the successful integration of indirect aspects such as teaching and research. The aim of this research was to develop additional insights into the benefits and challenges associated with a participatory approach to environmental management. The objective was to undertake a case study of the participatory approach to environmental management followed by Manchester Metropolitan University, UK. The approach reported here comprised four stages: (a) an environmental review, (b) a stakeholder analysis, (c) participatory meetings and workshops, and (d) synthesis. The outcomes of this approach were a comprehensive management system covering twelve environmental aspects, including teaching and research; a management group chaired by a lead stakeholder for each aspect; a high level environmental strategy board; and university wide networks in facilitating collaboration in sustainability teaching and research. Benefits of increased capacity and reduced conflict were greater than the effort required for overcoming challenges such as securing commitment from and providing support to stakeholders. This chapter will be useful to universities planning to

H. Tinker (✉)
Environment Team, Manchester Metropolitan University, Cavendish North Building, Cavendish Street, Manchester, M15 6BG, UK
e-mail: h.tinker@mmu.ac.uk

K. Tzoulas
School of Science and the Environment, Manchester Metropolitan University, John Dalton Building, Chester Street, Manchester, M1 5GD, UK

W. Leal Filho et al. (eds.), *Integrating Sustainability Thinking in Science and Engineering Curricula*, World Sustainability Series, DOI 10.1007/978-3-319-09474-8_31

implement ISO 14001 and/or those that already have an environmental management system and are wishing to expand the scope to include teaching and research.

Keywords
Environmental aspects · ISO 14001 · Stakeholder engagement · Teaching · Research · Higher education

1 Introduction

Universities and colleges across the world are increasingly implementing environmental management systems (Clarke and Kouri 2009; Disterheft et al. 2012; Noeke 2000). The two best known formally certified systems are the international ISO 14001 standard and the European Eco Management and Audit Scheme (EMAS; Disterheft et al. 2012). Implementing the ISO 14001 standard or EMAS requires organisations to identify the environmental aspects and impacts of their activities, products and services within a defined scope of their environmental management system (British Standards Institution 2004; EC 2009). Environmental aspects are the organisation's activities, products or services that can interact with the environment, and impacts are any consequent changes to the environment whether adverse or beneficial, wholly or partially resulting from any environmental aspect (EC 2009; British Standards Institution 2004). Universities can have direct and indirect environmental aspects (Sammalisto and Brorson 2006). Direct environmental aspects are associated with activities, products and services for which universities have direct management control (EC 2009). Direct aspects often include waste generation, emissions and discharges to air, water and land (Clarke and Kouri 2009). Indirect environmental aspects result from the interaction of an organisation with third parties (EC 2009). For universities indirect environmental aspects include investment practices and purchasing of products, goods and services (Clarke and Kouri 2009), as well as increasing knowledge capital through teaching and research activities (Von Oelreich 2004). Some authors have argued that because teaching and research activities are some of the largest indirect environmental aspects of universities they ought to be integrated in their environmental management systems (Clarke and Kouri 2009; Sammalisto and Brorson 2006; Simkins and Nolan 2004). However, the scope of environmental aspects that are covered in relevant management systems of universities varies significantly (Disterheft et al. 2012; Clarke and Kouri 2009; Sammalisto and Brorson 2006; Sammalisto and Arvidsson 2005; Simkins and Nolan 2004) and tends to overlook the indirect aspects of teaching and research (Disterheft et al. 2012; Clarke and Kouri 2009). Moreover, even when universities are including teaching and research in their environmental management system they may not be taking the practical steps for full integration into core business practices (Sammalisto and Arvidsson 2005). The drivers and barriers briefly outlined below may help explain the reluctance of universities to fully integrate teaching and research in their environmental management systems.

Bennett and James (1999) suggested that environmental management systems in universities may evolve over three stages, each with different drivers for change. First, driven mainly by cost reduction and legal compliance considerations, universities may focus on operations and direct environmental aspects. Second, driven by the need for stakeholder management, universities may start considering indirect aspects such as procurement or investments. Finally, factors such as stakeholder partnerships, lifecycle management and contributing to sustainable development drive universities to include teaching and research in their environmental management systems. Moreover, Alshuwaikhat and Abubakar (2008) outlined a list of fifteen international declarations that create a policy framework for universities to integrate teaching, research and campus operations in their contributions to sustainable development. However, international declarations may not apply to all universities and institutions may not consider them as operational priorities. Therefore, the drivers for universities to include teaching and research in their environmental management system are voluntary.

There are two initial barriers in integrating teaching and research in the environmental management systems of universities. First, environmental management of universities is usually coordinated by estates or facilities departments (Simkins and Nolan 2004), which may not be communicating with academic departments or research institutes. Second, environmental management coordinators in these departments may not have the necessary skills to engage with lecturers and researchers (Sammalisto and Brorson 2006). These barriers make it challenging for estates or facilities departments to instigate, drive and manage actions concerning teaching and research (Sammalisto and Brorson 2006). However, evidence suggests that barriers to integrating teaching and research can be overcome by using a participatory approach to developing and implementing environmental management systems in universities (Disterheft et al. 2012, 2014).

Disterheft et al. (2012) defined participatory approaches, in the context of universities, as the involvement of academic and support staff as well as students in institutional change processes. In their study of environmental management systems across forty seven European universities Disterheft et al. (2012) found that 60 % followed participatory approaches ranging from simple information sharing (low participation) to joint projects and collaborative visioning workshops aimed at empowering and building the capacity of stakeholders to take ownership of the required changes (high participation). Furthermore, they found that universities implementing the EMAS were more likely to follow a high participation approach than those implementing ISO 14001 (Disterheft et al. 2012). These findings suggest that the guideline requirements of different standards most probably influence the approach to environmental management that a university may take. However, the potential benefits and challenges of participation could also be important factors in determining environmental management approaches (Disterheft et al. 2012, 2014; Collins et al. 2005). Further understanding is needed of the potential benefits and challenges of participation. The aim of the research presented in this chapter was to develop additional insights into the benefits and challenges associated with a participatory approach to environmental management. The objective was to undertake a case study of the participatory approach to environmental management followed by Manchester Metropolitan University, UK.

2 Methods

Two reasons make Manchester Metropolitan University a case study that merits investigation. First, it achieved first position in the UK's People and Planet Green League Table in 2013. This is a system that ranks universities in the UK according to thirteen criteria (e.g. waste, energy, teaching, engagement and investment; People and Planet 2013). Second, Manchester Metropolitan University is working towards certification to the ISO 14001 standard and is using a bottom-up participatory approach to integrating teaching and research within its environmental management system. This makes Manchester Metropolitan University an unusual case study because European universities implementing ISO 14001 are mostly characterised by top-down approaches and limited participation (Disterheft et al. 2012).

3 Environmental Review

Environmental management starts with a review of the organisation's activities, products and services to identify which significant environmental aspects to include in the scope of its system (Simkins and Nolan 2004). The review was undertaken by the environment team at Manchester Metropolitan University, which was positioned within the operations department. The environmental review comprised audits of the university's activities, products and services and the associated environmental aspects (direct and indirect). The significance of each environmental aspect was then scored against three criteria i.e. whether the aspect (a) was associated with a legal obligation or existing voluntary commitment (if yes it scored fifteen points; if no it scored zero points); (b) presented a concern to the university or its stakeholders (if yes it scored ten points; if no it scored zero points); and (c) it's impact was estimated to being minimal (scored one point), minor (scored two points), moderate (scored three points) or major (scored four points). Significant aspects were defined as those scoring a minimum of three points (i.e. moderate impact, not related to a commitment or concern). The environmental review revealed twelve significant direct and indirect environmental aspects, which were then grouped into four themes reflecting the common functions of universities i.e. operations, administration, community and academic (the latter including teaching and research; Cortese 2003).

4 Stakeholder Analysis

For each environmental aspect a lead stakeholder was identified. Lead stakeholders had to meet at least two of three requirements i.e. have (a) good knowledge of, (b) strategic influence; or (c) top management responsibilities for, each particular environmental aspect. Potential lead stakeholders were identified by the environment team through a review of the roles and responsibilities of posts relevant to each environmental aspect. Once potential lead stakeholders were identified informal one to one meetings with the environment team were arranged to ensure their commitment and to help them build the business case for managing the relevant environment aspect.

5 Participation

Having secured commitment, formal one to one meetings were held between the environment team and each lead stakeholder. These meetings were designed to allow lead stakeholders to (a) develop ownership of, and (b) take responsibility for, the strategic direction and management of the environmental aspect(s) that they were concerned with. Each lead stakeholder was then asked to form a management group to develop the action plans and procedures for managing their environmental aspect(s). Stakeholders who were invited to join the management groups were identified by a snowball approach (i.e. first stakeholder identified the first group of other stakeholders, who then identified the second group and so on). In order to join a management group additional stakeholders had to meet two requirements i.e. (a) have knowledge of the particular environmental aspect, and (b) be involved in some part of its management. This way management group members as well as relevant consultees were identified from across the university.

In addition to the management groups two informal university wide networks were established by the relevant lead stakeholders and with support from the environment team. A sustainability research network was established to facilitate academic collaboration; and a sustainability teaching group to develop support tools and programmes for embedding sustainable development in curricula. The former involved researchers across the university and the latter academics from relevant departments of the university (business school, school of science and the environment, centre of excellence in teaching and learning). Top management commitment was secured through the establishment of an environmental strategy board comprising managers across the university with significance influence or control of relevant budgets. The environment team with help from the director of finance identified these managers, engaged them to secure their commitment and developed a proposal to the university's executive for the establishment of the environmental strategy board. Recognising the requirements of senior management commitment for ISO 14001 the university's executive approved the establishment of the board.

6 Synthesis of University's Environmental Management System

Each management group met individually and discussed and finalised its own terms of reference, objectives, targets, key performance indicators, registers of specific aspects and impacts and of legal compliance, and action plans. These contributions from each management group were then collected and synthesized by the environment team. The objectives, targets and key performance indicators were collated in a draft environmental strategy. The remaining documents formed the records for policy commitments and performance monitoring. The draft environmental strategy was finalised and agreed through a collaborative two hour workshop involving the environment team and all lead stakeholders. The final strategy was presented to and approved by the executive of the university. Finally, the environment team compiled the environmental aspects, policy commitments, strategy and performance monitoring records into the university's environmental management system.

7 Findings: Environmental Aspects

With the exceptions of biodiversity (Wright and Wilton 2012) and teaching and research (Disterheft et al. 2012; Clarke 2006) that are often not included in the scope, all remaining environmental aspects identified by the review are commonly addressed in the environmental management of universities (Table 1; Disterheft et al. 2012; Clarke and Kouri 2009; Sammalisto and Brorson 2006; Sammalisto and

Table 1 Direct and indirect environmental aspects

Environmental aspects	Direct	Indirect
Energy management[1]	•	
Emissions and discharges[1]*	•	
Water management[1]	•	
Waste management[1]	•	
Transport management[1]	•	
Capital programmes[1]**	•	
Buildings maintenance[1]	•	
Biodiversity management[1]	•	
Sustainable procurement[2]		•
Teaching sustainability[3]		•
Sustainability research[3]		•
Engagement[4]***		•

Themes 1 Estate and operations, *2* administrative, *3* academic, *4* community; *Notes* *includes to land, water and air; **includes contract specification and management; ***includes staff, students and neighbouring communities and is a cross cutting activity

Arvidsson 2005; Simkins and Nolan 2004). Four of the twelve environmental aspects were indirect, and the environment team had little or no control over these. These indirect aspects were procurement, teaching, research and engagement (Table 1).

8 Lead Stakeholders

Nine of the twelve lead stakeholders were identified within operational departments of the university and three from the academic community (Table 2). The deputy vice chancellor (student success) and the head of research (academic post at professorial level) were recognised as lead stakeholders for teaching and research respectively. At senior management level there was a lack of expertise in urban biodiversity, and consequently for this activity an academic with the appropriate research expertise was selected as the lead stakeholder. The remaining key stakeholders included two heads of services (property management and procurement) and three senior managers from the environment team (Table 2). These lead stakeholders, with appropriate support from the environment team, were responsible for developing the environmental management system for their respective environmental aspect.

9 Participation

Additional stakeholders that were involved in the development of the environmental management system, both as management group members and/or as consultees on particular aspects, ranged from estates and operation departments, to

Table 2 Lead stakeholders

Environmental aspect	Lead stakeholder
Energy management[1]	Head of property services
Emissions and discharges[1]*	Head of property services
Water management[1]	Head of property services
Waste management[1]	Manager of waste and recycling
Transport management[1]	Manager of the travel plan
Capital programmes[1]**	Head of property services
Buildings maintenance[1]	Head of property services
Biodiversity management[1]	Academic lead (biodiversity and green infrastructure)
Sustainable procurement[2]	Head of procurement
Teaching sustainability[3]	Deputy vice chancellor (for student success)
Sustainability research[3]	Head of research
Engagement[4]***	Manager of sustainability engagement

Themes *1* Estate and operations, *2* administrative, *3* academic, *4* community; *Notes* *includes to land, water and air; **includes contract specification and management; ***includes staff, students and neighbouring communities and is a cross cutting aspect

Table 3 Additional stakeholders

Other relevant internal stakeholders	Engaged in theme
Academics teaching relevant subjects across the university	3
Campus management and estates	1
Centre for excellence in teaching and learning	3
Communications and marketing	4
Continuing professional development	3
Deans of faculties	3
Deputy vice chancellor for strategic planning and research	3
Employability initiatives	4, 3
Health and safety	1
Human resources	2
Procurement and finance	1, 2
Key suppliers and contractors	1, 2, 3
Pro vice chancellor for students	3
Research and knowledge exchange	3
Research centres and groups	1, 3
Student halls management services	1
Students	1, 3, 4
Students union	1, 3, 4
Technical services teams (labs, workshops)	1, 2
Volunteering initiatives	4
External partners and local community	1, 2, 3, 4

Themes 1 Estate and operations, *2* administrative, *3* academic, *4* community; *Note* for environmental aspects related to each theme refer to Table 1

academic teaching and research departments, and the student union (Table 3). The environmental strategy board comprised representatives from estates and facilities, academic faculties, strategic planning, finance, and student success. The board's responsibilities included providing strategic support for the environmental management system; ensuring its integration with the university's strategic planning processes and core business; and overseeing implementation and reporting progress, opportunities and risks to the university's executive.

10 The Environmental Management System

The university's environmental management system shows the interrelationships between environmental aspects, policy commitments, strategy and performance (Fig. 1). For each environmental aspect, there are policy commitments, objectives with targets and key performance indicators, and action plans for delivering improvements. Progress on action plans and the key performance indicators are

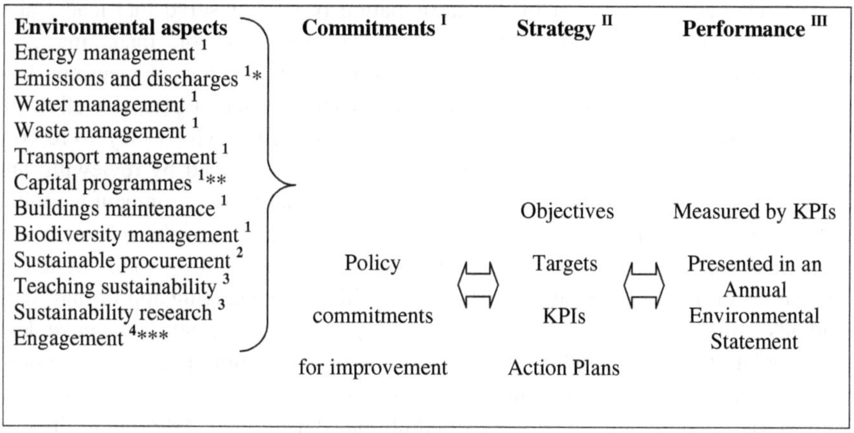

Fig. 1 The environmental management system. *Themes 1* Estate and operations; *2* Administrative; *3* Academic; *4* Community; *Notes* *includes to land, water and air; **includes contract specification and management; ***includes staff, students and neighbouring communities; (*I*) each environmental aspect has a number of policy commitments; (*II*) objectives, targets, key performance indicators (*KPIs*) and action plans for each environmental aspect form the strategy; (*III*) performance in achieving the targets is monitored annually through the key performance indicators; *arrows* indicate that there are two way influences between performance, strategy and commitments

monitored annually and reported in the annual environmental statement (Fig. 1). Performance monitoring and reporting informs the setting of new commitments, objectives and targets (this feedback loop is indicated by the two way arrows on Fig. 1).

11 Discussion: The Participatory Approach

Engagement of a wide range of stakeholders, commitment from senior management and executives, collaboration and shared ownership of an environmental vision and strategy are critical success factors for participatory processes in environmental management (Disterheft et al. 2014). The outcomes of the environmental review formed the foundation for developing the participatory approach followed by Manchester Metropolitan University. First, the four themes (i.e. Estate and operations, administrative, academic and community), in which the environmental aspects were grouped, helped clarify management responsibilities for each aspect. Second, the twelve environmental aspects of the university (i.e. energy, emissions, water, waste, transport, capital programs, buildings, biodiversity, procurement, teaching, research and engagement) determined the stakeholders that were engaged in the next stage of the process. So, the review resulted in a comprehensive description of the institution's functions (themes) and environmental aspects, which in turn informed the identification of all relevant stakeholders.

Two elements at the start of the participation process ensured that lead stakeholders developed ownership of, and took responsibility for, managing their respective environmental aspect(s). First, during the initial formal meetings lead stakeholders were engaged in identifying draft objectives and targets, as well as the priorities relating to managing the environmental aspect(s), which they were concerned with. These draft objectives and targets were developed by reviewing and discussing the environmental review documentation that the environment team had undertaken. This way the stakeholders developed ownership of the relevant environmental aspect(s) from the outset. Second, lead stakeholders were invited to finalise the draft objectives and targets by consulting and engaging appropriate staff, students and/or external organisations as they saw appropriate. This gave lead stakeholders the responsibility for both the strategic direction and management of the environmental aspect (i.e. objectives, targets and operational actions plans) as well as for engaging a wide network of additional stakeholders. Offering to the lead stakeholders ownership of, and responsibility for, managing the university's environmental aspect(s), and providing the necessary support from the environment team, ensured that stakeholders were empowered to influence change. This approach was particularly important for including teaching and research in the university's environmental management system because the environment team at the beginning of the process had limited working relationships with the academic community.

The management groups, the university wide research and teaching networks and the environmental strategy board were all established through a bottom-up approach (i.e. key stakeholders involving their networks). This bottom up-approach to engagement and participation was important to ensure that all lead and additional stakeholders across the university had ownership of the relevant environmental aspects. Importantly, the establishment of the environmental strategy board ensured top level commitment and addressed in advance the forthcoming changes to the ISO 14001 standard requiring the involvement of top management and the integration of environmental management into core business processes and strategy (International Standards Organisation 2013).

12 The Benefits of the Participatory Approach

Stakeholder participation in environmental management could bring about a number of benefits associated with inclusive decision-making (Narain Mathur et al. 2008). The participatory approach that Manchester Metropolitan University followed in developing its environmental management system highlighted five benefits. First, stakeholder engagement increased ownership and shared responsibility of the environmental management system, which facilitated policy and strategy development and delivery. Second, since the stakeholders were engaged in producing the environmental management policy commitments, objectives, targets and KPIs there was reduced conflict and resistance when these were implemented.

Third, the participatory approach facilitated the inclusion of teaching and research in the scope of the environmental management system. This is because it allowed relevant academic and non-academic stakeholders to come together and discuss shared goals and actions. Moreover, integrating teaching and research in the environmental management system not only contributed to the university's efforts to managing one of its largest indirect aspects (Clarke and Kouri 2009; Sammalisto and Brorson 2006); but it also allowed identifying relationships and integrating best practice between operations, academic and support departments. Fourth, collaboration, communication and exchange of information could encourage capacity building amongst a range of stakeholders (Disterheft et al. 2014). Capacity building was demonstrated in the case study reported here by the establishment of the management groups (one for each environmental aspect), the university wide teaching and sustainability networks, and the environmental strategy board. Finally, engaging the commitment of senior management provided opportunities for integrating environmental management into core business processes and strategy.

13 The Challenges of the Participatory Approach

Participatory approaches to environmental management present a number of challenges as well as benefits (Disterheft et al. 2014). In particular the Manchester Metropolitan University case study uncovered four important challenges. First, engaging many stakeholders often leads to a slow implementation process due to time required to build new relationships, increase knowledge and skills and discuss and arrive at a consensus (Disterheft et al. 2012, 2014). This challenge was resolved by anticipating and allowing for time delays during the planning stages of the environmental management system. Moreover, the more established the management groups and processes became, the less the time delays were. The second important challenge was securing ongoing key stakeholder interest and commitment. To overcome this challenge management support was secured at all levels, the core business case for each lead stakeholder was communicated clearly, and they were allowed to develop their own achievable and incremental performance objectives and targets. This was especially important for integrating the aspects of teaching and research.

Using quantitative key performance indicators presented a challenge for teaching and research. Quantitative indicators were not seen appropriate because these activities have important qualitative elements. This challenge was addressed by allowing the use of mixed qualitative and quantitative indicators. The final challenge presented by this participatory approach was maintaining momentum and interest amongst stakeholders, who had often competing priorities. To surpass this challenge ongoing, as well as additional ad hoc, communication and coordination of support from the environment team was provided. It remains important for Manchester Metropolitan University to maintain momentum and build upon the benefits of participation for moving its environmental management system from the implementing to the operating, and checking and correcting levels of the ISO 14001 standard. In particular, ongoing appropriate resources and support (e.g. sustainable development curriculum co-ordinator post) are required to ensure that policy commitments and targets for teaching and research are delivered. The insights developed by the case study reported here will be useful to (a) universities planning to implement ISO 14001 in incremental stages; and/or (b) universities that already have an environmental management system and are planning to expand the scope to include teaching and research. Additional research is required to explore how universities would have to re-align their environmental management systems to meet the proposed changes to ISO 14001 after 2015.

14 Conclusion

The approach that Manchester Metropolitan University followed for its environmental management is characterised by collaboration, shared ownership and empowerment in developing a strategy and system covering both direct and indirect environmental aspects including teaching and research. Teaching and research activities tend to be excluded from the scope of the environmental management system of most universities (Disterheft et al. 2012; Clarke and Kouri 2009). This reflects a missed opportunity to develop a holistic approach to environmental management and to focus resources on some of universities' biggest indirect environmental aspects (Clarke and Kouri 2009; Sammalisto and Brorson 2006). Integrating teaching and research in a university's environmental management system requires engaging all relevant internal stakeholders in policy and strategy development as well as in delivery and performance monitoring. The participatory approach that was followed by Manchester Metropolitan University increased ownership, captured expertise and enthusiasm, reduced conflict and increased the capacity of stakeholders and relevant staff to deliver environmental improvements. On the other hand, a participatory approach also presented a number of challenges including slowing implementation at the initial stages and needing ongoing and clear communication to secure commitment and maintain momentum. Despite the challenges Manchester Metropolitan University successfully established a new senior level environmental strategy board, university wide sustainability research and teaching networks, and a management framework and strategy with delivery mechanisms led by appropriate stakeholders and management groups. This demonstrates that the benefits of participation were greater than the effort required for overcoming the challenges.

Acknowledgments The authors would like to thank Dr. John Hindley, head of environmental strategy, Manchester Metropolitan University, UK, for providing resources and support for the case study reported here; Alex Hobbins, EcoCampus, Loreus Ltd., UK, for his insightful advice on implementing environmental management systems at universities; and all lead and additional stakeholders from Manchester Metropolitan University, UK, for their contribution to the development of the environmental management system reported here.

References

Alshuwaikhat HM, Abubakar I (2008) An integrated approach to achieving campus sustainability: assessment of the current campus environmental management practices. J Clean Prod 16:1777–1785

Bennett M, James P (eds) (1999) Sustainability measures: evaluating and reporting on social and environmental performance. Greenleaf Publishing, Sheffield, UK

British Standards Institution (2004) EN ISO 14001:2004 Environmental management systems-requirements with guidance for use. BSI Standards Ltd, s.l

Clarke A (2006) The campus environmental management system cycle in practice: 15 years of environmental management, education and research at Dalhousie University. Int J Sustain High Educ 7(4):374–389

Clarke A, Kouri R (2009) Choosing an appropriate university or college environmental management system. J Clean Prod 17:971–984

Collins E, Kearins K, Roper J (2005) The risks in relying on stakeholder engagement for the achievement of sustainability. Electron J Radical Organ Theor 9(1):1e19

Cortese AD (2003) The critical role of higher education in creating a sustainable future. Plan High Educ 31(3):15–22

Disterheft A, Caeiro S, Ramos M, Rosário A, Ulisses M (2012) Environmental management systems (EMS) implementation processes and practices in European higher education institutions—top-down versus participatory approaches. J Clean Prod 31:80–90

Disterheft A, Caeiro SA, Ulissess M, Leal Filho W (2014) Sustainable Universities- a study of critical success factors for participatory approaches. J Cleaner Prod (article in press). http://dx.doi.org/10.1016/j.jclepro.2014.01.030

EC (2009) Regulation No 1221/2009 on the voluntary participation by organisations in a Community eco-management and audit scheme (EMAS). European Council and European Parliament, Brussels

International Standards Organisation (2013) Environmental management systems—Requirements with guidance for use, Committee Draft-14001.2. International Standards Organisation, Geneva

Narain Mathur V, Price ADF, Austin S (2008) Conceptualizing stakeholder engagement in the context of sustainability and its assessment. Constr Manag Econ 26(6):601–609

Noeke J (2000) Environmental management systems for universities: a case study. Int J Sustain High Educ 1(3):237–251

People and Planet (2013) People and planet green league guide 2013: driving UK universities' transition to a fair and sustainable future. People and Planet, Oxford, UK

Sammalisto K, Arvidsson K (2005) Environmental management in Swedish higher education: directives, driving forces, hindrances, environmental aspects and environmental co-ordinators in Swedish Universities. Int J High Educ 6(1):18–35

Sammalisto K, Brorson T (2006) Training and communication in the implementation of environmental management systems (ISO 14001): a case study at the University of Gavle, Sweden. J Cleaner Prod 16(3):299–309

Simkins G, Nolan A (2004) Environmental management system in universities. Environmental Association for Universities and Colleges (EAUC), s.l

Von Oelreich K (2004) Environmental certification at Malardalen University. Int J Sustain High Educ 5(2):133–146

Wright TSA, Wilton H (2012) Facilities management directors' conceptualizations of sustainability in higher education. J Clean Prod 31:118–125

Author Biography

Helena Tinker is Environment and Energy Manager at Manchester Metropolitan University, UK. She is leading ISO 14001 implementation at the university. She has implemented EMAS and ISO 14001 in manufacturing companies and local authorities in the UK and has supported a number of local authorities across Europe achieve EMAS.

Dr. Konstantinos Tzoulas is a Lecturer in Environmental Management at the School of Science and the Environment at Manchester Metropolitan University, UK. His teaching focuses on sustainable development policy and urban and rural ecosystem management. He is the chairman of the university's Biodiversity Management Group.

Introducing Sustainability Values: The Case of Ibri College of Applied Sciences (Sultanate of Oman)

Sadiq Hussain

Abstract

Given that the implementation of sustainable development in higher education is a global trend, the particular importance of this study derives from the fact that Ibri College of Applied Sciences (CAS) was the very first campus to be established in Aldahira administrative region—one of the developing provinces of Oman. This study adds to Sustainable Development at Universities literature by analyzing the innovative approaches which enabled the topic of sustainability values to be introduced for the first time in the transnational Ibri CAS Information Technology (IT) Program curriculum and research. Finally outlined are some concrete Omani indigenous-knowledge steps for innovation methods and tools in collaboration with The Research Council (TRC), Public Establishment for Industrial Estates (PEIE) and the Industrial Innovation Centre (IIC) undertaken to allow Ibri CAS to integrate sustainability values in its research, innovation, practical projects and activities and also in Ibri regional endeavours. This paper will be useful to sustainable development education researchers, students working on graduation projects, project supervisors, project managers, educationalists, others interested in change management and the different levels of emphasis given to sustainability values at regional colleges in Oman.

Keywords

Sustainability · Oman · Information technology · Innovation · Projects

S. Hussain (✉)
Sultan Qaboos University, 32, 123 Al Khodh, Muscat, Sultanate of Oman
e-mail: profsadiq@yahoo.com

W. Leal Filho et al. (eds.), *Integrating Sustainability Thinking in Science and Engineering Curricula*, World Sustainability Series, DOI 10.1007/978-3-319-09474-8_32

1 Introduction: Sustainability Values and the Omani Public Higher Education System

1.1 Sustainability Values and the Context of Information Technology Innovation

Introducing sustainable development at universities (SDU) is an established research topic and worldwide there are many universities whose academics might be interested in one aspect or another of introducing a sustainability dimension in their studies. The state of the art in introducing sustainability values has been advanced in many dimensions: campus environment management systems (EMS) (Disterheft et al. 2012); offering students practical activities (Filho et al. 2010); academics introducing sustainability into their own courses (Hegarty et al. 2005); offering students ESD projects (Wals 2009); creating new knowledge (Lozano 2011); SDU-driven innovation facilitating the accumulation of indigenous knowledge (Hussain and Albarwani 2014; Amir 2013; Hammett 2013); offering students a general taught course (Lee 2014; Hussain and Albarwani 2014) and entrepreneurial campus (Mason 2014). Related to this last dimension is interesting to note that Huggins (2012) stated that place-based community social cohesion may stimulate entrepreneurship. Lee (2014) overviewed the pedagogy within the SDU literature, and Arabian-based sustainability examples are described in Baban and Al-Oun (2003), Luken and Castellanos-Silveria (2011), (Mezher et al. 2011) and Chang et al. (2013). Sarkis et al. (2013) identifies the progress in sustainability research as the most multidisciplinary of all academic topics covering individual action, UN policy, community ties, Information Technology Projects, and Technology and Innovation Management.

This paper introduces sustainability in a novel one-off regional local development setting that might inspire other academics to adopt a similar approach as well as to share effective student graduation project practices. The innovative elements of this study are that Ibri-based faculty initiatives have enabled the very first IT program students cohort to be introduced to sustainability values as part of their formal education. This entailed some courses being adapted, 'contextualized', and 'localized' to transform a transnational IT program whose curriculum is generated by non-teaching faculty at Otago University, New Zealand. Ibri-based faculty pioneered in introducing the modified innovative aspects on as-received IT program from the electronic access and use of Otago-made packages of Power Point lecture series and lab sessions suites. The significance of this study lies in its opening the door for replication elsewhere to introduce sustainability values at other institutions where the sustainability dimension is neither integrated nor embedded in the curricula. For a higher education institution in its infancy where no sustainability efforts have been previously undertaken and from the state of the art literature above the research question that follows is derived:

Research Question: *Can recognition and acceptance of the importance of innovation by students and faculty pioneer to create new sustainability knowledge and values through a set of local practical group project activity as part of formal education*?

The overall objective of this paper is to contribute to knowledge and deepen discussions on local and international approaches to introducing sustainability values in Oman-based higher education systems. The specific aim is to explore and analyze the case of introducing a sustainability dimension in the formal context of the transnational Software Major IT Program degree requirements some of whose modules are the software development graduation projects SFDV4001 and SFDV4004 at Ibri CAS campus, one of a network of six public regional CAS college campuses run by the Director General DG CAS, MOHE, Oman.

1.2 The Omani Public Higher Education System and Regional Higher Education Institutions HEI (e.g. Ibri College of Applied Sciences, CAS)

With the accession to the throne of His Majesty Sultan Qaboos in 1970, Oman has rapidly transitioned from a neo-tribal territory with no universities, no colleges, not even schools, to a modern Western-looking state emerging with a renaissance to satisfy national aspirations as well as being part of the global knowledge-based economy. Public higher education systems in Oman currently include the biggest and most prominent institution which is Sultan Qaboos University (SQU) established in 1986, whose sustainability values have recently been studied (Hussain and Albarwani 2014). The second public University of Science and Technology starts in 2016, and there are also joint tertiary education vocational training systems designed, regulated and funded by the government; that work with educational institutions of Germany, Holland, Scandinavia, for example. Recent international organizations educational studies have noted the rapidly developing modern systems in Oman compared with the 22 member-states of the Arab League, e.g. UNDP (2006), and World Bank (2008).

Unlike Ibri CAS, SQU does not offer an IT program and all the while the discipline of IT is changing in nature from an analytical tool into a strategic planning tool (Musango 2012). This is demonstrated by the fact that IT projects development formed an important part of the plan of action by the GCC council of ministers (of which Oman is a member) endorsing the UN MDG. MDG targets and indicators provide a framework to measure progress in the integration of environment sustainability and socioeconomic development (Khoja 2013). Powell (2012) notes that peoples well-being and a healthy environment are among the key indicators for sustainability in a country's regions, which is in line with the Omani public higher education system situating education (Ibri CAS with its pioneering IT program example) as a sustainability teaching and learning site (Hammett 2013), positioning IbriCAS to take good care of the region's educational, cultural and technology dimensions.

MOHE endorsed the UN International Decade of Education for Sustainable Development (IDESD). To further speed-up Oman-based undergraduate education

MOHE established in 2005 a new Directorate General (DG) for Colleges of Applied Sciences (CAS) and contracted international collaboration projects including that between DG CAS and Otago University, New Zealand for the CAS Information Technology (IT) program to be introduced for the first time at a public institution—as noted above SQU does not run IT programs. Stakeholders recognized at the outset that the structure and exact content of the new CAS project would need to evolve over time in the light of experience of the Omani students and the Oman-based lecture course delivery staff (not involved in the design of the Otago-based program). There are also water sustainability collaborations between Oman government and universities at the German Federal Republic on water sustainability in Oman (Kalbus private communication August 2013) and GCC-EU International Cooperation Network INCONET partnership with The Research Council (TRC) of Oman (Andersson private communication August 2013) undertaking greening innovation projects. With all this Oman brings a new dynamic to the sustainability community seeking a better-sustainability approach, and emerging as co-producers of scientific knowledge (Hussain and Albarwani 2014).

Ibri is the main town and is home to the very first college to be established in Aldahira province which is one of eleven administrative regions. This gave the college tremendous importance for the progress and human development of this friendly region. Students and tutors at Ibri do not have regular access to the kinds of educational, cultural and social amenities available in the larger cities. Aldahira province borders Saudi Arabia, has a relatively drier climate, and, by contrast, a most welcoming and social people in this desert of deserts known as the Empty Quarter. On the cultural dimension of sustainability, Aldahira is home to one of UNESCO's inscribed world heritage sites at Bat, near Ibri (Ministry of Heritage and Culture 2003).

2 Methodology

The specific objective of this study is to explore and analyze the case of introducing a sustainability dimension in the formal context of the transnational Software Major IT Program degree requirements. This objective is achieved by introducing a sustainability dimension in some modules namely the software development graduation projects SFDV4001 and SFDV4004. The IT software specialism was chosen because other IT majors—IT networks, and IT security—unlike the software major, do not offer student project modules. The following is a description of the objects/subjects to be studied comprising the population of Omani undergraduate students studying in Oman and the studied sample—the entire first Ibri CAS cohort of the IT (Software Major) students. This study is designed to target the larger population of Omani undergraduate students studying in Oman where Arabic is the language of instruction in public schools and students compete for HEI admission based on their score in the Ministry of Education general diploma examination. Once accepted and before registration all students have to sit an HEI-administered placement test that decides whether a student is allowed to register on arrival for a four-year academic program or has to study at the HEI for one semester or more to complete an Oman-wide HEI-run MOHE-designed foundation course to enable (low scoring placement test) students to develop their English language skills, and IT skills before proceeding in their HEI-based specialism studies. The network of six CAS College campuses serves regional cities and towns at Ibri, Nizwa, Sohar, Rustaq, Sur, and Salaleh and are witnessing a steady increase in the number of students entering from schools. These six CAS college campuses offer programs in Information Technology, Communication, International Business and Design while Sohar CAS offers in addition a fifth program in engineering through a partnership with Aachen University of Applied Sciences, Germany. These new CAS college campuses were housed in the infrastructure that formerly belonged to the network of colleges of education.

The population of the studied sample—the entire first Ibri CAS cohort of the IT (Software Major)—is inhabited by 49 students organised in 13 project teams with 2–5 students per team. The following are characteristics of the sample students: Age: 22–23, education: final year undergraduate, gender: two male and 47 female. These students are the best to be under observation and to collect the information needed to answer the research questions and fulfil the objective. The study period was the academic year 2009/2010 and timetable for completion of the project was the end of that academic year. In the first and second semester the general supervisor reached the students in the sample through meetings with student teams to discuss their graduation projects in the module SFDV4001 Software Projects A (first semester) and SFDV4004 Software Projects B (second semester). Also included were formal 4 h weekly contact hours of class room and lab life as instructor for lectures and lab sessions (additional role) in the first semester module INFT3003 on Ethics and Professional Practice and second semester module SFDV4002 on Software Projects Management. Meetings were also held with individual students at campus-based extra-curricular students life skills development workshops and being available in office at times suitable for students. Moreover, the general supervisor as head of IT department and full professor, kept in touch with all individual student teams supervisors comprising 7 tutors/supervisors, 3 of whom were PhD holders, 4 MSc holders. Through the lens of sustainability innovation methodology, the present study involves campus-based practice, research, observation, primary sources, and an exhaustive search on national and international sources of literature. Also involved are the emergent Oman countrywide innovation for sustainability policy with Ibri CAS as the pioneering target institution.

3 Results on Introducing Sustainability Values at Ibri CAS

This study makes use of the whole of the CAS hierarchy plus collaborators troika: TRC, IIC, and PEIE. In the innovation literature, two schools of Innovation Management research have been identified: first domain understands the causes of innovative behavior of students, where the student is used as the unit of analysis and the second domain is in areas of organizational theory and academic strategic management, where the institution e.g. MOHE, SQU, CAS, PEIE, IIC, TRC is used as a unit of analysis. In this study in addition to our troika PEIE, IIC, TRC, the whole CAS hierarchy are involved including DG CAS, PD IT and other PDs, Deans, Assistant Deans, HODs, 100-strong IT Faculty Academic Staff, LRC staff, Registry Staff, Administration and Finance Staff, Services Staff.

Faced with solving a problem or with understanding a task, students react by thinking, by applying judgment and by seeking information (Auger 1992). For software project students, the first two involve using information from the most readily available source, namely one's memory and going through lecture handouts and practice in lab sessions. For more information students may turn first to external sources starting with their supervisor and general supervisor. In this context it might be a good idea to remind students that having 'experts' so readily available with answers is a privilege of on-campus education which they might miss after they graduate. Second 'going and looking' for 'observation', and third online data banks held on web networks, books and journals. Through the Ibri students graduation projects shown in Table 1 this study seeks to link the innovation body of knowledge and the field of innovation management to the above context of introducing Omani innovation for sustainability.

3.1 Ibri Royal Camp and the Minister, Dr Rawya al Busaidi's Ibri CAS Campus Milestone Visit

In November 2009 at Ibri's Royal Camp, His Majesty Sultan Qaboos was accompanied by Her Excellency Dr Rawya Saud Al Busaidi, Minister of Higher Education. Excellency took this opportunity to visit Ibri CAS campus and the author invited Excellency to look at student project display—which Excellency Al Busaidi (2010) thought were excellent Ibri CAS sustainable development contributions as shown in the following officially archived quotation:

> I have received your report on the student projects which you mentioned during my visit to Ibri CAS. I read with interest the details of the very useful work your IT Department is undertaking with local organizations regarding the software development for electronic meter-reading; e-health services; and e-voting. I indeed am pleased to see that our students are involved in practical applied research projects where they are learning valuable new technical and generic skills that will be helpful to them later on the job. Local organizations will also benefit from the results of these projects, while faculty gains from experiential professional development. Thank you for this excellent contribution to the on-going development of IbriCAS. Signed Dr Rawya Saud Al Busaidi, Minister of Higher Education.

Also, in 2010 Ibri CAS IT project student teams won first and third place in the Omani National students projects competition and exhibition held at Sohar CAS campus. The Timeline for Ibri CAS development is shown in Table 2.

Table 1 Student project teams, titles, and supervision, Ibri CAS SFDV4004 software engineering project B (Academic year 2009–2010)

Student project teams Names of students	General supervisor Supervisor Project title	Notes
Team A Mariam Al-Whashi Ghaya Al-Shaibani Eman Al-Jadidi Dalal Al-Harmali	Prof. Sadiq Hussain, PhD Akram Pasha, MSc Ibri pharmacy software management system	Students, general supervisor, and supervisor visited the pharmacy at Ibri Hospital to gather data
Team B Khuloud Al Harrasi Sheikha Al Dohani Thuraiya Al Shabibi Laila Al Hakmani	Prof. Sadiq Hussain, PhD Akram Pasha, MSc Ibri hostel software management system	Ibri CAS had a hostel for female students only—the interest was to develop a more student-centred approach
Team C Nabila AL-zri Amna Al-Badi Nawal Al-Mamari Ruqaya AL-Abri Khalsa Al-Kharousi	Prof. Sadiq Hussain, Awny Abd Elhady, PhD Electricity e-billing and e-payment Software Management System	General supervisor met some local electricity kilowatt-hour meter readers and also visited Ibri agents for processing of electricity Bills for Public Authority for Electricity and Water (PAEW) to gather data
Team D Haneia Al Malki Aisha Al Hatmi Najeaa Al Jhafri Samia Al Sasdi	Prof. Sadiq Hussain, PhD Awny Abd Elhady, PhD Ibri Hospital Software Management System	Students, general supervisor, and supervisor visited the executive manager at Ibri Hospital to gather data
Team E Safiya Al Riamy Thuriaya Al Abri Sumaya Al Farsy Jameela Nasser Rashid-Al-Tobi Idris Ali Hamdan Al-Kharousi	Prof. Sadiq Hussain, PhD Amr Kamal, PhD capital market authority software management system	General supervisor visited Ibri Capital Market authority in Muscat to gather data
Team F Rahma Abdullah Al-Harthi Maryam Mohammed Al-Mujaini Laila Humaid Al-Abri Najwa Salem Al-Hatmi Nader Khalifa Al-Jadidi	Prof. Sadiq Hussain, PhD Amr Kamal, PhD Dental Clinic software management system	Students visited some private dental clinics at Ibri to gather data

(continued)

Table 1 (continued)

Student project teams Names of students	General supervisor Supervisor Project title	Notes
Team G Rahima Sulaiman Fatma Sulaiman Rahma Mubarak	Prof. Sadiq Hussain, PhD Shaker Jebari, PhD Ibri restaurant software management system	Students visited some Ibri restaurants to gather data
Team H Kawthar Mohammed Al Zahraa Sulaiman Raiya Hamdan Wafa Rasheed	Prof. Sadiq Hussain, PhD Shaker Jebari, PhD Nizwa real estate software management system	Students visited some Nizwa real estate agents to gather data
Team I Raba Mohsin Sulaiman Azza Hamed Obeid Al Amri Fathiya Hamood Hamed Al Sarmi Rasha Yaqoob Saif Al Rawahi	Prof. Sadiq Hussain, PhD Munesh Chauhan, MSc Ibri e-voting software management system	Students visited the Shura Council (elected chamber) in Muscat, one of Oman's two-tier people representative system, who were introducing e-voting for the first time in Oman
Team J Roqaiya Khalfan Bushra Sulaiyam	Prof. Sadiq Hussain, PhD Munesh Chauhan, MSc Ibri E-Time Table Software Management System	The students experienced scheduling problems at Ibri CAS– nterest in improving timetables with this software application
Team K Diana Mahmood Azhar Said Fetna Rasheed Salem Adhraa Sulaiman	Prof. Sadiq Hussain, PhD Naveen Vijayakumar, MSc Hotel information management system	The students visited some hotels in areas of cultural interest in the students own local home town
Team L Jameella Hamed Rahma Ali Jamill Fatma Nasser Ali Abeer Saleh Said	Prof. Sadiq Hussain, PhD Naveen Vijayakumar, MSc Ibri CAS Registration Software management system	Collaboration with Ibri CAS registry
Team M Narjes Abdullah Al Khlasi AsmaaMarhoun Al Belushi	Prof. Sadiq Hussain, PhD Rudy Prabowo, PhD Oman Castles software management system	The students visited some castles in areas of cultural interest in the students own local home town

Table 2 Ibri CAS timeline

Year	Topic
2005	Founding of the first public sector MOHE directorate general (DG) of Colleges of Applied Sciences (CAS) Appointment of first CAS Director General CAS signs contract with Otago University for curricula design of IT program Appointment of first Oman-based Information Technology Program Director Beginning of UN International Decade for Education for Sustainable Development (IDESD) First cohort student intake (foundation program) signalling the transformation of MOHE Ibri College of Education into Ibri CAS Appointment of Oman-based Ibri CAS faculty to teach the foundation year students (2005 cohort)
2006	Appointment of Oman-based Ibri CAS faculty to teach the year1 students (2005 cohort)
2007	Appointment of Oman-based Ibri CAS faculty to teach the year2 students (2005 cohort)
2008	Appointment of Oman-based Ibri CAS faculty to teach the year3 students (2005 cohort)
2009	Appointment of Oman-based Ibri CAS faculty to teach the final year (2005 cohort) His Majesty Sultan Qaboos Royal Camp at Ibri Her Excellency Minister of Higher Education Dr Rawya visits Ibri CAS IT students projects exhibition
2010	Graduation of first(2005) Ibri CAS IT program cohort
2011	End of CAS Contract with Otago University, New Zealand Graduation of the second (2008) cohort IT program Ibri CAS Beginning of the Oman 8th five year national development plan (2011-2015)
2012	Graduation of the third (2008) cohort IT program Ibri CAS
2013	Graduation of the fourth (2008) cohort IT program Ibri CAS
2014	Ibri CAS contributes a paper to WSSD-U-2014 Graduation of the Fifth (2008) cohort IT program Ibri CAS End of UN International Decade for Education for Sustainable Development (IDESD)
2015	Graduation of the sixth (2008) cohort IT program Ibri CAS
2016	Beginning of the Oman 9th five year national development plan (2016-2020) The new Oman University of Science and Technology opens its doors to students for the first time

4 Discussion and Conclusions

There has never been so many HEIs and at the start of the 21st century internationalization of higher education has highlighted important sustainability dimensions that are affecting the performance of countries' HEIs in the five UN-defined educational regions of the world. In the educational region of the 22 Arab States, and the GCC sub-region in particular for example, public universities are strategic national institutions that are shaping and are shaped by the sustainable development of their countries. Thus some HEIs, e.g. Ibri CAS campus, focused on advancing the IT graduate-recruiting knowledge-based industries dimension of sustainability. In this regional context Luken and Castellanos-Silveria (2011) examined the impact of industrial transformation on sustainability in developing countries, Baban and

Al-Oun (2003) paid attention to the impact of desert greening on sustainability, Mezher et al. (2011) explored sustainability knowledge in the Arab world, and Chang et al. (2013) studied the effects of international trade and industrial structure on sustainability. Another SQU pilot study is to explore using UCINET software packages for social network mapping techniques of Oman-based sustainable development researchers (Alhinae private communication February 2014).

Ibri CAS looks into developing innovation capability in the institution IT program itself as well as its students developing innovation capability of their own, acquiring knowledge on how to create new projects. Adoption of paradigm shifts in Omani IT innovation management techniques offers prospects for Oman actively participating in innovation globalization processes which enhance Omani national development. For instance one such opportunity was seized by the author in a discussion about the practical aspects of the World Wide Web, in a public lecture attended by students as well as face-to-face one-on-one during Sir Tim Berners-Lee December 2013 visit to Oman. The innovative topic aims to change the 'web world' for the better by experimenting with systems saving web pages with web page-URL embedded in the web page-save or the screenshot. At present, looking at the screenshot taken by a student, say, an academic observer, tutor, say, has no clue where the student's screenshot came from.

Ministry Of Commerce and Industry (MOCI) graduate startups are collaborations between MOHE, where her Excellency's involvement is evident as shown in the results section above and MOCI PEIE CEO also being active as member of CAS' Board of Trustees. This study has leveraged knowledge, expertise, and experience of Ibri CAS faculty and students in such a way to create complementary activity structure for learners, tutors, PEIE, IIC, TRC collaborators, seeking to expand the indigenous Omani knowledge base. This helps to narrow the gap between learning and teaching practice on the one hand, and on the other controlled research labs and field studies; enabled students and IT academics to become practitioners and achievers of Omani IT-driven sustainability capability; helping students to understand "what works, why, and how it works" with sustainability-driven innovation. Students acquire knowledge on how to create new projects, and how to identify and capitalize on opportunities to innovate based on qualitative evidence from several "best practice" organizations. Innovation management involves discerning and meeting the (human) needs of the student teams as well as transformation of knowledge into new processes; noting that what makes innovation tick with students is the prospect to be well-grounded in understanding the "why? i.e. innovation theory as well as the "how?" i.e. innovation practice.

The fact that in 2010 Ibri CAS IT software major student teams won first and third place in the Omani National CAS students IT projects competition as stated in the Results section above, refereed by a panel of external experts held at Sohar CAS campus is all the more significant because of the following observations. Sohar CAS campus is where the IT program director is based and consequently carries more prestige as the flagship college than other CAS colleges. Also at official IT faculty meetings Sohar colleagues gave the impression that they were very confident that their student teams would easily win the competition—they did not as it turned out—the honours went to Ibri CAS campus. Another professional merit to the credit of this Ibri CAS cohort is that as graduates they were offered top jobs and started working at the most prestigious enterprises e.g. TRC, Oman IT Authority, Oman IT Regulation Authority, Omantel (the biggest telephone and internet service provider in Oman) while other IT graduates preferred to study for MSc at SQU.

Beyond Ibri CAS campus, ESD in the GCC States has been one of the success stories of higher education in the conflict-prone Middle East with the GCC in its fourth decade epitomising regional integration into the global economy. It is common knowledge that such is the importance of sustainability education to the knowledge-based global economy, that the World Bank as part of the global soft governance has recently become one of the key publishers of international and comparative education studies, alongside UNESCO. The year 2014 marks the end of the IDESD and the coming year will usher in another decade in a transition marked by the emergence of newer sustainability-driven Arab education systems.

Traditionally, knowledge consumed in the GCC countries was produced in far off countries in the 'centre' by universities based in Europe and North America. Now GCC countries wish to move from the periphery to the centre through international collaborations to explore ways of generating indigenous knowledge as a valuable resource for local development in a global context.

In the temporal, cultural and spatial contexts of sustainability, in the recent millennial transition Europe and North America dominate; whereas in the earlier transition from first to second millennium, culture, geography, mathematics, and logic (the foundation of modern network and computer software), thrived along with other disciplines in Arab lands, while Europe was then 'under-developed' and America 'yet to be discovered'.

The conclusion is that based on the above statement of the research context, methodology, frameworks, results, and discussion, the recognition and acceptance of creativity by students and faculty can create new sustainability-knowledge and sustainability-values. This enables the higher education institution to meet the challenges of sustainability through student awareness and training to help the global efforts in a set of local practical project activities which may assist various groups as part of formal education.

Acknowledgments Sincere thanks are due to Walter Leal Filho and Ulisses M Azeiteiro for their comments on earlier versions of this paper.

References

Al Busaidi RS (2010) Formal letter of thanks from her excellency Dr Rawya Saud Al Busaidi Minister of Higher Education archived at The Minister's Office, Muscat: MHE MO 2010/2/267

Amir AF (2013) Understanding the regional innovation support systems in developing countries. Int Dev Plann Rev 35:41

Auger CP (ed) (1992) Information sources. Bowker, London

Baban SMJ, Al-Oun S (2003) Let the desert bloom: an overview of an attempt to promote sustainable development and environmental protection in the Jordanian Badia region. Int J Sustain Dev 11(3):159–170

Chang D-S, Yeh L-T, Chen Y (2013) The Effects of Economic Development, International Trade, Industrial structure and energy demands on sustainable development. Sustainable development, doi:10.1002/sd.1555, www.onlinelibrary.wiley.com

Disterheft A, Caeiro S, Ramos R, Azeiteiro UM (2012) Environmental management systems (EMS) implementation processes and practices in European higher education institutions—top-down versus participatory approaches. J Cleaner Prod 31:80–90. doi:10.1016/j.jclepro.2012.02.034

Filho WL, Pace P, Manolas E (2010) The contribution of education towards meeting the challenges of climate change. J Baltic Sci Educ 9(2):142–155

Hammett D (2013) Transition and the education of the new South African citizen. Comp Educ Rev 57(2):309–331

Hegarty B, Penman M, Nichols M, Brown C, Hayden J, Gower B, Kelly O, Moore M (2005) Approaches and implications of eLearning adoption on academic staff efficacy and working practice: An annotated bibliography. Ministry of Education: Author. Retrieved 25 Mar 2014, http://www.ascilite.org.au/conferences/sydney06/proceeding/pdf_papers/p156.pdf

Huggins R (2012) Entrepreneurship and community culture: a place-based of their interdependency. Entrepreneurship Res J 2(1):2–34

Hussain S, Albarwani T (2014) Sustainable development at universities in the Sultanate of Oman: the interesting case of Sultan Qaboos University (SQU). In: Leal W (ed) Transformative approaches to sustainable development at universities: working across disciplines. Springer, Berlin

Khoja T (2013) Working together for better health: technical cooperation among the states of GCC. Executive Board of the Council of GCC Health Ministers, Riyadh

Lee (2014) Manifestos for sustainable development: Student visioning of change through a geography course Transformative approaches to sustainable development across disciplines. Springer, Berlin (forthcoming)

Lozano R (2011) Creativity and organizational learning as means to foster sustainability. Sustain Dev. doi:10.1002/sd.540, www.onlinelibrary.wiley.com

Luken R, Castellanos-Silveria F (2011) Industrial transformation and sustainable development in developing countries. Int J Sustain Dev 19(3):167–175

Mason C (2014) Creating entrepreneurial campuses. Glasgow University Adam Smith Bussines School, Scottish QA Authority Report, Edinburgh

Mezher T, Noamani D, Abdul-Malak MA, Muddah B (2011) Analyzing sustainability knowledge in the Arab world. Int J Sustain Dev 19(6):402–416

Ministry of Heritage and Culture (2003) Bahla, Bat and Frankincense Trade Route in the World heritage list, Muscat

Musango (2012) Technology assessment of renewable energy sustainability in South Africa. University of Stellenbosch http://scholar.sun.ac.za. PhD, thesis

Powell (2012) IWRM platform for HEI to meet sustainability challenges. Int J Environ Educ Res:1–19. doi:10.1080/13504622.2012.704898

Sarkis J, De Bruijn T et al (2013) Guest editorial: sustainability in engineering management-setting the foundation for the path forward. IEEE Trans Eng Manage 60(2):301–314

UNDP (2006) Arab human development report: building a knowledge society. UNDP, New York

Wals (2009) Review of contexts and cultures for ESD, UNESCO, Paris

World Bank (2008) MENA development: the road not travelled for education reform. World Bank, New York

Author Biography

Professor Sadiq Hussain has spent decades in development technology learning, teaching, researching, consulting, and full-professor since 1986; spanning UK Cardiff University BP scholarship award, engineering BSc Honours; UK Warwick University, engineering PhD, UK Ministry of Technology scholarship award, research assistant, and Rootes Hall resident tutor (additional post); UK Brunel University, postdoctoral research fellow, Atomic Energy Research Establishment, Harwell, award; UK Durham University, electronic instrumentation lecturer, and lab demonstrator; In 2005 as Oman-resident appointed to represent Asia on the Internet Society World Advisory Council, Public Interest Registry (@pir.org, Reston, Virginia, USA; awarded Oman Minister of Higher Education's letter of thanks; Sultan Qaboos University visiting professor, Ibri College, IT professor, head of IT department, and Majan College IT professor and college research coordinator. In Iraq, awarded Iraq Republic President's Medal for distinguished national scholars (Wisam Al-Ulema) served at Baghdad University, University of Technology, Alnehrain University; research center director, deputy dean, head of department, board member Iraq Science Research Council. Visiting fellowships: USA Santa Clara University, California, NASA Ames research contract; Budapest Technical University, Hungary; Montpellier University, France; Knowledge Village Campus, Dubai, UAE.

Social Feasibility in the Adoption of Renewable Energy: A Case Study of REGSA Project

José Baltazar Salgueirinho Osório de Andrade Guerra, Youssef Ahmad Youssef, Leandro Piazza dos Santos and Ane Cristina Figueiredo Pereira de Faria

Abstract

Promoting Renewable Electricity Generation in South America (REGSA) is a project that aims at increasing the use of electrical energy obtained by means of renewable power generation in South America. It is a way to improve the environmental conditions, enhance energy security, and alleviate the poverty in the project areas. The REGSA project goals and objectives include: (i) to identify and disseminate the technical and socioeconomic potential for generating renewable energy power in the target countries; (ii) to raise awareness and provide support to the regional dialogue between public and private key actors and decision-makers about renewable electricity generation in the target countries; (iii) to support selected pilot communities in the development and implementation of viable renewable electricity generation projects; and (iv) to contribute to the polycentric integration of Brazil, Chile and Bolivia, as South-American countries, by developing experiences, which may replicate elsewhere in the region. In this context, the University of Southern

J.B.S.O. de Andrade Guerra (✉) · L.P. dos Santos · A.C.F.P. de Faria
Universidade Do Sul de Santa Catarina—UNISUL, Rua Trajano 219, Florianópolis, SC 88010-010, Brazil
e-mail: baltazar.guerra@unisul.br

L.P. dos Santos
e-mail: leandro.piazza@me.com

A.C.F.P. de Faria
e-mail: ane.faria@unisul.br

Y.A. Youssef
School of Business, Humber Institute of Technology and Advanced Learning, 3199 Lake Shore Boulevard West, Room H215, Toronto, ON M8V 1K8, Canada
e-mail: youssef.youssef@humber.ca

W. Leal Filho et al. (eds.), *Integrating Sustainability Thinking in Science and Engineering Curricula*, World Sustainability Series, DOI 10.1007/978-3-319-09474-8_33

Santa Catarina is developing a pilot module to establish a microhydroelectric system (MHS) in Rancho Queimado, a rural area in the State of Santa Catarina, southern Brazil. As a first step for the implementation of the MHS, a social feasibility study was conducted among the local population. The purpose of this survey was to measure the reaction of the population to the project that is going to generate energy for a local elementary school, and to raise awareness of the renewable power generation as well as support the dialogue between public and private key actors and decision-makers in this field.

Keywords
Energy · Renewable electricity generation · Clean power source · Microhydroelectric system

1 Introduction

Renewable energy has been pinpointed as a critical source for electricity generation enabling long-term sustainable development, where alternative and green energies play a decisive role on reducing the negative effects of a high concentration of carbon dioxide and other pollutants in the atmosphere. In a recent study, the Intergovernmental Panel on Climate Change IPCC (2014) warned that although research on climate change mitigation has grown significantly, there were not enough efforts dispended by the global community to minimize the impact of gas emissions on the ongoing global warming process.

It has been widely acknowledged by the scientific community that the global warming reality is affecting in an unprecedented way the climate and the life of the globe at the same time where demand for energy is growing to meet the basic requirement for social and economic development in the global stage (Edenhofer et al. 2012). For this reason, there is a need to increase the share of renewable energies (RE) into the current energy matrix.

In terms of renewable energy technologies and markets, Edenhofer et al. (2012) mentions different types that can supply electricity. Among the recommendations, are the RE technologies that can be deployed at the point of use in a decentralized architecture especially in rural and urban environments.

In South America, especially Brazil, hydroelectricity accounts for almost 80 % of electricity generation, even when questions have been raised about the negative impact of this source of energy on surrounding ecosystems. For this reason, there have been calls from environmentalists and civil society alike on moving toward alternative green sources of electricity generation to save the biodiversity of the Brazilian rainforest from the catastrophic environmental impact caused by the ongoing construction of hydroelectric plants, especially in the Amazon among other endangered areas such as the Atlantic coastal forest.

This paper sheds light on an innovative experience of a decentralized RE generation in a Brazilian rural area located in the Atlantic Coastal side of the southern Brazilian state of Santa Catarina, specifically in the city of Rancho Queimado as part of Renewable Electricity Generation in South America (REGSA). REGSA project is part of the thematic programme for Environment and Sustainable Management of Natural Resources including Energy (ENRTP) founded by the European Union. It provides the participant countries with the chance to promote renewable energy in rural areas and foster local job opportunities. The consortium of the REGSA project evolved from a current ALFA III project, the Joint European Latin-American Renewable Energy Project (JELARE), which supports a broader application of renewable energy by fostering labour market-oriented research and teaching approaches on the renewable energy sector at Latin American universities.

The REGSA project partners include the Hamburg University of Applied Sciences (HAW Hamburg), the University of Southern Santa Catarina (UNISUL), the Bolivian Catholic University (UCB) and the University of Chile.

The rural community of Rancho Queimado (State of Santa Catarina, Brazil), was selected as a REGSA pilot community following pre-established criteria. After the installation of an autonomous microturbine, a survey was conducted to measure the impact upon the community and assess people's awareness of the clean and renewable sources of energy.

The overall goal of this paper is to measure the acceptance of the REGSA micro hydroelectric power plant by the citizens of Rancho Queimado. Specific objectives included: (i) to know the opinion of the residents of Rancho Queimado on their electricity bills; (ii) to assess their level of awareness on renewable energies; (iii) to identify the importance given to renewable energies; (iv) to map their expectations regarding the MHS installation at the Roberto Schutz School; and (v) to measure their willingness to pay more for the use of renewable or clean energy.

2 The Rural Community of Rancho Queimado

Rancho Queimado is 60 km away from Florianópolis, the state capital of Santa Catarina, Southern Brazil. The municipality was colonized by the Germans in the late 18th century and is located at the base of the Serra do Mar, at an altitude of 810 m. Its geographic coordinates are latitude 27° 40′ 21″ South and longitude 49° 01′ 18″ West. Its population numbers 2,748 people, according to the 2010 Brazilian census, and has an area of 286,432 km^2 (IBGE 2010).

The main economic activities are based on agriculture, especially the cultivation of strawberries, the leading production in the state of Santa Catarina. The landscape is mountainous dotted with rivers, waterfalls and natural pools, which is why there is a large untapped hydropower potential (IBGE 2010).

A MHS was installed at Robert Schutz School in Rancho Queimado. The system will generate 10 kW of electricity, enough to supply the school's energy demand.

A survey questionnaire was administered to a sample of people living in Rancho Queimado to measure their acceptance of the proposed MHS. The field research was intended to identify the population's opinion on such an investment in this community. Information from the survey will help to evaluate the effectiveness and feasibility of the project.

3 Methodology

This study is engaged in a qualitative approach of applied nature, according to Gil (1999), to obtain deeper understanding on people's reaction to the establishment of a power plant project to solve specific problems of the local community.

With regard to the approach, Vianna et al. (2008) affirm that a research design is a roadmap that reduces the dichotomy between methodological procedure and research question. To that end, the authors emphasize that the development of research design is part of a methodological operation proposal in scientific research.

Regarding qualitative research, Godoy (1995) states that qualitative studies differ from each other in their form, method and goals. The author lists a number of essential features that characterize qualitative research such as the natural environment as a direct source of data and the researcher as the basic instrument; the descriptive nature; the inductive approach; and the meaning ascribed to things and life of people being a key concern of the researcher.

4 Data Collection and Sample Size

A semi-structured questionnaire was used in this study as a tool for gathering data, following the research classification system presented by Gil (1999). In addition, a literature review of the main topics of this study was also performed to provide a comprehensive background for understanding current knowledge and highlighting the significance of new research.

After the literature review a questionnaire was prepared and a pre-test was conducted with a reduced sample of the population within the community of Rancho Queimado. After the pre-test the researcher added two more questions: number 3 and 15 of the questionnaire. The survey was conducted and answered by residents of the community of Rancho Queimado. A filter question, as suggested by Rea and Parker (2000), was asked to identify whether the respondent lived in the town. The affirmative answer to this question was a prerequisite for the inclusion in the study; otherwise, the respondent was excluded. The interviews were conducted on the streets of Rancho Queimado, in the city center and in the district Taquaras, in October 2012.

To calculate the sample size, we determined a tolerable sampling error under a probabilistic approach as suggested by Haddad (2004). The sampling error was set at 7 %.

The formula used to determine the sample size without knowledge of the population size was:

$$n_0 = 1/E_0^2$$

where n_0 is a first approximation to the sample size, and E_0 is the tolerable sampling error.

Due to the large size of the population in question, the calculation was corrected to:

$$n = (N \times n_0)/N + n_0$$

where

N is the population size, and n is the number of elements in the sample.

Thus, for a sampling error of 7 % and a population of 2,748 inhabitants, we have:

$$n_0 = 1/E_0^2$$
$$n_0 = 1/(0.07)^2$$
$$n_0 = 1/0.0049 = 204.08$$
$$n = (2,748 \times 204.08)/(2,748 + 204.08) = 560,811.84/2,952.08 = 189.97$$

This meant that the minimum number of participants should be 190. Originally, 200 subjects were included, but after the filter question, 198 participants remained.

The second question asked whether the respondent had electricity at home, which was also a filter question, but it had no exclusion power, since it could only make the respondent unable to answer some questions, skipping to question 8. This was the case for two respondents, resulting in 196 who were able to complete the entire questionnaire.

The survey questionnaire had twenty open-ended and closed questions, as suggested by Rea and Parker (2000) and Triviños (1987). They were intended to identify the acceptance of the MHS establishment and the importance given to the issues of energy and sustainability, electricity consumption habits and socioeconomic profile of the respondents. A copy of the full survey questionnaire can be found in Appendix 1.

5 Findings and Discussion

From the 200 people interviewed, 198 were residents of Rancho Queimado. They responded affirmatively to question 1 and continued in the study. The answers that mention monetary value were based on the Brazilian currency, the Real—BRL.

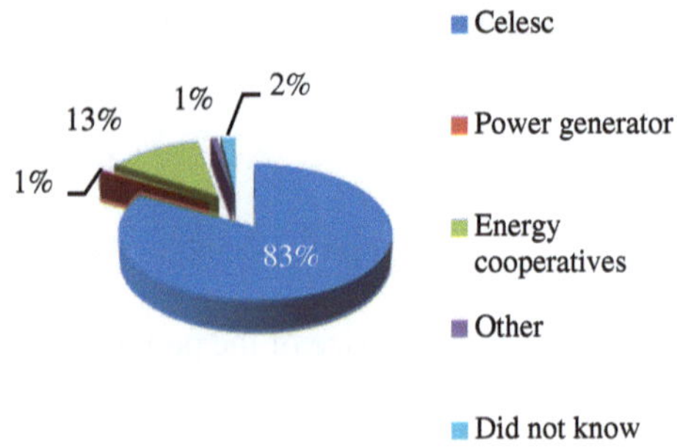

Fig. 1 Participants' knowledge about the power supplier that serves their homes

Question 2 asked whether the respondent had access to electricity at home. This filter question did not invalidate the interview. Respondents who answered negatively to this question were unqualified to answer the subsequent five questions related to their habits of energy consumption, skipping directly to question 8.

Celesc, a Brazil-based holding company involved in the electric power sector, is the leading supplier of energy in the municipality of Rancho Queimado, accounting for 83 % of the power supply. In addition, energy cooperatives are also very significant in the area, accounting for about 13 % of electricity generation.

Question 3 asked which of the following alternatives was the major home power supplier: Celesc, power generator, energy cooperative or other. Only 1 % of the respondents reported that they had a power generator at home. The use of power generators is suitable in Brazil, even though the 2010 census revealed that 98.2 % of rural households have electricity supplied by energy distributor companies, as shown in Fig. 1.

Question 4 asked what the monthly electricity expenditure was. Mean values were categorized in six levels and expressed in BRL: (a) up to 25; (b) between 26 and 75; (c) between 76 and 150; (d) between 151 and 300; (e) between 301 and 600; and (f) above 600. The vast majority of respondents reported that their monthly electricity bills corresponded to levels b (30 %) and c (46 %). A considerable proportion (15 %) of respondents paid between BRL 151 and BRL 300. Only 2 % paid less than BRL 25; 4 % paid between BRL 301 and BRL 600; and 1 % above BRL 600. Energy expenditure in rural areas of Brazil tends to be much lower than that found in Rancho Queimado. According to the Consumer Expenditure Survey conducted by IBGE in 2008 and 2009, the national average monthly electricity bill paid by households in rural areas is only BRL 29.90 (less than $15), as shown in Fig. 2.

The 2008–2009 Consumer Expenditure Survey conducted by IBGE points out that on average the cost of electricity corresponds to 2.1 % of household expenditures in rural Brazilian areas. Question 5 was intended to assess the population's perception of Rancho Queimado on residential electricity rates. Notably different perceptions could be observed, since 13 % of respondents considered their bills as "not expensive"; 16 % considered them "a bit pricey"; 31 % said the bills were

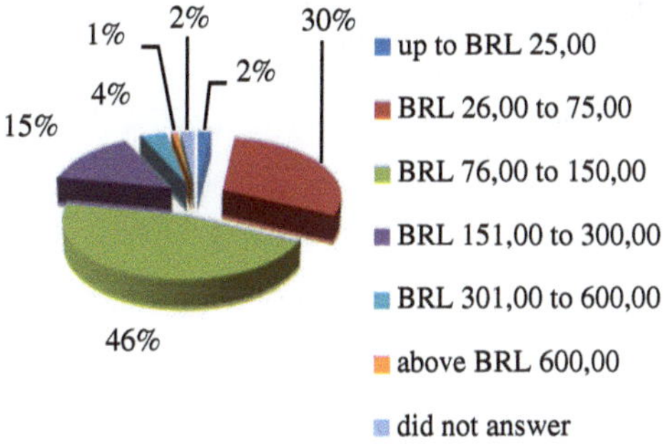

Fig. 2 Monthly cost of the electricity bills on average

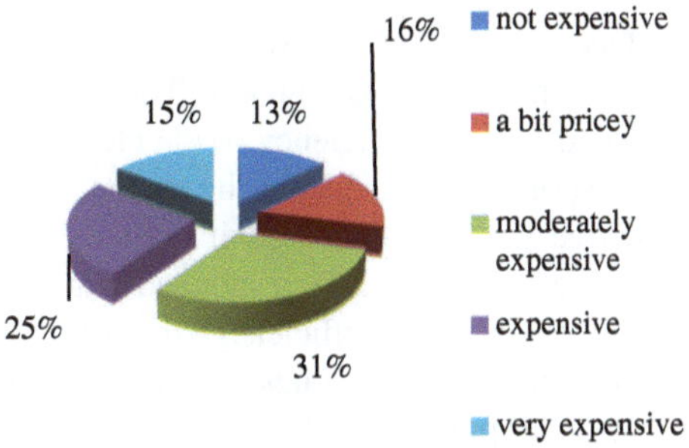

Fig. 3 Participants' opinions on the cost of the electricity bills

"moderately expensive"; 25 % considered them "expensive"; and 15 % found them "very expensive", as shown in Fig. 3.

With regard to the satisfaction of the population of Rancho Queimado in relation to the supply of electricity in the city, question 6 asked the surveyed population about the number of times per month, on average, there was a power failure in the city in 2012. None responded "never", which was the first alternative; 35 % of the respondents reported less than once a month; 42 % said from 1 to 3 times; 13 % from 3 to 6 times; 3 % from 6 to 9 times; and 2 % informed that it happened more than 9 times per month, as shown in Fig. 4. Despite being a small rural town, power outage was not frequent, which is somewhat surprising, since the quality of the electricity supply in Brazil still leaves much to be desired. According to the results presented by the National Electric Energy Agency (ANEEL), the average length of power outages in Brazil was 18 h and 42 min in 2011.

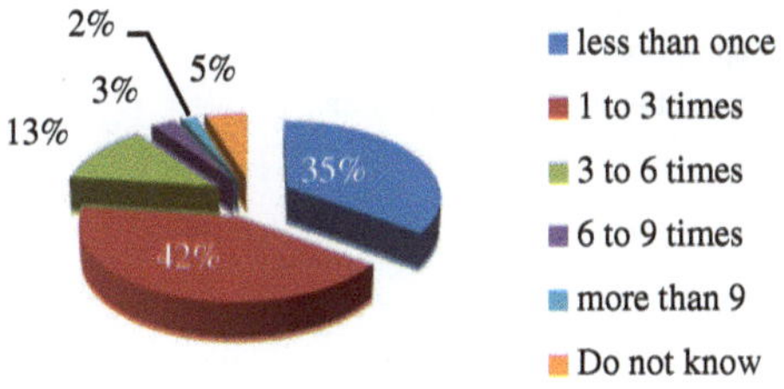

Fig. 4 Average power failure per month, as perceived by the participants

Fig. 5 Participants' opinion on the effectiveness of the power supply in Rancho Queimado

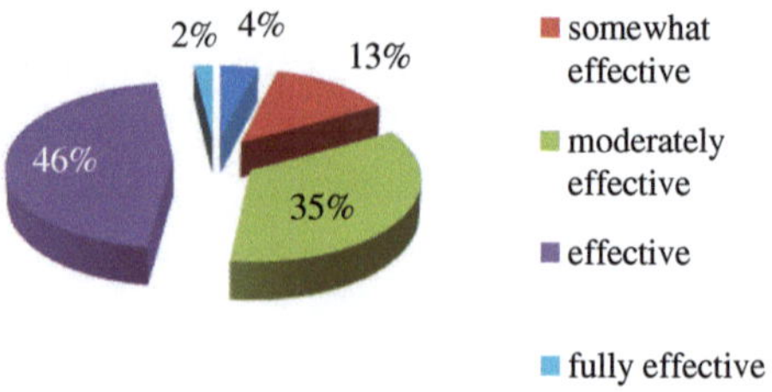

Question 7 was intended to assess the degree of satisfaction with the energy supply in Rancho Queimado. The majority of the townspeople appeared to be satisfied with the efficiency of the electric supply system of the city. In all, 46 % of the respondents considered the system "effective"; 35 % considered it "moderately effective"; 4 % considered it "not effective" and 13 % said it was "somewhat effective", whereas 2 % responded that the system was "fully effective", as shown in Fig. 5. The term "efficiency" describes, according to Hordeski (2005), the ability of equipment that operates in cycles or processes to produce the expected results.

Another purpose of this survey was to identify the townspeople's knowledge and opinions on issues related to sustainability and environment. Question 8 asked them if they had ever heard of clean and renewable energy generation. In response, only 37 % of the respondents said they had never heard about it, whereas the majority (63 %) gave an affirmative answer, which indicates that most of the population of Rancho Queimado is aware of the issue. (See Fig. 6).

Those who had heard about renewable energy were asked (question 9) to name clean or renewable power generation systems. 53 % of them pointed out wind power; 15 % mentioned solar energy; 5 % cited hydroelectric energy; 2 % indicated other sources; and 25 % did not answer, as shown in Fig. 7. The most noteworthy in these answers is the large number of respondents mentioning wind power as a clean or renewable source. Wind power is the conversion of kinetic energy contained in air masses in motion (wind). It converts kinetic energy that is present in the wind into more useful forms of energy such as mechanical energy or electricity. Its use occurs by converting translation kinetic energy into rotational kinetic energy, with

Fig. 6 Participants' awareness of clean or renewable energy generation

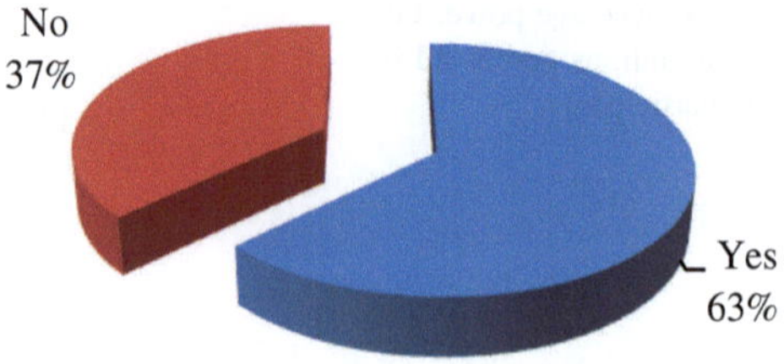

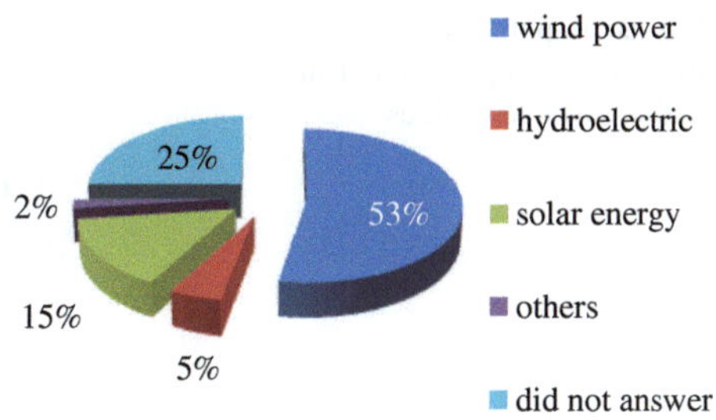

Fig. 7 Clean or renewable power systems mentioned by the participants

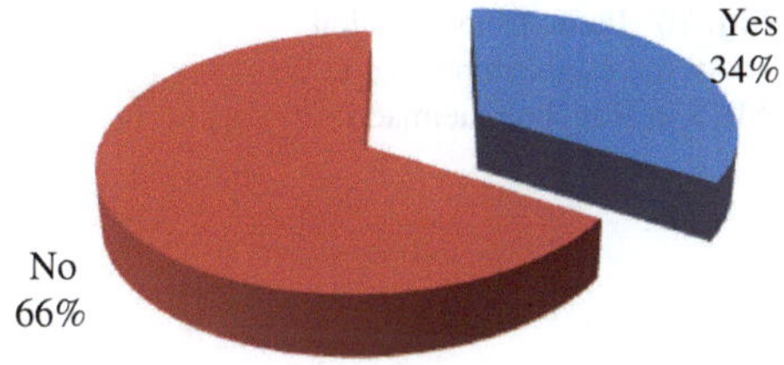

Fig. 8 Previous knowledge about micro-hydroelectric system (MHS)?

the use of wind turbines to generate electricity, or windmills to work as mechanical water pumping, for instance.

Question 10 asked whether the respondents had heard of the MHS, prior to this survey. Of the 194 valid answers, 34 % reported that they had already heard of it, whereas 66 % had no idea, as shown in Fig. 8. Clearly, the MHS is little known in the region, even though the establishment of the MHS at the Robert Schutz School has been officially announced to the community. Communication strategies should be used to disseminate the system among the local population.

Question 11 had great importance for the results of this study, since Brazil is the tenth largest consumer of electricity in the world, according to the International Energy Agency (IEA). Therefore, popular initiatives targeting the consumption of electricity from clean or renewable sources will generate a significant impact on global sustainable development. The survey participants were asked about the importance they gave to the electricity supply system in Rancho Queimado. It is remarkable to note that virtually the entire surveyed population gave importance to a sustainable energy generation system. Of the respondents, 47 % considered it "important" and 43 % considered it "very important"; 8 % believed it is "moderately important"; 2 % considered it "slightly important"; and 1 % considered clean power "not important" at all, as shown in Fig. 9.

Question 12 asked if they favoured or were against the implementation of the MHS in Rancho Queimado. The majority (81 %) favoured the MHS installation; only 3 % were against the idea, and 16 % did not answer, mainly because they did not have enough information to give an opinion or support the project, as shown in Fig. 10. Information about the benefits of the MHS implementation should be

Fig. 9 Importance given to the energy supply from clean and renewable sources

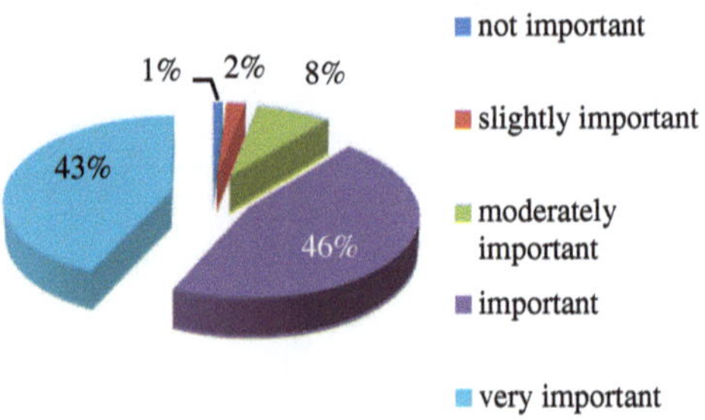

Fig. 10 Participants' opinion about the establishment of a MHS in Rancho Queimado

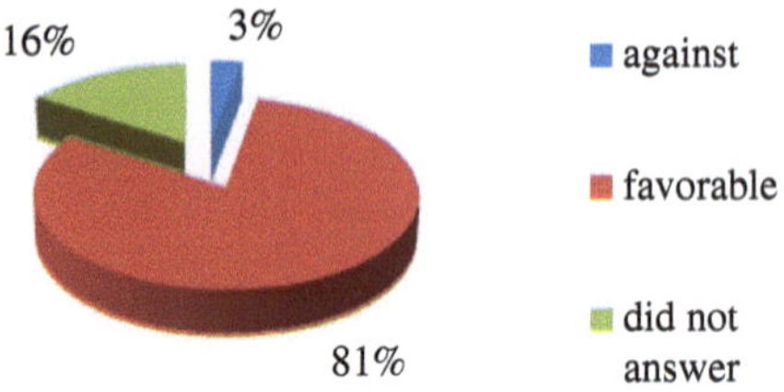

strengthened, since they are evident in rural areas that need power supply. According to the Department of Energy of the United States, the micro hydro-electricity system can be one of the simplest and most consistent forms of energy in a property.

Question 13 asked the surveyed population if the local authorities should support the implementation of the MHS in Rancho Queimado. The vast majority (83 %) of the respondents said they wanted the support of local authorities for the implementation of the MHS, as shown in Fig. 11. Clearly, the population wants government investment in sustainability projects, which is consistent with data from research commissioned by the Department of Energy and Climate Change (DECC) in England, in which 82 % of the British population supports multi-source renewable energy systems to ensure a reliable supply of energy in the country.

Fig. 11 Participants' opinion about the need for local authorities to support the implementation of MHS

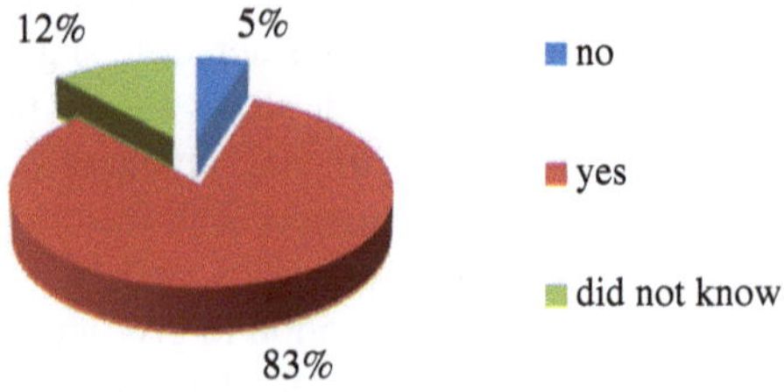

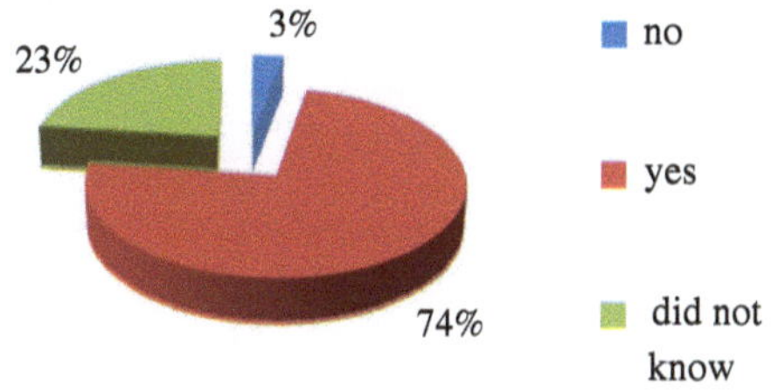

Fig. 12 Participants' opinion about the improvement of the quality of energy supply that the MHS prototype could bring to Roberto Schultz School

Question 14 asked if they believed the MHS installation at Robert Schutz School would improve the quality of the power supply in that facility. A large number (74 %) of respondents believed that the MHS would improve the power supply. Clearly, people have faith in sustainable initiatives. However, a significant number (23 %) said that they did not know if the MHS installation would improve the quality of the power supply at Robert Schutz School, stressing the need for more information on the issue, as shown in Fig. 12.

In question 15, the respondents were asked if they agreed to pay more for the electricity bill if it came from a clean and renewable power source. More than half (57 %) of the respondents said they agreed to pay more for clean or renewable energy, which shows that part of the population actually supports such initiatives, even though 32 % did not agree with higher bills, as shown in Fig. 13. It is known that the challenge of ensuring clean energy is often confronted with the question of cost, or the population's fear of an increase in electricity bills.

Questions 16–20 were intended to gather some socioeconomic information of the surveyed population. With regard to the number of people living in the same residence, the average was 3.25 persons per household, which can be considered low by the Brazilian standards. In the whole sample, the maximum number of persons per household was 7.

Question 17-collected information on gender of the participants. Out of the 198 participants who completed the questionnaire, 53 % were male and 47 % were female, as shown in Fig. 14.

In terms of age groups (question 18), 12 % of the respondents were under 18 years; 24 % were between 19 and 29 years; 19 % were between 30 and 39 years; 19 % were between 40 and 49 years; 14 % were between 50 and 59 years; and 12 %

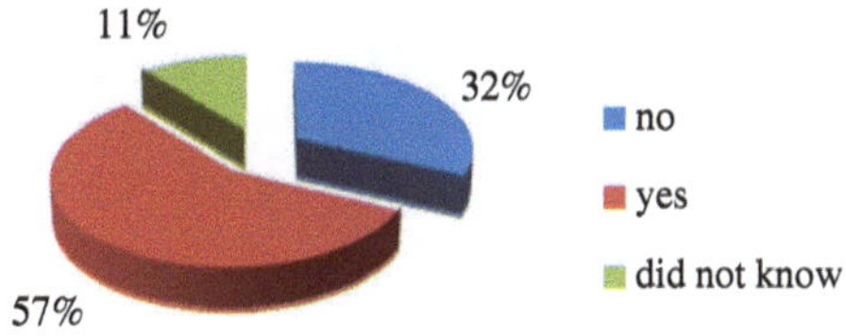

Fig. 13 Participants' willingness to pay a higher price for electricity from a clean, renewable source

Fig. 14 Gender of the surveyed population

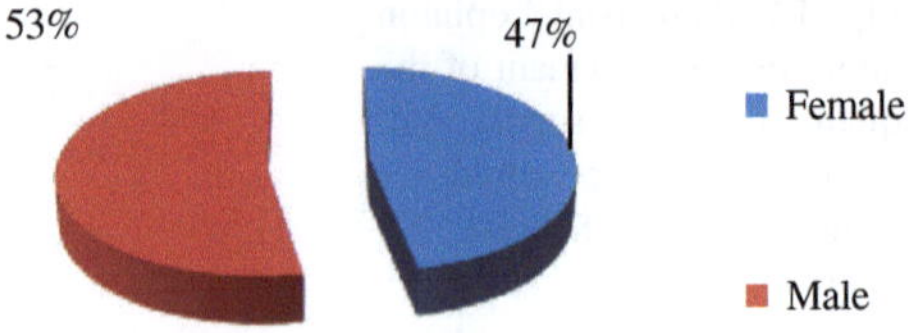

Fig. 15 Surveyed population age groups

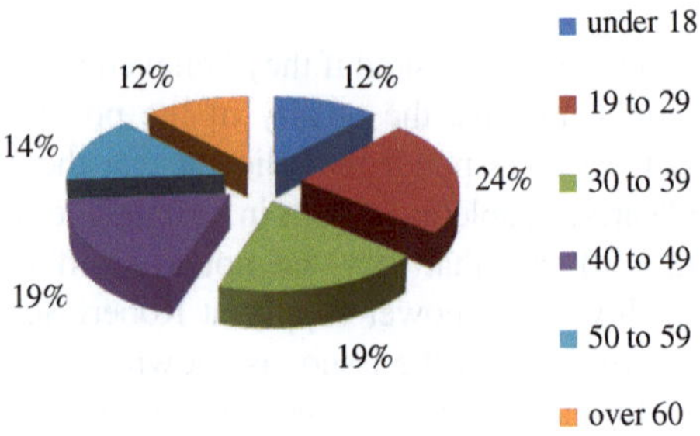

were over 60 years old, as shown in Fig. 15. The population of Rancho Queimado is well distributed among all age groups.

With regard to the education level of the survey population (question 19), out of the 194 respondents who gave valid answers, 10 % had incomplete primary education; 15 % had primary education; 10 % had incomplete secondary education; 27 % had secondary education; 13 % had incomplete higher education; 18 % hold a college degree; and 6 % hold a postgraduate degree, as shown in Fig. 16.

Fig. 16 Education level of the surveyed population

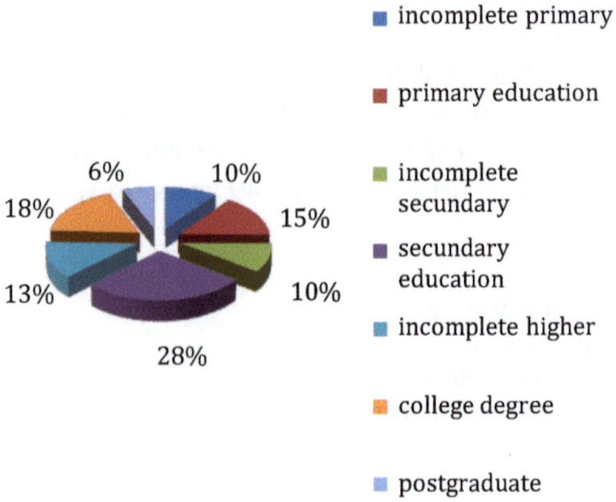

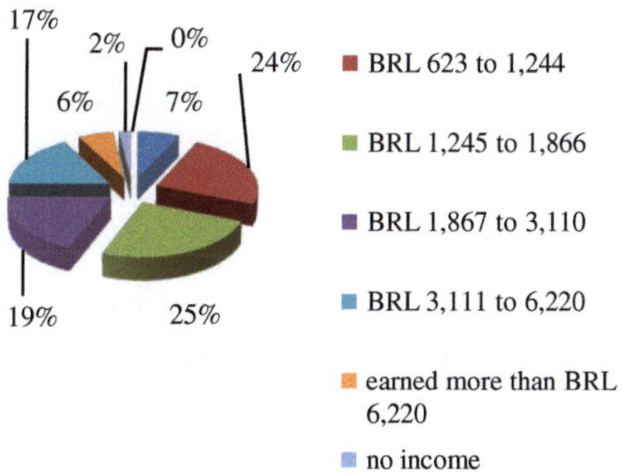

Fig. 17 Participants' household income per month (in BRL)

Question 20 asked about the household income of the surveyed population. Only 7 % earned up to BRL 622 per month, which is the minimum wage in Brazil (around $305); 24 % earned between BRL 623 and BRL 1,244; 25 % earned between BRL 1,245 and BRL 1,866; 19 % earned between BRL 1,867 and BRL 3,110; 17 % earned between BRL 3,111 and BRL 6,220; 6 % earned more than BRL 6,220, and 2 % had no income, as shown in Fig. 17.

It is interesting to observe the outcomes resulting from crosschecking some answers given to the survey questionnaire. Six cross-checkings of data were performed; three of them focused on energy supply systems available in Rancho Queimado, trying to understand the population's degree of satisfaction. The other three cross-checkings focused on the willingness of respondents to pay more for electricity from clean, renewable sources and socioeconomic profile of the surveyed population.

Since the electricity supply in Rancho Queimado is supplied by different systems, the first cross-cheking of data listed the different power supply systems used in Rancho Queimado with people's opinion of the price of the electricity bills. (Questions 3 and 5).

The population served by Celesc and by the energy cooperatives has varied opinions regarding the average values of the power bills paid monthly. The responses were well-distributed, almost uniformly throughout the options, but there was a slightly greater tendency to consider the monthly bill more expensive than cheap.

Of the respondents who used generators, one found the cost "not expensive" and another found it "moderately expensive". The occurrence of the response "other" to identify other sources of energy was cited as "expensive". The result from the first crosschecking of data can be considered satisfactory, since the cost of electricity is usually a limiting factor for power supply to communities in general. Rancho Queimado proved to be well supplied in terms of energy, with costs regarding average values of the electricity bills rated between reasonable and a little expensive (Fig. 18).

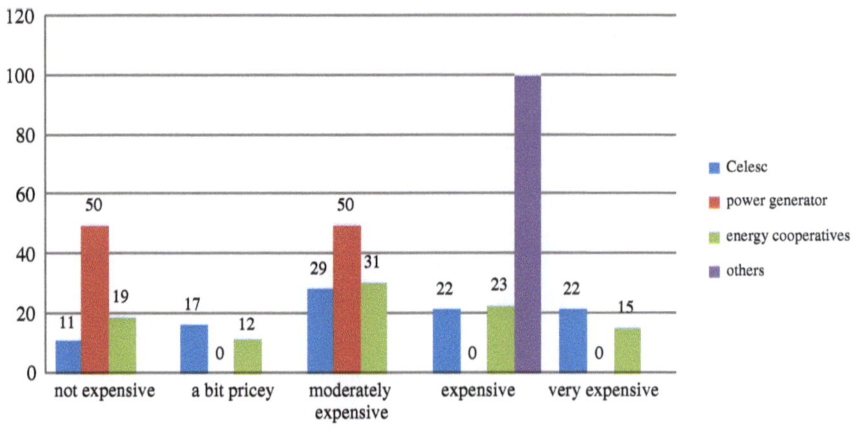

Fig. 18 Power supply sytems and the surveyed population's opinions on the monthly price of electriciy bills

The second crosschecking of data compared the respondents' opinions on the power supply system with the number of times per month that they had experienced power outage in 2012. (Questions 3 and 6).

The vast majority of the population served by Celesc had very few electricity failures per month. Forty-one percent of the respondents supplied by Celesc experienced power failures less than once a month, and 45 % had from 1 to 3 electricity outages per month, which shows that power supply by Celesc is very reliable. Among those supplied by energy cooperatives, 12 % experienced power failures less than once a month; 40 % suffered from 1 to 3 times per month; and 40 % had between 3 and 6 failures per month. The respondents who owned power generators reported that they had from 1 to 3 failures per month (Fig. 19).

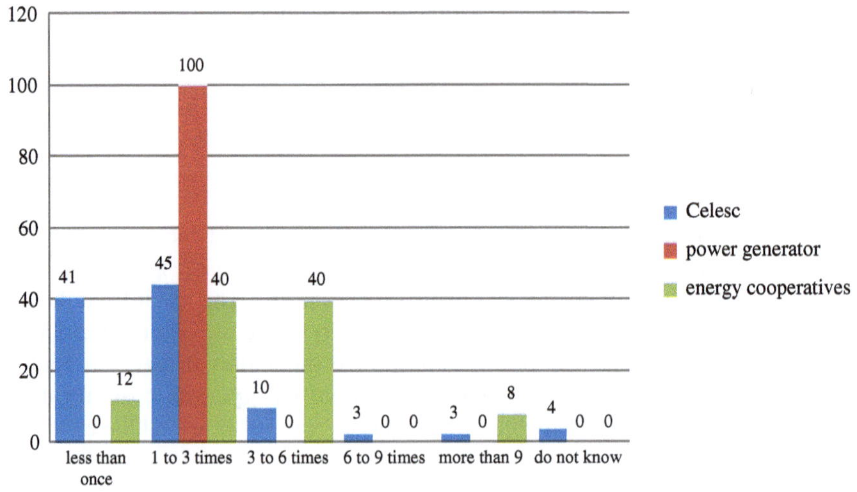

Fig. 19 Crosschecking data of power suppliers and electricity failures per month

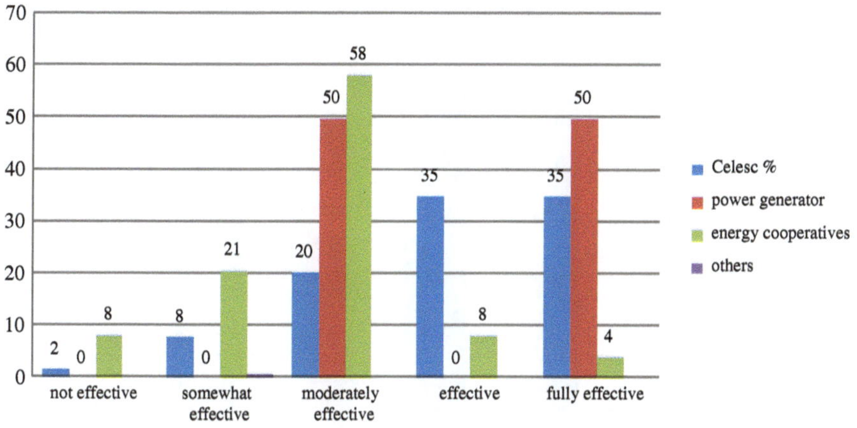

Fig. 20 Power supply systems compared with the respondents' opinion on the effectiveness of each supply system

The third crosschecking of data compared the power supply systems with the respondents' opinions regarding the efficiency of the electricity supply in Rancho Queimado. (Questions 3 and 7).

The respondents served by Celesc said that they were satisfied with the efficiency of the company. Twenty percent considered the electricity supplied by Celesc "moderately effective"; 35 % found it "effective"; and 35 % found it "fully effective", thus exceeding the level of satisfaction of those served by the energy cooperatives. Twenty-one percent of the respondents considered the electricity supplied by cooperatives "inefficient"; 58 % considered it "moderately effective." Of the respondents who owned generators, one considered it "moderately effective" and another considered it "fully effective" (Fig. 20).

The fourth crosschecking of data refers to the different age groups and their willingness to pay more for electricity from clean and renewable sources.

The Brundtland report states that it is very important to assess the impact of sustainable development projects for the different generations of people. Sustainable development is the kind of development that meets the needs of the present without compromising the needs of future generations (Brundtland 1987).

It can be clearly noted that the willingness to pay more for clean or sustainable energy is higher among young people compared to older generation. Among the age group under 18 years, 19 respondents answered "Yes" to the question, versus only 4 who answered "No". Among the respondents aged between 19 and 29 years, this number was 30 "Yes" versus 23 who replied "No". This was sharply reversed among older age groups. Among the respondents aged between 30 and 39 years, and also between 40 and 49 years, there were 18 "Yes" and 29 "No" answers in each age group. There was no "Yes" and 21 "No" answers among the age group between 60 and 69 years. However, among the age group between 50 and 59 years, there was a change in the acceptance decline curve, since 16 answered "Yes" and 15 "No" with regard to be willing to pay a higher bill for clean energy (Figs. 21 and 22).

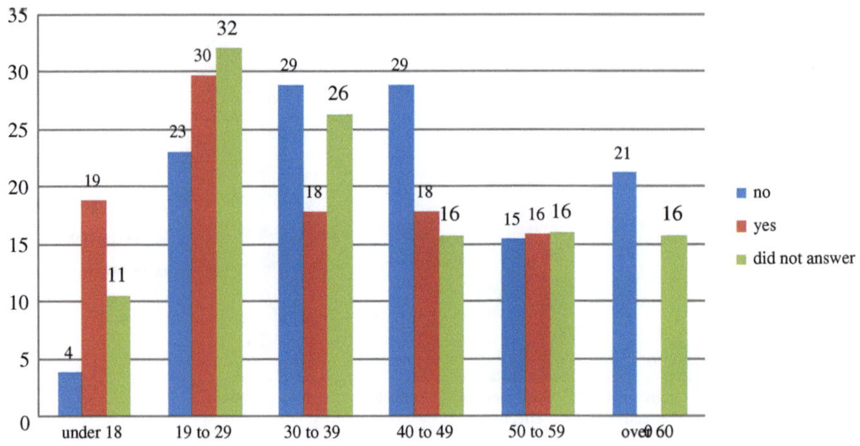

Fig. 21 Relationship between age groups and willingness to pay more for electricity generated by a renewable system

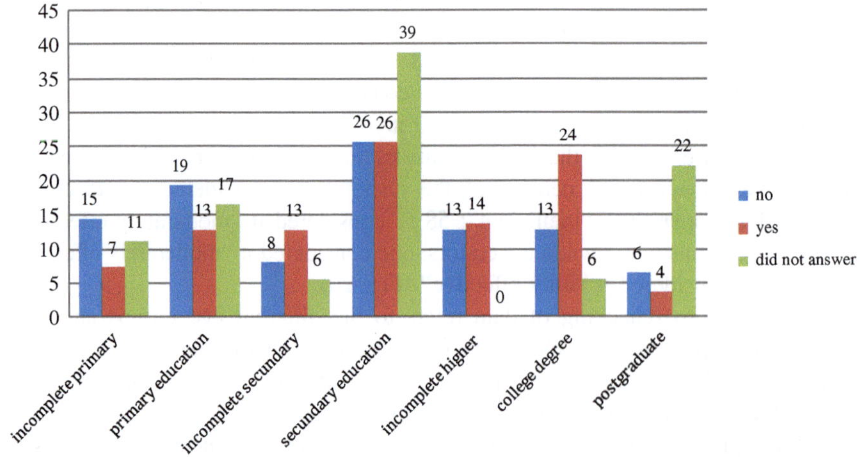

Fig. 22 Relationship between education level and willingness to pay more for electricity generated by a renewable system

The fifth crosschecking of data was related to the educational level of the respondents and their willingness to pay more for electricity from clean and renewable sources. There was a clear association between the education level of the respondents and a predisposition to invest in renewable and energy sources.

Finally, comparing the household income of the respondents and their willingness to pay more for electricity from clean and renewable sources, it could be noted that the higher the income, the greater the willingness to pay more for clean energy.

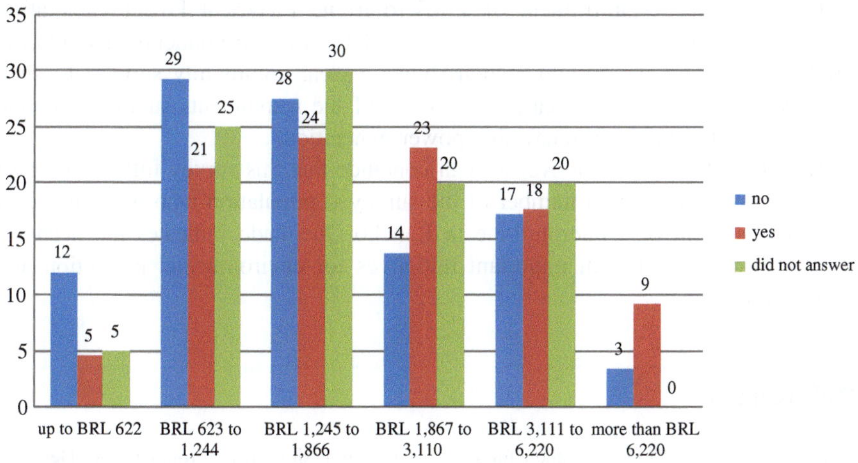

Fig. 23 Household income and willingness to pay more for electricity generated by a renewable system

Most respondents who answered that they were not willing to pay more for electricity from clean energy sources earned up to BRL 622, between BRL 623 and BRL 1,244, and from BRL 1,245 to BRL 1,866. The majority that answered, "Yes" to that question had an income between BRL 1,867 and BRL 3,110, and greater than BRL 6,220. People earning between BRL 3,111 and BRL 6,220 diverged from the rising trend of accepting a higher bill for clean energy, since 18 answered "Yes" and 17 gave a negative answer (Fig. 23).

6 Conclusion

This study was intended to measure the people of the town's degree of agreement with the MHS to be installed at Robert Schutz School in Rancho Queimado, Brazil. A twenty-question survey questionnaire was used to interview a sample of 198 subjects. Two were filter questions to figure out whether the interviewers were residents of Rancho Queimado and had electricity at home; thirteen questions were related to energy issues, and five questions addressed socioeconomic variables.

Of the 198 respondents living in Rancho Queimado, only 2 reported that they did not have electricity at home, which corresponds to 1.01 % of the respondents. This result is close to the expected, since 99.8 % of the households in the state of Santa Catarina are provided with electricity, according to the 2,000 Brazilian census.

Findings revealed that the majority of residential consumers paid between BRL 25 and BRL 300 per month for their electricity bills. Interestingly, 63 % of them have heard about renewable or clean energy, and 90 % considered important or very important to promote clean, renewable energies.

Regarding the establishment of a Microturbine Electrical Power Generation Project by REGSA, the survey revealed that 81 % of the respondents favoured the implementation of the system, which shows a clear community support to such initiatives. Findings also revealed that 83 % of the respondents supported public policies directed to clean, renewable power generation.

Based on the analysis above, we can conclude that this survey fulfilled its goal and objectives. The small number of the surveyed population who were aware of the establishment of a microturbine in Rancho Queimado indicates the need for greater disclosure of such important initiatives for environmental protection and sustainability.

References

Brundtland GH (1987) Our common future. World Commission on Environment and Development, Oxford University Press, Oxford

Edenhofer O, Madruga RP, Sokona Y (2012) Renewable energy sources and climate change mitigation: special report of the intergovernmanetal panel on climate change. Cambridge University Press, Cambridge

Gil A (1999) Métodos e Técnicas de Pesquisa Social, 5th edn. Atlas, São Paulo

Godoy AS (1995) Introdução à pesquisa qualitativa e suas possibilidades. Revista de administração de empresas 35(2)

Hadad N (2004) Metodologia de estudos em ciências da saúde. Rocca, São Paulo

Hordeski MF (2005) Emergency and backup power sources. CRC Press, Cleveland. http://www.nanotechdobrasil.com.br/eficiencia-energetica-conceitos-e-definicoes-mme/. Accessed 11 Nov 2012

IBGE. Instituto Brasileiro de Geometria e Estatística (Brazilian Institute of Geography and Statistics). http://censo2010.ibge.gov.br/. Accessed on 15 Jan 2013

IPCC. Climate change 214: mitigation of climate change. Retrieved at http://www.ipcc.ch/report/ar5/wg3/. Accessed on 19 Apr 2014

Rea LM, Parker RA (2000) Metodologia de Pesquisa. Editora Guazzelli Ltda, São Paulo

Triviños ANS (1987) Introdução à pesquisa em ciências sociais: a pesquisa qualitativa em educação. Atlas, São Paulo

Vianna WB et al (2008) Pesquisa quali-quantitiva em engenharia de produção e aspectos de cientificidade e validação. In: SEPROSUL. Bento Gonçalves, RS, Brasil

Authors Biography

José Baltazar Salgueirinho Osório De Andrade Guerra Dean and Professor at Unisul Business School (Universidade do Sul de Santa Catarina—UNISUL, Brazil). Director, at UNISUL, of two research projects: JELARE—Joint European-Latin American Universities Renewable Energies Project and REGSA—Promoting Renewable Electricity Generation in South America, both projects financed by the European Union (through ALFA III and the thematic program for environment and sustainable management of natural resources, including energy). Member of The Scientific Committees of the World Symposium on Sustainable Development at Universities

(WSSD-U-2012), a parallel event of Rio+20, and the Green Campus Summit 2013. Author and organizer of several books in the fields of Economics, International Relations Environment, Sustainability and Renewable Energy.

Youssef Ahmad Youssef Professor of the School of Business at Humber Institute of Technology and Advanced Learning in Canada. Active researcher on the subject of business, sustainability and renewable energies since 2004. Co-supervisor to two major Europe Aide' funded joint European and Latin American projects: JELARE and REGSA between 2008 and 2012. Founder and president of the Federation of Canadian Brazilian Businesses (FCBB), a not for profit organization that aims to foster sustainable innovation and entrepreneurial initiatives between Canada and Brazil. His research interests include informal entrepreneurship in emerging economies, renewable energies, entrepreneurship and sustainability and knowledge management, fields on which he has published several books and journal articles over his academic career.

Leandro Piazza dos Santos Holds a Bachelor in Management from Unisul Business School (UBS), Universidade do Sul de Santa Catarina (Unisul). At Unisul, Leandro is a volunteer researcher in the Project REGSA—Promoting Renewable Electricity Generation in South America.

Ane Cristina Figueiredo Pereira de Faria She is about to complete her Bachelor's degree in International Relations at Universidade do Sul de Santa Catarina (Unisul), Florianópolis. At Unisul, Ane is a volunteer researcher in the Project REGSA—Promoting Renewable Electricity Generation in South America.

An Innovative Model of Student-Led Sustainability in Higher Education. Green Dragons, City University London Students' Union

Maria Xypaki

Abstract

Education for Sustainable Development is about "equipping the younger generation with the broad knowledge needed to make informed decisions for the benefit of themselves and others, now and for the future" (UNESCO 2014). This paper examines how an innovative student-led sustainability project of extra-curricular activities with a bottom-up basis and a top-down motivation and capacity building support structure can deliver sustainability outcomes and inspire other institutions of higher education. Green Dragons project for City University London aims to establish pro-environmental behaviours and facilitate empowerment, while engaging academics, creating community links and affecting operations management. In order to test the effectiveness of this new model, qualitative and quantitative data were collected. The paper concludes with the actions needed by institutions like City University London for increasing the impact and for exploiting the full potential of Green Dragons project as a sustainability model that could be applied in different contexts and could deliver sustainable development in the curriculum, research, operations, community outreach and of course student participation and empowerment, contributing towards behaviour change for "a more sustainable future in terms of environmental integrity, economic viability, and a just society for present and future generations" (International Decade of Education for Sustainable Development, UN 2005).

M. Xypaki (✉)
City University London, Northampton Square, London EC1V 0HB, UK
e-mail: maria.xypaki.1@city.ac.uk

W. Leal Filho et al. (eds.), *Integrating Sustainability Thinking in Science and Engineering Curricula*, World Sustainability Series, DOI 10.1007/978-3-319-09474-8_34

Keywords
Education for sustainable development · Bottom-up approach · Pro-environmental behaviour · Empowerment · Capacity building · Innovation · Integration

1 Introduction: Need of Innovative Sustainability Models in Higher Education

Education was recognized as a fundamental prerequisite for the implementation of sustainable development in 1972, becoming more and more integral part of the governmental agenda in both industrialized and developing countries (IISD 2014). Dealing with sustainable development issues is complex and confirms the importance of acknowledging competing value positions and interests (Tytler 2012). Education for Sustainable Development (ESD) has been described as "an approach to teaching and learning based on the ideals and principles that underlie sustainability" including key issues such as human rights, poverty reduction, sustainable livelihoods, climate change, gender equality, corporate social responsibility and protection of indigenous cultures (UNESCO 2009).

Over the past 20 years, ESD has grown from an idea to a global movement; it has evolved in both maturity of understanding and in the variety of its implementation formats—from new corporate training programs to the reorienting of higher education degrees. Countries, provinces, states, regions and cities have undertaken massive strategies to use the potential of engaging their education, public awareness and training systems to improve life for all on this planet (Hopkins 2012). The most important result of these initiatives is that everyone recognizes the need to include the concept of sustainability at all levels of education (Wright et al. 2009; Arima 2009; Jones et al. 2010).

ESD has been so far approached through top-down initiatives, while in higher education the main focus is on operations management as well as on research and integration of sustainability into the curriculum "ranging from courses about sustainability to total curriculum redesign" (Lozano Garcia et al. 2006; Sterling and Scott 2008). However, there has been little progress so far. As Martin et al. (2005, p. 80) comment: "many universities have made a commitment to incorporating sustainability into higher education, but the practical implementation of these principles has proven to be more difficult and progress is frustratingly slow". It becomes clear that there is need for new models of sustainability in higher education with different priorities and approach that could progress faster and inform the existing literature and educators. This paper introduces an innovative student-led sustainability project of extra-curricular activities with a bottom-up basis and a top-down motivation and capacity building support structure that can serve as a model of sustainability in higher education. Green Dragons project, City University London Students' Union, was launched the summer of 2013 as part of Students' Green Fund (NUS 2013). This innovative project aims to primarily achieve pro-environmental behaviours and empowerment, while engaging academics, creating community links and affecting operations. The empowered students define the new sustainability structures, processes and operations.

The paper describes the genesis of the project, its main characteristics, and how its objectives fit well with the idea that ESD is about empowering the younger generation with sustainability skills (UNESCO 2014) and that building capacity in individuals is fundamental in integrating sustainability into education (O'Rafferty et al. 2014). It concludes with suggestions as to actions needed by institutions like City University London for exploiting the full potential of Green Dragons project and for increasing its impact as a sustainability model that meets the goals of the International Decade of Education for Sustainable Development (IDESD).

2 Education for Sustainable Development (ESD): A Historical Overview

Sustainable development is rising up the agenda of governments around the world, mainly due to the implications of increased CO_2 emissions and climate change aggravation with profound environmental, social and economic consequences (IPCC 2014). The first Earth Summit in 1972 identified education as fundamental for sustainable development. The idea of Education for Sustainable Development (ESD) germinated through the report of the World Commission on Environment and Development (1987) entitled "Our Common Future" (UNESCO 2005a). In 1992, the United Nations Conference on Environment and Development in Rio de Janeiro produced "Agenda 21" that further highlighted the role and importance of education as a tool to achieve sustainable development and identified four main areas: basic education, education for sustainable development programs (EDS), public awareness, training and understanding of the importance of environmental responsibility (Kommiyama and Taguchi 2006).

A few years later, the International Decade of Education for Sustainable Development (2005–2014) was established pursuing changes in behaviour for a more sustainable future in terms of environmental integrity, economic viability, and a just society for present and future generations (UN 2005; UNESCO 2005b, c). The World Conference on Education for Sustainable Development in 2009 further defined Education for Sustainable Development as "an approach to teaching and learning based on the ideals and principles that underlie sustainability" including key issues such as human rights, poverty reduction, sustainable livelihoods, climate change, gender equality, corporate social responsibility and protection of indigenous cultures (UNESCO 2009). Another landmark of ESD was United Nations Conference on Environment and Development in Rio de Janeiro in June 2012; Rio +20 (Paden 2012). The outcome document of Rio+20 titled "The future we want" (UN 2012; HEFCE 2013) recognises the significance of 'education at all levels'. It encourages institutions to adopt good practice, to teach sustainable development as an integrated component across disciplines, and to undertake research in this area.

The role of Education for Sustainable Development is to "equip the younger generation with leadership skills, management capabilities, and the broad knowledge needed to make informed decisions for the benefit of themselves and others, now and for the future, and to act upon those decisions" (UNESCO 2014). UNESCO argues that the Education for Sustainable Development should be interdisciplinary and holistic based on values determined on critical thinking and problem solving, including several methods of teaching, and be oriented to local participation (Calder and Clugston 2003).

3 ESD and Higher Education

It is difficult to imagine the individual and societal changes needed to ensure a sustainable future could occur without the assistance of institutions of higher education (Cortese 1999). With 17,000 universities in the world, higher education is a global enterprise operating collaboratively through the exchange of ideas, students and staff. The role of universities is significant not only because their research and teaching generates and transfers knowledge about sustainability, but because they educate future professionals to enable them to contribute to a sustainable future (Barth et al. 2007; Niu et al. 2010). Numerous studies in the last few years have described a variety of approaches to implementing education for sustainable development in higher education, ranging from courses about sustainability to total curriculum redesign (Lozano Garcia et al. 2006; Sterling and Scott 2008). Although there have been many initiatives to promote sustainability in higher education such as the European model "Copernicus-Campus", the results of these efforts are still unclear with many of the initiatives to promote the concept of sustainability in higher education, but with little impact on the actual education (Wright et al. 2009; Arima 2009; Jones et al. 2010).

Embedding sustainability in mainstream curricula is one of the main challenges as it needs to be supported by the development of "credible teaching materials which are fully contextualised and relevant" (O'Rafferty and Davison 2011). Haigh (2005) explains how sustainable development is missing from many degree programs because of the cross-disciplinary nature of the topic. The UK Higher Education Academy acknowledges four major barriers to the successful embedding of sustainability into many of the subject disciplines (Sterling and Witham 2008): overcrowded curriculum, perceived irrelevance by academic staff, limited staff awareness and expertise and limited institutional drive and commitment.

Another immense challenge is the broad skill set needed to practice more sustainable product design by design educators ranging from facilitation and people-centred skills to knowledge of manufacturing techniques, materials and environmental impacts (Richardson et al. 2005). Lozano (2006) believes that even though sustainable development is a radical innovation within universities, it is necessary to incorporate it incrementally to avoid resistance and unnecessary conflict. Leal Filho (Leal Filho 2010; Zint 2011) also supports that many universities perform activities in the area of environmental protection neglecting the fact that campus operations are important contributions, but not the only ones.

4 ESD and the Different Contexts

The relation between education and sustainable development is complex. Generally research shows that basic education is to nation's abilities to develop and achieve sustainable targets (Tilbury et al. 2002). The different contexts where Education for Sustainable Development is practiced differ greatly in terms of their socio-political, historical, socio-cultural, ecological, and economic realities that respectively affect the settings and priorities of their educational institutions (Kopnina and Meijers 2014). For instance, some types of ESD with a focus on social equity and health have been more appropriate in developing countries, whereas an ecological focus and behaviour change has been more appropriate in more affluent Western societies (Zarger 2010). Different contexts also mean different challenges and barriers for the implementation of ESD (Wright et al. 2009; Arima 2009; Jones et al. 2010). For example, some of the barriers identified in European countries include inflexible education systems, problems co-ordinating interdisciplinary education and the reconciliation of long-term sustainability objectives with short-term political expediency (UN 2005). In a completely different context such as China, the regional differences, the development issues and the social needs arisen due to the economic growth are the ones that mainly prevent the implementation of ESD (Niu et al. 2010).

5 Green Dragons: An Innovative Model of Student-Led Sustainability

Green Dragons (2013) is the first student-led sustainability project in City University London commissioned by the Higher Education Funding Council for England (HEFCE) and the National Union of Students (NUS 2013). City University London (2014) is an institution of higher education in the United Kingdom with 17,000 students from more than 150 countries. As part of Green Dragons project, City students are motivated by professional staff to deliver practical sustainability solutions in the campus and/or the local community by receiving funding, training and support. Students are encouraged to mainly address issues of energy and emissions reduction, waste and recycling, ethical procurement, elimination of retail plastic bottles, sustainable food and Fairtrade, food waste reduction and community engagement. A survey commissioned by City University in association with the NUS—entitled "Insight into student and staff environmental attitudes, behaviors and the impact of existing schemes"—clearly demonstrated that these were areas with significant student disengagement in the university. However, students are the ones who finally decide what sustainability issues they are going to address and in what way (bottom-up structure). Waste reduction, sustainable food, emissions reduction and community engagement have currently been the main focus of student leaders.[1]

The establishment of pro-environmental behaviours is one of the main objectives of Green Dragons project. Student leaders are expected to adopt pro-environmental behaviours by actually working on their sustainability project with the support of professional staff and academics; the direct and/or indirect engagement of other City students/staff in the leaders' projects is expected to facilitate a spillover of sustainable behaviors in the whole campus. This is practically achieved through the pledges each project should get from City students/staff in order for the project leaders to receive their delivery funding. This approach is consistent with DEFRA's 4Es model (Encourage—Enable—Engage—Exemplify) on changing behavior (DEFRA 2010). It also agrees with Lozano's (2006) approach suggesting that for successful behavior change incremental integration of sustainability within universities is needed to avoid resistance and unnecessary conflict.

Acknowledging that building capacity in individuals is an integral part of integrating sustainability into education (O'Rafferty et al. 2014) students are provided with weekly mentoring and training (empowerment) by professionals from across City University, i.e. Students' Union, Marketing, Energy and Environment, Careers and Enterprise teams in order to design and deliver their sustainability solutions (project plans). Students gain sustainability skills such as environmental knowledge, leadership, collaboration, project management, entrepreneurship, team building, communication and civic skills, while boosting their confidence and employability (Cohen 2007; Warburton 2003; Sterling and Thomas 2006; Barth et al. 2007; HEA 2013). Students are also being given the opportunity to practice sustainability skills ("skill-acts") as empowerment also involves the opportunities for developing and gaining skills, competence and confidence (Chavis and Wandersman 1990; Brady et al. 1995). This active learning that focuses on student engagement, interaction with topics and reflection on learning (Zeegers and Clark 2014; Martin et al. 2005) offers a balanced perspective on sustainability. Students also participate in another empowering process important for capacity building (Minkler 1990, p. 293), which is the creation of links (social capital) with internal and external stakeholders according to students' own initiatives and ideas (Putnam 1993). Finally, practical support such as resources, information and networking is being offered by professionals in the university based on the leaders' individual needs for the successful implementation of their sustainability projects (Stoker 2004, p. 214).

[1] For more details, visit Green Dragons official website: http://www.green-dragons.co.uk/.

6 Research Methodology

6.1 Qualitative and Quantitative Method

The methodology employed during this study was based on the collection of qualitative and quantitative data in order to test the hypothesis that a sustainability model for higher education such as Green Dragons (see Fig. 1) can integrate sustainability into higher education by primarily achieving pro-environmental behaviours and empowerment (sustainability skills). Qualitative investigation recognizes the importance of the subjective and its experiences (Silverman 2000, p. 20), while interviews are one of the major approaches in collecting data in qualitative research (Kvale 2007, p. 2). For this study, the 8 student leaders were asked open-ended questions (17–21 of March, 2014) that captured their beliefs and attitudes in regards to how their environmental behaviour and skills changed during their participation in Green Dragons project. In March, the 8 Green Dragons Leaders also completed a skills self-assessment questionnaire, also filled out at the beginning of their projects (November, 2013) that allowed for useful comparisons.

The skills self-assessment questionnaire (see Table 1) captured students' attitudes and behaviours (environmental awareness was also examined as a skill). The quantitative data collected address the limitations of the qualitative method such as the issues of anonymity, confidentiality and bias (Burns 2000, pp. 13–14). The findings have the precision of quantitative analysis (Burns 2000, pp. 9–10) supported by qualitative analysis, while students' beliefs, attitudes and behaviours are equally captured (Gollwitzer 1999; Ajzen 1971). The secondary objectives-outcomes (academic engagement, community links, operations) are investigated

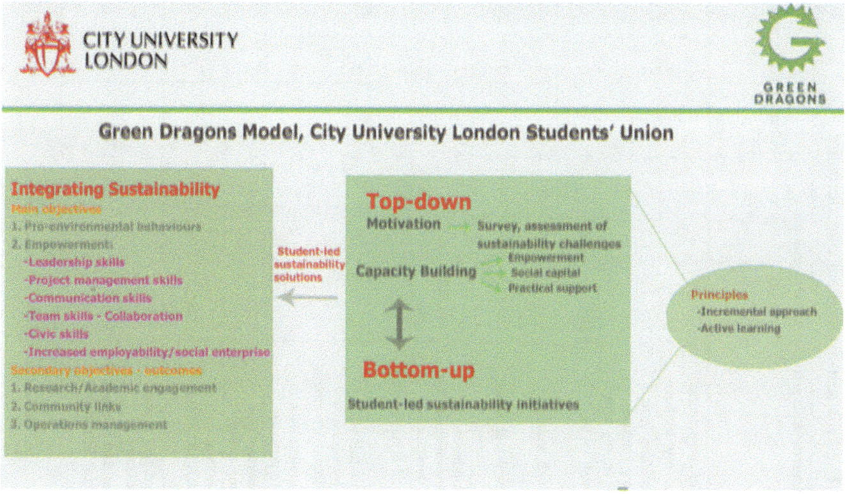

Fig. 1 Green Dragons model

Table 1 Skills self-assessment questionnaire

Green dragons skills self-assessment questionnaire	Strongly agree	Agree	Neither agree nor disagree	Disagree	Strongly disagree	Count	Hydrovillage leader	Meat free mondays leader	Smart sugar crystals leader	Waste no more leader	Restart project	PIEL leader	EASTRO leader	City green moniter
Skills and competences (November 2013)														
I meet all my deadlines	1	4	2	0	1	8	2	3	2	3	2	5	1	2
I am confident with prioritising multiple demands on my time	2	2	4	0	0	8	3	3	1	3	2	3	2	1
I work well under pressure	2	6	0	0	0	8	2	2	1	2	2	2	2	1
I work well without supervision	4	2	2	0	0	8	2	2	1	3	1	3	1	1
I work well with other people	2	5	1	0	0	8	2	2	1	3	2	1	2	2
I am confident with taking responsibility for making decisions	3	3	0	2	0	8	4	2	2	4	1	1	1	2
I am good at motivating others	1	2	4	1	0	8	2	3	1	4	2	3	3	3
I am confident with putting forward new ideas to improve a project	2	5	1	0	0	8	2	2	1	3	1	2	2	2
I am confident with managing projects	2	2	2	2	0	8	3	4	2	3	1	4	2	1
I am confident with organising events	1	2	4	1	0	8	3	3	1	3	2	3	2	4
I am confident with managing/planning a budget	2	4	0	1	1	8	5	2	1	2	2	4	1	2
I am confident giving presentations	1	3	1	2	1	8	1	4	2	3	2	5	2	4
I am confident at chairing or speaking in formal meetings	1	3	2	1	1	8	1	3	2	3	2	5	2	4

(continued)

Table 1 (continued)

Green dragons skills self-assessment questionnaire	Strongly agree	Agree	Neither agree nor disagree	Disagree	Strongly disagree	Count	Hydrovillage leader	Meat free mondays leader	Smart sugar crystals leader	Waste no more leader	Restart project	PIEL leader	EASTRO leader	City green moniter
I am confident with writing formal emails/reports	4	3	0	1	0	8	1	4	1	2	1	2	1	2
I am confident with talking with various people	2	3	2	1	0	8	3	3	1	2	2	2	1	4
I am confident with meeting new people	1	5	2	0	0	8	2	2	1	2	3	2	2	3
I am confident with building good working relationships with various people	1	4	2	1	0	8	3	2	1	3	2	4	2	2
I am confident with learning a new skill	3	5	0	0	0	8	2	2	1	2	1	1	2	2
I seek feedback from others to improve how I approach tasks	3	3	1	1	0	8	1	2	1	4	2	1	3	2
I am confident with providing solutions in difficult situations	2	3	2	1	0	8	3	2	1	4	1	2	2	3
I am confident with providing excellent customer service	1	3	3	1	0	8	2	2	2	3	3	1	3	4
I am confident with using microsoft office (word, excel, powerpoint, outlook)	7	0	0	0	1	8	1	5	1	1	1	1	1	1
For the sake of the environment, car users should pay higher taxes	1	1	3	2	1	8	4	1	3	5	2	3	3	4
I always try to recycle my paper coffee cups	4	3	0	0	1	8	5	1	1	2	2	1	1	2
I am happy getting involved in voluntary activities	4	4	0	0	0	8	1	2	1	1	1	1	2	2

(continued)

Table 1 (continued)

Green dragons skills self-assessment questionnaire														
	Strongly agree	Agree	Neither agree nor disagree	Disagree	Strongly disagree	Count	Hydrovil lage leader	Meat free mondays leader	Smart sugar crystals leader	Waste no more leader	Restart project	PIEL leader	EASTRO leader	City green moniter
I am confident that I know how to reduce my carbon emissions on a daily basis	4	4	0	0	0	8	2	1	1	2	2	2	1	1
Skills and competences (March 2014)														
I meet all my deadlines	1	5	2	0	0	8	3	2	2	2	2	3	1	2
I am confident with prioritising multiple demands on my time	2	3	1	2	0	8	3	4	1	2	2	4	2	1
I work well under pressure	2	4	1	1	0	8	2	4	1	2	2	3	2	1
I work well without supervision	4	3	0	1	0	8	4	2	1	2	1	2	1	1
I work well with other people	4	3	1	0	0	8	1	3	1	1	2	1	2	2
I am confident with taking responsibility for making decisions (March, 2014)	4	4	0	0	0	8	2	2	2	1	1	1	1	2
I am good at motivating others (March, 2014)	4	4	0	0	0	8	1	2	1	1	2	2	1	2
I am confident with putting forward new ideas to improve a project	5	3	0	0	0	8	1	2	1	1	1	1	2	2
I am confident with managing projects (March, 2014)	4	3	1	0	0	8	1	3	2	2	1	1	2	1
I am confident with organising events	3	3	2	0	0	8	1	3	1	3	2	1	2	2
	3	5	0	0	0	8	2	2	1	1	2	2	1	2

(continued)

Table 1 (continued)

Green dragons skills self-assessment questionnaire	Strongly agree	Agree	Neither agree nor disagree	Disagree	Strongly disagree	Count	Hydrovil lage leader	Meat free mondays leader	Smart sugar crystals leader	Waste no more leader	Restart project	PIEL leader	EASTRO leader	City green moniter
I am confident with managing/planning a budget (March, 2014)														
I am confident giving presentations (March, 2014)	5	3	0	0	0	8	1	2	2	1	2	1	1	1
I am confident at chairing or speaking in formal meetings (March 2014)	1	6	1	0	0	8	1	2	2	3	2	2	2	2
I am confident with writing formal emails/reports	4	3	0	1	0	8	1	4	1	2	1	2	1	2
I am confident with talking with various people (March, 2014)	3	5	0	0	0	8	ll	2	1	2	2	2	1	2
I am confident with meeting new people	2	5	1	0	0	8	1	2	1	2	2	2	2	3
I am confident with building good working relationships with various people (March, 2014)	2	6	0	0	0	1	1	2	1	2	2	2	2	2
I am confident with learning a new skill	3	4	0	1	0	8	1	4	1	2	1	2	2	2
I seek feedback from others to improve how I approach tasks (March, 2014)	2	6	0	0	0	8	1	2	1	2	2	2	2	2
I am confident with providing solutions in difficult situations	3	3	2	0	0	8	1	3	1	2	1	2	2	3
	2	3	2	1	0	8	1	3	2	2	3	4	1	2

(continued)

Table 1 (continued)

Green dragons skills self-assessment questionnaire														
	Strongly agree	Agree	Neither agree nor disagree	Disagree	Strongly disagree	Count	Hydrovillage leader	Meat free mondays leader	Smart sugar crystals leader	Waste no more leader	Restart project	PIEL leader	EASTRO leader	City green moniter
I am confident with providing excellent customer service (March 2014)														
I am confident with using Microsoft office (word, excel, PowerPoint, Outlook)	5	2	0	1	0	8	1	4	1	2	1	2	1	1
For the sake of the environment, car users should pay higher taxes (March 2014)	2	5	1	0	0	8	1	2	2	2	1	3	2	2
I always try to recycle my paper coffee cups (March 2014)	5	3	0	0	0	8	2	1	1	2	2	1	1	1
I am happy getting involved in voluntary activities (March 2014)	6	2	0	0	0	8	1	2	1	1	1	1	2	1
I am confident that I know how to reduce my carbon emissions on a daily basis (March 2014)	7	1	0	0	0	8	1	1	1	1	2	1	1	1

1 Strongly agree, *2* agree, *3* neither agree or disagree, *4* disagree, *5* strongly disagree

through examples as they occur from the current practice. The data presentation strategy is *Thematic* under the main and secondary objectives of Green Dragons project.

6.2 Limitations

Green Dragons is a newly established project. Therefore, more data need to be collected at the end of the project both in years 1 and 2 by repeating the same interview questions and by completing the same skills self-assessment questionnaire.

7 Integrating Sustainability into Higher Education

7.1 Findings

This chapter discusses the findings of the 8 student interviews and skills self-assessment questionnaires completed by Green Dragons Leaders at City University London in November 2013 and March 2014 respectively. The findings are discussed under pre-existing codes, which are the Green Dragons main objectives (pro-environmental behaviour and empowerment) and secondary ones (academic engagement, community links and operations). It becomes obvious that the students perceive their lifestyles as more environmentally friendly, while they have actually adopted more sustainable habits since their involvement in Green Dragons project. Green Dragons Leaders have also become more empowered by developing leadership, project management, communication, team and civic skills. As for the secondary objectives, a network of community partners has already being formed around the leaders' projects, academics are engaged as mentors in the research projects and students' sustainability solutions have also affected operations in the university.

7.2 Main Objectives

7.2.1 Pro-environmental Behaviours

In their interviews, the student leaders expressed the belief that their involvement in the project helped them adopt more sustainable habits. "Green Dragons remind me of the "green" theme anytime. I walk longer every day to reduce the emissions, recycle more than before, and switch off the monitor of my computer. I am also trying to persuade others to do that" (City Green Monitor Leader). The students also learnt more about sustainability. "This project has taught me a lot about sustainability… it is a great experience" (Smart Sugar Crystals Leader).

The skills self-assessment questionnaire data are represented in column charts. Column chart (1) shows that 5 students strongly agreed with recycling their cups in

March in comparison to 4 students who strongly agreed in November. No students disagreed in March, whereas 1 student strongly disagreed in November. Based on the data of Table 1, 2/8 students recycle more since the beginning of the project and 6/8 students recycle the same. Column chart (2) shows that 5 students in March strongly agreed that car users should pay higher taxes in contrast with just 1 in November. Table 1 shows that 6/8 students have a more pro-environmental behaviour and 2/8 have the same behaviour. The findings of column chart (3) show that 7 students in March strongly agreed with knowing how to reduce their carbon emissions in comparison to only 2 in November. All students knew how to reduce their carbon footprint in March. Table 1 shows that 3/8 students have developed more pro-environmental behaviours, while 5/8 haven't seen any difference in their behaviour. Overall, students have developed more pro-environmental behaviours during their participation in the project (11 positive changes in total).

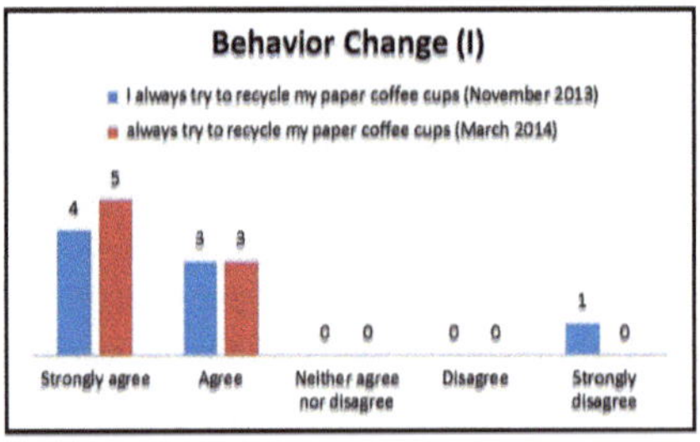

Column Chart 1

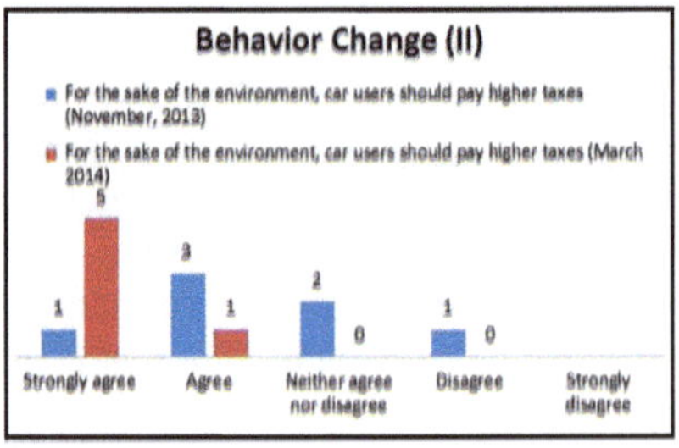

Column Chart 2

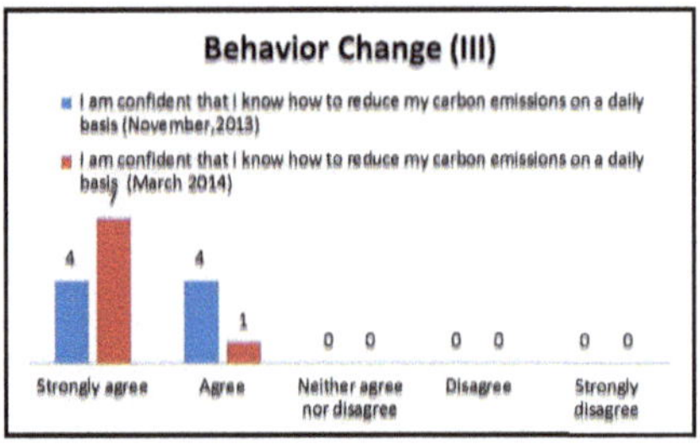

Column Chart 3

Besides the pro-environmental changes in the leaders' behaviour, it is important to mention that more than 2,500 students/staff members have been involved in the project both directly and indirectly (see Table 2). Even though there are currently no data to prove a wider behaviour change in the university at this phase of the project, the current impact remains promising.[2]

---

[2] The "Insight into student and staff environmental attitudes, behaviours and the impact of existing schemes" survey will be repeated at the end of Green Dragons project in order to accurately measure behaviour change in City University London.

Table 2 Green Dragons Impact (November 2013–March 2014), updated last: 30th of March, 2014

Impact of Green Dragons project, City University London Students' Union November, 2013–March, 2014
1949 Unique page views of Green Dragons microsite
1080 Pledges on Green Dragons microsite
105 Facebook "likes"
118 Twitter followers
1,101 students engaged across the challenge: • 40 directly engaged • 1,061 indirectly engaged
20 staff engaged across the challenge: • 18 directly engaged • 2 indirectly engaged

7.2.2 Empowerment

Green Dragons Leaders supported in their interviews that they have more confidence to manage their own sustainability projects since they took part in Green Dragons project. "I am now doing something about an issue that was concerning me, I feel in control. The project has given me more confidence to do something positive [...] it is good for me as well" (Waste No More Leader). All the students supported that their participation in the project has equipped them with project management, communication and team skills, while increasing their confidence and employability opportunities."It made me much more employable and confident to follow ideas, meet a lot of people, develop management and business skills" (Hydrovillage Leader). PIEL Leader also said "I learnt new skills such as blogging. I am aware about how people interact in a business sense, how to write formal emails, how to network with different people" and Meat Free Mondays Leader emphasised gaining time management skills. "I am doing something that I am feeling positive about. I learnt how to deal better with my time... This project is developing so many crucial skills such as communication, time and money management".

Leadership Skills

Based on the data from the skills self-assessment questionnaire, column chart (4) shows that 4 students strongly agreed in March that they are good at motivating others in comparison to just 1 in November. However, 2 students agreed in March in comparison to 4 students that agreed in November. In March, no students disagreed about their ability to motivate others. Table 1 shows that 6/8 students feel more empowered to motivate others since their Green Dragons project began, while 2/8 students feel as empowered as when the project first started. Column chart (5) shows that 4 students in March highly agreed that they are confident with taking responsibility for making decisions in comparison to 3 in November. All students were confident in making decisions in March. Table 1 shows that 2/8 students feel more empowered to take responsibility for making decisions. Column chart (6) shows that 5 students strongly agreed that they are confident to suggest new ideas in March (column chart 3) in comparison to only 2 in November. All students were confident in March. 3/8 students feel more empowered and 5/8 feel as empowered in November as in March regarding decision-making. The data above show that Green Dragons project has helped students develop their leadership skills (11 positive changes in total).

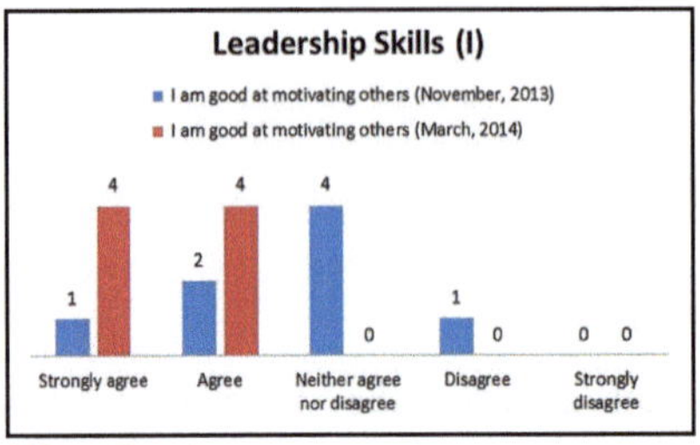

Column Chart 4

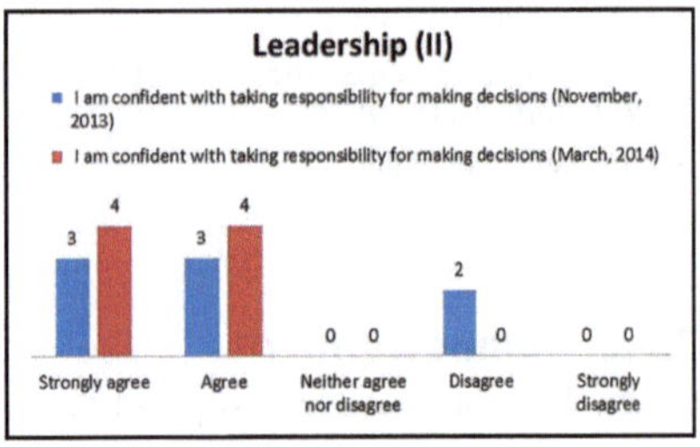

Column Chart 5

Column Chart 6

Project Management Skills

In November, 2 students strongly agreed that they have the confidence to manage projects (column chart 7); this figure doubled in March. 4/8 students feel more empowered as far as project management is concerned, whereas 4/8 students feel as confident as when the project first started. According to column chart (8), 3 students strongly agreed that they can manage/plan a budget in November, whereas this number doubled again in March. All students felt confident to plan/manage a budget in March. In total, 3/8 students feel more confident to manage/plan a budget and 5/8 students have the same confidence as when the project started. There have been a total of 7 positive changes in relation to how empowered students feel to manage a project since November.

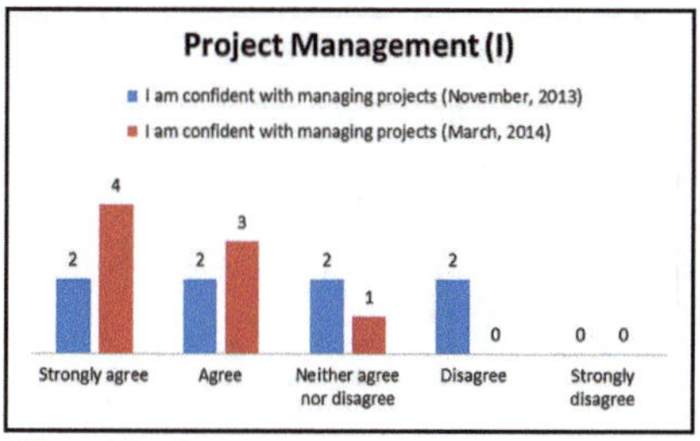

Column Chart 7

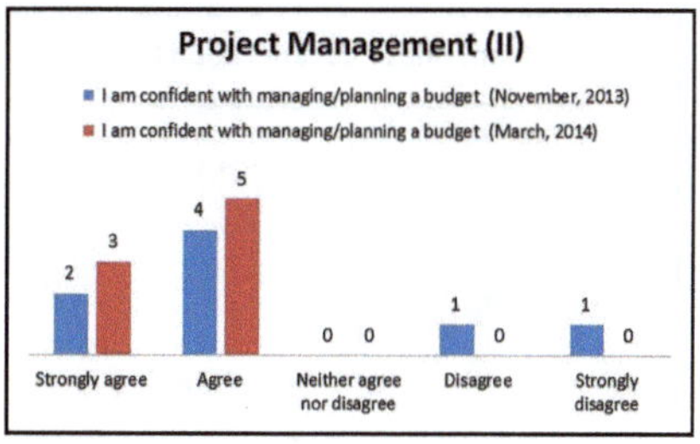

Column Chart 8

Communication Skills

In March, 5 students strongly agreed that they have the confidence to give presentations, while only 1 student expressed such confidence in November. All students had such confidence in March (column chart 9). 5/8 students felt that they had better communication skills after their participation in Green Dragons project. In November, 2 students felt very confident to talk with various people, while 3 students felt that level of confidence in March (column chart 10). According to the data of Table 1, 3/8 students are more empowered in terms of communicating with various people. Overall, there have been 8 positive changes in students' communication skills for the period November 2013–March 2014.

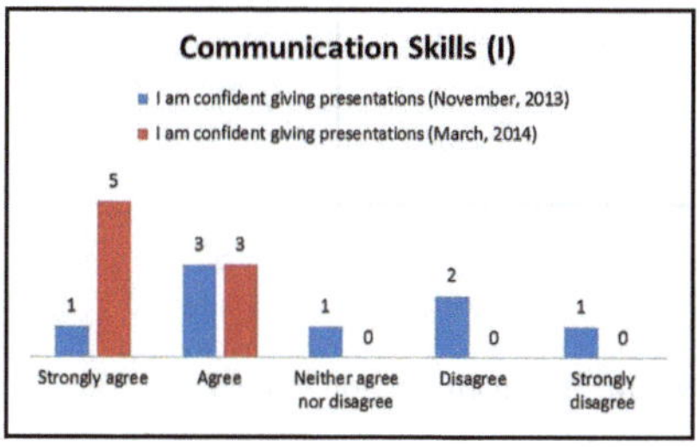

Column Chart 9

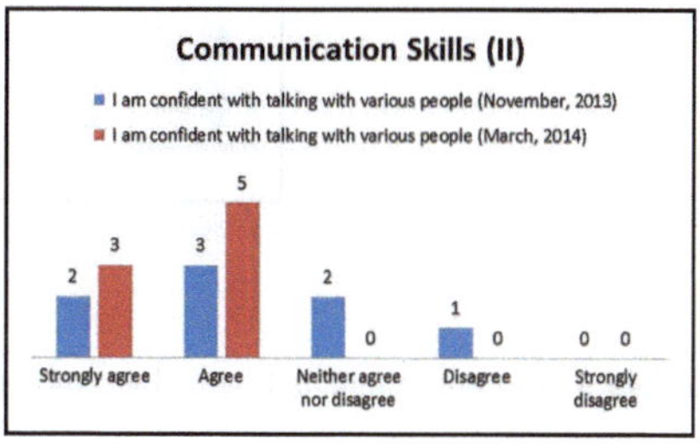

Column Chart 10

Team Skills—Collaboration

At the beginning of the project, 5 students agreed to have the confidence to build good working relationships with various people, while 6 students felt the same in March (column chart 11). In that same month, there was no student disagreeing having this ability. According to Table 1, 3/8 students felt more confident to build good working relationships with various people. Even though in November, only 3 students agreed that they sought feedback to improve the way they approach tasks (column chart 12), 3 more students developed this approach in March. Based on Table 1, 2/8 students seek more feedback from others, whereas 1/8 seeks less feedback. Overall, there have been 5 positive changes in students' team skills and a negative one.

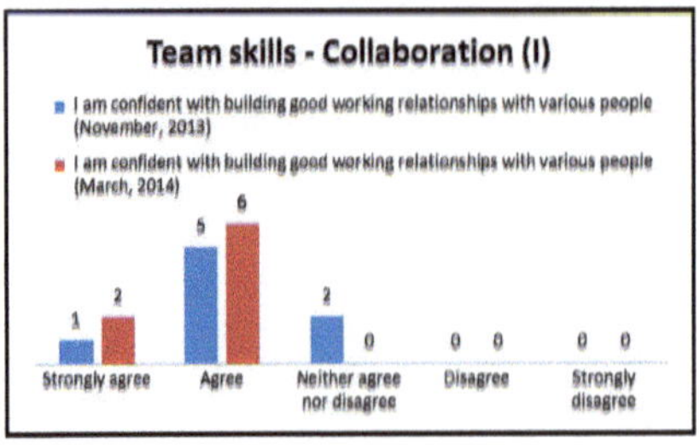

Column Chart 11

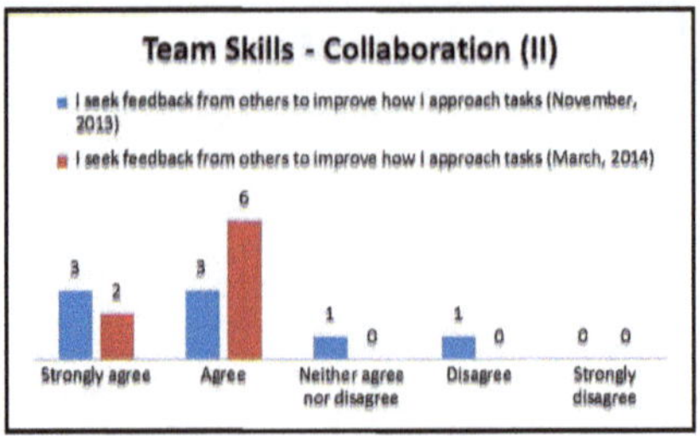

Column Chart 12

Civic Skills

In November, 4 students strongly agreed that are happy getting involved in voluntary activities whereas in March, 2 more students expressed the same opinion (column chart 13). Table 1 shows that 2/8 students expressed a more positive attitude regarding participation in the wider community and 6/8 feel the same. Overall, there have been 2 positive changes (Table 3).

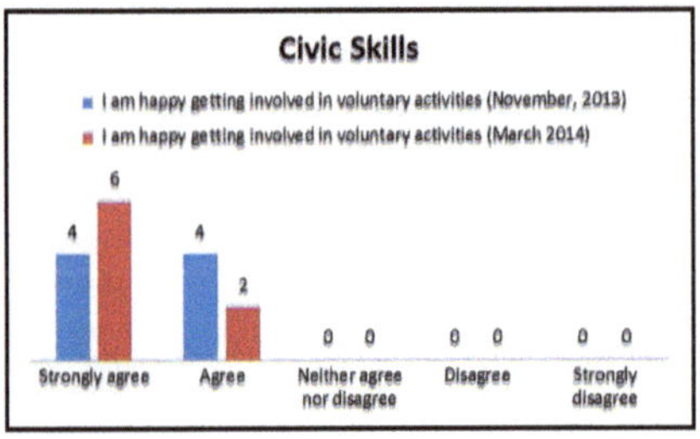

Column Chart 13

Table 3 Pro-environmental behaviour and empowerment, Green Dragons model

Summary table: pro-environmental behaviour and empowerment, Green Dragons model		
	Positive changes	Negative changes
Behaviour change	11 in total	–
Empowerment skills	33 in total	1 in total
Leadership skills	11 in total	–
Project management skills	7 in total	–
Communication skills	8 in total	–
Team skills—collaboration	5 in total	1 in total
Civic skills	2 in total	–

7.3 Secondary Objectives-Outcomes

7.3.1 Academic Engagement and Research

City Green Monitor and Smart Sugar Crystals Leaders have linked their projects with their PhD research; they closely work with their tutors and other academics from City University London.

7.3.2 Community Links

There is an increasing network of community partners around Green Dragons project. For example, Meat Free Mondays' project is being mentored by the Vegetarian Society (Vegsoc 2014), Waste No More has partnered with CRISP charity (CRISP 2014) and Client Earth (2014) has created links with PIEL team.

7.3.3 Operations

Waste No More Leader introduces new reuse services in the University, while City Green Monitor Leader is currently running experiments for creating the first eco-lab in City campus.

8 Conclusion

Can a student-led sustainability project of extra-curricular activities significantly contribute to the integration of sustainability into higher education by empowering students and building new sustainability structures, processes and operations based on students' ideas? A qualitative and quantitative evaluation of the above findings confirms that students participating in the project actually developed pro-environmental behaviours and sustainability skills. Green Dragons project has clearly empowered the student leaders with an impact on more than 2,500 City University students. It has also encouraged new procedures and structures, i.e. Green Dragons steering group meetings bring together different teams from across the university to work towards sustainability, new partnerships have been established with the community, academics are engaged in Green Dragons research projects and new environmentally-friendly operations are being introduced. Finding the right communication tools for effective engagement, partnering with strategic stakeholders, tailoring unique engagement approaches based on the portfolio of each partner, understanding internal politics and overcoming people's natural resistance to change through insistence and consistency have been the key actions to address limited staff awareness, the lack of institutional drive (Sterling and Witham 2008) and clashing values and mentalities.

However, Green Dragons project hasn't yet exploited its full potential. For the 8 months of the project, the more than 2,500 students engaged shows a significant impact, but taken that about 15,000 City students haven't been engaged directly or indirectly means that a large part of the student body neither has participated in sustainability skill-acts/training and mentoring nor has it influenced the creation of new sustainability capacity building structures. In order to fully exploit the potential of Green Dragons project, especially after the end of the two-year NUS funding, it should be officially recognized and promoted by the senior management of the university. For example, by being included in the University's Vision 2016 (City University London 2014), more streams of funding could be encouraged, new internal and external collaborations could be pursued and integration between the different university teams could be easily achieved. The new joint-up processes, structures and communication channels as well as the enriched skill sets available to student leaders could result in greater student engagement as well as in more effective top-down support and empowering opportunities. Academic involvement in the project has so far been limited to research. By strategically targeting and linking academics to the existing Green Dragons projects could lead to the development of "credible teaching materials which are fully contextualised and relevant" (O'Rafferty and Davison 2011). Sustainability could then be integrated into the curriculum through extra-curricular activities defined and "legitimized" by students.

Green Dragons project is an innovative student-led sustainability project of extra-curricular activities with a bottom-up basis and a top-down motivation and capacity building support structure that encourages behaviour change for "a more sustainable future in terms of environmental integrity, economic viability, and a just society for present and future generations" (International Decade of Education for Sustainable Development). Green Dragons can serve as a model of holistic integration of sustainability into higher education which due to its bottom-up structure can be applied to different contexts. Targeted communication, strategic partnerships, engagement of senior management and academics, adequate funding and joint-up thinking can address initial resistance to change and unveil the full potential of Green Dragons as a bottom-up model that empowers the sustainability leaders of tomorrow and delivers sustainable development in the curriculum, research, operations and community outreach.

References

Ajzen I (1971) Attitudinal vs. normative messages: an investigation of the differential effects of persuasive communications on behaviour. Sociometry 34:263–280

Arima A (2009) A plea for more education for sustainable development. Sustain Sci 4:3–5

Barth M, Godemann J, Rieckmann M, Stoltenberg U (2007) Developing key competencies for sustainable development in higher education. Int J Sustain High Educ 8(4):416–430

Brady EH, Verba S, Schlozman LK (1995) Beyond SES: a resource model of political participation. Am Political Sci Rev 89(2):271–294

Burns R (2000) Introduction to research methods, 4th edn. SAGE Publications, London

Calder W, Clugston RM (2003) International efforts to promote higher education for sustainable development. Plan High Educ 31:30–44

Hopkins C (2012) Twenty years of education for sustainable development. J Edu Sustain Dev 6 (1):1–4

Chavis DM, Wandersman A (1990) Sense of community in the urban environment: a catalyst for participation and community development. Am J Community Psychol 18(1):55–81

City University London (2014) The city vision 2016, http://www.city.ac.uk. Last accessed 24 Feb 2014

Client Earth (2014) Client earth homepage, http://www.clientearth.org/. Last accessed 24 Feb 2014

Cohen B (2007) Developing educational indicators that will guide students and institutions toward a sustainable future. New Dir Inst Res 134:83–94

Cortese AD (1999) Education for sustainability—the university as a model for sustainability. Second Nature, Boston. Second Nature, http://secondnature.org/pdf/snwritings/articles/univmodel.pdf. Last accessed 26 Feb 2014

CRISP (2014) CRISP homepage, http://www.crispej.org.uk/. Last accessed 02 Mar 2014

DEFRA (2010) Understanding and influencing behaviours: a review of social research economics and policy making in Defra. Department for Environment, Food and Rural Affairs, http://archive.defra.gov.uk/evidence/series/documents/understand-influence-behaviour-discuss.pdf. Last accessed 02 Mar 2014

Green Dragons (2013) Green dragons homepage, http://www.green-dragons.co.uk/. Last accessed 27 Feb 2014

Gollwitzer PM (1999) Implementation intentions: strong effects of simple plans. Am Psychol 54:493–503

Haigh M (2005) Greening the university curriculum: appraising an international movement. J Geogr High Educ 29(1):31–48

HEA (2013) Student attitudes and skills for sustainable development. Higher education academy, www.heacademy.ac.uk/resources/detail/sustainability/2013_student_skills_final_report. Last accessed 05 Mar 2014

HEFCE (2013) Sustainable development in higher education, consultation on a framework for HEFCE. Higher education funding council for England, http://www.hefce.ac.uk/pubs/year/2013/201331/name,83723,en.html. Last accessed 05 Mar 2014

IPCC (2014) IPCC homepage. Intergovernmental panel on climate change, http://www.ipcc.ch/. Last accessed 05 Mar 2014

IISD (2014) National strategies for sustainable development. International Institute for Sustainable Development, http://www.iisd.org. Last accessed 05 Mar 2014

Jones P, Selby D, Sterling S (eds) (2010) Sustainability Education: perspectives and practice across higher education. Earthscan, London

Kommiyama H, Taguchi K (2006) Sustainability science: building a new discipline. Sustain Sci 1 (1):1–6

Kopnina H, Meijers F (2014) Education for sustainable development (ESD), exploring theoretical and practical challenges. Int J Sustain High Edu 15(2):188–207

Kvale S (2007) Doing interviews. SAGE Publications Ltd., London

Leal Filho W (2010) Teaching sustainable development at university level: current trends and future needs. J Baltic Sci Edu 9(4):273–284

Lozano R (2006) Incorporation and institutionalization of SD into universities: breaking through barriers to change. J Clean Prod 14(9–11):787–796

Lozano Garcia FJ, Kevany K, Huisingh D (2006) Sustainability in higher education: what is happening? J Clean Prod 14(9–11):757–760

Martin S, Brannigan J, Hall A (2005) Sustainability, systems thinking and professional practice. J Geogr High Edu 29(1):79–89

Minkler M (1990) Improving health through community organization. In: Glanz K, Lewis FM, Rimer BK (eds) Health behaviour and health education: theory, research and practice. Jossey-Bass Publishers, San Francisco, 257–287p

Niu D, Jiang D, Li F (2010) Higher education for sustainable development in China. Int J Sustain High Educ 11(2):153–162

NUS (2013) 25 students' green fund projects. National Union of Students, http://www.studentsgreenfund.org.uk/. Last accessed 27 Feb 2014

O'Rafferty S, Davison J (2011) Interdisciplinarity, design thinking & sustainable development: strategies for UK higher education. Higher Education Academy, York

O'Rafferty S, Curtis H, O'Connor F (2014) Mainstreaming sustainability in design education—a capacity building framework. Int J Sustain High Educ 15(2):169–187
Paden M (2012) Rio+20 conference set for 20–22 June 2012. J Edu Sustain Dev 6:5–7
Richardson J, Irwin T, Sherwin C (2005) A scoping report for the sustainable design forum, design council, London
Putnam DR (1993) The prosperous community. Am Prospect 4(13):35–42
Putnam RD (1995) Bowling alone: America's declining social capital. J Democracy 6(1):65–78
Silverman D (2000) Doing qualitative research. SAGE Publications. London
Sterling S, Thomas I (2006) Education for sustainability: the role of capabilities in guiding university curricula. Int J Innov Sustain Dev 1(4):349–370
Sterling S, Scott W (2008) Higher education and ESD in England: a critical commentary on recent initiatives. Environ Educ Res 14(4):399–412
Sterling S, Witham H (2008) Pushing the boundaries: the work of the Higher Education Academy's ESD project. Environ Educ Res 14(4):399–412
Stoker G (2004) Transforming local governance: from Thatcherism to new labour. Palgrave MacMillan, Basingstoke
Tilbury D, Stevenson B, Fien J, Schreuder D (2002) Education and sustainability responding to global challenges. IUCN—The World Conservation Union, Switzerland
Tytler R (2012) Socio-scientific issues, sustainability and science education. Res Sci Educ 42 (1):155–163
UN (2005) UN Decade of Education for sustainable development 2005–2014. United Nations http://unesdoc.unesco.org/images/0014/001416/141629e.pdf. Accessed 27 Feb 2014
UN (2012) The future we want. United Nations www.un.org/en/sustainablefuture. Accessed 27 Feb 2014
UNESCO (2005a) United Nations decade of education for sustainable development (2005–2014), Framework for the International Implementation Scheme, 32 C/INF.9. United Nations Educational, Scientific and Cultural Organization http://unesdoc.unesco.org/images/0013/001311/131163e.pdf. Accessed 27 Feb 2014
UNESCO (2005b) Guidelines and recommendations for reorienting teacher education to address sustainability. United Nations Educational, Scientific and Cultural Organization Education for Sustainable Development in Action. Technical Paper, 2, 74
UNESCO (2005c) United Nations Decade of Education for Sustainable Development (2005–2014): International Implementation Scheme. United Nations Educational, Scientific and Cultural Organization. Education Sector, Paris. ED/DESD/2005/PI/01, 31
UNESCO (2009) World Conference on Education for Sustainable Development. United Nations Educational, Scientific and Cultural Organization,www.esd-worldconference-2009.org. Accessed 27 Feb 2014
UNESCO (2014) Education for Sustainable Development. Canadian Commission for United Nations Educational, Scientific and Cultural Organization, http://unesco.ca/en/home-accueil/esd-edd. Accessed 24 Feb 2014
Vegsoc (2014) Vegetarian Society Homepage, Vegetarian Society https://www.vegsoc.org/. Accessed 27 Feb 2014
Warburton K (2003) Deep learning and education for sustainability. Int J Sustain High Educ 4 (1):44–56
Wright JS, Habit E, Adlerstein S, Semrau DJ (2009) Sustainability education through an interdisciplinary international case study. Sustain Sci 4:29–36
Zarger RK (2010) Learning the environment. In: Lancy D, Bock J, Gaskins S (eds) The anthropology of learning in childhood. AltaMira Press, Lanham. pp 341–369
Zint M (2011) Evaluating education for sustainable development programs. In Leal Filho W (ed) World trends on education for sustainable development. Peter Lang, Frankfurt. pp 329–348
Zeegers Y, Clark FI (2014) Students' perceptions of education for sustainable development. Int J Sustain High Educ 15(2):242–253

Author Biography

Maria Xypaki City Green Challenge Coordinator, Maria Xypaki, received her degree in political science from Panteion University in Athens, Greece, where she first became interested in sustainability and international development. After working for research institutes on environmental security and energy diplomacy, she pursued her master's in Bartlett School of Planning, UCL. There she explored the implementation of triple bottom line sustainability in an urban context. Prior to join City University London, she worked for not-for-profit organizations on community development, environmental behaviour change, Corporate Social Responsibility and social innovation projects. Maria would like to thank all of her students for their contribution to this paper as well as for their promise for a more sustainable future.

Revealing Complexity in Educating for Sustainability: An Update on the Work of the Roundtable on Environment and Sustainability

Paul A. Barresi, Will J. Focht, Michael A. Reiter, Richard C. Smardon, Michael Humphreys, Kimberly D. Reiter and Steven A. Kolmes

P.A. Barresi (✉)
Environmental Management Program, Southern New Hampshire University,
2500 North River Rd., Manchester, NH 03106-1018, USA
e-mail: p.barresi@snhu.edu

W.J. Focht
Department of Political Science, Oklahoma State University, 228 Murray Hall,
Stillwater, OK 74078, USA
e-mail: will.focht@okstate.edu

M.A. Reiter
Department of Integrated Environmental Science, Bethune-Cookman University,
640 Dr. Mary McLeod Bethune Blvd., Daytona Beach, FL 32114-3099, USA
e-mail: reiterm@cookman.edu

R.C. Smardon
Department of Environmental Studies, SUNY College of Environmental Science
and Forestry, 1 Forestry Drive, Syracuse, NY 13210, USA
e-mail: rsmardon@esf.edu

M. Humphreys
B. J. Moore Center for Integrated Environmental Science, Bethune-Cookman University,
640 Dr. Mary McLeod Bethune Blvd., Daytona Beach, FL 32114-3099, USA
e-mail: humphreysm@cookman.edu

K.D. Reiter
Department of History, Stetson University, 421 North Woodland Blvd., DeLand
FL 32723, USA
e-mail: kreiter@stetson.edu

S.A. Kolmes
Department of Environmental Studies, University of Portland,
5000 North Willamette Blvd., Portland, OR 97203, USA
e-mail: kolmes@up.edu

W. Leal Filho et al. (eds.), *Integrating Sustainability Thinking in Science and Engineering Curricula*, World Sustainability Series, DOI 10.1007/978-3-319-09474-8_35

Abstract
The Roundtable on Environment and Sustainability was convened to provide academics and others with a collaborative forum in which to pursue their commitment to providing students with the tools needed to craft holistic strategies for meeting sustainability challenges. In the first effort of its kind, the Roundtable is developing a framework for designing, recognizing, and assessing academic programs that prepare students to engage in the holistic, adaptive management of the interactions of the systems at the human-environment interface needed to support both stewardship of the natural environment and long-term improvement in the quality of life for human individuals and communities. The Roundtable tentatively has identified as pedagogically optimal at least eight supradisciplinary skills, fourteen supradisciplinary perspectives on sustainability challenges, and twelve supradisciplinary contexts for applying the two. Revealing the complexity of the set of interactions among human and environmental systems is the most important pedagogical touchstone of the Roundtable's approach. This revealed complexity is partly a function of the conceptual resolution of the interacting systems into sets of subsystems, and has at least five dimensions. Viewing sustainability challenges from the supradisciplinary perspectives helps students to acquire a holistic understanding of the set of interactions among systems crucial to meeting those challenges.

Keywords
Complexity · Environment · Pedagogy · Systems · Sustainability

1 Introduction

The sustainability concept increasingly has become the aspirational guiding light of environmental higher education, notwithstanding the lack of a firm consensus on its precise meaning and scope. A consensus on the competencies that graduates of environmental degree programs should be able to demonstrate remains equally elusive, despite a growing recognition that programs seeking to impart those competencies must transcend a traditional, discipline-based framework (see Caldwell 1983; Orr 1990; Wilke 1995; Soule and Press 1998; Maniates and Whissel 2000; Wiek et al. 2011). Surveys of environmental programs in US colleges and universities sponsored by the Council of Environmental Deans and Directors (CEDD) of the National Council for Science and the Environment have established baselines for understanding the evolving landscape of environmental higher education (Vincent 2010; Vincent et al. 2012). Some progress also has been made toward developing a broadly applicable framework for externally reviewing environmental degree programs (see Smardon 2011). If these programs are to produce graduates capable of crafting practical strategies for meeting the pervasive sustainability challenges of the modern world, however, much work remains to be done on pedagogical approaches, curriculum design, program structure, administration, and other issues.

The Roundtable on Environment and Sustainability was founded with this formidable task in mind. The Roundtable is an ongoing, inclusive collaboration of academics and others committed to providing students in environmental degree programs with the tools needed to craft practical management strategies for meeting the pervasive sustainability challenges that exist at the human-environment interface. The Roundtable's proposals build on discussion groups organized by the Interdisciplinary Environmental Association, CEDD's work, and other sources,[1] and are in use or are being considered for use in sustainability-focused programs at several universities (e.g., Reiter et al. 2009). The Roundtable hopes that its proposals ultimately will serve as a template for the design, recognition, and assessment of sustainability-focused environmental programs throughout the academy.

2 The Relevant Academic Domain

The Roundtable has proposed "Environment and Sustainability" as the academic domain of relevance to sustainability-focused environmental programs. The goal of this supradisciplinary field is the holistic, adaptive management of the interactions of the systems at the human-environment interface needed to support both stewardship of the natural environment and long-term improvement in the quality of life for human individuals and communities. This goal recognizes the extent to which the interactions among human systems and the systems that environ humanity shape the fate of both. It also implies that neither stewardship of the natural environment nor long-term improvement in the quality of life for human individuals or communities can be achieved without an approach to managing the interactions among those systems that is both holistic and adaptive (Fig. 1). This goal could be achieved in the context of any of the three general approaches to environmental education—the *systems science* approach, the *policy and governance* approach, and the *adaptive management* approach (Vincent 2010)—that currently prevail in the US (Reiter et al. 2012, p. 111).

3 Achieving Disciplinary Synthesis

The more complete the disciplinary synthesis achieved by a given environmental program, the more effective that program is likely to be in achieving the goal of the Environment and Sustainability field (Table 1). *Supradisciplinary* approaches are

[1] As of this writing, the Roundtable has convened eight times since March 2009 at locations throughout North America. Reiter et al. (2011, 2012) offer comprehensive accounts of the Roundtable's work through its seventh meeting in September 2011. This chapter summarizes the results of those earlier meetings, and offers a detailed account of the Roundtable's subsequent work. Parts of this chapter have been summarized and adapted from the earlier publications. As a comparison of the earlier publications with this chapter will reveal, the Roundtable's name, the conceptual boundaries of the academic domain on which its work focuses, and the details of its approach to education within that domain have evolved over time.

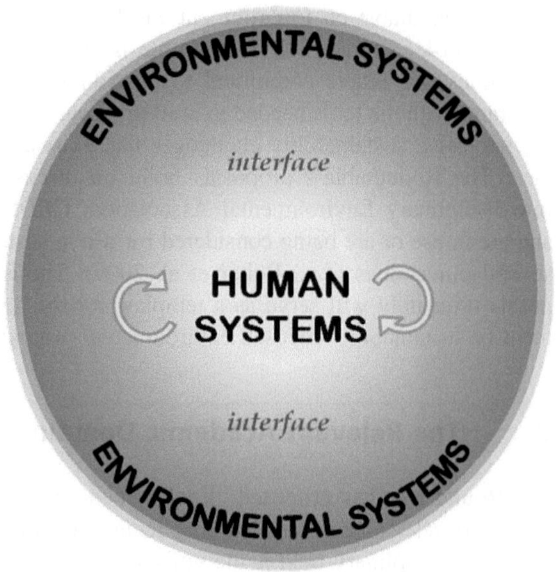

Fig. 1 The Roundtable's conception of the global ecosystem, showing the interface between human and environmental systems on which the Environment and Sustainability field focuses. The fuzziness of the interface recognizes that in an age that many scientists have dubbed the Anthropocene (see, e.g., Steffen et al. 2011), few if any systems that environ humanity are wholly non-human

likely to be the most effective, although *interdisciplinary* approaches are likely to be the most practical in most academic settings given the formidable institutional barriers to achieving supradisciplinarity that exist throughout the academy. *Pluri-disciplinary* approaches could play a valuable transitional role in the evolution of a program toward the more highly effective approaches.

Moreover, the Roundtable tentatively has identified as pedagogically optimal for achieving the goal of the Environment and Sustainability field at least eight supradisciplinary skills, fourteen supradisciplinary perspectives on sustainability challenges, and twelve supradisciplinary contexts for applying the two (Table 2). Different institutions or degree programs at the same institution might choose to prioritize certain skills or perspectives over others, or to devote more time and energy to different application contexts. Viewing sustainability challenges from the supradisciplinary perspectives helps students to acquire a holistic understanding of the set of interactions among systems crucial to meeting those challenges. The elaboration of the finer points of how this aspect of the Roundtable's pedagogical approach could be implemented, as well as the explication of and development of learning outcomes associated with the skills, perspectives, and application contexts, remain on the Roundtable's agenda.

4 Program Structure

Some program structures are likely to be more conducive than others to the more holistic approaches to education in the Environment and Sustainability field (Table 3). Most current environmental programs have a Conical or Inverted Conical

Table 1 Approaches to disciplinarity and their implications for the likely effectiveness of programs in achieving the goal of the Environment and Sustainability field

Approach	Connotation	Symbol	Disciplinary perspective	Example	Pedagogical implications for achieving the goal of the field
Unidisciplinary	One, single, alone		Views sustainability from a single discipline's perspective	History, geology	Unlikely to be effective
Multidisciplinary	Many, several		Invokes many disciplinary perspectives applied separately	Chemistry + biology	Unlikely to be effective
Transdisciplinary	Across		Recognizes points of intersection among otherwise separate disciplines	Historical archeology, geophysics	Unlikely to be effective
Pluridisciplinary	*Belonging to many*		*Recognizes some overlap among disciplines, but preserves the distinctions among them*	*Cultural psychology*	*Moderately likely to be effective*
Interdisciplinary	Among		Views sustainability from a perspective synthesized from multiple disciplines, but applied as a coherent whole	Natural ecology, American studies	Highly likely to be effective
Supradisciplinary	Above, over, beyond		Ignores disciplinary boundaries entirely	Sustainability	Most likely to be effective

Table 2 Pedagogically optimal supradisciplinary skills, perspectives on sustainability challenges, and contexts for applying the two in environmental degree programs aimed at achieving the goal of the Environment and Sustainability field

Skills	Perspectives	Application contexts
Analysis	Governance	Problem-solving
Synthesis	Risk	Research
Discernment	Ecologic	Conflict management
Conceptualization	Optimization	Advocacy
Communication	Civic	Planning
Collaboration	Ethical	Adaptive management
Creativity	Justice	Design
Imagination	Entrepreneurial	Assessment, mitigation, and restoration
	Cultural	Education
	Aesthetic	Modelling
	Physiologic	Introspection
	Dialectic	Field experience
	Synergistic	
	Participatory	

structure, although the Double Conical, Hourglass, and Extensional structures would be more conducive to the implementation of the more holistic pedagogical approaches.

5 Administration

Roundtable participants agreed that the optimal administrative status for an Environment and Sustainability program would be that of a department with its own promotion standards, tenure requirements, and programmatic learning outcomes. Perhaps the most important administrative constraint on the effectiveness of such a program would be whether administrators above the department or program level recognize the value of, and are willing to reward and to provide resources for pursuing, holistic approaches to sustainability-focused education.

6 Program Recognition

The Roundtable has proposed that any formal effort to recognize Environment and Sustainability programs as such should rely on a truth-in-advertising-based certification approach rather than a value-judgment-based accreditation approach. The multifaceted scope of the Environment and Sustainability field means that graduating students with different strengths and weaknesses from different Environment and Sustainability programs—or even from the same Environment and Sustainability

Table 3 Structural options for Environment and Sustainability programs and their implications for the implementation of the more holistic pedagogical approaches, whether interdisciplinary (ID) or supradisciplinary (SD)

Structure	Connotation	Symbol	Pedagogical implications
"Conical"	Broad base, then a narrowing of perspective		Least conducive to the more holistic approaches (ID or SD)
"Inverted Conical"	Narrow base, then a broadening of perspective		Least conducive to the more holistic approaches (ID or SD)
"Double Conical"	*Narrow base, then a broadening of perspective, then a re-narrowing of perspective (e.g., through the specialized application of knowledge and skills in a capstone course)*		*More conducive to the more holistic approaches (ID or SD)*
"Hourglass"	*Broad base, then a narrowing of perspective, then a re-broadening of perspective (e.g., through the synthesis of knowledge and skills in an integrative capstone course)*		*More conducive to the more holistic approaches (ID or SD)*
"Extensional"	Simple to complex, but always broad in perspective		Most conducive to the more holistic approaches (ID or SD)

program—is a virtue rather than a flaw because it corresponds to the diversity of sustainability challenges faced by students, employers, and the world as a whole. In this respect, the Environment and Sustainability field contrasts sharply with unidisciplinary fields, where a more or less monolithic consensus exists with respect to what every graduate of a program in the field should know and should be able to do across a relatively narrow range of competency categories. Institutional program assessment efforts will need to be correspondingly broad and flexible without losing sight of either the capacity for positive criticism or what the varied programmatic designs are intended to emphasize.

7 The Pedagogical Role of Complexity

Environment and Sustainability degree programs—whatever their names or the substantive contexts in which they pursue the goal of the field—seek to prepare students to engage in the holistic, adaptive management of the interactions of the systems at the human-environment interface needed to support both stewardship of the natural environment and long-term improvement in the quality of life for human individuals and communities. They do so in part by revealing to students the complexity of the set of interactions among the systems at the human-environment interface of relevance to the goal of the field.

8 Complexity Defined

Complexity in this sense is a measure of the number and diversity of the interactions among the systems at the human-environment interface of relevance to the goal of the Environment and Sustainability field in a given subject matter context, and thus also is a measure of the number and diversity of the systems themselves. Complexity increases as the number or diversity of the interactions among systems increases, and thus increases as the number or diversity of the systems themselves increases. This complexity is only partly an emergent quality of the systems and their interactions, however. As a standard of measurement, complexity also is partly a function of which interactions the human observer chooses to measure, which in turn is a function of how the observer chooses to parse the internal structure of—and to delimit the external boundaries of—the set of systems the interactions of which the observer purports to be measuring. Thus, complexity in the Environment and Sustainability context is simultaneously both an objective measure and a subjective measure of the number and diversity of the interactions among systems at the human-environment interface of relevance to the goal of the field.

8.1 The Complexity of Relevance to Program Recognition

The complexity of relevance to the recognition of programs in the Environment and Sustainability field is the complexity of the set of interactions among the systems at the human-environment interface revealed to students by those programs. This *revealed complexity* serves as a principal measure of the degree to which any given program approximates the field's pedagogical ideal. By definition, a program that reveals more of the complexity of the set of interactions among the systems at the human-environment interface of relevance to the field's goal in a given subject matter context approximates that ideal more closely than a program that reveals less of that complexity.

9 The Role of Systemic Resolution

The complexity of the set of interactions among the systems at the human-environment interface revealed by a given program in the Environment and Sustainability field is partly a function of *systemic resolution*, which is the conceptual resolution of a system into a set of subsystems. The degree of systemic resolution increases as each of these subsystems is resolved conceptually into its own set of subsystems, and so on, through any number of degrees of conceptual resolution. By analogy to the phylogenetic field of cladistics, a system and the subsystems into which it is resolved conceptually through any number of degrees of resolution together comprise a *systemic clade* (Fig. 2). The subsystems that reside at the same degree of conceptual resolution either within a systemic clade or across systemic

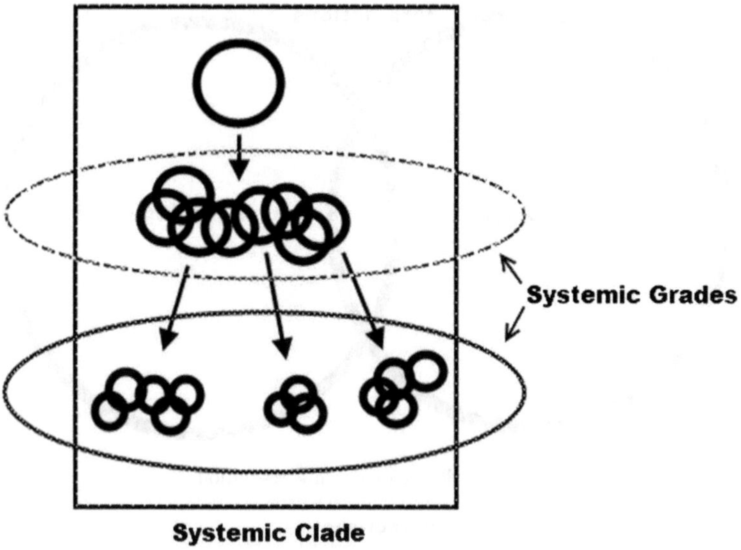

Fig. 2 Illustrating the distinction between a systemic clade (the systems and subsystems with the *rectangle*) and systemic grades (the systems/subsystems within each *ellipse*) generated by systemic resolution. The *circles* represent individual systems or subsystems

clades together comprise a *systemic grade* (Fig. 2). The degree of systemic resolution is maximized when the process of conceptually resolving systems into sets of subsystems generates a grade of subsystems that is incapable of further conceptual resolution. All else being equal, a program that conceptually resolves the systems at the human-environment interface of relevance to the field's goal in a given subject matter context to a higher degree reveals more of the complexity of the set of interactions among those systems than a program that conceptually resolves those systems to a lower degree (Fig. 3). Thus, the former approximates the field's pedagogical ideal more closely than does the latter, at least insofar as the additional complexity revealed is relevant to the goal of the Environment and Sustainability field.

9.1 Aspects of Systemic Resolution

The attributes of the systems being conceptually resolved are partly a function of the world as it is, but also partly are imposed on the world by the human observer. The Roundtable has identified at least five possible aspects of the systemic resolution process, which are as follows.

Systemic resolution in its phenomenological aspect is the conceptual resolution of a system into a set of phenomenologically defined subsystems, the phenomenological categories of which differ qualitatively from but are subsumed by the

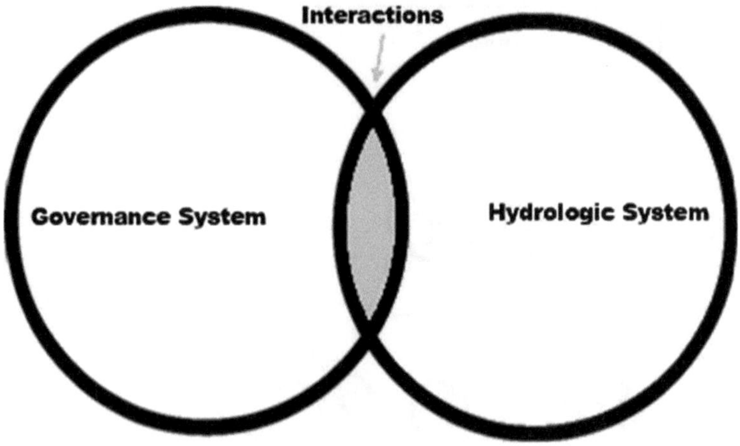

Lower degree of systemic resolution

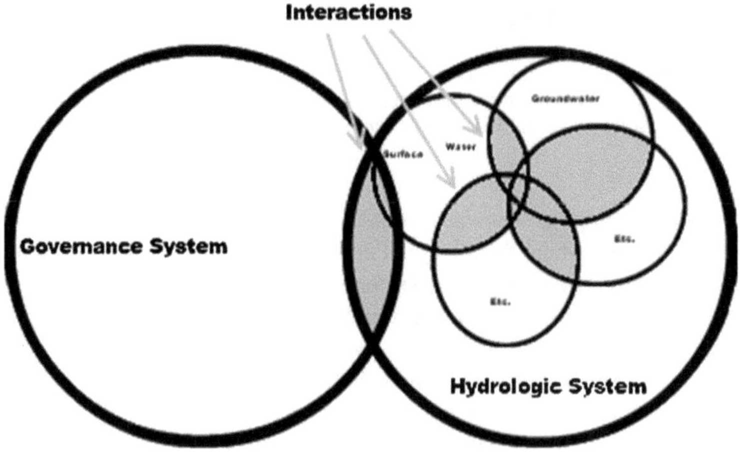

Higher degree of systemic resolution

Fig. 3 A conceptual illustration of the way in which a program with a higher degree of systemic resolution reveals more of the complexity of the set of possible interactions among systems than a program with a lower degree of systemic resolution

phenomenological category of the parent system, and also differ qualitatively from each other. For example, the conceptual resolution of undeveloped territory within the jurisdiction of a given local governance system into a likely commercial area and a likely residential area, or the conceptual resolution of a hydrologic system untapped by humans into a potable hydrologic subsystem and a non-potable hydrologic subsystem, would illustrate systemic resolution in its phenomenological aspect. Given that phenomenological understandings are intrinsically the product of interactions between the human observer and the phenomena being observed, the phenomenological categories to which different observers can assign systems and the subsystems of which those systems are comprised can vary widely.

Systemic resolution in its functional aspect is the conceptual resolution of a system into a set of functionally defined subsystems, the functions of which differ qualitatively from but are subsumed by the function performed by the parent system, and also differ qualitatively from each other. For example, the conceptual resolution of a governance system into a political subsystem, an administrative subsystem, and a legal subsystem, or the conceptual resolution of a hydrologic system used by humans into a hydrologic subsystem used for irrigation purposes and a hydrologic subsystem used for drinking purposes, would illustrate systemic resolution in its functional aspect.

Systemic resolution in its spatial aspect is the conceptual resolution of a system into a set of spatially defined subsystems, the spatial scopes of which differ qualitatively from but are subsumed by the spatial scope of the parent system, and also differ quantitatively from but are qualitatively the same as each other. For example, the conceptual resolution of a national governance system into a set of regional governance subsystems, or the conceptual resolution of a regional hydrologic system into a set of local hydrologic subsystems, would illustrate systemic resolution in its spatial aspect.

Systemic resolution in its cyclical aspect is the conceptual resolution of a system into a set of temporally defined subsystems that differ qualitatively from each other, the qualitatively different features of which recur in accordance with a cyclical pattern. For example, the conceptual resolution of a democratic governance system into a set of subsystems that captures the qualitatively different forms assumed by the parent system during the phases of a recurring election cycle, or the conceptual resolution of a hydrologic system into a set of hydrologic subsystems that captures the qualitatively different forms assumed by the parent system during the different seasons of the year, would illustrate systemic resolution in its cyclical aspect.

Systemic resolution in its transformational aspect is the conceptual resolution of a system into a set of temporally defined subsystems that differ qualitatively from each other, and either have become manifest in the past—or are likely to become manifest in the future—in accordance with a non-recurring pattern. For example, the conceptual resolution of a governance system into a set of governance subsystems that captures the qualitatively different forms assumed by the parent system in the past—or likely to be assumed by the parent system in the future—before, during, and after a successful national independence struggle would illustrate systemic resolution in its transformational aspect. Similarly, the conceptual resolution of a hydrologic system into a set of hydrologic subsystems that captures the qualitatively different forms assumed by the parent system in the past—or likely to be assumed by the parent system in the future—before, during, and after the construction of a dam, would illustrate systemic resolution in the same aspect.

10 Dimensions of Revealed Complexity

The revealed complexity of the set of interactions among systems at the human-environment interface of relevance to the goal of the Environment and Sustainability field can be measured along at least five dimensions that correspond to the five aspects of systemic resolution. Each dimension of revealed complexity is partly a function of the degree to which the interacting systems have been resolved conceptually in the corresponding aspect.

The *phenomenological dimension of revealed complexity* is a measure of the extent to which a given program reveals interactions among systems assigned to different phenomenological categories, both within and across systemic clades and grades. All else being equal, the revealed complexity of the set of interactions among systems increases along its phenomenological dimension as the number or diversity of systems assigned to different phenomenological categories revealed by the program increases, either within or across systemic clades and grades. For example, a program that revealed the interactions among the likely commercial and likely residential areas of which a set of interacting, undeveloped territories within the jurisdiction of a set of local governance systems is comprised, both within and across territories, would reveal more complexity than a program that revealed merely the interactions among the undeveloped territories per se. Similarly, a program that revealed the interactions among the potable and non-potable hydrologic subsystems of which a set of interacting hydrologic systems untapped by humans is comprised, both within and across hydrologic systems, would reveal more complexity than a program that revealed merely the interactions among the untapped hydrologic systems per se.

The *functional dimension of revealed complexity* is a measure of the extent to which a given program reveals interactions among systems with different functions, both within and across systemic clades and grades. All else being equal, the revealed complexity of the set of interactions among systems increases along its functional dimension as the number or diversity of systems with different functions revealed by the program increases, either within or across systemic clades and grades. For example, a program that revealed the interactions among the political systems, administrative systems, and legal systems of which a set of interacting governance systems is comprised, both within and across governance systems, would reveal more complexity than a program that revealed merely the interactions among the governance systems per se. Similarly, a program that revealed the interactions among the hydrologic subsystems used for irrigation and the hydrologic subsystems used for drinking purposes of which a set of interacting hydrologic systems used by humans is comprised, both within and across hydrologic systems, would reveal more complexity than a program that revealed merely the interactions among the hydrologic systems per se.

The *spatial dimension of revealed complexity* is a measure of the extent to which a given program reveals interactions among systems of different spatial scopes, both within and across systemic clades and grades. All else being equal, the revealed complexity of the set of interactions among systems increases along its spatial dimension as the number or diversity of spatially defined systems revealed by the program increases, either within or across systemic clades and grades. For example, a program that revealed the interactions among the national governance systems, regional governance systems, and local governance systems of a set of interacting nation-states, both within and across nation-states, would reveal more complexity than a program that revealed merely the interactions among the nation-states per se. Similarly, a program that revealed the interactions among the local hydrologic systems of which a set of interacting regional hydrologic systems is comprised, both within and across regional hydrologic systems, would reveal more complexity than a program that revealed merely the interactions among the regional hydrologic systems per se.

The *cyclical dimension of revealed complexity* is a measure of the extent to which a given program reveals interactions among systems in the qualitatively different forms assumed by those systems in the different phases of recurring cycles, both within and across systemic clades and grades. All else being equal, the revealed complexity of the set of interactions among systems increases along its cyclical dimension as the number or diversity of systems in different phases of the relevant cycles revealed by the program increases, either within or across systemic clades and grades. For example, a program that revealed the interactions among a set of democratic governance systems in the qualitatively different forms assumed by those systems during all of the different phases of their respective election cycles would reveal more complexity than a program that revealed merely the interactions among the governance systems in the forms assumed by those systems during the nomination phases of their respective election cycles, or that revealed merely the interactions among the governance systems common to all phases of their respective election cycles. Similarly, a program that revealed the interactions among a set of hydrologic systems in the qualitatively different forms assumed by those systems during all seasons of the year would reveal more complexity than a program that revealed merely the interactions of the hydrologic systems in the forms assumed by those systems only during their respective dry seasons, or that revealed merely the interactions among the hydrologic systems common to all phases of their respective seasonal cycles.

The *transformational dimension of revealed complexity* is a measure of the extent to which a given program reveals interactions among temporally defined, qualitatively different, non-recurring systems, both within and across systemic clades and grades. All else being equal, the revealed complexity of the set of interactions among systems increases along its transformational dimension as the number or diversity of temporally defined, qualitatively different, non-recurring systems revealed by the program increases, either within or across systemic clades and grades. For example, a program that revealed the interactions among a set of governance systems in the qualitatively different forms assumed by those systems in the past—or likely to be assumed by those systems in the future—before, during, and after successful national independence struggles would reveal more complexity than a program that merely revealed the interactions among those governance systems before those struggles, or that merely revealed the interactions among the systems that were common—or are likely to be common—to the colonial, revolutionary, and post-colonial eras of the respective governance systems. Similarly, a program that revealed the interactions among a set of hydrologic systems in the qualitatively different forms assumed by those systems in the past—or likely to be assumed by those systems in the future—before, during, and after the construction of a dam, would reveal more complexity than a program that merely revealed the interactions among those hydrologic systems in the forms assumed by the systems after the construction of the dam, or that merely revealed the interactions among the systems that were common—or are likely to be common—to the dam pre-construction, dam construction, and dam post-construction eras of the respective hydrologic systems.

11 Conclusion

The Roundtable on Environment and Sustainability is an ongoing, inclusive collaboration of academics and others committed to providing students in environmental degree programs with the tools needed to craft practical management strategies for meeting the pervasive sustainability challenges that exist at the human-environment interface. The Roundtable has proposed "Environment and Sustainability" as the relevant academic domain. The goal of this supradisciplinary field is the holistic, adaptive management of the interactions of the systems at the human-environment interface needed to support both stewardship of the natural environment and long-term improvement in the quality of life for human individuals and communities. Environmental programs that pursue this goal do so in part by revealing to students the complexity of the set of interactions among systems at the human-environment interface. Complexity in this context is simultaneously both an objective measure and a subjective measure of the number and diversity of the interactions among systems at the human-environment interface of relevance to the goal of the Environment and Sustainability field. It is partly a function of the conceptual resolution of the systems at that interface into sets of subsystems. The Roundtable has identified at least five possible aspects of the systemic resolution process—the *phenomenological aspect*, the *functional aspect*, the *spatial aspect*, the *cyclical aspect*, and the *transformational aspect*. The revealed complexity of the set of interactions among systems at the human-environment interface of relevance to the goal of the Environment and Sustainability field can be measured along at least five dimensions that correspond to these five aspects of systemic resolution. Each dimension of revealed complexity is partly a function of the degree to which the interacting systems have been resolved conceptually in the corresponding aspect.

Notwithstanding the Roundtable's recent work on the pedagogy of revealing complexity in educating for sustainability, many important topics remain on its agenda, including elaboration of the content and implications of the lists of pedagogically optimal skills, perspectives, and application contexts in Table 2, and the development of a means of discerning where any given environmental program falls within the Environment and Sustainability field in accordance with a truth-in-advertising approach to program recognition. The Roundtable welcomes participants from across the academic spectrum and throughout the world who share the Roundtable's goals. Parties interested in participating in the Roundtable's effort should contact any of the three Roundtable co-chairs, who are the first three named authors of this paper.

References

Caldwell LK (1983) Environmental studies: discipline or metadiscipline? Environ Prof 5(3/4):247–259

Maniates MF, Whissel JC (2000) Environmental studies: the sky is not falling. Bioscience 50 (6):509–517

Orr DW (1990) Environmental education and ecology literacy. Educ Dig 55(9):49–53

Reiter MA, Coggins PC, Howse ME (2009) Designing an integrated interdisciplinary environmental science curriculum using an IRMA chart: an example from Bethune-Cookman University. In: Leal Filho W (ed) Umweltbildung, Umweltkommunikation und Nachhaltigkeit [Environmental education, communication and sustainability]: vol 31: Sustainability at Universities: opportunities, challenges and trends. Peter Lang Scientific, Frankfurt am Main, Germany, pp 153–162

Reiter MA, Focht WJ, Barresi PA, Bumpous S, Smardon RC, Reiter KD (2011) Making education for sustainability work on your campus: the roundtables on environmental systems and sustainability. In: Leal Filho W (ed) Umweltbildung, Umweltkommunikation und Nachhaltigkeit [Environmental education, communication and sustainability]: vol 32. World trends in education for sustainable development. Peter Lang Scientific, Frankfurt am Main, Germany, pp 61–76

Reiter MA, Focht WJ, Barresi PA, Gill S, Smardon RC, Baker SL, Reiter KD, Fitch E, Rolfe T, Bumpous S (2012) Making education for sustainability work on campus: the proposals of the roundtable on environmental systems and sustainability. In: Leal Filho W (ed) Umweltbildung, Umweltkommunikation und Nachhaltigkeit [Environmental education, communication and sustainability]: vol 33. Sustainable development at Universities: New Horizons. Peter Lang Scientific, Frankfurt am Main, Germany, pp 109–116

Smardon RC (2011) Developing a national framework for internal/external review of undergraduate environmental studies/environmental science programmes. Interdisc Environ Rev 12 (3):244–257

Soule ME, Press DE (1998) What is environmental studies? Bioscience 48(5):397–405

Steffen W, Grinevald J, Crutzen P, McNeill J (2011) The Anthropocene: conceptual and historical perspectives. Philos Trans R Soc 369(1938):842–867

Vincent S (2010) Interdisciplinary environmental education: an exploration of field identity and core competencies. National Council for Science and the Environment, Council of Environmental Deans and Directors, Washington, DC, 86 p

Vincent S, Bunn S, Sloane L (2012) Interdisciplinary environmental and sustainability education on the Nation's Campuses 2012: curriculum design. National Council for Science and the Environment, Council of Environmental Deans and Directors, Washington, DC, 59 p

Wiek A, Withycombe L, Redman C, Mills SB (2011) Moving forward: on competence in sustainability research and problem solving. Environment 53(2):3–12

Wilke R (1995) Environmental literacy and the college curriculum: colleges and universities have a challenge to meet. EPA J 21(2):28–30

Authors Biography

Paul A. Barresi is Professor of Political Science and Environmental Law at Southern New Hampshire University in Manchester, New Hampshire, USA, where he coordinates the Environmental Management program. His scholarly work has been published or is awaiting publication in three languages on four continents.

William J. Focht is Associate Professor of Political Science and a past Director of the Oklahoma Water Resources Research Institute, the Environmental Institute, and the Environmental Science Graduate Program at Oklahoma State University in Stillwater, Oklahoma, USA.

Michael A. Reiter is Professor and Chair of the Department of Integrated Environmental Science and Director of the B.J. Moore Center for Integrated Environmental Science at Bethune-Cookman University in Daytona Beach, Florida, USA, past President of the Interdisciplinary Environmental Association, and Associate Editor of the international journal *Interdisciplinary Environmental Review*.

Richard Smardon is a State University of New York (SUNY) Distinguished Service Professor and Professor of Environmental Studies at the SUNY College of Environmental Science and Forestry in Syracuse, New York, USA. He has been director of the Division of Environmental Science and Chair of the Department of Environmental Studies, and has reviewed some 15 environmental science and studies programs in the US and Canada.

Michael L. Humphreys is Associate Professor of Ethics at Bethune-Cookman University in Daytona Beach, Florida, USA. His primary academic interests include restorative justice, restoration ethics, and ecological sustainability. Dr. Humphreys's scholarship in environmental ethics is supplemented by his prior service in the US Coast Guard, where he served as Chief, Port Operations at the Marine Safety Office in Portland, Maine. His duties included environmental preparedness and response.

Kimberly D. Reiter is Associate Professor of History at Stetson University in DeLand, Florida, USA, and Chair of the Advisory Board of the Interdisciplinary Environmental Association.

Steven A. Kolmes is Director of the Environmental Studies Program and occupant of the Rev. John Molter, C.S.C., Chair in Science at the University of Portland in Portland, Oregon, USA. His interests are in the areas of salmon recovery planning and combining ethical and scientific analyses in environmental policy discussions. He has served on federal and state scientific advisory panels and on the Steering Committee for "The Columbia River Watershed: Caring for Creation and the Common Good".

Embedding Sustainability Across the Curriculum via Collaboration: A Review of Progress in the University of the West of England

Georgina K. Gough and James W.S. Longhurst

Abstract

Students, the Higher Education Academy (HEA), the Quality Assurance Agency (QAA) and the Higher Education Funding Council for England (HEFCE) have all recommended that sustainability be given greater importance in the curricula of programmes delivered by HEIs in the UK. The University of the West of England (UWE), Bristol is responding to these calls by working with staff and students to enhance their education for sustainable development (ESD) provision. Recent curriculum changes have been undertaken with sustainability as a guiding principle. The University's approach recognises that for some programmes the focus is upon education about sustainability but for all programmes it must include education for sustainable development. The process which has been undertaken to achieve these developments will be presented. This includes engagement of senior managers within faculties and departments and targeted support for academics, via a range of individual and group activities. Dedicated staff and a collaborative approach have been critical to the success of these efforts. This paper will be useful for academics and managers from any discipline who are interested in ensuring that students develop an awareness of sustainability principles regardless of their programme of study.

Keywords

Higher education · Sustainability · Curriculum · Collaborative working

G.K. Gough (✉) · J.W.S. Longhurst
Faculty of Environment and Technology, University of the West of England, Bristol, BS16 1QY, UK
e-mail: Georgina.Gough@uwe.ac.uk

W. Leal Filho et al. (eds.), *Integrating Sustainability Thinking in Science and Engineering Curricula*, World Sustainability Series, DOI 10.1007/978-3-319-09474-8_36

1 Introduction: Sustainable Development in the Higher Education Sector

The relationship between education and sustainable development has been described as 'complex' (McKeown 2002) and problematic. Nonetheless universities across the world have worked decisively to achieve sustainable management of their estates and have made significant progress in engaging with the ideas of Education for Sustainable Development (ESD) in their curricula. Enabling ESD to find its place in the often crowded curricula of a university can be mandated by policy and, whilst this is necessary to signal institutional importance it is rarely sufficient and is unlikely to achieve the meaningful widespread engagement required. An approach based on developing a shared understanding of the necessity for engagement and developing collaborative engagement pathways is more likely to have long term benefits.

There is increasing recognition of the important role that the higher education sector can and should play in assisting global society to adopt a sustainable development pathway. In the UK there have been unprecedented policy developments in 2013 and 2014 concerning the nature and place of ESD in UK higher education. These developments include the new Higher Education Funding Council for England (HEFCE) Sustainable Development Framework (HEFCE 2014) and the Quality Assurance Agency (QAA) and Higher Education Academy (HEA) guidance document on ESD (QAA and HEA 2014). HEFCE in their new framework have formulated an ambitious vision for the role and impact of the English higher education sector which is for

> universities to be leaders in society's efforts to achieve sustainability—through the understanding, skills and attitudes that students gain and put into practice, through research, knowledge exchange and community involvement as well as through their strategies and operations that bring all these together (HEFCE 2014).

Achieving this ambition will require collaborative action within and between institutions as well as a wide range of stakeholder and community interests.

The QAA and HEA guidance covers all of the UK and is intended to provide further contextual support for Chap. B3 of the UK Quality Code for Higher Education which specifically refers to ESD as an academic quality theme that crosses subject boundaries (QAA 2013).

> In addition to subject-specific content, higher education providers consider the way their strategic approach reflects themes that cross subject boundaries. These themes reflect topics which may be considered to have a broad relevance to the purposes of higher education and its wider context in society. Where the themes are embedded within the curriculum and form an integral part of a programme of study, learning and teaching activities are designed to take them into account (QAA 2013).

The QAA and HEA guidance and the HEFCE Framework are important drivers for institutional engagement. These are reinforced by the results of the NUS longitudinal surveys of student attitudes to and skills for sustainable development which show a very strong level of expectation for institutions to act in a sustainable manner and to infuse the curricula with ESD (Drayson et al. 2013). The NUS survey data provides important evidence to suggest that students expect higher education to enhance their knowledge, skills, capacities and capabilities in the area of sustainable development. National and institutional engagement with ESD will be further highlighted by the launch of the UN Education for Sustainable Development Global Action Programme in November 2014 (UN ESD Global Action Programme, online), as the UN Decade of Education for Sustainable Development draws to a close.

Most UK HE institutions are engaged to varying degrees with the bodies listed above. Subsequently, the experience of the University of the West of England (UWE), in attempting to transform its curriculum to meet the expectations of these various bodies, should be of wide interest. Given the very recent emergence of the guidance documents, this paper is timely and the approach which UWE has taken has not, to the authors' knowledge, been widely taken previously and so should provide innovative ideas.

2 Education for Sustainable Development: Definitions and Characteristics

UNESCO has offered a definition of ESD: "Education for sustainable development means enabling students to develop the knowledge and understanding, skills and attributes needed to work and live in a way that safeguards environmental, social and economic wellbeing, both in the present and for future generations" (UNESCO, undated). This is extended by a specification of the key characteristics of ESD, including expectations that it "uses a variety of pedagogical techniques that promote participatory learning and higher-order thinking skills; promotes lifelong learning; is locally relevant and culturally appropriate; engages formal, non-formal and informal education; addresses content, taking into account context, global issues and local priorities; builds civil capacity for community-based decision-making, social tolerance, environmental stewardship, an adaptable workforce, and a good quality of life; and, is interdisciplinary".

Drawing on both the Brundtland definition of sustainable development (Brundtland 1987), and its 2005 recalibration, the QAA and HEA Guidance defines ESD as follows (QAA and HEA 2014).

> Education for sustainable development means enabling students to develop the knowledge and understanding, skills and attributes needed to work and live in a way that safeguards environmental, social and economic wellbeing, both in the present and for future generations.

The QAA and HEA Guidance encourages students to consider what the concepts of global citizenship and environmental stewardship mean in the context of their own discipline; think about issues of social justice and equity, and how these relate to ecological and economic factors, develop a future-facing outlook, learning to think about consequences of actions, and how systems and societies can be adapted to ensure sustainable futures. It further articulates expectations that a graduate with ESD skills, knowledge, capabilities and competencies will be able to demonstrate. These will be acquired by students through engagement with all elements of their curriculum.

3 Curriculum: Content, Change, Collaboration and ESD

According to Fraser and Bosanquet (2006, p. 272) "the curriculum is conceptualised as:

- Category A: The structure and content of a unit or module (subject);
- Category B: The structure and content of a programme of study;
- Category C: The students' experience of learning;
- Category D: A dynamic and interactive process of teaching and learning."

Maher (2004, p. 47) discusses changes in the conceptualisation of curriculum from a teaching to learning focus observing that "the adoption of a 'learning paradigm' in HE puts the learner at the heart of the educational process, a proposition that appeals to both teachers and students alike". This evolution of the scope of the term curriculum has been one catalyst for curriculum change within the HE sector.

The process of curriculum change is considered in many contexts. Of particular relevance here are discussions of drivers for, aims of and processes for curriculum change. Changes and developments within disciplines, professions, industry and society are established drivers of curriculum. Further drivers of curriculum change can include new developments in the subject knowledge base, new professional body expectations, developments in the QAA's Code of Practice such as new Subject Benchmark Statements, as well as institutional intentions to maintain distinctive and attractive programmes of study.

Working with colleagues is a necessity when redesigning curricula within modules and programmes. A coherent package of learning would not be achievable if academics were to design their modules in isolation from others. As such, collaboration within programme and discipline teams is standard practice within HE. However, collaboration involving colleagues and others outside of these teams is less common. Pasterczyk (1986), for example, writes of the importance of collaborating with and communicating details of curriculum changes to support staff, such as librarians in order to ensure adequate learning resources to support delivery of new curricula. The literature on co-creation offers further insight into the value of collaboration for all collaborators. Ind and Coates (2013, p. 86) note that work on co-creation has largely developed in the context of customers and organisations, "to counter charges that co-creation exploits consumers and other stakeholders who gift their time and intellect for the benefit of organizations, it needs to move beyond the co-opting lens and engage stakeholders in a reciprocally useful way". Some commentators observe that this is now the situation higher education finds itself in the wake of the new fee arrangements. Indeed, Nordensvard (2011, p. 157) notes that the "dominant metaphor for the students has lately been that of the consumer" and explores the implications such a position has on the students' attitudes to and engagement with learning, their degree, their educators and their university. Nordensvard (2011) further argues that "a citizenship perspective is far more compelling than a consumer perspective since it can open another set of roles which can give us a better understanding of the complexity of education" (ibid.). The notion of citizenship is very much at the heart of current discussions within the HE sector about the role HE institutions should play in developing global citizens of the future, and about the place of education for sustainable development within all spheres of the higher education curriculum.

The recent suite of policy drivers from QAA, HEA and HEFCE have been considered above and it is recognised that these drivers of and for change must contend with a range of other change actions that seek to influence the shape of the curriculum. However, as alluded to earlier, the drivers for enhancing curriculum through greater ESD provision are currently at an unprecedented level in terms of number and influence, and the literature reflects academic consideration of the relationship between sustainable development and HE. Gacel-Ávila (2005, p. 123) calls upon universities to develop graduates that are well-rounded, global citizens. *"One of the basic and fundamental functions of a university should ...be the fostering of a global consciousness among students, to make them understand the relation of interdependence between peoples and societies, to develop in students an understanding of their own and other cultures and respect for pluralism". The implementation of HE curriculum change in response to these developments, has focused as much on the teaching and learning processes and delivery mechanisms, and on the student experience, as it has on content.*

4 ESD and the Curricula of the University of the West of England

Traditionally universities have been concerned with the relationship of their estate and resource management practices to the ideas of sustainable development. However, there is increasing recognition that the major impact of the HE sector on future sustainability arises from the education received by students (Longhurst and Gough 2013). Hence, the University of the West of England has systematically addressed the place of education for sustainable development in the curricula and is progressively infusing consideration of these issues into the formal and informal curricula of all undergraduate and post graduate provision. A policy framework expressed in the University's Strategy 2020 (UWE 2014a), the former Sustainability Strategy and the new Sustainability Plan (UWE 2014b) provides the context in which these initiatives are implemented. However, it is understood that the strategy and plan, whilst necessary, are not on their own sufficient. Collaborative action between staff, between staff and students and between the university and its stakeholders are equally necessary to achieve successful implementation and ultimately, normalisation of ESD across the institution.

Strategy 2020 incorporates a clear commitment to sustainable development and in particular to the role that the curricula of the university have in delivering its sustainability ambitions though the development of "graduates ready and able to realise their full potential, make a positive contribution to society and their chosen field of employment or further study, and play their full part in the development of a sustainable global society and knowledge economy" (UWE 2014a).

The Sustainability Plan operationalises the ambitions of Strategy 2020 and outlines the University's sustainability aims for the period 2013–2020. It defines the mechanisms by which sustainability is embedded within the University's day to day operation and effective decision making. The Plan contributes to UWE's Strategy 2020 priorities through its associated action plans which will deliver ready and able sustainability literate graduates and which locates ESD within the context of the discipline of study. It recognises that all disciplines must engage with ESD but that the extent and nature of engagement can differ. In particular the Plan contains three important ambitions which, if they are to be achieved, will require extensive staff engagement and collaborative action:

1. 'The University will enable all students to explore sustainable development in the context of their discipline';
2. 'All Departments… will ensure ESD is embedded at the programme level';
3. 'The University will ensure that its staff are trained and developed in their understanding of the underpinning concepts.'

An ESD Action Plan provides a more detailed exposition of the aims and objectives laid out in the Sustainability Plan. The Action Plan is developed, owned and implemented by the Knowledge Exchange for Sustainability Education (KESE) group. The approach taken recognises that for some programmes the focus will be upon education about sustainable development but for all programmes it must include education for sustainable development.

5 Curriculum Change for Sustainability at UWE, Bristol

Curriculum changes at UWE occur at times dictated by the institution's review processes. Programmes and modules are reviewed as part of the annual Quality Management and Enhancement process. Over a 6 year cycle all provision is subjected to a Periodic Curriculum Review in which the currency, quality and suitability of the provision is assessed. In addition, professionally accredited provision is periodically reviewed for the purpose of assuring compliance with the expectation of the accrediting body. New modules can be proposed at any time and are approved by a Curriculum Approval Panel. From 2007 ESD has been increasingly represented in the curriculum of many departments at UWE. The Quality Assurance

Agency and Higher Education Academy guide academics to consider the place of ESD within modules and programmes noting that "At one level, this could be fulfilled via provision of a relatively superficial introductory lecture on sustainable development, origins and definitions of the concept and a contextual positioning of the economic, social and environmental impacts of professions in relevant disciplines to the programme of study" (QAA and HEA 2014). However, it is recognised within the growing literature on ESD, that a much deeper engagement with sustainable development is required if student are to graduate with the knowledge, skills and attributes that will enable them to meet the expectations of the HEFCE Framework (2014), the QAA and HEA Guidance (2014), or the UNESCO essential ESD characteristics (UNESCO, undated).

At UWE, there is a relatively long history of providing ESD content to students across a range of programmes and disciplines. Over the past few years the institution has been implementing a range of activities to broaden, and more importantly deepen its ESD provision. These activities fall into three primary categories, all with collaboration at the core.

1. Development of quality management and enhancement processes to encourage consideration of sustainability in the curriculum.
2. Increased visibility of existing ESD provision to champion natural internal leaders and provide examples of good practice and
3. Support for academic and professional staff to engage with ESD regardless of their discipline of study or area of work.

The rationale behind this holistic and comprehensive approach is that institutions must lead by example. To be most effective at producing sustainable literate global citizens, universities must themselves be staffed by sustainably literate, global citizens.

6 Development of Quality Management and Enhancement Processes to Encourage Consideration of Sustainability in the Curriculum

Collaboration has been crucial to the successful engagement of relevant professional staff in the discussions and subsequent inclusion of ESD into the quality management and enhancement processes of the university. In order to ensure that ESD was considered in the design and approval of new modules and programmes the University commissioned a guide for staff on the various ways in which ESD should and could be considered: in the design of new programmes and modules; in the annual review of programmes and modules and in periodic curriculum review. In addition to guidance for the developers of curricula it also provides advice on issues for approval panels to consider. Tables 1, 2 and 3 present key issues considered in the UWE approach to Quality Management in relation to ESD. Throughout these

Table 1 ESD and quality management at UWE, Bristol

Quality management issue	Summary of advice and guidance
New or changed programmes	The design team for a new programme should consider the place of ESD in the proposed development and make explicit in their documentation how ESD has been considered in the curricula Learning outcomes with an ESD context should be highlighted and assessments of ESD attributes made clear. The role of the informal curricular including PAL and volunteering should be made explicit
New modules and changes to a module	Within either the development of a module or the process of module change a module team should consider the opportunity to enhance the visibility and place of ESD within the changed curriculum. Learning outcomes with an ESD context should be highlighted and assessments of ESD attributes made clear. Module designers should consider the contribution to ESD that will arise from group work, projects, field visits, external speakers, work placements, PAL activities and extra curricula activities
Programme and module approval processes	Academic Standards and Quality Committees (AQSC) and Curriculum Approval Panels (CAPs) should confirm that the programme or module under consideration has reflected upon the opportunity to position ESD considerations within its specification. A rationale should be provided to explain how ESD has been considered in the design process. Examples of enhancement or good practice such as constructive alignment of learning outcomes and assessments of ESD attributes in the programme or module should be recorded
	ASQC and CAP minutes should confirm that provision under consideration has clearly identified the contribution of the provision to the students understanding and knowledge of core SD issues in the programme or module

processes, core issues of global citizenship, social justice and equity, environmental stewardship and a future facing outlook, the ESD aims of the institution and the graduate outcomes articulated in the QAA/HEA ESD Guidance document are identified as reference points.

The collaborative activities involved in developing the guidance on ESD in the Quality Management and Enhancement Framework extended beyond just quality management staff and lead ESD staff. An internal Knowledge Exchange for Sustainability Education (KESE) group provided advice on issues to consider, acted as a steering group and approved the final draft prior to its approval by Academic Quality and Enhancement Committee and adoption by the University. The ESD quality management guidance document supports staff in meeting the new ESD reporting requirements of programme and curriculum review and approval processes.

One of barriers identified by Heads of Department as preventing further staff engagement in ESD provision was a sense of this subject matter being perceived as irrelevant or peripheral to their discipline or area of work. Two strands of collaborative activity have been put into place to help overcome this barrier. Firstly, work

Table 2 ESD in annual reporting processes

Annual department monitoring Report	The report should consider the extent to which the provision of the Department is supporting the university's ESD ambition, in particular the ambition to ensure that every student is exposed to the ideas of sustainable development within the context of their programme of study. Enhancement activities designed to make ESD more visible within the Department should be presented. Staff development requirements and other actions and resources needed to support ESD enhancement activities should be considered and appropriate recommendations made
Programme annual report	This is the key reporting mechanism from which the extent and depth of ESD engagement in the curricula can be gauged. Programme Managers should carefully consider the nature and extent of ESD in the delivery of the programme and present recommendations for enhancement in the following year. These can include changes to learning outcomes, assessments of ESD characteristics, employability outcomes and related matters. Staff development requirements and other actions and resources needed to support ESD enhancement activities should be considered and appropriate recommendations made
	The Programme report should consider the role played by the formal and informal curricula in delivering ESD within the programme. In particular the report should make clear how core ESD issues have been included in the provision. Enhancement actions in respect of these four areas should be presented
	The contribution of PAL and volunteering opportunities should be evaluated alongside the opportunities provided by the UWE Futures Award. The contribution of dissertation or project modules, group work activities, and employability enhancements should all be evaluated for their contribution to ESD. The contribution of curriculum enrichment activities such as external speakers or field visits should be reported
Annual module report	The module report should consider the contribution of the module to meeting the ESD aims of the university, the expectations of the QAA and HEA guidance and Chap. B3 of the Quality Code. Module reports should consider *Inter Alia* the contribution to ESD arising from group work, projects, field visits, external speakers, work placements, PAL activities and extra curricula activities present within the module

has been undertaken to raise the profile and visibility of existing ESD provision. Secondly, a variety of support mechanisms are being developed for staff to engage with in order to explore the relevance of sustainability in their own professional and personal contexts.

Table 3 Consideration of ESD in periodic curriculum review

Periodic curriculum review	Periodic Curriculum Review provides the opportunity for the discipline to reflect upon its engagement with the university's ESD ambitions and to consider how to enhance visibility and activity relating to ESD in the provision. Opportunities to consider enhancements to constructive alignment of learning outcomes and assessments, and the role of informal curricula activities supportive of ESD should be considered as part of the review. Recommendations for ESD actions should be included in the Critical Evaluation Document (CED)
	Examples of ESD being used to enhance curricula and other examples of good practice should be noted. Reports and Programme Action Plans should document the presence of ESD to date and the intended enhancements of the provision in order to demonstrate that students will have the opportunity to develop understanding and knowledge of the role and importance of core ESD issues in their programme

7 Visibility of Existing ESD Provision

In 2008, a baseline study revealed that there were 210 modules being delivered at UWE with visible sustainable development content. Furthermore, 67 staff were identified as delivering ESD within the institution. It was found, though, that much of this activity was largely unrecognised. Since that baseline survey, the visibility of ESD provision has been increased internally, largely through the activities of the KESE and through annual all staff ESD events. This has helped to support and champion natural internal leaders and provides examples of good practice across nearly all of the University's academic departments. The use of a variety of mechanisms such as centralised online visibility to help create a sense institutional ESD identity and promoting a sense of cohesiveness to activities, but also promotion within groupings which academics more readily associate with, particularly departmental and programme teams has been very beneficial. The intension is for all academics to recognise that sustainability does relate to them, to see that other people 'like them' are engaged with ESD and to begin to see how others are delivering ESD (both in relation to content and in teaching and learning approaches). All of these outcomes will help academics to find their own appropriate engagement 'space', thus developing a culture of ESD across the institution.

8 Support for Academic and Professional Staff to Engage with ESD Regardless of Their Discipline of Study or Area of Work

Annual ESD all staff events have been important in engaging staff in ESD work. Equally important has been direct engagement with Heads of Department to discuss barriers to further ESD provision and to develop solutions. This in turn has led to staff development activity including bespoke sessions for Departments, clusters and programme teams as well as inclusion of ESD in the university's Academic Development Programme and in induction activities. These activities are regularly reinforced by discussions of ESD with faculty executives and deliberative committees of the university. In all of these activities, collaboration between staff charged with ESD delivery across the institution and staff in various discipline and role groupings has been vital in identifying local issues and in gaining traction across staff groups.

9 Factors for Success of Recent Accelerated Change

Activity designed to help locate and acknowledge existing provision has been helpful in normalising the place of ESD in the curricula. This has been backed up by strong central support for the sustainability agenda from Strategy 2020, which emphasises sustainability in relation to graduates of UWE as well as creating a sustainable place to live and work. Of equal importance has been the development of centralised reporting and approval processes. The work of local 'champions' has been very important to the success of the institutional approach and has allowed organic growth of ESD provision, albeit with an encouraged sense of urgency.

Central to these developments is the clear recognition of and respect for 'local' issues and a commitment to the proposition that one size does not fit all. The approach has recognised that there are programme specific, discipline specific, department specific and individual specificities that need to be considered in order to achieve real curriculum change for sustainable development. Thus it follows that academics, in considering why, how or where ESD should be embedded in the curricula, must be given time and freedom to identify the best ways and places for achieving the ESD goals the institution has set itself.

10 Conclusion: Collaboration Expedites Transformative Processes for Curriculum Change for Sustainability

In response to the expectations of students and the policy drives of the Higher Education Academy, the Quality Assurance Agency and the Higher Education Funding Council for England, universities are giving greater attention to the provision of sustainability in the curricula of programmes delivered by HEIs in the UK (Drayson et al. 2013; QAA/HEA; 2014; HEFCE 2014). The University of the West of England, Bristol has responded to these calls by working collaboratively to enhance its ESD provision across the curricula. This experience has enabled the following lessons to be learnt:

- Capitalise on the opportunities which existing curriculum change schedules and processes provide to review and enhance, where necessary, ESD provision. This minimises additional workload for both academic and professional staff.

- Ensure engagement of senior managers (at institutional, faculty and department level) in promoting sustainability in the curriculum and in curriculum change activities. This is important for coherency of activity and consistency of message.
- Provide targeted support for academics, via both individual and group activities, such as workshops, seminars and even one-to-one appointments. These should be made relevant to groups with which people identify, such as department or programme team. Each institution will need to identify relevant groupings.
- Utilise the passion and influence of staff already dedicated to ESD to demonstrate possibilities for ESD provision in a variety of disciplines and contexts. This will help to drive a collaborative approach which is proving to be critical to the success of efforts at UWE.

In summary, the drive to embed ESD in UWE's curricula has begun with a strong sustainability focus in the institution's strategy, policies and plans, including a specific sustainability plan, with ESD components. Together, these provide context and send signals which are amplified by senior staff. Successful implementation of the ESD plan, however, requires collaboration between academic staff, quality management professionals, teaching and learning support staff, students and others, facilitated by dedicated ESD leads. Ultimately, this will lead to the establishment of a sustainability culture across the institution, with a truly sustainability-focused curriculum.

References

Brundtland GH (1987) Report of the world commission on environment and development: our common future. United Nations, Geneva

Drayson R, Bone E, Agombar J, Kemp S (2013) Student attitudes towards and skills for sustainable development York: Higher Education Academy. Third report in a longitudinal study of students' views (full 2013 report and executive summary). Available at http://www.heacademy.ac.uk/resources/detail/sustainability/2013_student_skills_final_report. Last accessed 26 Mar 2014

Fraser SP, Bosanquet AM (2006) The curriculum? That's just a unit outline, isn't it? Stud High Educ 31(3):269–284

Gacel-Ávila J (2005) The internationalisation of higher education: a paradigm for global citizenry. J Stud Int Educ 9(2):121–136

HEFCE (2014) Sustainable development in higher education. A Framework for HEFCE. HEFCE, Bristol, p 28

Ind N, Coates N (2013) The meanings of co-creation. Eur Bus Rev 25(1):86–95

Longhurst JWS, Gough GK (2013) Education for sustainable development at the University of the West of England, Bristol. Paper presented at influence, implement, engage: the triple imperative for ESD in HE. HEA thematic seminar series 2013/14, 29th November 2013, University of the West of England, Bristol. Available at http://www.heacademy.ac.uk/events/detail/2014/Seminars/Themes/GEN876_UWE. Last accessed 24 Mar 2014)

Maher A (2004) Learning outcomes in higher education: implications for curriculum design and student learning. J Hospitality Leisure Sport Tourism Educ 3(2):46–54

McKeown R (2002) Education for sustainable development toolkit version 2. Available online at http://www.esdtoolkit.org/esd_toolkit_v2.pdf. Last accessed 27 Mar 2014

Nordensvard J (2011) The consumer metaphor versus the citizen metaphor in education policy. In: Molesworth M, Scullion R, Nixon E (eds) The marketisation of higher education and the student as consumer. Routledge, Abingdon, pp 157–169

Pasterczyk CE (1986) Involvement in curriculum change. Coll Res Libr 47(1):7–15

QAA and HEA (2014) Education for sustainable development. Guidance for UK higher education providers. QAA, Gloucester

QAA (2013) UK quality code for higher education part B: assuring and enhancing academic quality Chapter B3: learning and teaching. Available at http://www.qaa.ac.uk/Publications/InformationAndGuidance/Pages/quality-code-B3.aspx. Last accessed 28 Mar 2014

UNESCO (undated) Education for sustainable development. Available at http://www.unesco.org/new/en/education/themes/leading-the-international-agenda/education-for-sustainable-development/education-for-sustainable-development/. Last accessed 28 Mar 2014

United Nations Education for Sustainable Development (ESD) (undated) Global action programme. Available at http://www.unesco.org/new/en/unesco-world-conference-on-esd-2014/esd-after-2014/global-action-programme/. Last accessed 29 Mar 2014

University of the West of England, Bristol (UWE) (2014a) Strategy 2020. Available at: http://www1.uwe.ac.uk/aboutus/visionandmission/strategy.aspxv. Last accessed 28 Mar 2014

University of the West of England (UWE), Bristol. (2014b) Sustainability strategy and policy. Available at: http://www1.uwe.ac.uk/aboutus/visionandmission/sustainability/sustainabilityaction/sustainabilitydocuments.aspx. Last accessed 29 Mar 2014

Author Biography

Dr. Georgina Gough is Senior Lecturer in Education for Sustainable Development at UWE, Bristol. She received her doctorate in environmental interpretation from the University of the West of England in 2008. She lectures on sustainability in business and on tourism and supports academics in developing their ESD engagement. Professor James Longhurst is Assistant Vice Chancellor, Environment and Sustainability at UWE, Bristol. He chairs the Higher Education Academy's Education for Sustainable Development Advisory Group and also chaired the joint Quality Assurance Agency and Higher Education Academy's development group preparing the guidance for UK higher education providers on Education for Sustainable Development.

Developing a Holistic Approach to Implementing Education for Sustainable Development Across a Higher Educational Institute: A Longitudinal Study of the University of the West of England, Bristol

James W.S. Longhurst, Georgina K. Gough, Marcus Grant and Sarah Hills

Abstract

Over the last 20 years the University of the West of England (UWE) has developed policy, strategy and plans to address its environmental and sustainability impacts. The University identifies its role in education for sustainable development (ESD) as being about the contribution that can be made through learning and teaching. Key to this is the development of skills, knowledge and values through programmes of study that support the goal of reconciling human needs with the environmental limits of the planet. Actions undertaken to implement this approach include the commitment for all students to be exposed to the ideas of sustainable development through their programme of study. The challenges experienced whilst undertaking such actions are discussed in the context of organisational leadership and change processes.

Keywords

Education for sustainable development · UWE · Holistic · Change management

1 Introduction

This paper presents a longitudinal analysis of the journey of UWE towards becoming a sustainable university. We distinguish here between the "what" of a sustainable university i.e. how the principles of sustainable development manifest in

J.W.S. Longhurst (✉) · G.K. Gough · M. Grant · S. Hills
Faculty of Environment and Technology, University of the West of England, Frenchay Campus, Bristol, BS16 1QY, UK
e-mail: James.Longhurst@uwe.ac.uk

W. Leal Filho et al. (eds.), *Integrating Sustainability Thinking in Science and Engineering Curricula*, World Sustainability Series, DOI 10.1007/978-3-319-09474-8_37

strategy, campus infrastructure, operations and learning, teaching and research and the "how" of sustainability i.e. the processes through which a university makes the journey towards sustainability. The primary focus in this paper is on organisational leadership and change processes, but reference is made to examples of sustainability outcomes to illustrate the connection between those processes and institutional change. This journey is analysed through the lens of theories of sustainability leadership and organisational transformational change. It is important to note that universities are complex entities, not simply machines or cultures or political systems or instruments of domination (Morgan 1986) but exhibit dynamic combinations of all of the qualities suggested by each of these organisational metaphors. If organisations are complex and dynamic, and sustainable development is arguably even more so, then bringing the two together within the context of a university is "a classical example of supercomplexity in action" (Temple 2010, p. 105).

2 Sustainability Leadership

There is a substantial literature on sustainability leadership. And many of the qualities of good sustainability leadership discussed within it tend to align with the broader bodies of theory on leadership and change management (e.g. CPSL 2011). The sustainability literature emphasises the transformational nature of sustainability leadership (Blowfield 2013) with practices rooted in a sense of purpose and values aligned to sustainable development (Parkin 2010; Birney et al. 2008; Galea 2004). There is also emphasis on 'organisational intelligence', social and emotional intelligence (Parkin 2010) distributed leadership and the capacity to facilitate collaboration and create the conditions in which pro-sustainability innovation and 'intrapreneurship' (Blowfield 2013, p. 98) can thrive. Not surprisingly, the ability to approach sustainability change from a systems perspective is also emphasised (Henrik et al. 2006; Senge 2008; Parkin 2010; CPSL 2011).

3 Sustainability in Higher Education

Much of the literature on higher education and sustainability has focussed on education for sustainable development, with a particular emphasis on the case for, progress in and methodologies for integrating sustainability into learning and teaching (for example, Orr 1992; Lozano 2006; McMillin and Dyball 2009; Tilbury 2011). There is also a growing literature on the sustainability management of the university and its operations (for example, Brown (2010); Sharp and Pollock (2012); Posner and Stuart 2013; Swearingen White 2014). Temple (2010) highlights some of the difficult choices that universities will face if they want to make serious attempts to reduce their carbon footprint. This is complemented by the literature on leadership, organisational change and organisational learning (Scott and Gough 2006; Temple 2010; Mader 2012; Mader et al. 2013; Barth 2013; Cebrian et al. 2013; Wright and Horst 2013). Recent literature emphasises the need

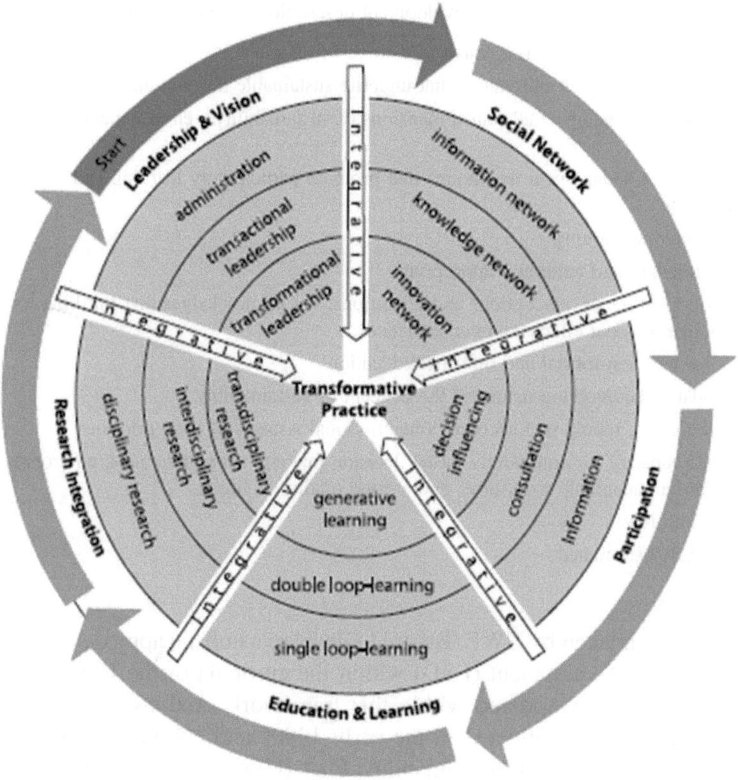

Fig. 1 Graz model for integrative development (Mader 2012, p. 83)

for holistic, transformative and integrative approaches to leadership and change for sustainability in the higher education sector. Mader (2012) identifies five interconnected dimensions of pro-sustainability change: Leadership and Vision, Social Network, Participation, Education and Learning, and Research Integration, which he suggests can be used to assess "transformative performance" towards sustainability in higher education. For each of these dimensions he identifies the qualitative parameters within which transformational practice takes place (Fig. 1). Scott et al. (2012) on turnaround leadership for sustainability education identify ten distinguishing characteristics of effective leadership, many of which align with previous studies of leadership, change management and pedagogic excellence in higher education and other sectors. Many of the issues that they identify could apply equally to the broader challenge of reorienting a whole university towards sustainability including the emphasis on transformation, working across silos, developing a clear and shared understanding of key terms, aligning organisational systems and structures to support key objectives and sustainability leaders, tracking and improving performance, using incentives, focus on strategies to engage the disinterested and employing effective change leaders.

Table 1 Education for sustainable development description according to UNESCO

Education for sustainable development
is based on the principles and values that underlie sustainable development
deals with the well-being of all four dimensions of sustainability—environment, society, culture and economy
uses a variety of pedagogical techniques that promote participatory learning and higher-order thinking skills
promotes lifelong learning
is locally relevant and culturally appropriate
is based on local needs, perceptions and conditions, but acknowledges that fulfilling local needs often has international effects and consequences
engages formal, non-formal and informal education
accommodates the evolving nature of the concept of sustainability
addresses content, taking into account context, global issues and local priorities;
builds civil capacity for community-based decision-making, social tolerance, environmental stewardship, an adaptable workforce, and a good quality of life
is interdisciplinary

Source UNESCO, undated

The approach taken by UWE, Bristol to develop a holistic approach to Education for Sustainable Development (ESD) within the curricula of the University over a 20 year period are considered within this framework. Actions commencing with environmental policies initiated in the early 1990s and culminating in 2013/2014 academic year in the new Sustainability Plan (UWE 2014c) are discussed. The Sustainability Plan mandates the curricula to provide exposure and opportunity for all students to engage in and understand the meaning of sustainable development within the context of their discipline of study (Gough and Longhurst 2014; Longhurst and Gough 2013, 2014a, b). As part of the comprehensive approach, a range of co-curricula activities providing ESD opportunity are available to all UWE students and particular emphasis is placed upon the sustainable development of the campus locations to provide ESD opportunities. UWE recognises the definition and description of ESD provided by UNESCO and this is presented in Table 1. The UWE approach to ESD seeks to embed this definition in policy and action (Longhurst and Gough 2014a, b).

4 A Comprehensive Approach to Education for Sustainable Development at UWE, Bristol

UWE is a large multi-site institution with approximately 30,000 students, some 3,000 staff, and an annual budget of £220 million(UWE, 2014a). The purpose of the University is to advance knowledge, inspire people and transform futures and recognises that it has an obligation to society to provide leadership in the field of sustainable development(UWE, 2014b). UWE sees this as being exercised through the generation and wise use of knowledge and through our public engagement and knowledge exchange activities.

In common with many universities UWE has addressed the sustainability of its campus locations. An important step in this process was the development a dedicated Sustainability Team, formerly Environmental Management Team in the Facilities Service in the 1990s. This team is led by the University's Sustainability Manager and takes responsibility for initiating, managing and monitoring those aspects of the sustainability performance related to waste management, energy management, travel planning and biodiversity amongst others. These initiatives are ensuring sustainable management of the current and future estates.

UWE recognises that the education of students, future graduates, represents the biggest opportunity for a university to encourage awareness and action for the development of a more sustainable future. Hence, UWE has systematically addressed the place of ESD in the curricula. The university is infusing sustainable development into the formal curricula of all undergraduate and post graduate provision. To do this UWE has established a policy framework and from this plans and strategy to support sustainable development have evolved. These have been implemented into the business of the university both in the educational role and in the management of our estate. This approach is overseen by a high level University Sustainability Board, chaired by the Deputy Vice Chancellor, and action is devolved to services and faculties to embed sustainable development into their business decision making. Success factors here include strategic leadership from the Vice Chancellor and senior managers, establishment of a cross university research and knowledge exchange group and a multidisciplinary staff group tasked with promoting education for sustainable development and overseeing its implementation into the curricula.

UWE's systematic engagement in ESD can be traced to the Sustainability Board commissioning a baseline survey of Education for Sustainable Development at UWE (Grant 2008). This consisted of two elements. First, a quantitative census of all provision at modular level, to assess where and to what extent content associated with the teaching of sustainable development was present in modules across all undergraduate and M level provision. Second, a qualitative interview sample of 26 lecturers drawn from 15 schools or departments to test the breadth of understanding and terminology used in the delivery of content associated with the teaching of sustainable development. An analysis of the findings from this survey enabled a list of strengthens and weaknesses of ESD provision to be drawn-up and discussed. This in turn resulted in 12 recommendations, which became the basis for the first ESD action plan. These were presented to, and endorsed by the Sustainability Board in 2008.

As such the study proceeded through simple twin track iterative process; both, scoping what ESD activities are in place and at the same time developing a suitable defining and descriptive framework for what to include as ESD practice. The quantitative survey used several text searching approaches to identify elements of ESD in the modular descriptions and outlines in the validation documents across the whole of UWE. This search yielded 210 potential modules. The module specification for each of these was then hand searched resulting in finding 136 modules that were directly relevant to ESD teaching, 38 of these were deemed as core ESD,

Table 2 Knowledge exchange for sustainability education group terms of reference and reporting relationships

KESE terms of reference and reporting relationships
Purpose
To support the delivery of the university's sustainability strategy and learning teaching and assessment strategy
Objectives
To provide strategic direction to, and oversight of, activities contributing to education for sustainable development in the curriculum
To develop networking and institutional learning opportunities in ESD
To periodically refresh our understanding of the progress of ESD nationally and internationally through benchmarking with other institutions
To identify opportunities for new programmes and/or short courses in ESD
To provide periodic updating of institutional developments and practices in the form of reports to the sustainability board, academic standards and quality committee and other UWE fora
To provide advice and support for the training and development of UWE staff for and in ESD

having comprehensive coverage of sustainable development. 61 were deemed as specialist, in that they assume some basic knowledge of sustainable development or they are in a route way such that the very specialist knowledge they provide is what would be most valued for ESD in that discipline. The final 38 were all candidate modules for ESD, but were deficient in specific aspects, which could easily be addressed. The qualitative work found that aspects of ESD were evident in almost all departments but in most cases weakly developed in terms of reflecting a coherent approach.

In 2008, largely unrecognised and with limited institutional policy support, elements of ESD related teaching were fairly widespread. Between them the 26 interviewees were able to immediately identify and name a total of 67 members of staff who were also delivering elements of ESD. This work was conducted as action research, in that the interviews and the subsequent cross institution discussions were also used to deliberately deepen the understanding of ESD at UWE. In 2007/2008, in order to promote and encourage academic engagement in education for sustainable development and to respond to the findings of Grant (2008) a cross university group, Knowledge Exchange for Sustainability Education (KESE), was created. The group is charged with promoting, encouraging, challenging and leading the ESD agenda in UWE. Some 70 staff are engaged in this network drawn from the four faculties of the university. In meeting the responsibility placed upon it KESE has developed a university wide ESD Action Plan. Table 2 presents the terms of reference of KESE as mandated by the Sustainability Board.

The university's approach to ESD is to recognise it as a change management process with all the attendant problems of change management. Progress is being made from a baseline expressed in the Environmental Policy of the 1990s, recodified in the Sustainability Strategy 2007–2012 (see UWE 2014c) and now superseded by the new Sustainability Plan (see UWE 2014c) for the period 2013–2020. This new plan was approved by Academic Board and the Board of Governors and signed off by the Vice Chancellor, the Chair of the Board of Governors and the President of UWESU. This tripartite ownership and responsibility for the implementation of the plan is a symbolic statement of the distributed responsibility for leadership of the change process in support of a more sustainable university.

UWE is committed through the plan to ensuring our students—future graduates—are exposed to the ideas of sustainable development in their programme of study (Longhurst and Gough 2014a, b). Whilst it is important that ESD is contextualised within the discipline the student is studying UWE also provides opportunities for students through co curricula volunteering and other public engagement activities in the field of sustainable development. Internship and placement opportunities are available to enable real world learning in enterprises of various types and sizes. A particular feature of the UWE approach is the dedicated Green Internship Scheme with an annual Green Intern of the Year award. These activities provide an adjunct to the disciplinary engagement of the programme of study and support integration and application of disciplinary knowledge.

5 Outcomes of Institutional ESD Activity

Each academic year KESE and the Sustainability Board audits progress in meeting the KPI of all students being exposed to ESD in their programme of study. In 2011/2012 academic year 71 % of UK domiciled students engaged with sustainable development within the context of their discipline. In academic 2012/2013 around 90 % of our UK based students had had this opportunity. There is still progress to make but overall this is a satisfying performance in respect of inputs to the curricula. Clearly, further work is needed to judge the extent to which students acquire ESD skills, knowledge and attributes from this engagement. Departmental annual surveys reveal that there is a widespread engagement of academic programmes in the exploration of sustainable development and its meaning in the discipline. This engagement is either introductory or specialised according to the specific needs of the individual discipline or programme. As first noted by Grant (2008), and subsequently noted by annual surveys (see Gough 2011; Longhurst 2011, 2012, 2013), each Department of the University contains modules and programmes of study exposing students to some or the entire context of sustainable development.

The annual survey of progress in the departments (Gough 2011; Longhurst 2011, 2012, 2013) enables a degree of benchmarking between years and Departments and allows a synoptic evaluation of progress with ESD at UWE. Its findings help KESE to understand local issues and Departments to think about current issues and future opportunities. KESE, in evaluating the outcomes of these surveys, reflects and reports on achievements against the Sustainability Strategy and ESD Action Plan, identifies resource and other gaps in support, particularly in the form of staff development and curricula resources for staff, and considers the extent to which the ESD sessions in the Academic Development Programme require enhancement to meet the needs identified in the survey. Table 3 presents a

Table 3 Proportion of departmental programmes considering ESD

	Percentage (%) of programmes which consider ESD				
	0–20 %	21–40 %	41–60 %	61–80 %	81–100 %
2011	57 % (8)	14 % (2)	0 % (0)	7 % (1)	21 % (3)
2013	13 % (2)	6 % (1)	6 % (1)	13 % (2)	64 % (10)

Source Gough (2011), Longhurst (2011, 2013)

Table 4 Departmental stage of development in embedding sustainability into the curriculum?

	2011 (%)	2012 (%)
Interested stage Willingness to include sustainability in the curriculum	36	0
Incipient stage A small number of staff are teaching sustainability and a few programmes include consideration of sustainability	50	56
Emergent stage The programme team is working towards engaging with the sustainability agenda across most or all of its programmes and modules	7	25
Advanced stage Teaching staff and students immersed in the sustainability agenda in all aspects of the programme	7	19

Source Gough (2011); Longhurst (2011, 2012)

comparison of Departmental programmes providing ESD context for all students in 2011 and 2013. As can be seen, in 2013, 64 % of programmes considered ESD, and these programmes represent 69 % of students. Table 4 provides a further illustration of the development of departmental engagement in the provision of ESD. In 2011 36 % of departments were classified as being in the Incipient Stage of engagement and only 7 % had achieved the Advanced Stage. By 2012 the percentage of departments classified as Incipient had fallen to zero whilst the Advanced Stage now totalled 19 %.

In the annual survey Heads of Departments note a range of drivers for change including institutional policy and strategy, staff interest, professional body expectations, developments in the disciplines and broad societal pressures. These drivers are not uniformly distributed across the departments as the following quotes illustrate.

> It is already embedded in virtually all modules therefore we are not looking to take it any further. Currently, sustainability is there in rhetoric and practice but there is scope for improvement. We need to keep ensuring that what we do is what we say we do (UWE Head of Department 2013).
>
> Sustainability is important to me in relation to well being of staff and students, including health and safety, in the broadest sense. The culture within the department is also important, regarding energy use, resource use, safe workplaces, etc. Students are not pushed to be sustainable (UWE Head of Department 2013).
>
> At the moment this is unimportant (to our professional bodies) but as skill set scoping exercises begin, it will grow in importance (UWE Head of Department 2013).

Table 5 Whole institution ESD staff development days 2011, 2012, 2013

Year of ESD staff development day	Aim of day	Themes addressed
2011	Sustainability + Education	A springboard for innovation, student enrichment and partnership
2012	Education for sustainable development in practice	Practical insight, inspiration and networking Providing a platform for debate Showcasing and sharing approaches Inspiring our next steps
2013	Space for the exchange of ideas for enhancing sustainability awareness amongst staff and students	A snapshot of progress. Supporting education for sustainable development activity

> This is turning into a key feature of the University and the department recognises pressure is coming (UWE Head of Department 2013).

An important part of the ESD Action Plan is the provision of an annual Education for Sustainable Development staff development day (see Longhurst 2011, 2012, 2013). Table 5 illustrates the themes covered in the last 3 staff development days. As part of its commitment to sharing practices in ESD the university hosted the first of the 2013/2014 HEA thematic seminars on Education for Sustainable Development (Longhurst and Gough 2013).

6 Broadening the Range of ESD Opportunities for Students

The university recognises that some students will want to explore sustainable development at their own pace outside of the formal curriculum. In order to address this need the university is developing a student resource called Keys to Sustainability (Gough and Longhurst 2014). This will be available to all UWE students during induction in academic year 2014/2015. The resource will be delivered on-line via Blackboard and students will be able to work through it at their own pace. On completion they will be awarded a certificate, have completed a component of the UWE Futures Award and had the outcome noted as part of their Higher Education Achievement Record.

As part of its broad scale commitment to ESD the university hosted an Erasmus Integrated Programme (IP) in the summer of 2013. Entitled "Exploring Hopeful Pathways towards Preferable Local and Global Futures" the IP brought over 50 students, both UG and PG, and 17 academic staff from 9 EU countries to Bristol to work with UWE staff and students, exploring issues and challenges in sustainability (UWE 2013).

In pursuit of its sustainability goals UWE works closely with the UWE Students' Union. The university is supporting the Students' Union to develop a cadre of 70 Student Green Leaders across the Departments of UWE. The Green Leaders are supported by a co-ordinator within UWESU. The university and Students' Union provide training, resources for projects and, in exchange, the university expects the Green Leaders to challenge and encourage us to do better in meeting our new Sustainability Plan's aims and objectives. They are, and UWE wants them to be, critical and noisy friends.

UWE's research and knowledge exchange activities support the goal of sustainable development. This work is led by a cross university initiative linking sustainability, health and environment with important contributions to new knowledge in the interlinked domains of public health, community development, carbon management, climate adaptation, environmental technology, air quality management, ecosystem services, flood risk management and water security. An innovative cross-university MSc Sustainable Development in Practice provides a specific contribution to education about sustainable development counterpointing the university's curriculum orientation towards education for sustainable development.

UWE is committed to sharing its practices and the challenges we face within the sector. We are active in the Higher Education Academy's Education for Sustainable Development Advisory Group and in the joint HEA—Quality Assurance Agency's ESD Development group which produced a reference point for use with the QAA Code of Practice (QAA and HEA 2014).

Mader (2012) proposed five interconnected dimensions of pro-sustainability change: Leadership and Vision, Social Network, Participation, Education and Learning, and Research Integration, which he suggests can be used to assess "transformative performance" towards sustainability in higher education. Table 6 provides a chronology of some of the key UWE developments in pursuit of sustainability from 1994 to 2014. Each of these is assessed against Mader's dimensions of pro-sustainability change. From this table it can be seen that 11 of the UWE developments relate to Leadership and Vision, 11 to Social Network, 11 to Participation, 18 to Education and Learning and 3 developments relate to Research Integration. Recognising the interconnectedness of these dimensions 12 of the 28 developments relate to more than one dimension. The initial phases of UWE's activity principally relate to the Leadership and Vision dimension with the dimension Education and Learning increasingly prominent from 2008 onwards and developments addressing multiple dimensions being relatively common across the chronology.

Table 6 UWE chronology of developments assessed against Mader's five interconnected dimensions of pro-sustainability change

UWE chronology of development	Mader (2012) dimension addressed
Environmental Policy 1994 periodically updated	Leadership and vision
Sustainability board established 2004	Leadership and vision
Sustainability Action Plan 2005	Social network, participation
Institute for Sustainability, Health and Environment established 2007	Research integration
Sustainability strategy with ESD goals 2007	Leadership and vision, social network, participation, education and learning, and research integration
Baseline survey and report of ESD in the modular scheme 2008	Education and learning
Knowledge exchange for sustainability education (KESE) established 2008	Social network, participation, education and learning
KESE Action Plan 2009	Social network, participation, education and learning
KESE Annual Report to LTAC and sustainability Board 2010	Leadership and vision
Head of Department ESD survey 2011	Social network, participation, education and learning
Assistant Vice Chancellor, Environment and Sustainability appointed 2011	Leadership and vision
Sustainability + Education Conference 2011	Social network, participation, education and learning
KESE Annual Report 2011	Leadership and vision
Sustainable Development Resource Pack 2011/2012	Education and learning
ESD Session in the Academic Development Programme 2011/2012 onwards	Education and learning
MSc Sustainable Development in Practice 2012	Education and learning
New ESD modules in Department of Education 2012	Education and learning
Education for Sustainable Development in Practice Staff Development day 2012	Social network, participation, education and learning
KESE Annual Report 2012	Leadership and vision
First L/SL in Education for Sustainable Development appointed 2012	Leadership and vision, education and learning
Head of Department survey 2013	Social network, participation, education and learning
Education for Sustainable Development Day 2013	Social network, participation, education and learning
KESE Annual Report 2013	Leadership and vision
Environmental Policy Review 2013	Leadership and vision

(continued)

Table 6 (continued)

UWE chronology of development	Mader (2012) dimension addressed
ESD and Quality Management Framework. A Guide for Staff 2014	Education and learning
Education for Sustainable Development Day 2014	Social network, participation, education and learning
New Sustainability Plan 2013/2014–2020	Leadership and vision, social network, participation, education and learning, and research integration
Keys to sustainability—available start of 2014/2015	Education and learning

7 Concluding Comments

This paper has outlined the comprehensive approach of the university in responding to the challenge of creating a more sustainable future. The university is preparing its students for the challenges of the 21st century and no bigger challenge confronts them than that of sustainable development. In order to thrive and excel in this environment the disciplinary attributes of a graduate must be blended with those that are the result of engagement with, and understanding of, the challenges and opportunities posed by sustainable development.

This paper has analysed the approach of UWE through the lens of Mader's (2012) interconnected dimensions of pro-sustainability change. Whilst the university can demonstrate that it has made progress on the journey, particularly in its goal of exposing all students to the ideas of ESD in their discipline, there remains much to do. The university can demonstrate elements of integration in and between the five dimensions but a culture shift is still required perhaps best summed up by a respondent to the 2013 HoD survey.

> You will never gain momentum unless it is part of everything you do. It is not currently worked on explicitly by all programme teams. There is still scope for developing all staff knowledge and awareness. You are not advanced unless it is so ingrained that it is automatic for all staff to make the connection (UWE Head of Department 2013).

This paper provides a longitudinal study of the approach of one UK university, UWE, Bristol, to develop a comprehensive engagement with ESD in the curricula. It describes progress and barriers towards the institution's goal of providing an ESD experience for all students. Success factors have been identified, of these factors institutional leadership is critical to legitimise engagement. However, it is not sufficient on its own as staff and student interest and enthusiasm are equally necessary if the goals are to be achieved. This longitudinal study can provide a template for other universities considering the introduction of ESD into the institutional mission. Whilst limited to a single university this paper does provide the basis for future comparative studies of institutional approaches to the comprehensive introduction of ESD into university curricula.

References

Barth M (2013) Many roads lead to sustainability: a process-orientated analysis of change in higher education. Int J Sustain High Educ 14(2):160–175

Birney A, Salazar C, Morgan J (2008) How do we enable systems change for a one planet future? J Corp Citizensh Summer 2008:23–36

Blowfield M (2013) Business and sustainability. Oxford University Press, Oxford

Brown GH (2010) Sustainability in higher education: Management and governance challenges. Perspect: Policy Pract High Educ 14(4):103–104

Cambridge Programme for Sustainability Leadership (2011) A journey of a thousand miles: the state of sustainability leadership in 2011. Cambridge Programme for Sustainability Leadership, Cambridge

Cebrian G, Grace M, Humpris D (2013) Organisational learning towards sustainability in higher education. Sustain Account Manage Policy J 4(3):285–306

Galea C (2004) Teaching business sustainability. Greenleaf, Sheffield

Gough GK (2011) Education for sustainable development at UWE. Heads of Department survey. University of the West of England, Bristol, 19 p

Gough GK, Longhurst JWS (2014) Raising sustainability awareness in a Higher education institution: engaging the whole institutional community. Paper presented at on sustainability. Tenth international conference on environmental, cultural, economic and social sustainability, University of Split, Croatia, 22–24 Jan 2014

Grant M (2008) Education for sustainable development baseline study: the state of play at UWE. Report for the Sustainability Board. University of the West of England, Bristol, vol 1, 86 p; vol 2, 69 p

Henrik N, Macdonald J, Broman G, Yamamoto R, Robèrt K-H (2006) Sustainability constraints as system boundaries: an approach to making life cycle management strategic. J Ind Ecol 10 (1–2):61–77

Longhurst JWS (2011) Sustainability + education. A springboard for innovation, student enrichment and partnership. Presentation at annual ESD staff development day, University of the West of England, Bristol, 10 May 2011

Longhurst JWS (2012) Education for sustainable development in practice. Presentation at annual ESD staff development day. University of the West of England, Bristol, 29 June 2012

Longhurst JWS (2013) Education for sustainable development at UWE: space for the exchange of ideas for enhancing sustainability awareness amongst staff and students. Presentation at annual ESD staff development day. University of the West of England, Bristol, 24 Apr 2013

Longhurst JWS, Gough GK (2013) Education for sustainable development at the University of the West of England, Bristol. Paper presented at influence, implement, engage: the triple imperative for ESD in HE. HEA thematic seminar series 2013/2014, University of the West of England, Bristol, 29 Nov 2013. Available at http://www.heacademy.ac.uk/events/detail/2014/Seminars/Themes/GEN876_UWE. Last accessed 19 Mar 2014

Longhurst JWS, Gough GK (2014a) Implementing a comprehensive institution-wide approach to education for sustainable development. A case study of the University of the West of England, Bristol. Paper presented at University Leadership for Sustainability, University of Gloucestershire, 10–11 Jan 2014

Longhurst JWS, Gough GK (2014b) Implementing a comprehensive institution-wide approach to education for sustainable development. A case study of the University of the West of England, Bristol. Paper presented at on sustainability. Tenth international conference on environmental, cultural, economic and social sustainability, University of Split, Croatia, 22–24 Jan 2014

Lozano R (2006) Incorporation and institutionalization of SD into universities: breaking through barriers to change. J Clean Prod 14(9–11):787–796

Mader C (2012) How to assess transformative performance towards sustainable development in higher education institutions. J Educ Sustain Dev 6:79–82

Mader C (2013) Sustainability process assessment on transformative potentials: the Graz Model for integrative development. J Clean Prod 49:54–63
Mader C, Scott G, Abdul Razak D (2013) Effective change management, governance and policy for sustainability transformation in higher education. Sustain Account Manag Policy J 4 (3):264–284
McMillan J, Dyball R (2009) Developing a whole university approach to educating for sustainability. Int J Sustain High Educ 3:55–64
Morgan G (1986) Images of organization. Sage, Thousand Oaks
Ny H, MacDonald JP, Broman G, Yamamoto R, Robért K (2006) Sustainability constraints as system boundaries: an approach to making life-cycle management strategic. J Ind Ecol 10 (1–2):61–77
Orr D (1992) Earth in mind: on education, environment and the human prospect. Island Press, Washington
Parkin S (2010) The positive deviant: sustainability leadership in a perverse world, 1st edn. Earthscan, London
Posner S, Stuart R (2013) Understanding and advancing campus sustainability using a systems framework. Int J Sustain High Educ 14(3):264–277
QAA and HEA (2014) Education for sustainable development. Guidance for UK higher education providers. Available in Online http://www.qaa.ac.uk/en/Publications/Documents/Education-sustainable-development-Guidance-June-14.pdf
Scott G, Gough S (2006) Sustainable development within UK higher education: revealing tendencies and tensions. J Geogr High Educ 30(2):293–305
Scott G, Tilbury D, Sharp L, Deane L (2012) Turnaround leadership for sustainability in higher education. Learning and Teaching Excellence Division, DEEWR, Australian Government, Canberra
Senge P (2008) The necessary revolution: how individuals and organizations are working together to create a sustainable world. Nicholas Brealey, London
Sharp L, Pollock Shea C. (2012) Institutionalizing sustainability: achieving institutional transformation from the inside. In: Martin J and Samuels J (Eds). Sustainable university: Green goals and new challenges for higher education leaders. John Hopkins University Press, Baltimore
Swearingen White S (2014) Campus sustainability plans in the United States: where, what and how to evaluate? Int J Sustain High Educ 15(2):228–241
Temple P (2010) Sustainability. Persp: Policy Pract Higher Educ 14(4):105–107
Tilbury D (2011) Higher education for sustainability: a global overview of commitment and practice. In: Anon (ed) Higher education in the world 4: higher education's commitment to sustainable development: from understanding to action. Palgrave MacMillan, London, pp 1–21
UNESCO (undated) Education for Sustainable Development Available at http://www.unesco.org/new/en/education/themes/leading-the-international-agenda/education-for-sustainable-development/education-for-sustainable-development/. Last accessed 18 March 2014
University of the West of England, Bristol (2013) Erasmus Integrated Programme news item http://info.uwe.ac.uk/news/uwenews/news.aspx?id=2585. Last accessed 19 March 2014
University of the West of England, Bristol (2014a) About the University http://www1.uwe.ac.uk/aboutus. Last accessed 19 March 2014
University of the West of England, Bristol (2014b) Strategy 2020. http://www1.uwe.ac.uk/aboutus/visionandmission/strategy.aspx. Last accessed 19 March 2014
University of the West of England, Bristol (2014c) Sustainability strategy and policy http://www1.uwe.ac.uk/aboutus/visionandmission/sustainability/sustainabilityaction/sustainabilitydocuments.aspx. Last accessed 18 March 2014
Wright T, Horst N (2013) Exploring the ambiguity: what faculty leaders really think of sustainability in higher education. Int J Sustain High Educ 14(2):209–227

Authors Biography

Prof. James Longhurst is Assistant Vice Chancellor, Environment and Sustainability at UWE, Bristol. He chairs the Higher Education Academy's Education for Sustainable Development Advisory Group and the joint Quality Assurance Agency and Higher Education Academy's development group which prepared the guidance for UK higher education providers on ESD published in June 2014.

Dr. Georgina Gough is Senior Lecturer in Education for Sustainable Development at UWE, Bristol. She lectures on sustainability in business and on tourism and supports academics in developing their ESD engagement.

Marcus Grant is the Deputy Director of the WHO Collaborating Centre for Health Urban Environments. He undertook the initial baseline survey of UWE's ESD provision. Marcus is a member of the KESE group and delivers the ESD sessions to the Academic Development Programme for new staff.

Dr. Sarah Hills is the Programme Manager for the MSc Sustainable Development in Practice at the University of the West of England, Bristol. Sarah supports the introduction of ESD across UWE and is a member of the KESE group.

Environmental Education for Staff at the University of São Paulo, Brazil: Capillarity and Critical Environmental Education Put into Action

Daniela Cássia Sudan, Ana Maria de Meira, Marcos Sorrentino, Fernanda da Rocha Brando Fernandez, Rosana Louro Ferreira Silva, Laura Alves Martirani, Denise Maria Gândara Alves, Paulo Ernesto Diaz Rocha, Silvia Aparecida Martins dos Santos, Taitiany Karita Bonzanini, Maria Estela Gaglianone Moro, Welington Braz Carvalho Delitti and Marcelo de Andrade Romero

Abstract

This article addresses an experience report regarding an ongoing socio-environmental education program for 17,000 administrative staff members on the seven campuses of the University of São Paulo (USP) in Brazil, from the critical and emancipatory perspective of environmental education. The program is being run by the superintendent of environmental management at USP. The objective of the program is to educate all USP staff about the environment; increase the inclusion of sustainability in university management; and encourage sustainable actions in the university. Capillary architecture, in which a group known as "people who learn by participating" (PAP; the acronym in Portuguese is used in its original form in this article because of its resonance) makes a

D.C. Sudan (✉)
Environmental Management Superintendence, University of São Paulo (USP), Av. Bandeirantes, 3900, Ribeirão Preto-SP CEP 14040-900, Brazil
e-mail: danisudan@usp.br

A.M. de Meira · M. Sorrentino · F. da Rocha Brando Fernandez · R.L.F. Silva · L.A. Martirani · D.M.G. Alves · P.E.D. Rocha · S.A.M. dos Santos · T.K. Bonzanini · M.E.G. Moro
Group for Research and Studies on Environmental Education at SEM, University of São Paulo, São Paulo, Brazil

W.B.C. Delitti · M. de Andrade Romero
Environmental Management, University of São Paulo, São Paulo, Brazil

W. Leal Filho et al. (eds.), *Integrating Sustainability Thinking in Science and Engineering Curricula*, World Sustainability Series, DOI 10.1007/978-3-319-09474-8_38

commitment to mobilize other groups in their workplace through theoretical and practical courses, was used to ensure that all staff members could be reached. The initial group (PAP-1) is responsible for developing the political-pedagogical project and preparing, encouraging, and mentoring PAP-2 leaders. Subsequently, this group will offer courses in their workplaces to PAP-3 (estimated staff of 4,590), who will involve other university students (PAP-4), totaling 17,000 people at various levels. Some of the preliminary program elements were: (i) ten planning meetings with experts (PAP-1) to improve the project and its strategies; (ii) presentation and approval of the program by the university authorities; (iii) selection of socio-environmental leadership groups on the campuses; (iv) 22 h of education for PAP-2 on environmental education, civilization crisis, climate change, culture industry and consumerism, public policies, and good practices at universities; and (v) dissemination of the initiative through USP media. The results of this unprecedented educational effort are expected to contribute to other institutions of higher education and to make the University of São Paulo a space of coherence and reference in the socio-environmental field.

Keywords
Environmental education · Sustainability · University · Capillary · Staff

1 Introduction: Sustainability in the University and Staff Training

One of the major challenges in today's world is to create global socio-environmental sustainability through the contribution of every single institution and territory worldwide. Institutions of higher education in general, and universities in particular, can and should take on important roles in this process. Sorrentino and Nascimento (2010) emphasized the inseparable connection between the utopian ideal of sustainable societies and efforts in environmental education, in which universities play a dual role: the education of university staff per se and contributing to the education of society regarding environmental issues.

Hence, investment in the socio-environmental education of staff is required in order to allow them to take on the role of educators, and also in order to enable their collaboration in the daunting challenge of sensitizing faculty and the university community as a whole so that they also can join efforts in this field. It is necessary to keep in mind that because of the rather focused education received by faculty and the university community, they often find it difficult to understand the complexity of the socio-environmental field and the cross-departmental and multidisciplinary perspectives required in critical environmental education.

The objective of this paper is to present the preliminary results of an ongoing socio-environmental education program for employees at the University of São Paulo (USP), based on the critical and emancipatory perspectives of environmental education. More specifically, we aim to describe and reflect upon the theoretical and methodological framework of the educational program, its main content, practices, and connections, as well as the expected outcomes.

USP has seven campuses located in the State of São Paulo in the cities of São Carlos, São Paulo, Piracicaba, Pirassununga, Ribeirão Preto, Bauru, and Lorena. According to the "2012 Statistical Abstract of USP," the university comprises a total territory of 76,314,505.94 m^2, with a developed area of 1,821,970 m^2. The faculty has 5,860 members, and there are 58,303 students enrolled in 249 undergraduate courses. USP also offers 239 graduate courses: 332 master's courses with 13,836 students; and 309 doctoral courses with 14,662 students. The university staff of 16,839 people is the target population in this socio-environmental education program (USP 2014).

The program has been run by the superintendent of environmental management (SEM) at USP since 2013, and is based on critical and emancipatory perspectives on both environmental education and action-research (Órgão Gestor da Política Nacional de Educação Ambiental 2006).

To develop this program, in March 2013, the SEM created the Group for Research and Studies on Environmental Education in March 2013. The group comprises of experts from several campuses, and aims to conceptualize and structure the education plan, which is continuously reviewed and improved for collective development.

Considering the participative, permanent, connected, and continuing perspectives of the program, it has several main objectives: to collaborate in the socio-environmental education of the university community; to contribute to the internalization of sustainability into university management; to promote a change in the organizational culture based on agreed-upon socio-environmental values; and to offer support for USP staff in broadening their view/perception/analyses and possibilities of socio-environmental action in their own work and in other areas within this community.

2 Theoretical and Methodological Premises: Capillary Environmental Education

The premises of the program are based on the concept of critical and emancipatory environmental education in the methodological processes of capillary architecture and participant-action-research (Viezzer 2005).

The critical and emancipatory perspective of environmental education is essentially characterized by: a complex understanding of environmental issues; a critical attitude in view of the challenges of the civilization crisis; an understanding that children are not the main target for environmental education; a belief that democracy, dialogue, and participation are essential for building sustainability; and a search for changing realities that are contrary to collective well-being both locally and globally, based on individual and collective changes, as well as structural changes in public policies. This approach stimulates the engagement of individuals and communities in processes that aim to change the current model of society, seeking to build other interactions among individuals, culture, work, and nature (Carvalho 2001; Lima 2005; Sawaia 2001; Loureiro 2005; Sorrentino 2003, 2005, 2010; Layrargues 2006).

Alves et al. (2010) addressed five aspects pertinent to this proposed program, presenting important interconnected parameters for thinking, feeling, and acting in environmental education: community, identity, dialogue, power of acting, and happiness. The parameter of community involves recovering community meaning by creating collective environmental educators in the dialogic perspective, and developing the power of communication by strengthening individual and collective identities. To consolidate an identity, whether individual or collective, it is necessary to provide elements to search for a global identity that can serve as the basis for dealing with socio-environmental issues on both the global and micro-local scale. The third aspect—dialogue—is understood in this paper as the flow of meaning (different from debate, consensus, and discussion), which allows for building something new and sharing something collective, without competitiveness and eagerness to win. It should be noted that, according to Alves et al. (2010), there is a distinction between the I-Thou relationship, which is the encounter between the essence of beings, and the I-It relationship, which is fundamentally utilitarian. The power of action, the fourth aspect under consideration, is discussed mainly using the Espinosa framework (Baruch Spinoza, 1632–1677); its essence is participation, immanent/constituent and inseparable from the human condition. Each person must awaken the capacity to identify "the environmental issues, become involved, and commit towards making whatever decisions are necessary" (Alves et al. 2010, p. 10). The authors also highlighted the happiness issue—the idea of a utopia for all—as a fundamental aspect to be considered in the environmental education process, which is intended to be emancipating. In view of the various definitions and studies of happiness, it is important to consider that the human condition in the present degrading and unequal society has a deep impact on people's feelings and their search for happiness. Aspects related to happiness (psychological well-being, health, education, standard of living, etc.) must be considered, including when creating indicators and planning goals.

The challenge of the USP socio-environmental education process is to promote "good encounters" that allow "subjects to share and dialogue about their experiences, and thus strengthen the power of action" (Alves et al. 2010, p. 28).

Following this line of thought, Sorrentino and Nascimento (2010) postulated a number of important questions on environmental issues and sustainability within the university scope:

> What role (or roles) does the University play in the field of sustainability and environmental education? And, on the other hand, what is the role of sustainability and environmental education at universities? What roles should the public policies play for these institutions to work more determinedly in the field of sustainability and environmental education? (Sorrentino and Nascimento 2010, p. 17).

According to these authors, first, the participants and society who maintain the university must define the direction to be followed in the complex context of contemporaneity. Second, but equally important, they should determine how it should be done. To talk about building sustainable societies through education implies the need to delve simultaneously into knowledge and action proposals that involve the formulation and implementation of public policies, and the improvement of teaching and learning methods and techniques that allow such studies, debates, and learning (Sorrentino and Nascimento 2010).

Considering the target audience of approximately 17,000 public workers (administrative staff) from the USP campuses who are from various sectors and work areas, this educational program uses capillary architecture in a horizontal network of multipliers. This system of capillary education refers to two connected and complementary meanings: "participant-action-research" and "people who learn by participating", or simply PAPs, comprising a learning, interpretative, and affective community of life and meaning that is in a place for good and humanizing encounters and for praxis (Ferraro and Sorrentino 2005).

The superintendent of environmental management (SEM) and the Group for Research and Studies on Environmental Education were included in the capillary network as a PAP-1 collective educator (25 people), playing the role of developing the political-pedagogical project and creating and monitoring (on-site and at a distance) a group of agents and commission members, the PAP-2s (135 people). The latter are committed to teaching courses at their workplace to PAP-3s (4,590 people), and the PAP-3s are in turn committed to developing educational action connected to environmental management, and involving other public workers (the PAP-4s), implementing the education of a total of 17,000 people with different levels of knowledge regarding the environment (as illustrated in Fig. 1).

The whole process is being recorded by a support team, in both print and audiovisual files. These files are stored and shared through email and on a USP virtual platform, comprising a collection of data made available to all program members.

3 Content and Practices of USP Staff of the Socio-environmental Education Program

The socio-environmental education program for USP staff is structured on three interconnected axes: the availability and problem-posing of contents; praxis pedagogy, creating dialogical and reflexive processes; the production of new knowledge in education action; and the creation of interpretive and learning communities.

For the availability and problem-posing of the contents, the PAP-1 created a "learning menu", with contents that are fundamental for environmental education as well as sustainability practices, which must be developed with the PAP-2s, PAP-3s, and PAP-4s on all campuses.

The learning menu, according to Tonso (2005), should propose activities in a variety of formats that develop the recreational, affective, and aesthetic meanings of educators-learners, beyond technical and objective information. It presents items of different natures, including informative (providing several types of content and knowledge) and formative (allowing creation of methodologies, values, perceptions,

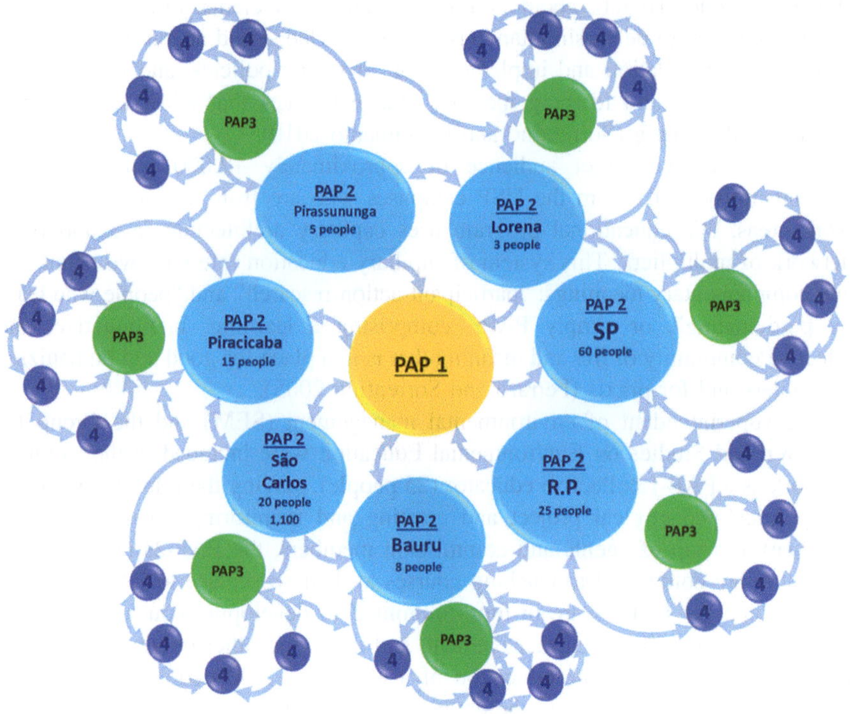

Fig. 1 Illustrative image of capillary architecture

and attitudes of the educator being trained). The learning menu also assumes an identity of its own, according to the specific demands of the local community, in this case, each campus.

Though approached through different modules, the topics in environmental education, communication, and public policies are, in fact, cross-sectional to every module and practice. The contents are addressed at different depths, according to the duration of the educational actions, with PAP-2s and PAP-3s receiving contents in the highest number and greatest depth.

Each group has access to: a *conceptual repertoire*, which comprises a set of information on society and the environment, sustainability, and environmental education; a *situational repertoire*, with resources to make participative socio-environmental diagnoses; and an *operational repertoire, in which each PAP will have to develop a sustainable educational practice in the workplace that could have a positive impact on environmental management at the university* (SGA/USP 2014).

The second methodological axis, praxis pedagogy, brings in the theoretical-practical dialectics of Paulo Freire (1921–1997). It is based on the principle that knowledge is a continuous exchange of experiences accumulated in life. The educator is, at the same time, also a learner open to the teaching-learning process in the search for sustainable solutions and practices. This attitude is particularly important to building sustainable societies as we move away from collective unawareness of deeper answers to contemporary questions, which culminated in the current unsustainability. Based on this pedagogical approach, the actor-subject becomes the protagonist-agent of his or her own choices and actions, although those choices and actions are agreed upon in the groups in which they participate. In other words, within the groups there is a foreseen collective attitude of collective learning and implementation of shared solutions. This is, therefore, the third methodological axis of the present study: interpretative and learning communities.

Inspired by the Paulo Freire culture circles and interpretative and learning communities (Avanzi and Malagodi 2005; Brandão 2005), this collective educator is a collegiate enterprise that gathers people with support from their institutions to work in an educational program of participant-action-research on the environment, citizenship, and quality of life in a given territory (Ferraro and Sorrentino 2005).

Within the PAP theoretical and methodological axis and similar to its capillarity proposal, the National Environmental Fund financed the project "Making Utopia Viable", developed by the São Paulo Collective Educator, Araraquara, Jaboticabal and Region (CESCAR)/Brazil, in the period between 2007 and 2008. This project implemented an education program for environmental educators, with the participation of 90 people and 39 partner institutions from 12 different municipalities of the State of São Paulo. As part of the proposal, PAP-1 s belonged to the Environment Board's Managing Office of the Ministry of the Environment, PAP-2 s were the educators from CESCAR's partner institutions, and PAP-3 s were the participants in the education program who developed educational interactions with PAP-4 s in their municipalities.

During the periods of reflection carried out during the process, participants reported that this course had been different from others they had previously attended or experienced, due to the innovative attitudes and spaces for dialogue which had been created (Oliveira 2011). Santos (2010) analyzed CESCAR's constitution and observed that, despite the inherent difficulties of the PAP theoretical and methodological axis, such as some lack of participation, attempting to reconcile schedules, waiting for and respecting each person's response time, which implies an excessively lengthy process, the results were satisfactory. According to assessments by the participants, not only did they obtain the theoretical basis they had been looking for, but they were also able to form PAP-4 s and provide continuity in the education process. Participants said that they were able to understand the process and realize that learning done in this manner is more significant. They also stated that they were able to put what they had learned into practice during the interaction projects with PAP-4s, and learn from them as well.

Another example of a participative process which lends itself to comparison and exchange of experiences is the creation of the University Environmental Education Program (PUEA), which has been developed on the Luiz de Queiroz/USP campus, located in the municipality of Piracicaba, São Paulo, Brazil. This project aims at making environmental education intrinsic to research, teaching, outreach programs and university management. The process was created with the involvement of all campus sectors through meetings, seminars, and talking circles, and by sharing experiences and welcoming initiatives. Within 5 years (from 2009 to 2013), the campus was able to write up a document (Cooper 2009) and have it approved by its highest bodies (congregation and councils). Implementing the process has been a challenge and a great learning experience, for it implies a revision of all the practices and relationships within the university. The process has been gradually implemented on the campus and is part of the Participative Socio-Environmental Master Plan for the Luiz de Queiroz campus (Escola Superior Luiz de Queiroz 2013).

Other practical experiences in this line of work within the University of São Paulo have also strengthened groups, supported joint actions, created spaces for speaking and making decisions, and led to positive outcomes, which can be found in publications and articles concerning events in this area (Meira et al. 2009; Sorrentino et al. 2003; Sudan et al. 2007, 2009).

4 Connections with Educommunication as Part of the Education Process

Allied with the perspective of critical and emancipatory environmental education, educommunication, as an educational practice, stimulates the production of communication materials in a participative and dialogical way, aiming to encompass both groups and communities with the perspective of forming social communication networks and affirming human rights, as well as the values of social justice and sustainability.

According to the USP Center for Communication and Education, educommunication is defined as:

> A group of actions aimed at broadening the communicative coefficient of educational actions, which can be formal, non-formal, and informal, by increasing the expression skills of educational community members, as well as their competence in handling information technology, so as to build open and democratic communicative ecosystems, thus ensuring equal opportunity of expression for the whole community (NCE/USP apud. Brasil 2014).

The principles of educommunication involve, besides dialogical and participative processes as the basic foundation for all educational and communicative action, interactive production and dissemination of environmental education contents through media, as well as firmly establishing environmental education among communication professionals, and the perspective of partnered communication with the media aiming at developing public awareness of sustainability (Brasil 2005, 2008).

Thus, the program foresees the production of communication materials by the PAP-2, aiming at their dissemination in digital environments linked to this educational program, as well as artistic actions and initiatives, including the possibility of developing information and environmental signal systems. With the purpose of guiding and broadening the reach of these activities, a proposition was made to produce journalistic material in the form of *press releases* following a standard template, aiming at facilitating transit, publication, and dissemination of the activities developed by the media, both inside and outside the university.

Besides the communication material, the program continuously supports the production of didactic materials, considering audiovisual methods and the use of distance tools to record every action, image, report, and discussion forum among the PAPs.

Considering that the program aims at reaching a large number of people, in addition to on-site meetings, it was resolved that communication technologies, as well as distance education, should be included at some point. As described by Valdívia (2008), there is an acknowledgement of innumerable benefits that can be added by information and communication technology to education, regardless of the dominant pedagogical model. They increase opportunities to learn at any time and any place; promote more fluid and permanent relationships between students and knowledge; open learning opportunities throughout life; create learning communities between students and faculty; increase opportunities for professional development; and permit the creation of new means of communication among those involved in the educational program.

Considering the target audience, it would be possible to use distance learning processes to offer participants multiple opportunities for interaction, mediation, and expression of feelings, provided by the flow of information, the diversity of discourse, and the availablity resources, which can be in print, audio, or visual forms, as well as by flexibility of time and schedules. Furthermore, virtual learning environments contribute to the development of activities that emphasize collective production, because education processes that develop at a distance are not solitary. In fact, technological tools and resources result in the subjects being more closely involved, and contribute to awareness of the need to build autonomy regarding one's own educational process. Individuals can depend on their own initiative and ability to organize the times when they interact virtually, as well as their involvement in the proposed activities and readings. A virtual learning environment using the Moodle platform is under construction, and can be accessed by program participants who are affiliated with the university.

5 Current Context of the Program and the Expected Outcomes

The socio-environmental education program for employees at USP aims to consolidate: participative socio-environmental mappings and diagnoses; moderation of problem-solving activities; permanent, connected, and continuing education; inter- and cross-disciplinary dialogue; and management-related sustainable processes, including water, energy, waste, and consumption management, among others. Every process is followed by participating research teams, who support the action-reflection-action process.

In 2013, the following took place: (i) Ten meetings with experts (PAP-1) to improve the program and make strategic connections; (ii) presentation and approval of the program by university authorities on the Bauru, Pirassununga, São Carlos,

Ribeirão Preto, and Piracicaba campuses; (iii) communication with university authorities to select/indicate the socio-environmental leadership groups at the campuses; (iv) three educational meetings with the 180 PAP-2s (leaders from the seven USP campuses) and invited experts, totaling 22 h of education covering six topics (environmental education, civilization crisis, climate change, culture industry and consumerism, public policies, and good sustainability practice at universities); (v) dissemination of the initiative through USP media (IPTV, TV USP, USP News Agency, and others); (vi) coordination of the work of six fellows (grants provided by the USP Dean's Office for Culture and Outreach), three trainees, and three educators hired for the program.

Complementarily, the "Sustainability is" project is being developed, which aims at promoting the feeling of belonging, identity, and the affirmation of values in this educational process by posting remarks and photos by program participants in partners' online spaces, regarding their perspectives on sustainability and the related practices.

The expected outcomes from the program are: (i) the creation by PAP-2s, of 25–78 on-site short-term courses (between 10 and 40 h) that take local situations into consideration; (ii) 870 simultaneous actions in environmental education and management on the campuses; and (iii) publication of educational material and audiovisual productions, combining all the experiences and outcomes of the process.

Being aware of the difficulties associated with the program, the participants in this project understand that the university must act as an example and model for society regarding the implementation of environmental management policies and processes. In this sense, USP has advanced over the last few years, utilizing a cross-sectional approach to elaborating its guidelines, programs, and actions aimed at environmental sustainability in the fields of research, teaching, and outreach.

The participative education process described herein aims at increasing the environmental perception of all USP staff through the development of their critical sense by changing their own everyday attitudes and behaviors. This process should awaken a stronger commitment to the preservation, recovery, and care of the environment and quality of life in the university's working environment and in society in general. Attaining this purpose will certainly take the university to a permanent sustainable level, thereby sensitizing and stimulating society and mobilizing movement toward perceiving environmental issues and adopting sustainable practices in daily life.

6 Discussion of the Initial Results: Commitment Indicators Within the Institution

We can affirm that the goals of the educator-environmental intervention described in the present article environmentally educating all university staff; increasing the inclusion of sustainability in university management; and encouraging sustainable actions in the university—were fully accomplished within the first year of activity. In other words, the education process began with the education of the first learning circles (PAP-1 and PAP-2s), involving coordination by both the university coordination each campus. Also, the institution leaders became aware of the necessity for continuing the education process. At the end of 2013 the university underwent a competitive electoral process for university president and new management; the new regime was very critical of the previous administration and put several projects on hold, but did not suspend this project but continued the program in its entirety. Based upon this, we can affirm that the program passed its first sustainability test.

In order to reach all staff, capillarity architecture has been adopted, in which "people who learn by participating" (PAPs) "research their own participative action". This means that actions that have already been implemented, such as the education program for the first two circles of participative learning and the presentation of this descriptive-analytical paper by PAP-1 at an international scientific event, can also be considered indicators of the commitment that has been made to mobilize other groups in the workplace. This mobilization was conducted through monitored theoretical and practical groups, constituted as self-managed and self-analytical processes of knowledge production, based on collective action and participation.

7 Final Considerations: Challenges and Prospects for the University

The university should be an example and model for society regarding policy implementation and management processes in the environmental field. In this context, USP has advanced over the last few years, transversally creating guidelines, programs and actions whose objective is environmental sustainability in the fields of research, teaching and outreach programs.

The participative education process described here seeks to raise the level of environmental perception of all USP staff, by developing critical thinking displayed through changes of attitude and behaviors experienced in each person's daily life. This process should awaken in individuals a greater commitment to conservation, recovery and care of the environment and quality of life, both in the university work environment and in society in general. If this goal is reached, it will surely take the university to a permanent level of sustainability, sensitizing, mobilizing and encouraging society to become aware of environmental issues and to adopt routine sustainable practices.

Taking into consideration USP's scientific leadership and the dialogue which has already been established with other universities, for example, with the University Network of Environmental Education Programs (RUPEA) and at national and international events which are currently taking place coordinated by the university, it is possible that the impact of this study and intervention aimed at environmental education will move beyond the walls of the institution and have an impact beyond the actions that were planned for its own staff.

Acknowledgments This study presents a collective project by the University of São Paulo, and, therefore, we thank all the other members of the Group for Research and Studies on Environmental Education who were unable to participate in the production of this article, though they have been collaborators in the Environmental Education Program for Staff of the University of São Paulo since 2013, namely: Antonio Vitor Rosa, Maria Angélica Penatti Pipitoni, Patricia Gabryela Moreira, Patricia Di Vitta, Patricia Christina Silva Leme, Tamara Gomes, Fabricio Rossi, Aldo Ometto, Carolina Goes.

References

Alves Denise MG et al (2010) Em busca da sustentabilidade educadora ambientalista. AmbientalMente Sustentable, Revista Científica Galega-Lusófona de Educación Ambiental, A Coruña 1(9–10):7–35. ANO V

Avanzi MR, Malagodi MAS (2005) Comunidades Interpretivas. Encontros e caminhos: formação de educadoras (es) ambientais e coletivos educadores. Brasília: MMA, Diretoria de Educação Ambiental 1:93–102

Brandão CR (2005) Comunidades Aprendentes. Encontros e caminhos: formação de educadoras (es) ambientais e coletivos educadores. Brasília: MMA, Diretoria de Educação Ambiental 1:83–92

Brasil (2005) Ministério do Meio Ambiente. Programa de Educomunicação Socioambiental. Série Documentos Técnicos 2. Brasília: Órgão Gestor da Política Nacional de Educação Ambiental. http://www.daep.com.br/coletivos/adm/download/dt_2_programa_educomunicacao_socioambiental_4a_versao_maio_final.pdf. Accessed 1 Dec 2014

Brasil (2008) Ministério do Meio Ambiente Programa Nacional de Educação Ambiental (ProNEA). Educomunicação socioambiental: comunicação popular e educação. Brasília: MMA, 50 p

Brasil (2014) O que é educomunicação? Ministério da Educação. Mídias na Educação. http://webeduc.mec.gov.br/midiaseducacao/material/radio/radio_basico/inicio_oqueeeducomunicacao.htm. Accessed 16 Mar 2014

Cooper M (Org.) (2009) Plano Diretor Socioambiental Participativo do Campus "Luiz de Queiroz". Universidade de São Paulo. Piracicaba, 432 p. Disponível em: <http://www.esalq.usp.br/biblioteca/PDF/plano_diretor_socioambiental.pdf>. Acesso em 22 de abril de 2014

da Lima GFC (2005) Crise ambiental, educação e cidadania: os desafios da sustentabilidade emancipatória. In: Loureiro FB, Layrargues PP, Castro RS de (orgs). Educação ambiental: repensando o espaço da cidadania. Cortez, São Paulo, 264 p

de Carvalho ICM (2001) Qual educação ambiental? Elementos para um debate sobre educação ambiental e extensão rural. Agroecologia e Desenvolvimento Rural Sustentável, Porto Alegre 2 (2):43–51

dos Santos SAM (2010) Políticas públicas de formação de educadores ambientais: análise do processo de constituição do Coletivo Educador de São Carlos, Araraquara, Jaboticabal e Região—CESCAR. Tese (Doutorado)—Programa de Pós-Graduação em Ciências da Engenharia Ambiental, Escola de Engenharia de São Carlos, Universidade de São Paulo, São Carlos, 208 p

Escola Superior de Agricultura "Luiz De Queiroz" (2013) Programa Universitário de Educação Ambiental. Piracicaba. SP, 35 p

Ferraro LA Jr, Sorrentino M (2005) Coletivos Educadores. In: Ferraro LA Jr (org.). Encontros e caminhos: formação de educadoras (es) ambientais e coletivos educadores. Brasília: MMA, Diretoria de Educação Ambiental 1:57–69

Layrargues PP (2006) Muito além da natureza: educação ambiental e reprodução social. In: Loureiro CFB, Layrargues PP, de Castro RC (orgs.). Pensamento complexo, dialética e educação ambiental. Cortez, São Paulo, pp 72–103

Loureiro CFB (2005) Educação ambiental e movimentos sociais na construção da cidadania ecológica e planetária. In: Loureiro FB, Layrargues PP, Castro RS de (orgs). Educação ambiental: repensando o espaço da cidadania. Cortez, São Paulo, 264 p

Meira AM et al (2009) Construcción de indicadores de sostenibilidad para el Programa USP Recicla—Universidade de São Paulo—Brasil. In: VII Congreso de Educación Ambiental para el Desarrollo Sostenible, 2009, La Habana. Anais do VII Congreso de Educación Ambiental ara el Desarrollo Sostenible, La Habana

Oliveira HT (2011) Pesquisa-Ação-Participante ou pessoas que aprendem participando: de qualquer maneira, pap. In: Oliveira HT, Santos SAM, Dominguez IGPE, Kunieda E (organizadoras). Cadernos do Cescar—Educação Ambiental—Caderno 1—"Os fundamentos e as políticas públicas de Educação Ambiental na constituição do Coletivo Educador de São Carlos, Araraquara, Jaboticabal e Região"—Projeto Viabilizando a Utopia (ViU) 2005–2011. São Carlos: Gráfica e Editora Futura, 100p. Disponível em: http://www.CESCAR.ufscar.br/material_didatico/CESCAR_caderno_1.pdf. Accessed 15 Apr 2014

Órgão Gestor da política Nacional de Educação Ambiental (2006) Documentos Técnicos. ProFEA: Programa Nacional de formação de educadoras(es) ambientais. Por um Brasil educado e educando ambientalmente para a sustentabilidade. Sério Documentos Técnicos—8. Brasília, DF, 46 p

Sawaia BB (2001) Participação e subjetividade. In: Sorrentino M (Org.) Ambientalismo e participação na contemporaneidade. São Paulo: EDUC; FAPESP, p 255

Sorrentino M (2005) Desenvolvimento sustentável e participação: algumas reflexões em voz alta. In: Loureiro FB, Layrargues PP, Castro RS de (Orgs) Educação ambiental: repensando o espaço da cidadania. Cortez, São Paulo, 255 p

Sorrentino M et al (2003) Programa USP Recicla: como construir uma gestão compartilhada? São Paulo, Brasil First Environmental Education Congress. Espinho, Portugal: Abstract Book, 20–24/maio/2003

Sorrentino M, Nascimento EP do (2010) Universidades e políticas públicas em Educação Ambiental. Educ. Foco, Juiz de Fora, 14(2):15–38, set 2009/fev 2010

Sudan D et al (org.) (2007) Dá pá virada: revirando o tema lixo. Vivências em educação ambiental e resíduos sólidos. São Paulo: Programa USP Recicla/Agência USP de Inovação

Sudan DC et al (2009) Educación ambiental y gestión de residuos en la Universidad de São Paulo —Brasil: 15 años de actuación, muchos puntos para reflexión. In: Congreso de Educacion Ambiental para el Desarrollo Sostenible, 2009, La Habana. Anais da VII Convencion Internacional sobre Medio Ambiente y desarrolllo. La Habana

Tonso S (2005) Cardápio de aprendizagem. In: Ferraro Júnior LA (org.) Encontros e caminhos: formação de educadoras (es) ambientais e coletivos educadores. Brasília: MMA, Diretoria de Educação Ambiental 1:47–56

USP—Universidade de São Paulo (2014) Superintendência de Gestão Ambiental. GT Educação. http://www.sga.usp.br/?page_id=1287. Accessed 16 Mar 2014

USP—Universidade de São Paulo—Anuário Estatístico (2012). https://uspdigital.usp.br/anuario/AnuarioControle. Accessed 23 Mar 2014

Valdivia IJ (2008) Las políticas de tecnología para escuelas en América Latina y el mundo: visiones y lecciones. Santiago de Chile, Naciones Unidas, 320 p

Viezzer M (2005) Pesquisa—Ação—Participante. In: Ferraro LA Jr (Org.) Encontros e caminhos: formação de educadoras(es) ambientais e coletivos educadores. Ministério do Meio Ambiente. Diretoria de Educação Ambiental. Brasília: MMA/Diretoria de Educação Ambiental, pp 279–294

Authors Biography

Daniela Cássia Sudan holds an undergraduate degree in Biological Sciences, Master's of Education and is currently a doctoral student in Education. She has been an Environmental Educator at USP since 2001. She has experience in the field of Education, with emphasis on Environmental Education, and works primarily with the following themes: social-environmental sustainability, shared management and minimization of solid residue, collective educators and the educating of new environmental educators.

Ana Maria de Meira holds an undergraduate degree in Forest Engineering and Agricultural Sciences, Master's in Forest Science and a Doctorate in Science. She is an Environmental Educator at USP. She has experience in the fields of residue management, environmental planning, environmental education and forest resources.

Marcos Sorrentino holds an undergraduate degree in Biological Sciences, Master's and Doctorate in Education. He is also a university professor and an environmentalist. Faculty member of the Forest Science Department of ESALQ/USP since 1988 and coordinates the Education and Environmental Policies Laboratory. He was the director of environmental education for the Brazilian Ministry of the Environment from 2003 to 2008.

Fernanda da Rocha Brando Fernandez holds an undergraduate degree in Biological Sciences; specialized in environmental management; Master's and Doctorate in Science Education. Faculty member of the Department of Biology of the School of Philosophy, Sciences and Literature, USP—Ribeirão Preto, and coordinates Epistemology and Didactics Laboratory. Works with teaching, research and outreach programs in the field of environmental education and sustainability.

Rosana Louro Ferreira Silva holds an undergraduate degree in Biological Sciences, Master's in Ecology and Doctorate in Education in the science education line of research. Faculty member of the Zoology Department of the Bioscience Institute of USP in the field of biology education. Has worked with environmental education research since 2001 and coordinates teaching, research and outreach projects in the field.

Laura Alves Martirani holds an undergraduate degree in Art Education, specialized in Computer Graphics, Master's in Arts and Doctorate in Education. Professor of Communication and Education for the Department of Economy, Administration and Sociology at the Higher School of Agriculture Luiz de Queiroz, USP—Piracicaba. Faculty member of the Undergraduate Agrarian and Biological Sciences Program and Applied Ecology Graduate Program of CENA and ESALQ/USP.

Denise Maria Gândara Alves holds an undergraduate degree in Agricultural Sciences, Master's in Biochemical Physiology of Plants and Doctorate in Biological Sciences. Research collaborator for the Laboratory of Education and

Environmental Policies. She works with the principles and values of environmental education and complex indicators of socio-environmental sustainability.

Paulo Ernesto Diaz Rocha holds an undergraduate degree in Biological Sciences, specialized in Environmental Theory and Praxis, and in Agricultural Biodynamics and Art Education, Master's in Botany and Doctorate in Development, Agriculture and Society. She was an educator for the USP Recycling Program and currently works with university outreach programs and popular education with the Technological Incubator for Popular Cooperatives—ITCP/USP, working with culture, environment, solidarity economy, cooperativism and self-management.

Silvia Aparecida Martins dos Santos holds an undergraduate degree in Ecology, Master's and Doctorate in Environmental Engineering Sciences. Works for the Center of Scientific and Cultural Diffusion (CDCC/USP) since 1992, where she is responsible for the Biology and Environmental Education Sector, working primarily in the technical sector of scientific diffusion and educational material.

Taitiâny Kárita Bonzanini holds an undergraduate degree in Biological Sciences and Education, Master's and Doctorate in Education for Science. Faculty member of the Department of Economics, Administration and Sociology for the Luiz de Queiroz Higher School of Agriculture, USP—Piracicaba, working in the fields of biology education, agricultural sciences education and fundamentals of education for the blended sciences undergraduate program offered by USP in partnership with Univesp.

Maria Estela Gaglianone Moro holds an undergraduate degree in Veterinary Medicine, Master's in Genetics and Animal Improvement, Doctorate in Zootechnology. She is a faculty member of the Zootechnology Department of the College of Zootechnology and Food Engineering, USP–Pirassununga. She lectures the following courses: apiculture, wild animal health, alternative husbandry and cat and dog nutrition and feeding. Responsible for the Socio-Environmental Master Plan for USP's Pirassununga campus. Campus Vice-Mayor from 2009 to March 2014.

Welington Braz Carvalho Delitti holds an undergraduate degree in Biological Sciences, Master's and Doctorate in Biological Sciences in the field of Botany. Experienced in the field of ecology, with emphasis in functional ecology, working primarily with the following themes: nutrient cycling and ecological restoration. Superintendent of Environmental Management at USP from 2012 to 2013 and currently Vice-Director of the Institute of Biosciences.

Marcelo de Andrade Romero holds a degree in Architecture and Urbanism, specialized in Philosophy and Philosophy of Education, Master's in Environmental and Urban Structures and Doctorate in Renewable Energies and Environmental and

Urban Structures. Professor at the Architecture and Urbanism College of USP. Director of the Architecture and Urbanism College of USP from 2010 to 2014, and is currently a Member of UIA—International Union of Architects—WP Group Architecture for a Sustainable Future and Superintendent of Environmental Management at USP.

Values, Behaviors, and Economic Outcomes: Sustainability in the Teaching of Economics

Madhavi Venkatesan

Abstract

Increasing global awareness of natural resource depletion, heightened weather-related volatility attributable to climate change, and the subsequent emergence of multi-disciplinary sustainability programs in higher education has pronounced the void in the behavioral explanatory science of economics to address the values that have yielded the economic and environmental outcomes observable in prevailing sustainability discussions. Economic theory, models, and analysis are central to a discussion of how individuals interact not only with one another but also with the environment. Further, the implicit inclusion of economics in the daily behavior of individuals and nations strongly influences the observations of global sustainability, including economic equity and social and environmental justice issues. This paper addresses an unmet need specific to introductory economics course curriculum, bridging the gap between consumption dependent growth and sustainability and explicitly addressing the creation of rational economic agent thinking as an outcome of introductory economics. Following a discussion of the present state of introductory economics instruction, the paper provides a replicable and novel starting point for the introduction of sustainability in the introductory economics curriculum, inspiring the development of students as "sustainable" rational economic agents.

Keywords

Education · Economics · Sustainability · Curriculum

M. Venkatesan (✉)
Department of Economics, Bridgewater State University, 117 Hunt Hall,
131 Summer Street, Bridgewater, MA 02324, USA
e-mail: mvenkatesan@bridgew.edu

W. Leal Filho et al. (eds.), *Integrating Sustainability Thinking in Science and Engineering Curricula*, World Sustainability Series, DOI 10.1007/978-3-319-09474-8_39

1 Introduction

Present models of economic growth primarily focus on the role of expenditures as captured in the commonly cited economic indicator, gross domestic product (GDP), where GDP is defined as the sum of final goods and services sold within a country's natural borders. Noting that a country's expenditures are referred to as "aggregate expenditures" and that the majority of spending is specific to consumption or consumer spending, especially in the United States where this spending category is nearly two-thirds of annual GDP (other expenditure categories for GDP include investment spending, government spending and foreign spending as proxied by net exports), there exists a significant relationship between consumer expenditures and macroeconomic growth, justifying the standard acceptance of consumption-based expenditures as being a significant driver of economic expansion. Given the consumption and growth relationship, consumption values and behaviors have a significant impact on economic outcomes as well as other parameters including the environment and social and economic equity, where the later are defined as relating to disparities between groups within a country, as well as across countries.

Following a discussion of the impact of consumer-led growth on sustainability parameters: the environment, economic and social equity, this paper provides an explicit linkage between the measure of economic progress in universal use, GDP, and the degradation to common global resources, connecting the endogeneity present between the modeling of economic growth and the values and behaviors that support the outcome of the very same growth. A discussion of the present teaching methods specific to introductory macroeconomics provides the foundation for an innovative, replicable, and grant-funded case study of introducing sustainability alongside the conscious development of the student of macroeconomics as a rational economic agent. The curriculum variants discussed are not in widespread use and at present, there are no standard textbooks for the instruction of Principles of Macroeconomics that explicitly include sustainability and provide sustainability-based economic parameters for alternative evaluation to standard economic growth as presently and singularly enumerated in GDP. Additionally, there is no standard text targeted to the creation of the rational economic agent as part of the study of Principles of Economics. The outcome of the case study reviewed is presented in the summary and conclusion of the paper where outcomes of the sustainability inclusive curriculum are captured in a qualitative assessment of student reaction and absorption of sustainability as a value and behavior catalyst.

2 The Present State of Introductory Economics Instruction

The course "Principles of Macroeconomics" is typically taught as one of a two part introductory requirement leading to the formal study of Economics and other undergraduate business majors. The course focuses on the aggregate factors that define economic progress as this concept relates to the "standard" expenditure-based variable of quantifying growth: gross domestic product (GDP). The course incorporates elements of Microeconomics to establish the foundation for the assumptions of individual and firm behavior (Fagg, 1981)where such behavior is guided by a "rational" agent assumption and the underlying premise of rational decision-making is defined as maximizing return while minimizing cost.

Costs are assumed to be priced through efficient market assumptions and are therefore simplified as being indicative of market pricing with limited consideration devoted to the potential for systemic under-enumeration of costs resulting from informational, regulatory, or exploited variations. Additionally consumption is not evaluated as a value that is enabled through pricing distortions, rather consumption, along with other expenditure components, investment, government spending, and net exports, are discussed as "given" parameters of economic growth.

As Nelson (1995) points out, economics evaluates efficiency with respect to the "use of resources to maximize production and consumption, not by the moral desirability of the physical methods and social institutions used to achieve this end." Therefore sustainability does not enter standard economic thinking. Reduction in consumption in the current period is only addressed through the lens of an increase in consumption in a later period. Nelson notes, "The possibility that consumption should be reduced because the act of consumption is not good for the soul, or is not what actually makes people happy, has no place within the economic value system."

In their analysis of the teaching of the Principles of Macroeconomics, Knoedler and Underwood (2003) found that "the myopic teaching style of economics is attributable to both economic instructors and to the leading texts in use for the instruction of Principles of Macroeconomics courses." Furthermore, they state that the standard teaching of economics did not provide access to alternative approaches, for example, normative economics was found to be "demonstrated by a quick example as something to be avoided at all costs, unlike the exercise in positive economics."

The seeming lack of attention to values and behavior as captured within standard expenditure based macroeconomic modeling has distanced students from the tangibility of economics, limiting their understanding of the explanatory potential of economics and the application of economics as both a cause and a remedy of unsustainable practices. As noted by O'Hara (1995), understanding of economics is essential to sustainability and an understanding of the social and ecological context related to observable sustainability issues is requisite in the evaluation of a solution, "sustainability challenges us to recover the links between social and ecological contexts."

To the extent that economic growth has been and continues to be a national goal both within and without the United States, there is an inherent endogeneity between the current expenditure-based teaching of Principles of Macroeconomics and the observable natural resource degradation, and economic and social inequities. The focus on implicit consumption-led growth limits the potential for sustainability, as myopic consumer-led, immediate gratification-oriented growth (utility and profit maximization) as captured in the modeling of economic activity is inconsistent with the quantified evaluation of the externalities caused by production to meet consumption demand (Czech 2000). Boran (2006) points out, albeit subtly, economic modeling is limited in its ability to adequately assess and address inequities that may arise as a result of standard economic analysis. Specifically, the increase in pollution that arises due to an inability to properly quantify the environmental burden of ambient emissions at a specific point in time.

The practice and teaching of economics, implicitly influences and reinforces consumption-led growth while contributing to the development of economic agent behavior. In turn, the learned economic literacy, in essence, contributes to the scientific approach to evaluating economic outcomes. Therefore, if the discipline does not include values and behavioral outcomes related to sustainability, it can be stated to be a contributing element of the present-day outcome of a lack of sustainability.

3 Rationale for Explicit Integration of Sustainability in Economics

The present multi-disciplinary emergence of sustainability is a result of anthropogenic attribution of the increasing speed of climate change and environmental degradation (Lovejoy 2014). Adverse impacts to global resources may not have been properly noted or addressed during the most recent period of human-led environmental modification as defined by the period from the late 17th century to the present. Further, the environmental impact presently observed, was most likely attributed to the myopic desire and attainability of human-centric, higher quality of life standards accessible through changing economic structures as addressed by early economists and political philosophers, Adam Smith and Jeremy Bentham. As noted by Nelson (1995), "the present shift from efficiency to sustainability no doubt reflects in part the moral disappointments of the twentieth century, relative to the hopes for economic progress that were widely shared at the beginning."

Efficiency has been simplified in economic modeling to account for "process efficiency," which through the most recent period has been evaluated as present costs relative to revenue generation. Costs articulated in standard evaluation have rarely been holistic or inclusive of non-quantifiable components to production; therefore, costs as described have not included items that were not readily quantifiable or were created as a result of natural resource utilization, degradation, or replenishment. Common assets such as water, air, and land were not included in production assessments; furthermore, damages resulting to any of these common elements were not articulated unless indirectly addressed as being a part of social and governmental regulation.

Anthropomorphic impacts related to production and consumption, inclusive of profit and utility maximization have been significant. To a large extent this is attributable to the lack of inclusion of the inter-relationship between the environment and human activity, as is captured in the concept of sustainability. Explicit attribution of sustainability in economics serves to increase awareness of the consumption and production linkage to the adverse externalities being currently faced, by also promoting, within a classroom setting, the endogeneity of values to economic outcomes. By focusing on the economic, social, and environmental outcomes from an expenditure–based view relative to the quality of life outcome attributable to a sustainability value paradigm, students are able to evaluate their own values, including the attribution of their personal values, and to then develop awareness of alternative value structures that promote balanced growth along the lines of economic equity, and social and environmental justice, all of which comprise the concept of sustainability.

4 Principles of Macroeconomics: Sustainability Integration Case Study

As part of a semester-long Principles of Macroeconomics course and following student introduction to basic macroeconomic concepts and related market mechanisms (movement along, shift of supply and demand; general understanding of consumption led growth, and the behaviors or values embodied within the framework of growth as it is currently defined and evaluated through GDP), students were introduced to the aggregate expenditure function[1] and the explicit need for consumption in the determination of economic growth.

Following the introduction of the foundation of macroeconomic variables and concepts, students were asked to fill out a questionnaire, where the questionnaire focused on the student's understanding of the relationship between behaviors and economic outcomes through the lens of the values that were of significance to the student. Students were provided with the opportunity to comment on the rationale for the values expressed in their actions. Sustainability values were not directly addressed, though students were asked if they had taken a course in sustainability or that incorporated sustainability. The questionnaire provided a baseline for reviewing any changes in student value orientation as a result of the explicit introduction of sustainability.

Students were then introduced to the concept of consumerism and were made aware of the endogeneity between marketing, advertising, consumption, credit access, and GDP growth through detailed discussion of the aggregate expenditure function, as well as monetary and fiscal policy mechanisms. Using standard aggregate demand and supply models, students were introduced to the impact of excess demand on natural resource utilization, exploitation, and development; they were provided with an understanding of non-articulated and under-enumerated

[1] *Aggregate expenditure function*: $Y = C + I + G + (X - M)$; where Y identifies the total amount expended in a given economy on goods and services over a given period; C is consumer expenditures, which in the United States accounts for two-thirds of Y; I is investment spending attributable to businesses; G is government spending; and $(X - M)$ is next exports and is reflective of foreign spending on domestic goods and services.

costs and asked to assess the benefits and costs of trade from both a profitability and environmental-social welfare perspective. Further, students were introduced to the role and purpose of the central bank and the government in the target goal of monetary and fiscal policy to maintain full employment as defined as the employment level consistent with the maximum aggregate expenditure level found at the optimal economy-wide production capacity (full production capacity) at a given point in time.

The process of instruction and assessment of sustainability factors related to the dependence of consumption on credit; trade and exploitable, quantifiable cost differentials; natural resource use and cost of degradation and replenishment; marketing and consumer demand; and the tragedy of the commons as related to the exploitation of common resources. All of the mentioned topics were addressed and evaluated as a class and on an individual student basis.

Weekly readings related to a macroeconomic topic and its sustainability correlate were assigned from a variety of sources, including popular press articles and books. Each week, four students were selected to be discussion group leaders, with a responsibility to facilitate weekly in-class sustainability focused discussion related to an aforementioned macroeconomic topic and corresponding readings. The group leaders were required to develop at minimum two questions specific to the macroeconomic and sustainability-based reading assignment for the week and the questions were distributed to all students one class period prior to scheduled class discussions. Students were expected to address all questions prior to class and to formally submit answers for two questions via an electronic journal. The questions, discussion, and journal entries were meant to provide students with an assessment of the role that implicit values have in individual decision-making and to increase awareness of how implicit social values impact individual consumption behavior and thereby eventually culminate in economic outcomes.

The integration of sustainability overlapped the standard course curriculum. Weekly sustainability topics followed and countered implicit values within the subject matter of the course, giving students an ability to understand the present macroeconomic phenomenon and then evaluate the same through a sustainability lens. The focus of macroeconomic and corresponding sustainability discussion was consumer-based consumption as an integral and targeted component of aggregate expenditures and corresponding GDP growth.

The students' weekly journals, discussion participation, and initial questionnaire submission provided a qualitative capture of the students' individual evolving and static economic values and behaviors. The responses detailed in these vehicles though the course of the semester were compared against the exit questionnaire; the chronology of responses was used to assess the impact to decision-making and economic rational agent creation, resulting from taking a course in introductory macroeconomics that incorporated sustainability.

5 Summary and Conclusion: Values, Behaviors and Economic Outcomes

Though not exhaustive in the mechanics implemented, the curriculum revision shared in this paper, reflects an innovative approach to the introduction of sustainability in introductory economics curriculum. As noted, the curriculum introduced incorporates a required self-assessment based journal. As a result, student tracking of own-value led consumption behavior and the evolution of the value/behavior relationship through the course of the term, provided a qualitative assessment of the outcome of the introduction of sustainability to both values and related behaviors as well as the development of the student into a sustainable rational economic agent.

The results of the grant-based project indicate what would be expected, understanding of implicit values in conjunction with the global issues resulting from the same values fosters modification of behaviors, potentially culminating in the development of sustainable values and modified economic outcomes, as well as the potential for the establishment of revised and universally adopted metrics to assess economic growth.

Student discussions and responses provided clear evidence that as awareness to issues including but not limited to the adverse health and environmental impact related to plastic use and disposal, to water access and sanitation increase, modification of behavior in conjunction with an increased awareness of economic drivers such as, marketing, to both promote and align with consumption behaviors, and access to consumer credit, to promote inter-temporal consumption, can be augmented. Student commentary included a cessation in the purchase of plastic bottles and use of recyclable and reusable containers, to reduction in the use of water for showers and daily hygienic activities, to an awareness of the adverse environmental, social justice, and credit impact of the purchase of excess clothing as a result of marketing induced need or want.

The underlying rationale for the introduction of sustainability in a Principles of Macroeconomics course was the premise that most individuals are not rational economic agents; rational economic agent behavior is created or taught through economic literacy, which can be considered as an implicit outcome of the introductory economics course objective. Rational agent behavior is defined as including an understanding and conscious incorporation of cost-benefit analysis (utility maximization) and other economics-based parameters in individual decision-making. As a result, most decision-making is fostered by limits that are consciously imposed, awareness of sustainability in consumption decisions can therefore, modify utility decisions by including the impact to the greater environmental good in consumption decisions. For example, in a consumer-oriented society, marketing and advertising as well as convenience drive behavior, where as, when the concept of holistic evaluation in decision making is presented, such that the individual thinks beyond just immediate consumption to incorporate the waste of consumption and the externalities resulting as a by-product of production—both of which are often either neglected or under-enumerated in price, consumption behavior can be augmented. In summary, in the same manner that the rational agent is "created," the "sustainable rational agent" can emerge.

The project in essence was designed with an expectation of making the course material more relevant to student interests (Colander, 2005; Dillard, 1982) by challenging assumptions of behavior to increase student critical thinking to evolve individual assessment of values, thereby fostering the development of the sustainable rational economic agent. The basis for the integration of sustainability is congruous with recommendation from Knoedler and Underwood (2003), "The alternative set of economic principles offers a foundation for a principles course that provides a richer understanding of the real economy…Whether the instructor of Principles chooses to build his or her course exclusively around the alternative principles or instead uses them as counterpoints to introduce a multi-paradigmatic and thoughtful survey of major issues, we are certain that students will be more engaged in the subject matter while continuing to increase their capacity for critical thinking. After all, economics is the business of ordinary life, and it is time that we return to that subject matter in our principles courses."

The curriculum exercise shared in this paper provides a significant step forward with respect to the explicit introduction of sustainability into Principles of Macroeconomics curriculum, where next steps could include at minimum a replication of the curriculum components as described, to the development of an introductory economics text targeted at the explicit creation of the sustainable economic agent. Additionally, advocated modification in the teaching of economics would include the development of a singular introductory economics course targeted solely on the development of the sustainable economic agent, which would be followed by sustainability infused Principles of Microeconomic and Principles of Macroeconomics course(s). A course as advocated would be beneficial to the establishment of economic literacy across all academic concentrations, given that many universities have established core requirements in economics.

References

Boran I (2006) Benefits, intentions, and the principle of fairness. Can J Philos 36(1):95–115

Colander D (2005) What economists teach and what economists do. J Econ Educ 36(3):249–260

Czech B (2000) Economic growth as the limiting factor for wildlife conservation. Wildl Soc Bull 28(1):4–15

Dillard D (1982) Rewriting the principles of economics. J Econ Issues 16(2):577–585

Fagg J (1981) The fundamental principles of economics. J Econ Issues 15(4):937–942

Knoedler JT, Underwood DA (2003) Teaching the principles of economics: a proposal for a multi-paradigmatic approach. J Econ Issues 37(3):697–725

Lovejoy S (2014) Scaling fluctuation analysis and statistical hypothesis testing of anthropogenic warming. Clim Dyn

Nelson RH (1995) Sustainability, efficiency, and god: economic values and the sustainability debate. Annu Rev Ecol Syst 26:135–154

O'Hara SU (1995) Sustainability: social and ecological dimensions. Rev Soc Econ 53(4):529–551

Author Biography

Dr. Madhavi Venkatesan has been engaged in economics and sustainability from the perspective of a public company financial communications executive and an academic instructor. She has been an advocate of the revision of introductory economics to incorporate sustainability values and to explicitly evaluate embedded consumerism in the defining of economic growth and progress. Dr. Venkatesan is active in the development of an economics curriculum that provides students of introductory economics with an understanding of the relationship between values and behavior, and ultimately, observable economic outcomes and has recently conducted grant-funded research on the impact of implicit values on economic outcomes. She has a Ph.D. in Economics from Vanderbilt University and was a Mellon post-Doctoral fellow at Washington University in St. Louis.

Elaboration and Implementation of a Waste Management Policy at the University of São Paulo (USP)

M. Cooper, Adriano C. Pimenta, Aline Mellucci, Ana Maria de Meira, Arthur R. Silva, Maria Cecília H.T. Cavalheiro, Clara M. Zorigian, Daniela C. Sudan, Danilo Vitorino, Elizabeth T. Lima, Fabrício Rossi, Glauco A. Tavares, Juliana Monti, Maria Estela Gaglianone Moro, Patricia B. Di Vitta, Patrícia G. Moreira, Patrícia C.S. Leme, Simone B.J. Simonelli, Tamara Gomes, Vera Gandra C. Albuquerque, Welington B.C. Delitti and Marcelo Romero

Abstract

The University of São Paulo (USP) is the largest university in Brazil and has developed independent environmental actions related to solid waste until 2012. This year, an Environmental Management Superintendence (SGA in Portuguese) was created in order to construct a Solid Waste Policy to the University. This policy was constructed in an innovative way, since it required a participatory process. The need for more articulation gave rise to the Environmental Management Superintendence (SGA in Portuguese) in 2012, so as to improve sustainability on the *campi* and change USP into a sustainability model. SGA has formed a work group on wastes under a participative process whose priority

M. Cooper (✉)
Department of Soil Science, University of São Paulo, Piracicaba, SP 13418-900, Brazil
e-mail: mcooper@usp.br

M. Cooper
Environmental Management Office, University of São Paulo, Av Pádua Dias, 11, Piracicaba, SP 13418-900, Brazil

A.C. Pimenta · A. Mellucci · A.M. de Meira · A.R. Silva · M.C.H.T. Cavalheiro · C.M. Zorigian · D.C. Sudan · D. Vitorino · E.T. Lima · F. Rossi · G.A. Tavares · J. Monti · M.E.G. Moro · P.B. Di Vitta · P.G. Moreira · P.C.S. Leme · S.B.J. Simonelli · T. Gomes · V.G.C. Albuquerque · W.B.C. Delitti · M. Romero
Working group of residues, University of São Paulo, Rua da Praça do Relógio, 109 Bloco K, 3° andar, sala 309, Cidade Universitária, SP 05508-050, Brazil
e-mail: uspverde@usp.br

W. Leal Filho et al. (eds.), *Integrating Sustainability Thinking in Science and Engineering Curricula*, World Sustainability Series, DOI 10.1007/978-3-319-09474-8_40

is to develop strategies and procedures for the elaboration and implementation of a Waste Management Policy (PGR/USP in Portuguese), as a response to the specificities of the National Solid Waste Policy (PNRS in Portuguese)—Law n. 12.305/10 to the university needs. The PGR/USP contemplates the elaboration of Waste Management Plans for all units at USP and in order to plan it, a course for the staff, the elaboration of the plans and a data base, integrated with the University Corporative System, and indicators to evaluate and improve the management are foreseen. This study analyses the elaboration process and the implementation of the PGR so as to identify the opportunities to promote sustainability at universities by taking waste management as a starting point.

Keywords
Higher education institution (HEI) • Waste management • Environmental management

1 Introduction

According to Tauchen and Brandli (2006), Higher Education Institutions (HEIs) can be compared to small urban clusters that develop teaching, research and extension activities besides others related to their financial, feeding and housing operations. Moreover, HEIs use isolated environmental practices aimed at law compliance, with no concern over the promotion of preventive practices.

Overall, the waste generated in a university environment is characterized by a complex and heterogeneous matrix, due to the particularities of its activities, which challenges the academic community to promote the right waste management for all sectors. Therefore HEIs must use the management of their waste as part of the academic management to develop and implement policies related to the aspects and impacts of their activities (Conto 2010).

In Brazil no specific law or inspection has been defined for the waste generated at HEIs. A way for HEIs to show their compromise with environmental issues is by adopting institutional policies and implementing environmental management systems that meet their specificities. Therefore the involvement of many sectors, especially senior managements is fundamental for an effective undertaking of actions (Giloni-Lima and Lima 2008; Silva and Mendes 2009).

Corrêa (2009) emphasizes that the University has the advantage of being built by an academic community composed of many knowledge areas, which makes the development of waste management policies stronger.

According to Tauchen and Brandli (2006), the implementation of waste management policies raises an awareness process at all levels and activities (managers, professors, students and staff) and contributes to a higher environmental quality of the institutions.

Corrêa (2009) apud Corrêa (2012) states that the development of a waste management policy at HEIs, with a focus on environmental education, promotes an articulated involvement in the university environment and enables the establishment of awareness from the parts involved and their responsibilities as waste generators under the context of their activities in many sectors, triggering a continuous and effective educational process.

Due to the scarcity of research into the waste generation process at universities, most solutions are achieved when the waste has been produced, which is one of the obstacles that hinder the participation of the community. Besides, the actions of the agents regarding waste generation in the teaching, research and extension activities are not clear, because of the scarcity of interaction among these three areas (Conto 2010).

This paper reports on the initial phase of the implementation process of PGR/USP. This policy was constructed in an innovative way, since it required a participatory process. The perspectives identified will provide knowledge on the context of the institution, so that discussions on the topic of sustainability at universities will be promoted.

2 The Building Process of the Waste Management Policy at USP

2.1 History

The University of São Paulo (USP) is comprised of eight *campi* and a community of 120,000 people including professors, students and staff members. USP has promoted many environmental actions, including projects and programs related to waste management. However, such actions are poor in articulation and integration.

In 2009, USP approved a Proposal for an Environmental Policy. In 2010, when the PNRS was launched, USP defined a coordinator for Environmental Management with the rectory and, in 2012, the creation of an institutional body responsible for the university environmental management. The SGA/USP was then established.

The SGA goals are to promote environmental sustainability on all USP *campi*, preserve their natural resources, promote a healthy and safe environment and the rational use of resources on the *campi*, and educate focusing on sustainability, and build, in a participative way, a sustainable University, changing USP as a model of sustainability to society.

2.2 Strategic Aspects

The first stage of planning for the waste management was the organization of a Work Group formed by specialists in waste issues on each USP *campus*. The priority of the group named “Work Group on Wastes at USP” (GT Resíduos da USP, in Portuguese) is to develop, in a participative way, strategies and procedures for the elaboration and implementation of a Waste Management Policy for the University, so as to adjust the specificities of PNRS to the University context.

In the second phase of planning, the participation of the academic community was promoted by a workshop[1] organized by SGA. The event had approximately 110 representatives of different categories of all USP *campi,* and those directly involved could contribute to the preliminary version of the document, which was then submitted for posterior alterations.

The preliminary version of PGR/USP was analyzed by the University's legal body to be implanted in 2014, after the regulation of the University's senior management.

The PGR/USP has established the preparation of Waste Management Plans in all its units, therefore the staff members will undergo a process of participatory training.

The waste management plans will enable the SGA to understand the state-of-the-art of the waste management at USP, as well as conduct planning, decision making, necessary investments and definition of indicators that will guide the adjustments and improvements regarding waste management on *campi.*

As part of the shares of SGA, an integrated database has been developed for the corporate system of the institution so as to control the inputs used, such as generation, storage, treatment and disposal of the waste generated in our units, and minimize environmental impacts through activities at the University (teaching, research, extension and university management).

3 Preliminary Analysis of the Implementation of the Waste Management Policy at USP

The preliminary analysis refers to the identification of difficulties encountered by USP units from the current scenario found, so as to encourage the incorporation of new correction strategies.

Such difficulties include:

- Academia's resistance behaviors. In general, as the habits and attitudes previously established and systematized in a community or group of people over a long period (USP), have led individuals to the repetition of the same actions and convenience in their attitudes, such situations have been aggravated due to the bureaucratic procedures that often prevent the fluidity of the work (Nicolades 2006);
- Lack of awareness on the part of the university community in the process of building, deploying and maintaining the environmental policy, due to the difficulty in inserting collaborators in participatory processes and budget limitations;

[1] Workshop named "Initiatives for the building of the Waste Management Policy of USP" held in April 25, 2013, with around 110 representatives of all USP *campi.*

- A decentralized and fragmented structure of institutions, which hampers the recognition of responsibilities and changes in the management throughout the process, as they are generally reluctant to support actions under development and prefer to create new strategies (Nicolades 2006);
- Difficulty in inserting the environmental dimension in the different sectors of the institution and the educational and research projects, which reflects a resistance of teachers to incorporating environmental issues in their discussions and prioritizing the allocation of institutional funds for investments. This point is directly linked to the need for ongoing environmental education programs to become part of the logistic planning of the institution as a whole (Nicolades 2006);
- Waste or mismanagement of funds received by funding agencies and research which are generally used for the storage of products;
- Lack of encouragement by funding agencies that prioritize valuing publications only in the area of expertise of the professional. This situation is associated with the unawareness of the importance of studies on Waste at Universities. This multidisciplinary theme should be an integral part of research activities, which may be directly or indirectly related to the projects, since everyone is somehow a waste generator. Induced by such bodies, priority becomes contradictory because projects on the subject at universities may provide data that guide the best use of available funds;
- Lack of commitment of the administration and university community with actions and practices that aim at sustainability. Such a lack is directly linked to the fact that the theme may not be incorporated into everyday issues (Nicolades 2006);
- Lack of time from the people engaged due to increasing demands in their activities, as the overload of activities with intellectual production in the academic area. Teachers also assume administrative positions and the structure of the university often limits new hires (Nicolades 2006).

The problems listed were observed in other situations in which structural changes had been introduced in the routine within the university environment without prior training/orientation. Below are some perspectives on the implementation of an institutional policy of waste management at USP. These perspectives may result in effective actions and catalysts for the process in question, in which the SGA/USP as an institutional body can encourage and support:

- Continuous and permanent environmental education so as to mobilize students and professors to develop appropriate technologies for the recycling of waste and recovery activities with the use of techniques, such as forum discussions (Dias in Conto 2010);
- Creation of a database of publications related to environmental programs and actions involved at University (Conto 2010);
- High costs between with waste management. All units must know where and how their waste is being sent and processed (Teixeira in Conto 2010);

- Development of theses, dissertations, monographs, internships and research projects on environmental issues and in particular with the University as a focus of study (Conto 2010);
- Stimulus for actions that minimize waste generation and think about the possibility of creating a specific research sector in cleaner technologies adapted to the university context. Such an action aims at decreasing the costs of treatment and disposal of waste for the units and the University;
- Inclusion of courses in the curricular program so that undergraduate students become aware of the "Sustainability at the University" theme, involving the issue of "Waste at Universities" (Conto 2010);
- Reevaluation of projects and programs on education and extension so that they can be developed with the least environmental impact. Every project, throughout its development, should be committed to the environment and provide measures to prevent waste generation;
- Preparation and insertion of an efficient system of Waste Management by taking into account the context of sustainability, which also expands the social and economic university context and includes a proposal for the development a sustainability report for the university. The current scenario of preparing reports for existing platforms has not been fully adapted to the reality of an HEI. Another challenge lies in the selection of a good indicator for the analysis of environmental impacts at the University;
- Encouragement by funding agencies to research projects of related fields associated with "Waste Management at Universities".

The use of a computerized system is expected to positively influence the organization of information from the identification of the local generator, centralization of data and diagnoses of waste generation and its current forms of management (storage, disposal or final disposition) such as system will be standardized and enable the sharing of responsibilities for waste management at USP.

The platform will also enable the development of waste management plans and creation and monitoring of indicators on waste management in the Units and on the USP *campi*.

As the case in question was partially analyzed because it is the initial stage of the establishment of the policy, the need for an educational approach that promotes the participation of all stakeholders and is based on the experiences of questioning and problem solving in the university environment could be verified.

4 Conclusions

Despite the difficulties reported, the University of Sao Paulo has had a breakthrough in a short time because it has initiated the process of institutionalization of Waste Policy and created an environmental agency responsible for centralizing and support within the University in a single management.

Initiatives were individualized according to the reality of the *campi* and units, with no centralization of the information and support by an institutional body. Presently, they can be seized and taken as gained knowledge through mechanisms of dissemination, such as meetings, training, forums, among other actions and address the technical solutions to educational work. Those involved in discussions from their experiences in teaching, research, extension and service provision activities can also be encouraged.

Just as the PNRS has brought a new look on the society regarding its waste, the implementation of a policy on waste management for a university by SGA may also bring the same focus to the university, from the moment that all will be mobilized for the development of a diagnosis and a survey of the current situation. Mechanisms that enable the development of a new approach regarding waste management in the units will be created, hence a greater commitment to environmental issues addressed in the proposed Environmental Policy.

The implementation of the Waste Management at the University will provide knowledge on the context of the institution. The results will become effective when the university community has incorporated the environmental dimension into its policies, plans and activities, including the question from an articulated dialogue among all university communities seek solutions to environmental problems (Corrêa 2009).

Only through a collective effort and combining individual and group experiences, actions will promote the necessary changes in attitudes at the university.

Other mechanisms for the dissemination of knowledge on Waste HEIs can be introduced, for example, through the creation of educational corporate videos, dissemination programs on the USP radio, newspapers, lectures and forums on the subject.

References

Corrêa LB (2009) Construction of policies for waste management at an institution of higher education in the perspective of environmental education. Rio Grande: Federal University of Rio Grande. http://www.ufrgs.br/sga/SGA/materialdeapoio/textos/textosapoio/links/tese%20luciara%20bilhalva%20correa.pdf. Accessed 07 Dec 2013

Conto SM org (2010) Gestão de resíduos em universidades: Uma complexa relação que estabelece entre heterogeneidade de resíduos, gestão acadêmica e mudanças comportamentais. EDUCS, Caxias do Sul, RS; 401 p

Giloni-Lima PC, Lima VA (2008) Gestão integrada de resíduos químicos em instituições de ensino superior. Química Nova 31(6):1595–1598. http://www.scielo.br/pdf/qn/v31n6/a53v31n6.pdf. Accessed 6 Dec 2013

Nicolades A (2006) The implementation of environmental management towards sustainable universities and education for sustainable development as ethical imperative. Int J Sustain High Educ 7(4):414–424

Silva ER, Mendes LAA (2009) The Role of Universities in Building Environmental Sustainability: a proposed model of Integrated Waste Management. Revista ADVIR da UERJ 23:78–85. http://www.eng.uerj.br/publico/anexos/1278655730/ArtigodeElmoeLuizAntoniopublicadonaRevistaAdvir_n23_2009.pdf. Accessed 6 Dec 2013

Tauchen J, Brandli LL (2006) A gestão ambiental em instituições de ensino superior: modelo para implantação em campus universitário. Revista Gestão & Produção 13:503–515. http://www.scielo.br/pdf/gp/v13n3/11.pdf. Accessed 4 May 2013

Vaz CR, Fagundes AB, Kachba YR, Oliveira IL, Kovaleski JL (2010) Sistema de Gestão Ambiental em Instituições de Ensino Superior: uma revisão. GEPROS. Gestão da Produção, operações e sistemas 3:45–58. http://revista.feb.unesp.br/index.php/gepros/article/viewFile/327/314. Accessed 5 May 2013

Authors Biography

Prof. Dr. Miguel Cooper In addition to a bachelor's degree in Agronomic Engineering Dr. Miguel Cooper holds a master's degree and a Ph. D. in Agronomy, specializing in Soil Science and Plant Nutrition (University of São Paulo). He is currently working as an Associate Professor at the University of São Paulo. He is also a Technical Advisor to the Superintendent of Environmental Management in this University. He has experience in Agronomy with emphasis on soil conservation and recovery of degraded areas, mainly in the following: soils, soil genesis, micromorphology, soil erosion and soil functioning.

Dr. Adriano C. Pimenta received his academic education (chemistry) at the University of São Paulo and obtained a Ph.D. degree in 2006. Dr. Pimenta is currently working at the Laboratory of Chemical Waste at the University of São Paulo (Ribeirão Preto, São Paulo, Brazil).

Aline Mellucci holds a Bachelor's degree in biology and an MBA in environmental management technologies. She is currently working at the Physical Plant of the University of São Paulo (São Paulo, São Paulo, Brazil).

Dra. Ana Maria Meira received her academic education (engineering) at the University of São Paulo and obtained a Ph.D. degree in 2010. Dr. Meira is currently working as an Environmental Educator at the University of São Paulo (School of Agriculture Luiz de Queiroz ESALQ-USP, Piracicaba, São Paulo, Brazil).

Dr. Arthur Roberto Silva received his academic education (chemistry) at the Federal University of São Carlos and obtained a Specialist in Environmental Management degree in 2007. He is currently working at the Laboratory of Chemical Waste at the University of São Paulo (School of Agriculture Luiz de Queiroz—ESALQ-USP, Piracicaba, São Paulo, Brazil).

Dra. Maria Cecília H. T. Cavalheiro received his academic education (chemistry) at the Federal University of São Carlos and obtained a master degree in 1995 and a Ph.D. degree in 1999. Dr. Cavalheiro is currently working at the Laboratory of Chemical Waste at the University of São Paulo (São Carlos, São Paulo, Brazil).

Clara M. Zorigian is a lawyer. She is currently working as an advisor at the Office of Environmental Management (University of São Paulo, São Paulo, São Paulo, Brazil)

Daniela C. Sudan holds a Bachelor's degree in biology and a Master degree in Education. She is currently working as an Environmental Educator at the University of São Paulo (Ribeirão Preto, São Paulo, Brazil).

Dr. Danilo Vitorino dos Santos received his academic education (chemistry) at the São Paulo State University and obtained a Ph.D. degree in 2009. Dr. Santos is currently working at the Laboratory of Chemical Waste at the University of São (Ribeirão Preto, São Paulo, Brazil).

Elizabeth T. Lima holds a Bachelor's degree in geography and a Masters degree in Environmental Science. She is currently working at the Physical Plant of the University of São Paulo (São Paulo, São Paulo, Brazil).

Prof. Dr. Fabrício Rossi received his academic education (agronomic engineering) at the Viçosa Federal University and obtained a Ph.D. degree in 2009. Dr. Rossi is currently a Professor at the Faculty of Animal Science and Food Engineering (FZEA), University of São Paulo (Pirassununga, São Paulo, Brazil)

Dr. Glauco Arnold Tavares received his academic education (chemistry) at the Methodist University of Piracicaba and obtained a master degree in 2000 and a Ph.D. degree in 2004. Dr. Tavares is currently working at the Center for Nuclear Energy in Agriculture of the University of São Paulo (Piracicaba, São Paulo, Brazil).

Juliana Monti holds a Bachelor's degree in biology.

Dra. Moro received his academic education (veterinary medicine) at the São Paulo State University and obtained a Master degree in 1991 and a Ph.D. degree in 1996. Dr. Moro is currently a Professor at the Faculty of Animal Science and Food Engineering (FZEA), University of São Paulo (Pirassununga, São Paulo, Brazil).

Dra. Patricia Busko Di Vitta received his academic education (chemistry) at the University of São Paulo and obtained a Ph.D. degree in 2001. Dr. Di Vitta is currently working at the Laboratory of Chemical Waste at the Institute of Chemistry of the University of São Paulo (São Paulo, São Paulo, Brazil).

Patricia Gabryela Moreira holds a Bachelor's degree in biology and an MBA in environmental management technologies. She is currently working at the Institute of Biosciences of the University of São Paulo (São Paulo, São Paulo, Brazil).

Dra. Patrícia C. S. Leme received his academic education (biology) at the Federal University of São Carlos and obtained a Master Degree in 2000 and a Ph.D. degree in 2008. Dr. Leme is currently working is as an Environmental Educator at the University of São Paulo (São Carlos, São Paulo, Brazil) and at the Laboratory of Chemical Waste at the Institute of Chemistry of the University of São Paulo

Simone B. J. Simonelli holds a Bachelor's degree in Engineering. She is currently working at the College of Dentistry of Bauru of the University of São Paulo (Bauru, São Paulo, Brazil).

Prof. Dra. Tamara Gomes received his academic education (agronomy) at the São Paulo State University and obtained a Master degree in 1998 and a Ph.D. degree in 2001. Dr. Gomes is currently a Professor at the Faculty of Animal Science and Food Engineering (FZEA), University of São Paulo (Pirassununga, São Paulo, Brazil).

Vera Gandra C. Albuquerque holds a Bachelor's degree in Chemistry and Safety Engineering. She is currently working as an occupational engineer at the University of São Paulo (São Paulo, São Paulo, Brazil).

Prof. Dr. Welington B. C. Delitti received his academic education (biology) at the São Paulo State University and obtained a Master degree in 1982 and a Ph.D. degree in 1984. Dr. Delitti is currently a Professor at the Institute of Biosciences of the University of São Paulo (São Paulo, São Paulo, Brazil).

Prof. Dr. Marcelo de Andrade Romero received his academic education (architecture) at the Brás Cubas University. He obtained a Master degree in 1990 and two Ph.D. degrees (1993, 1994). Dr. Romero is currently a Professor at the Faculty of Architecture and Urbanism of the University of São Paulo (São Paulo, São Paulo, Brazil). He is also the Manager at the Office of Environmental Management.

Incorporating an Ethos of Sustainability into the Conceptualisation and Development of an Academic Building: The Kings Cross Development in London

Fayyaz Vellani and Naveed Nanjee

Abstract

The Aga Khan Development Network (AKDN) is the largest private development network in the world. Although its work encompasses health, education and social, economic and cultural development, all of the work within this network increasingly reflects a commitment to sustainability. One of its key forthcoming initiatives is the King's Cross Development in London, which will host a number of institutions including the Aga Khan University's (AKU) Institute for the Study of Muslim Civilisations and the Institute of Ismaili Studies; both recognised UK higher education institutions. This chapter highlights design elements of the project, which showcase its commitment to incorporating an ethos of sustainability into the conceptualisation and development of the main academic building within the project site. These are: an in-built recognition and appreciation of the environmental and aesthetic value of green spaces in the form of gardens, as well as the inclusion of the highest environmental building specifications for a construction project on this scale.

Keywords

Sustainable development · Ethics · Built environment · Cultural sustainability

F. Vellani (✉)
University of Pennsylvania, Walnut Street, Philadelphia, PA 19104, USA
e-mail: fayyaz@writing.upenn.edu

N. Nanjee
University of Edinburgh, Edinburgh, UK
e-mail: naveed.nanjee@gmail.com

W. Leal Filho et al. (eds.), *Integrating Sustainability Thinking in Science and Engineering Curricula*, World Sustainability Series, DOI 10.1007/978-3-319-09474-8_41

1 Introduction

> Inspired by Islamic ethics, humanistic ideals and the philosophy of The Aga Khan Development Network, the University is committed to building an environment that fosters intellectual freedom, distinction in scholarship, pluralism, compassion and humanity's collective responsibility for a sustainable physical, social and cultural environment (Aga Khan University, in Vellani and Nanjee 2012).

According to the above mission, humans are responsible for taking care of the earth and this is in keeping with Islamic teachings. These teachings embody the Qur'anic notion that humans are *khalīfah*, or stewards of the earth—inheriting it and providing for its perpetuation for future generations. The centrality of environmental stewardship to the Aga Khan University (AKU), as a central responsibility and goal rather than an afterthought, dates back to its inception. Founded in 1983, the university includes sensitive considerations for a broad range of environments—human, social, physical and cultural—from its very inception and this is in harmony with the wider Aga Khan Development Network (AKDN), within which it exists. As Kassam (2003: 493) notes, "The AKDN's goal of relieving ignorance, disease and deprivation is profoundly relevant to environmental issues."

The university was designed to blend in organically with its physical surroundings as embodied in the choice of building materials for example, and to appropriately meet the educational and social development needs of Pakistanis in unobtrusive and intellectually sophisticated ways. Located at the Karachi campus is AKU's Institute for Educational Development[1] (IED). This Institute aspires to be "A Role Model for Teaching Excellence", being part of a university-wide Network of Teaching and Learning and Quality Assurance and Improvement (Kassam Khamis 2013). As noted by some, the IED's teacher education programme is "innovative", particularly in the context of the developing world, as its

> outcomes, aspirations and changes adopted by teacher education graduates closely aligned to the aims and expectations of the programme. This suggests a high degree of effectiveness of the programme, which is based on innovative educational practices, pedagogies and the promotion of children-centered and active learning methods as well as the empowerment of graduates to promote needs-based changes to support school improvement (Khamis and Sammons 2007: 572).

More specifically, the IED aims not just to provide students with Master's degrees but also to ensure that graduates have significant impact on the poverty-ridden areas surrounding the university.[2] For example, graduates of IED contribute to the teaching at Prince Aly Boys School, which serves "poor, destitute and orphan children in low-income areas of Karachi" (ibid). Alumni of IED's Master of Education programme have launched a comprehensive school improvement

---

[1] http://www.aku.edu/collegesschoolsandinstitutes/ied/Pages/home.aspx.

[2] http://www.aku.edu/Pages/home.aspx.

programme. In seeking to create equality of opportunity for those less able to access quality education, this initiative exemplifies the Qur'anic notion of *'adl.*

The exemplification of these principles—rooted in Qur'anic ideals and embodied through the socially-conscious actions of the founders of the AKU—are highlighted in two main ways in this chapter: (Sect. 2) through the use of building specifications which meet or exceed the latest standards of environmental design and (Sect. 3) through the use of garden spaces as strategic elements of design aimed and environmental and cultural sustainability.

2 Building Specifications

The number of people estimated to live in urban environments is expected to double from approximately from 3.4 billion in 2009 to 6.4 billion by 2050 (UN 2009). This astounding growth of urban environments poses significant challenges to the energy supply, water collection and distribution, waste management, and mobility within cities, among others. Sustainability is an issue relevant to architecture and urban planners alike as His Highness the Aga Khan elucidates, "we are beginning to see that buildings and places can be created in ways that reconcile—with elegance—the different demands of sustainability" (in Nanji 1994: 122). Going further, he calls for action, urging architects to consider sustainability as core and integral to their practice: "because we share the burden of stewardship of the earth, please ask how the design and technology of buildings can minimise the call on non-renewable resources" (ibid). This section will highlight the key building strategies of the AKDN proposed buildings: R1, Q1, T5, and P2 within the larger King's Cross Project. It is maintained that even though these buildings incorporate both passive and active technologies optimizing energy efficiency, it is not the scope of this paper to analyse the particularities of such mechanisms. However, what will be discussed is that the AKDN network of buildings incorporates a broader understanding of sustainability to include cultural, religious, and artistic ideals.

A plethora of literature has highlighted the importance of sustainability in architecture and design, as the building is the single largest contributor the world's greenhouse gas emissions (UNEP 2011). Incorporating sustainability into the fabric of architectural design of buildings is a critical element of reducing carbon emissions globally (Sayigh 2014). The UK's national sustainability policy highlights the importance of local buildings to take account climate change not just in short term building requirements to limit carbon emissions, but also considerations for long term sustainability strategy. The National Planning Policy Framework (NPPF) requires Green Infrastructure (GI) to have a fundamental role in planning and consideration to drive a strategic approach for the country. In this way, the NPPF has promoted energy-efficient buildings, which positively contributes to the reduction of carbon emissions. The policy objectives of energy efficiency and carbon emissions reductions have had a large influence in planning and development the King's Cross Project.

The King's Cross Project is considered the largest area of urban redevelopment in Europe and is committed to sustainability (Smith and Allen 2013). Overall, the King's Cross Project has set targets of reducing carbon emission through multi pronged approach, which consists of energy efficiency technologies, use of renewable energy, supply efficiency, and pedestrian promoted areas. The buildings highlighted in this chapter plan to connect to the King's Cross Combined Heat and Power (CHP) Energy Centre, which provides low carbon heating and electricity to these building and the rest of the development. This will mitigate much of the high cost of heating buildings and promote low carbon technologies. The CHP system is projected to provide for 95 % of the heat demand for the development for the entire King's Cross Project (King's Cross n.d.). Further study will be needed to assess whether or not energy usage throughout the network of buildings will be minimized and decrease the long-term operational costs with the generation of renewable energy onsite.

Furthermore, the King's Cross Project is seen as a predominantly pedestrian area focused around the largest public transportation hubs in the UK for domestic and international travel. More than 150,000 people everyday travel through King's Cross and St. Pancras International stations, which amount to 28 million annually (Network Rail 2013). The nature of the number of travellers promotes a network of open pedestrian spaces to ensure mobility. This ethos of pedestrian access is paramount in the sustainability strategy of the buildings considered and the King's Project as a whole. Thus, these strategies incorporate sustainable design elements to mitigate carbon emissions and further London's target for aim to be carbon neutral.

Furthermore, the design targets for all buildings part of the King's Cross Project are to achieve Building Research Establishment Environment Assessment Method (BREEAM) Very Good' or better with the hopes of achieving 'Excellent' criterion described in the Outline Planning Permission. As boldly outlined in the 2011 London Plan and the UK National Building Regulations have set standards requiring residential buildings by 2016 and non-domestic buildings by 2019 to be carbon neutral. In its efforts to minimize carbon dioxide emissions all development proposals are recommended to be lean, clean, and green. These regulations specifically affect the King's Cross Development, as the entire development is to reduce carbon emissions by at least 25 % (London Plan 2011). These baseline targets are driving the highest standards of sustainable design and construction throughout the project. Yet, minimizing ecological impact, improved energy efficiency, and limiting carbon emissions is only one facet of the AKDN buildings' sustainability initiative. The R1, Q1, T5, and P2 network of buildings feature sustainability from a technical capacity, but also sustainability in the sense of 'conceptual overlays' which embodies religious, cultural, and artistic ideals that permeate meaning (Levitt 2005).

To illustrate this notion of sustainability, which refers to a broader complexity, this chapter highlights the particular attention to external forms used in the T5 building whose purpose is to house approximately 198 students (see Fig. 1). The architects designed the building in a hollowed ellipsoid form to resemble an Islamic

Fig. 1 Rendering of T5—student accommodation

fort that was of particular relevance to Shia Ismaili history (King's Cross Development Forum 2013). Through, this integrative approach the building not only incorporates passive and active energy efficiency technology in its efficiency but also embodies a preservation of culture heritage through its very design. Furthermore, His Highness the Aga Khan expands the notion of sustainability as he explains, "because the resources we pass on to future generations are cultural as well as material, I urge you to ask how better to recognise and honour the requirement that both be enriched" (in Nanji 1994: 122).

Incorporating the preservation of culture into the sustainability strategy is more self-evident in the use of gardens throughout the network of buildings. In the R1 building there are proposed to be seven inter-linked gardens where the T5 building there are to be two gardens, each representing different regional styles creating a consistent pluralistic series of gardens throughout the AKDN buildings within the King's Cross Project. The significance of these gardens will be further explored in the subsequent sections.

3 Garden Spaces

> Architects have a critical and subtle role to play in moving the world towards a sustainable society (His Highness the Aga Khan, in Nanji 1994: 121).

In the context of art and architecture in the Muslim world, beauty is conferred great reverence. As Clark (2004: 7) notes, it "is not simply a surface decoration but is a reflection of a deep knowledge and understanding of the natural order" and of the all-pervading awareness of a higher being, representing the transcendence and immanence of Allah, the Arabic term for God. Islamic notions of appreciating and venerating the beauty to be found in the natural world can be found in the scripture's representations of gardens; spaces exemplifying a lofty experience of pleasure bestowed on humankind (see Qur'an 9:72, for example). Clark (2004) mines the concept of paradise as a garden, finding it pre-dates Islam, Christianity and Judaism and originates, in her view, with the Sumerians, some 4,000 years before the Common Era.

Art historians point to the role of art in Islam as a conduit to hope: "it should remind us of what it means to be human, of our place in the universe and our role, as is said in Islam, as God's vice-regent (*khalifat 'Allah*) on earth" (Clark 2004: 7). The notion of *khalifat 'Allah* is a kind of caretaking role bestowed on humans by the divine; we are seen as stewards of the natural world, referenced by many as an allusion to our environmental responsibility (Abdul-Matin 2010; Vellani and Nanjee 2012). Gardens, therefore, hold a unique place in Islam as both man-made and divinely inspired; they represent humans' impressions of the natural world but, of course, plant life is seen as a divine creation and gardens thus hold a special place in material and spiritual realms of the daily life of a Muslim. Gardens, in a sense, represent the intersection and co-mingling of these two realms. They also serve as potent reminders of the temporary nature of human existence and the concomitant promise of the afterlife. As Clark (2004: 7) notes:

> The Quran is a sacred presence to Muslims and its references to nature, like its description of the paradise gardens, are worth considering carefully when looking at the meaning of the Islamic garden.

Far from being removed from the vagaries of everyday life, design approaches to Islamic gardens pay attention to the smallest of details, concerning themselves with an alignment with the divine at the broader level while not ignoring the usual, the commonplace, and the mundane. Garden spaces appear in formal, ornamental settings as well as in the midst of the urban chaos of the built environment (Clark 2004). Furthermore, a garden designed with traditional principles in mind, such as the *chaar-bagh* or four-fold garden, can serve as an embodiment of the spiritual values of its designers and users, not simply as a decorative space. The *chaar-bagh* is traditionally constructed around a fountain or pool, from which flow four streams, symbolically reaching outwards in the directions of the earth (North, East, South, West) (ibid).

This section of the chapter: (Sect. 3.1) traces the roots of Islamic approaches to garden design and (Sect. 3.2) considers the design approaches to be taken in relation to the London Kings Cross campus of the Aga Khan University.

3.1 Tracing the Roots of Islamic Gardens

Clark (2004) and other art historians (Ettinghausen 1976; Lehrman 1980) posit that the oeuvre of formal garden design, inspired by the Islamic principles relating the reverence for beauty and nature noted above, also significantly influenced European gardens from the 16th to 18th centuries, in terms of both approaches to their design and the pragmatics of material use and planting—and from as early as the 15th century in Spain, given the presence of the Umayyad Caliphate in that country.

As Islam originated in the Arabian Peninsula, and rapidly gained geopolitical influence in surrounding areas, two very specific forms of design emerged as the archetypal gardens, both incorporated from the Persian tradition, and propagated as design hallmarks: the *chaar-bagh*, or the four-fold garden, and the *gulistan*, literally translated from *Farsi* as rose garden, but including broader forms of decorative gardens. The *chaar-bagh*, based on the envisaged gardens of paradise alluded to in the Qur'an, has itself become a symbol of beauty, and thus of God in Muslim traditions of art and architecture. Starting in the 7th century, shortly after the death of the Prophet, when Islam spread to areas beyond Arabia, many of which were home to ancient civilisations, including Persia, with established traditions of royal pleasure gardens and hunting grounds, travelling Muslims learned about these beautiful traditions, replicated them in the building of cities and palaces, and imbued them "with a whole new spiritual vision" (Clark 2004: 23). Ettinghausen (1976: 6) goes further, highlighting the symbiotic relationships blossoming between travelling Muslims and the cultures they encountered:

> The institution of the royal pleasure garden already existed in pre-Islamic times in the Near East. This applies especially to Iran. Thus, in Islam, there exists both a sacred, visionary, and a secular, hedonistic tradition, each centered on a special garden of the highest beauty. These two traditions strongly influenced each other and the descriptions of both were couched in the most poetic wording, often rich in meaningful metaphors (in MacDougall and Ettinghausen 1976).

Ettinghausen notes that these forms of garden emerged in response to their 'ecological conditions' (1976: 6). As he and others note, those travelling Muslims encountered a perfect design form for expressing their loftiest notions of paradise in the form of the *chaar-bagh* garden: these four-fold gardens were not just functioning physical spaces but also spaces intended to facilitate deep aesthetic enjoyment (Ettinghausen 1976; Lehrman 1980). More importantly, these gardens are meant to evoke humans' higher intellectual and spiritual sensibilities.

Following this period of expansion and the evolution of the concept of gardens as spaces of contemplation for Muslims, they were designed in relation to their geographical locations and surrounding climatic and cultural environments. For example, the Mughals, who maintained and cultivated an imperial presence in South Asia for more than three centuries, incorporated traditional elements of landscaping and decorative art from Turkish, Persian, Central Asian and Indian cultural spheres into their gardens (Jellicoe 1976).

3.2 Design Approaches to the Kings Cross Campus

The gardens to be designed and constructed as integral elements of the Aga Khan University's campus in the Kings Cross Development will be of considerable aesthetic and environmental merit. In all, the unique concentration will include seven gardens including on the roof, as part of inner courtyards, and outer balconies, including one inspired by the traditional *chaar-bagh* fourfold Persian garden design. In terms of usage, the plans are for the students from the Aga Khan University's Institute for the Study of Muslim Civilisations and the Institute of Ismaili Studies to be able to access gardens in the Atrium, the Atrium Terrace, the Library Terrace and the Sky Park on the roof of the main academic and administrative building. The researchers, faculty and staff will be able to use the aforementioned gardens plus a miniature Sky Court, a Terrace which will form part of large meeting room and an Anteroom Garden on the top floor of the building. These gardens will make landscape and architectural reference to the Al Hambra Palace and Complex in Spain, the mosques at Isfahan, Iran, the medieval fortress city of Aleppo, Syria, the picturesque Srinagar, in Kashmir, India, and one of the central touchstones of the Aga Khan Development Network's cultural legacy: the city of Cairo, Egypt.

The carefully considered, historically and culturally sensitive approach to the design of these garden spaces is not merely a natural result of the well-documented rigorous concern with design and architecture of the management of the Aga Khan University, and the wider Aga Khan Development Network, but is also in tandem with broader Muslim goals about the importance of intentionality in all efforts undertaken. Indeed, Clark's (2004) observations on intentionality offer some glimpses into the inspiration behind this fastidious planning. For example, there is a hadith—a saying of the Prophet Muhammad (peace be upon him)—which says that if anyone has intended a good deed and has not carried it out then God writes it down as a full good deed, but if he has intended it and has carried it out then God writes it down 'as from ten good deeds to seven hundred times, or many times over'" (Clark 2004: 14). Clark (2004) therefore places much store in the intention or the spirit behind creating a garden.

Indeed, it works both ways—a person with pure intentions and good thoughts will transform his or her environment into a pure space and, a beautifully ordered space is thought to help humans achieve inner tranquility. Gardens are also spaces of contemplation—a respite from the chaos of urban life—and much emphasis is placed on the role of 'aql—intellect, in Islam. Having tranquil spaces in which to reflect hearkens back to the mystical branch of Islam (Sufism), and indeed, to the world's oldest university, Al-Azhar, founded in Cairo in the 11th century by the

Fatimids, a Shia sect, of whom the current Aga Khan is an ancestor. As a tribute to the rich intellectual and cultural heritage of the Fatimids, the AKDN undertook a massive project of creating Al-Azhar Park in Cairo, the largest park of its kind in the city, in the historic district of the Darb Al-Ahmar, and in the process excavating 12th century Ayyubid walls. As His Highness the Aga Khan articulates:

> In much of Islamic architecture you find a sense of spirituality. You find that spirituality not only in religious buildings. If you think of the history of landscape architecture and you relate that to references to heaven in the Koran, you find very strong statements about the value of the environment, the responses to the senses, to scent, to noise, music or water. I think that in a number of spaces in the Islamic world, which are not religious buildings, there is a heightened sense of spirituality. You do not treat these spaces as theological spaces; you treat them as spaces that aim to give you a sense of spiritual happiness, of spiritual enjoyment. In a funny way, Azhar Park has some of that. We carried out surveys on visitor reactions and a large percentage of visitors to this park in Cairo talk about spirituality (in Jodidio 2007: 41).

Just as Muslims throughout history incorporated design elements from the multiplicity of cultures they encountered into their ornate gardens—the recently revamped grounds at the UNESCO World Heritage site of Humayun's Tomb in India, sponsored by the Aga Khan Trust for Culture being a prime example of this —likewise the gardens in the Kings Cross Development will use local materials where possible, incorporate geographical contingencies such as light, climate, location and the growing season, and aesthetic considerations of the surrounding environs and the wider cultural and architectural landscape of London. The discreet approach to the placement of gardens on the Aga Khan University's campus in London—on rooftops and in semi-private courtyards above the hustle and bustle of the Kings Cross pedestrian walkways—is in keeping with the idea of the subtle, hidden forms of beauty alluded to in scripture:

> It is in the nature of paradise to be hidden and secret, since it corresponds to the interior world, the innermost soul - *al-jannah*, meaning 'concealment' as well as garden, echoing the *hortus conclusus* of the medieval monastic garden (Clark 2004: 24).

Indeed, the traditional house of a prosperous Muslim in an Arab country is laid out in a similar manner to a chahar bagh garden. Lehrman (1980) takes the view that although most gardens in the Islamic world were only accessible to a relatively small number of privileged rulers and aristocrats, the more humble courtyard, however, was accessible to most members of the public. He notes that "all mosques and many dwellings, no matter how simple, possessed such a space" (ibid: 217).

Traditionally, these humble courtyards were also very private. The windows typically faced into buildings rather than out onto streetscapes. Similarly, the Kings Cross Development will feature balconies facing into the atrium of the building, as well as aspects of the paradise garden—a soothing respite from urban life and even from the built environment of the campus itself, fulfilling a need to "escape from daily stress" (Lehrman 1980: 217) which remains as relevant today as it did centuries ago in the Middle East. Clark notes: "This distinction between the public and private domains become one of the principles of traditional Islamic architecture and, by extension, the traditional Islamic four-fold garden" (2004: 25).

In the traditional Islamic house, splendid examples of which can be found in Damascus and Cairo, the courtyard represents the heart, symbolically reminding humans of the inward (*batin*), and contemplative search for meaning. Conversely, the contemporary house represents the outward, worldly (*zahir*) aspect of human life on earth. There were more prosaic reasons for building courtyards as well: "courtyards were common to dwellings throughout most of the Islamic world, owing as much to earlier living traditions and climate as to any specifically Muslim requirements" (Lehrman 1980: 21). Perhaps it is this deliberate use of design to create an inner sanctum—such a rare approach in the construction of university buildings in the contemporary higher education sector but sometimes found, as in the example of the Skirball Cultural Center in Los Angeles, designed by Moshe Safdie. Lehrman (1980: 21) evokes a beautiful picture of such a sanctum in a traditional house (*bait*) in an Arabic city:

> It may be in the middle of the bustling medina (old city) – of Marrakech for example – but when the door to the street is shut the visitors enter a totally different world: it is immediately quiet since the high, thick stone walls keep out the noise and bestow a kind of muffled silence on the interior, not dissimilar to entering a church. The gentle murmur of a fountain in the centre draws the visitor in, contributing to the atmosphere of inward reflection.

The intricate design and elaborate selection process for world-class architects—Japan's Fumihiko Maki being among the most frequently commissioned for the landmark projects of the Aga Khan Development Network—point not only to the desire of the network to be on the leading edge in terms of aesthetics and sustainability. In addition, several sources allude to the deep sensitivity of Maki to the values of his clients, promulgating a kind of cultural sustainability, beyond the obvious need for environmental sustainability. Seeking to expand notions of sustainability, His Highness the Aga Khan points to the role of architecture and design which is socially, environmentally and culturally sustainable:

> The development of sustainable environments draws not only on the physical resources of a society but also upon the cultural and artistic values that shape the values of that society. Buildings, spaces, settlements and cities are the embodiment and bearer of those values (in Nanji 1994: 121).

4 Conclusion

Embracing a design aesthetic which deeply enmeshes the multiple layers of intellectual and spiritual symbolism alluded to earlier in the chapter—of the human search for meaning—the managers of the Aga Khan University have designed the Kings Cross campus of their university with a clear desire to inspire a unique experience of inner contemplation in their building and garden spaces.

Thus, the Kings Cross Development is a carefully concerted effort on the part of the Aga Khan University, which hearkens back towards Islam's glorious cultural and architectural past whilst simultaneously looking forward. In keeping with the highest level of BREEAM standards and promoting London's vision of carbon neutrality the work of the Aga Khan Development Network is expanding notions of sustainability, to create an inspiring educational environment for its postgraduate students and members of faculty. Given the historical centrality of the city of Cairo to these rich cultural allusions—founded by the Fatimids, predecessors to the current Shia Ismaili Muslim community—including Al-Azhar, the world's oldest university, it is certainly conceivable that the managers of the Aga Khan University wish to recreate spaces which will inspire future generations of thinkers and intellectuals, including those who will strive to expand notions of what is required by humans to foster their own sustainability.

References

Abdul-Matin I (2010) Green deen: what islam teaches about protecting the planet. Berrett-Koehler, San Francisco

Burton E, Jenks MM, Williams K (eds) (2000) Achieving sustainable urban form. E & FN Spon, London

Clark E (2004) The art of the Islamic garden. Crowood Publishing, Marlsborough

Jodidio P (2007) Under the eaves of architecture, the Aga Khan: builder and patron. Prestel, Munich

Kassam Khamis T (2013) AKU networks of teaching and learning and quality assurance and improvement. Office of the Provost discussion paper, found at: http://www.aku.edu/aboutaku/Provostoffice/Documents/2013%2011%2005%20Networks%20of%20QA%20and%20Improvement%20and%20TLDP.pdf

Khamis A, Sammons P (2007) Investigating educational change: the Aga Khan University Institute for educational development teacher education for school improvement model. Int J Edu Dev 27(5):572–580

King's Cross Development Forum (2013) Camden town hall meeting of May 16, 2013. Found at: https://www.camden.gov.uk/ccm/navigation/environment/planning-and-built-environment/major-developments/king-s-cross-central/;jsessionid=D05F126C027747B25FEC16753BF2BD48

King's Cross (n.d.) The energy centre and CHP plant. Retrieved 26 Mar 2014, from http://www.kingscross.co.uk/the-energy-centre-at-kings-cross

Lehrman J (1980) Earthly paradise: garden and courtyard in Islam. Thames and Hudson, Over Wallop

Levitt B (2005) Veiled sustainability: the screen in the work of Fumihiko Maki [Speaking of Places]. Places 17(2)

MacDougall E, Ettinghausen R (eds) (1976) The Islamic garden. Trustees for Harvard University, Washington DC

Mulej M (n.d) Plan B 3.0: mobilizing to save civilization. PLAN B 3.0; Mobilizing to save civilization. Syst Res Behav Sci 27(1):113–123

Nanji A (ed) (1994) Building for tomorrow: the Aga Khan Award for architecture. Academy Editions, London

Sayigh A (ed) (2014) Sustainability, energy and architecture: case studies in realizing green buildings. Academic Press, Oxford

Smith P, Allen J (2013) King's cross & St Pancras green infrastructure audit. Land Use Consultant Ltd, London
UNEP (2011) Green economy report, Towards a green economy: pathways to sustainable development and poverty eradication, introduction—introduction: setting the stage for a green economy transition, found at: http://www.unep.org/greeneconomy/Portals/88/documents/ger/ger_final_dec_2011/Green%20EconomyReport_Final_Dec2011.pdf/
UN (2009) Urban and rural areas in world urbanization prospects: 2009 revision. United Nations Department of Economic Social Affairs, found at: http://www.un.org/en/development/desa/population/publications/pdf/urbanization/urbanization-wallchart2009.pdf
Vellani F, Nanjee N (2012) Sustainable development in higher education in the Muslim World. In: Leal W (ed) Sustainable development at Universities: new horizons. Environmental education, communication and sustainability vol 34. Peter Lang, Frankfurt

Authors Biography

Fayyaz Vellani is a Senior Critical Writing Fellow and College House Fellow at the University of Pennsylvania. He holds degrees from the University of London, New York University and the University of Waterloo. Dr. Vellani previously served as Associate Director of the Critical Writing Program at Penn and as Head of the Department of Graduate Studies at the Institute of Ismaili Studies in London. His book, *Understanding Disability Discrimination Law through Geography*, was published by Ashgate in 2013.

Naveed Nanjee is originally from Sugar Land, Texas and is currently studying at University of Edinburgh for a LLM in Global Environment and Climate Change Law. Prior to this, Naveed worked with Dr. Vellani as part of the Graduate Program in Islamic Studies and Humanities at the Institute of Ismaili Studies in London. Naveed completed his undergraduate degree with high honours from Vanderbilt University in Nashville, Tennessee. He double majored in Cognitive Studies and created his own interdisciplinary major called International Development and Sustainability. At Vanderbilt, in collaboration with various faculty members, he laid the foundation for an academic curriculum on sustainability, which was presented at the Cumberland Project Conference in 2010. Naveed spent a semester abroad in South Africa studying development at Stellenbosch University and interning at the Sustainability Institute.

Implementing a Holistic and Student Centered Outreach Programme Towards Integrated Sustainable Development of the Campus—A Case Study of a Residential School from South India

Nandhivarman Muthu, Bhaskaran Shanmugam, Poyyamoli Gopalsamy and Golda A. Edwin

Abstract

For several decades now, many institutions are doing their best to strive towards a sustainable future through education, research and out reach. According to UNESCO, the goal of education is to make people wiser, more knowledgeable, better informed, ethical, responsible, critical and capable of continuing to learn. Education is also the means to disseminate knowledge and develop skills, in order to not only bring about desired changes in behaviours, values, lifestyles but also to promote public support towards the continuing and fundamental changes that will be required if humanity is to alter its course towards sustainability. To meet ever growing global environmental challenges, every school, colleges and universities must prepare today's students in all disciplines, and it cannot be implemented/achieved without students' involvement. This article presents a case of a residential school in the developing part of the world that transforms itself into a model, self sufficient sustainable campus through constant motivation sustained by the result of their positive actions towards the sustainable development of the campus. While the green campus initiatives are more likely to focus solely on ecological issues (downplaying social and economic issues), this initiative revolves around the continually evolving and adaptive sustainability model that

N. Muthu (✉) · P. Gopalsamy · G.A. Edwin
Department of Ecology and Environmental Sciences, Pondicherry University, Puducherry, India
e-mail: m.nandhivarman@gmail.com

B. Shanmugam
Association for Promoting Sustainability in Campuses and Communities, Puducherry, India

N. Muthu
23, Kamaraj Salai, Thattanchavady 605009, Puducherry, India

W. Leal Filho et al. (eds.), *Integrating Sustainability Thinking in Science and Engineering Curricula*, World Sustainability Series, DOI 10.1007/978-3-319-09474-8_42

depends on the integrated knowledge from all the stakeholders in the school (lowest to the highest echelons) by bringing all of them to work as one single group with a vision on school sustainability. Emphasizing more on this, the Green Campus Initiative (GCI) is successfully evolved and implemented within 100 working days. This pilot scale Green Campus Initiative was evolved, based on the research done at Pondicherry University, and implemented as the student centered outreach programme, encompassing water-food-energy-ecosystems nexus, with an integrated systems approach. The various measures undertaken by the campus are discussed in brief and how they helped transform the campus into a Green Campus. The lessons learnt will be very useful for replication of similar experiments elsewhere in India or in any other developing country.

Keywords
Green campus · Sustainable development · Campus environmental sustainability

1 Introduction

Even though the origins of the terms 'sustainable development' or 'sustainability' goes back many decades, it was substantially enhanced since the Brundtland Report (1987), which is also known as Our Common Future. Following this event, India became part of 187 countries that agrees to carry out an important commitment towards SD by signing the Rio Declaration during 1992 UN Conference on Environment and Development. The goal towards a sustainable future have taken root when India became a part of the UN General Assembly resolution for establishing the UN Decade of Education for Sustainable Development (UN DESD, 2005–2014) in the year 2005, emphasizing Agenda 21, that education is an essential tool and an indispensable element for achieving sustainable development. Over the past two decades, across the globe, the term 'sustainable development' or 'sustainability' has found its way into the International conventions and agreements, government and non-government agencies/enterprises, universities, higher educational institutions, schools, NGO's, CBO's, etc. More importantly, the modern views of sustainable development are not just limited to ecological considerations; rather the economic, social, cultural and political dimensions are also being embraced (UNESCO 1997).

The World Summit on Sustainable Development, held at Johannesburg in 2002, affirmed and confirmed the commitment to the 'full implementation' of Agenda 21, along with the other Millennium Development Goals by emphasizing education as an indispensable element for sustainable development. Till today the majority of the international declarations, agreements and action plans have one thing in common: 'they have never been fully implemented until now' (Leal Filho 2012).

These and other similar authoritative international documents and commitments, such as the Tbilisi Declaration, the Dakar Framework for Action, and Local Agenda 21, have pushed the education sector to focus for change towards sustainability since the 1970s. Initially, the concepts of sustainability were well received in the western countries and Universities in the United States and Europe started a movement centered on the concept of campus environmental sustainability. A common value shared among the green universities is their ultimate goal of taking sustainable development for human society as a direction in developing higher education through incorporation of appropriate steps into university administration, development plans, teaching and research, environmental maintenance and student life (Henderson and Tilbury 2004; NTNU 2014)

The drive to attain the "GREEN CAMPUS" status is to keep water, food, energy as the core components, and to serve as a model for local/regional environmental sustainability where all the processes and operational functions of the campus are closely knit, providing educational and practical value to the institution and to the surrounding environment. As a general observation of institutional campuses across India, their largely unplanned growth (from ecological/social dimensions) over a period of time have resulted in unsustainable resource extraction and utilization, resulting in the degradation of campus quality in terms of environment and life. Each and every campus is endowed with a measurable stock of environmental assets such as top soil; flora and fauna; fertile and non-cultivable land; ground and surface water; clean air, etc. However many campuses have already lost the quality of basic life support systems like water, soil and air. Unless the resources are managed sustainably, it will result in the loss of ecosystem services and the ecological carrying capacity of the campus. It is the responsibility of every individual of the campus to embrace the Principle of Intergenerational equity "as members of the present generation, we hold the earth in full trust for future generations", which in recent years emphasized by many bilateral and multilateral agreements, declarations and resolutions for Sustainable Development.

Education, in short, is humanity's best hope and most effective means in the quest to achieve sustainable development (UNESCO 1997). In the United States and Canada, institutions of higher education are also implementing plans and taking actions that aim to achieve socio-ecological sustainability within their confines. With 22 million people engaged in higher education throughout this region, the potential for transformation is immense. Known as the Campus Sustainability Movement (CSM), it is expanding across the United States and Canada as campuses work towards greening their operations, as well as tackling the deeper issues of sustainability within their institutions. These initiatives are taking the much-needed first steps towards sustainability, as well as sparking conversations and increasing awareness around sustainable development (Shriberg 2001; Henderson and Tilbury 2004).

This article presents a case of a residential school, Jawahar Navodaya Vidyalaya in Puducherry, India that transforms itself into a model, self sufficient sustainable campus through constant motivation sustained by the result of their coordinated actions. This pilot scale Integrated Green Campus Initiative (GCI) model was based on the lessons learnt from the research done at Pondicherry University, and implemented as student centered outreach programme, encompassing water-food-energy-ecosystems nexus. Two researchers (NM and GE—among the authors of the present study) from Pondicherry University, came forward to donate fully for this project to set an example that, "*sustainable development can be achieved only when others are benefitted by selfless attitudes and attributes through life giving cooperation and coordination*". In this residential school, "*this, GCI, is a program that plans, formulates, designs and implements a package of sustainable solutions by the campus community to reduce the environmental impact, enhance the campus sustainability and to protect the health and well-being of the surrounding community and ecosystem*". The agenda of GCI is fundamentally based on the following principles outlined in the international agreements.

- **Precautionary principle 15 of Rio Declaration**: "where there are threats of serious or irreversible damage, lack of full scientific certainty shall not be used as reason for postponing cost effective measures to prevent environmental degradation"
- **Principles of Inter-generational Equity**: "as members of the present generation, we hold the earth in full trust for future generations"
- **Art.3 of the UN Conference on Climate Change**: "The parties should protect the climate system for the benefit of present and future generations of humankind, on the basis of equity and in accordance with their common but differentiated responsibilities and respective capabilities".
- **Principle 3 of the Rio Declaration**: "the right to development must be fulfilled so as to equitably meet developmental and environmental needs of present and future generations".

This GCI is expected to provide an additional momentum, for the efforts taken by the Government fostering MDG's and Decade of Education for Sustainable Development (2005–2014), and with much focus on what educational institutions can achieve in this regard for the region/country and also for their campus environment sustainability.

2 Project Site

The project location is at Jawahar Navodaya Vidyalaya (JNV), a residential central school located at Puducherry which is one of the Union territories of India, situated on the coromandal coast, 160 kms. south of Chennai (Fig. 1). JNV is a fully residential co-educational institution established in the year 1986–1987 under Ministry of Human Resource Development, Government of India. The GCI program is a part of the "Green Campus Initiative Co-operative Movement" (GCICM) originally initiated by the Association for Promoting Sustainability in Campuses and Communities (APSCC), an organization committed to promote sustainability in every area among the educational institutions and local communities. APSCC holds the responsibility to analyse leading researches and build a tailor made framework which fits into the Indian scenario, to suit the local conditions in the implementation

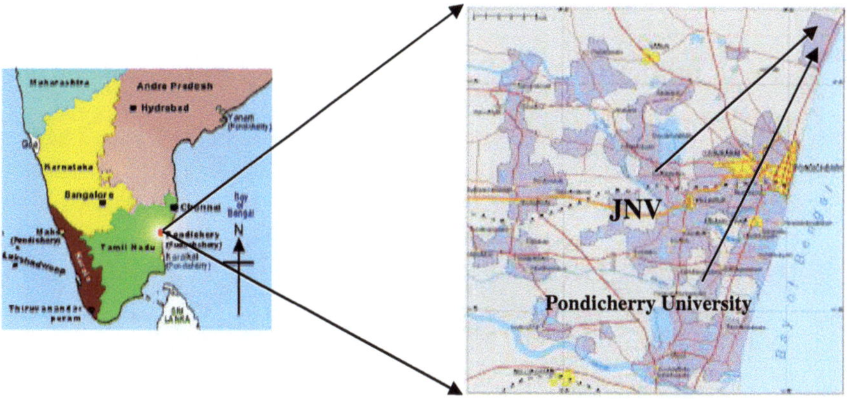

Fig. 1 Map showing the regions of Puducherry, Pondicherry University and JNV

of various programs and projects. The association provides a forum for information exchange and expertise assessment in related field at the national and international level and develops innovative programs and distributes educational materials, for learning, training, networking and capacity building in the field of Education and Sustainable Development (ESD). The GCICM was formed to achieve the objectives of campus sustainability by creating awareness, mobilizing resources and setting up an environment of opportunity to develop and implement tailor made programs to enrich the sustainability quotient of the campus. These are primarily achieved by taking into consideration the local contexts by empowering the student community and involving them through practical learning and networking. The different initiatives implemented as part of this program are listed in Fig. 2.

3 Vision

The vision of this green campus initiative by APSCC is to transform JNV into a self-sufficient sustainable campus.

4 Objective

The main objective of this research oriented initiative is to serve as a model for other schools and institutions to adopt this green campus initiative, for their campus sustainability. The project has a novel integration of the following ten components (Fig. 2) in its first phase.

Fig. 2 Ten integrated components of green campus initiative

5 Research Methodology

- Literature review
- Identification of current environmental issues
- Identification of components to address such issues
- Novel integration of components
- Appropriate site selection for the implementation of the components
- Identification of approaches needed for successful implementation
- Analysis
- Outcome

6 The Integrated Components of GCI

6.1 Identification of Components

To meet ever growing global environmental challenges, every school, colleges and universities must prepare today's students in all disciplines to identify the key strategies for achieving the goals of integrated sustainability. Every campus have burning issues related with waste water, solid waste, and unhealthy foods, and it is better that the components are identified at the earliest to address the need immediately. The following are the ten components which were evolved for the "Green Campus Initiate" at JNV, nine components addressing the issues related with water, energy, and food with the tenth component being the governance part of the GCI (Fig. 3). Each component was integrated with another to optimise the sustenance of the frame work fostering water security, food security, energy security and green economy in the campus.

1. Water sustainability
   i. Constructed wetland for water reclamation and reuse
   ii. Xeriscape for Biodiversity Conservation and Habitat Restoration
   iii. *H*ügelkultur—Raised Garden Beds for Ground Water Conservation
2. Energy Sustainability
   i. Anaerobic Digester for Biogas Generation from food/garden wastes
   ii. *Jatropha curcas* Linnaeus for Pollution Control and Renewable Energy (bio diesel) using campus grey water
3. Food sustainability
   i. Windrow Composting—Japanese "Bokashi" and Indian "Indore" Method
   ii. Bio-fertilizer and Bio-pesticide Production
   iii. Green House for Organic Olericulture, and Hydroponics
   iv. Climate Resilient Organic Farming
4. Governance
   i. Office of Sustainability

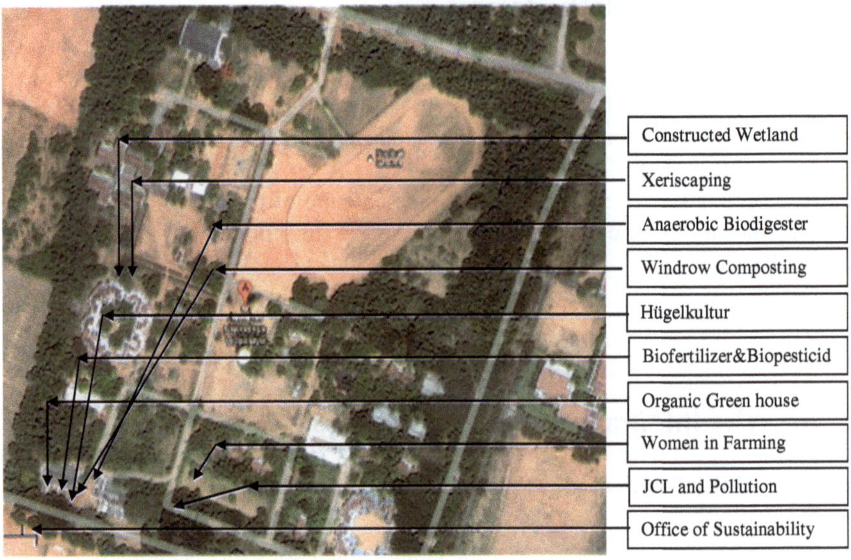

Fig. 3 Location of the ten components of GCI at Jawahar Navodaya Vidyalaya (JNV)

6.2 Selection of Appropriate Site for the Implementation Integrated Components of GCI

1. Water Sustainability

 i. **Constructed wetlands for water reclamation and reuse**: Horizontal flow Constructed wetlands system serves as a sustainable solution for water treatment. A constructed wetland is basically a bio filtration system which mimic the natural pond/lake ecosystems significantly removing a considerable amount of physic-chemical and biological pollutants from grey water before it seeps into the aquifer or flows into the river, or natural wetland (Brix 1993). This will promote a healthier ecosystem and more importantly it destroys pathogens, bacteria, and non-biodegradable toxins. Environmentalists, with ariel view have referred to wetlands as nature's kidneys (Wei et al. 2013), and viewing from the aquifer, we refer the wetlands to the lungs of aquifer. Native macrophytes such as *Arundo donax* is used as one of the species in the wetland along with well studied *Typha* species (Golda et al. 2012). The challenge is to implement the strategies through a collaborative approach that excite and engage the campus communities to deliver environmental sustainability for the present and future generations. Figure 4 shows the volunteering students to build and establish the system. The campus wastewater is economically and efficiently treated and reused for not only enhancing the landscape quality and local ecology but also serving as a practical environmental lab for the students and faculties. Constructed wetland treatment system therefore becomes a potential exemplar of ecological technology to fit

Fig. 4 Student volunteers participating in the building of the Constructed Wetland, and Xeriscape with Principal, and APSCC green action team

the goals of a sustainable campus and the conservation of this fragile water resource by every citizen is of prime importance (cgwb.gov.in; www.pdywaterinfo.org).

ii. **Xeriscape**: Xeriscape refers to landscaping and gardening in ways that reduce or eliminate the need for potable water for irrigation. It originated in Colorado and eventually the idea spread throughout the world to utilize water efficient landscape designs to save water (Hilaire et al. 2008). In GCI, we have integrated xeriscape with reclaimed water application. Xeriscape reduced the consumption of ground water by 100 %; facilitate more rainfall/moisture retention; reduce the need for fresh water; with organic inputs eliminate the need for synthetic alternatives; restoration of top soil; enhancement of soil beneficial micro flora and fauna; biodiversity conservation and habitat restoration; attract butterflies, honey bees, migrants and native species; sequesters carbon and mitigate climate change (UNEP 2013). As a motivation factor, every student contributed to the xeriscape with a suitable plant (Fig. 5).

Fig. 5 Student volunteers participating in the building of the Constructed Wetland, and Xeriscape with Principal, and APSCC green action team

iii. **Hügelkultur for ground water conservation**: Since Peninsular India is prone to frequent cyclones and storms, a large number of trees would be uprooted/damaged during such disasters (Shanmugasundaram et al. 2000). On December 29th, 2011, when the cyclone Thane hit the region, about 70 % of trees fell in Puducherry including the JNV campus. Anchoring roots of the trees usually spread where soil conditions provide nutrients and moisture, while usually 85 % of them are within the top 500 mm of soil. Due to poor watering and let to live un attended, much of these trees fell due to poor anchoring root system and they are left to rot without any use. In order to reuse the resource effectively, we adopted hügelkulture, as one of the GCI components. It is a practice of making raised garden beds filled with rotting wood. It is in effect creating a "nurse log", a fallen tree which, as it decays, provides ecological facilitation to seedlings. Benefits of hügelkultur garden beds include water retention and warming of soil. Buried wood becomes like a sponge as it decomposes, able to capture water and store it for later use by crops planted on top of the hügelkultur bed. As a result, the fertile top soil retains the moisture through the group of micro biota which will sustain the health of the roots (Samples 2013).

2. Energy Sustainability

While the conservation of energy cannot be implemented without student involvement, energy recovery from kitchen waste, contributing a fraction of green power profile could lead to a significant decrease in carbon footprint and GHG emissions. Focusing on energy sustainability, energy recovery by anaerobic digester and energy production by JCL, is strategically adopted. Students volunteered in the set up of the Anaerobic Digester and establish fencing using the felled tree logs available within the campus (Figs. 6 and 7).

i. **Energy Recovery**: "ABCD- Hybrid Biomethanation plant", developed by biogas development and training centre (BDTC), Puducherry, is installed. BDTC was formed in December 2011 by the APSCC in order to further the research and training in the area of biogas generation and storage. Since then,

Fig. 6 Student volunteers participating in the building the 7 m^3 Anaerobic Digester, and fencing by felled tree logs available within the campus

Fig. 7 Student volunteers participating in the building the 7 m^3 Anaerobic Digester, and fencing by felled tree logs available within the campus

BDTC actively promotes programs on biogas generation, synchronizing with the initiatives undertaken by The Ministry of Non-Conventional Energy Sources (MNCE) and The Ministry of New and Renewable Energy (MNRE) of the Govt. of India. One installed 7 m^3 plant produces biogas that substitutes approximately 6 cylinders per month with the potential reduction in petroleum gas by 10.34 %. The advantages and environmental benefits of this initiative are 25–30 % of petroleum gas is supplemented; elimination of pathogenic loads by 90 %; prevents cross contamination of soil during monsoon; sustainable management of spent slurry result in bio-fertilizer production. More importantly, a sum of Rs. 7,864/- per month could be saved on the cooking gas Indane and the slurry can replace Rs 5,000/month of chemical fertilizer, notwithstanding the associated ecological co-benefits (Duan et al. 2014).

ii. **Energy Production**: In spite of all these measures, there is an ample amount of waste water yet to be treated. Such waste water can be diverted to the cultivation of energy crop such as *Jatropha curcas* Linnaeus (JCL), which is principally promoted by the National Planning Commission, Government of India (Indian Programs 2014). Since JCL is energy crop, researchers and scientists conclude that energy crops will have high water footprint and should neither be irrigated with fresh water nor be planted near water bodies. In resonance with this concept, JCL plantation is integrated with grey water treatment and application. The advantages and environmental benefits of growing energy crops includes—reducing the waste water load on municipal sewer system; helps to close the water cycle; easily adapted and can be cultivated on lands adjacent to sewer systems which are not suitable for food crops (Rajaona et al. 2012).

3. Food Sustainability

The concept of organic farming through compost/vermicompost is gaining importance world-over to mitigate the damage caused by the increasing and indiscriminate use of synthetic fertilizers and pesticides which often resulted in soil fatigue, and gradual deterioration of soil health and productivity

Fig. 8 Student volunteers collecting biomass to prepare the windrow compost yard

Fig. 9 Student volunteers collecting biomass to prepare the windrow compost yard

(Padmavathy and Poyyamoli 2013). As part of this GCI, the biogas spent slurry is mixed with the campus biomass, consisting mainly of leaf litters in the ratio 1:5 and composted by windrow composting. The environmental benefits of composting includes improved soil porosity, balanced pH, attraction and feed for earthworms; increased soil fertility and soil health by increasing organic matter in soils; supporting vigorous growth of beneficial micro flora and fauna thereby increasing productivity; creating better environment by reducing ecological risks (Sinha et al. 2010). Figures 8 and 9 shows the student volunteers collecting biomass to prepare the Windrow compost yard.

i. **Bio-fertilizer**: Conventional farm compost will have low nutrient content needing synthetic supplement (Padmavathy and Poyyamoli 2013). In order to overcome this situation, vermicomposting, of composted biogas spent slurry with campus organic wastes were used to produce fortified bio-fertilizer. This could provide invisible ecosystem services by serving as versatile natural bioreactors to harness energy and destroy soil pathogens, even assimilating heavy metals during their life cycle. The driving forces behind this component is to recover organic material and to return it back to the field for maintaining the natural cycle of the recycling process, leading towards habitat restoration and biodiversity conservation (Garg et al. 2006; Suthar 2010; Nandhivarman et al. 2012).

ii. **Greenhouse for Organic Horticulture**: Greenhouse cultivation is evolved to create favorable micro-climates, which favors the crop production, possible throughout the year. Greenhouses may be used to overcome shortcomings in the growing qualities of a piece of land, such as a short growing season or poor light levels, and they can thereby improve food production in marginal environments (Olsen et al. 1999). As a component of GCI, the plants such as tomato, onion, ginger and gourd varieties were cultivated through application of biofertilizers and biopesticide resulting in 100 % organic produce. In order to replace the synthetic pesticides that are used for the kitchen garden, Panchagavya, a fermented organic bio pesticide was tried as an alternative. It is an organic plant liquid fertilizer cum pesticide and the term means the blend of five products obtained from traditional cows (dung, urine, milk, curd and ghee). It is also a traditional method in India that is used to safeguard plants and soil microorganisms and to increase plant production. Its application cures infested plants of diseases and increases the immunity of the plants and other organisms in the ecosystem. It activates the photosynthetic system for enhanced biological efficiency, enabling synthesis of maximum metabolites and photosynthates. It also contributes to less irrigation and ensures drought hardiness by forming a thin oily film on leaves and stem, thus reducing evaporation (Sangeetha and Thevanathan 2010; Ali et al. 2011; Padmavathy and Poyyamoli 2013).

iii. **Climate Resilient Organic Farming**: In the study of ecology, resilience is the capacity of an ecosystem to respond to a disturbance by resisting damage caused through stochastic events such as fires, flooding, windstorms, insect population explosions, anthropogenic activities (that culminates in climate change) and recovering quickly. In the context of emerging climate change challenges, we need to ensure that agriculture contributes to addressing food security (availability, access, utilization, stability/sustainability issues), development, livelihoods, health and climate change issues (adaptation and mitigation) (IAASTD 2009). One such innovative approach is climate-smart agriculture (CSA), an emerging paradigm (World Bank 2010; Branca et al. 2011; 2012; IFAD, 2012; Scherr et al. 2012). Examples of some of the key specific strategies of sustainable CSA (Stabinsky and Ching 2012) are: bio-intensive organic farming (Moore 2010; FAO 2011), low external input, sustainable agriculture agro-ecological and bio-dynamic production systems (Altieri 1995). While garden-based nutrition-education programs for youth are gaining in popularity and are viewed by many as a promising strategy for increasing preferences and improving dietary intake of fruits and vegetables (O'Brien and Heim 2009) in developed countries, it is not yet widely practiced in India. Thus, we have tried to emulate such hidden opportunities in the JNV campus by adopting such practices on a pilot scale to grow vegetables needed for the school kitchen. This program is again sustained by the volunteering efforts of a dedicated team of students (Fig. 10).

Fig. 10 Student centered climate resilient organic farming

As part of this initiative, we have integrated organic farming to serve as an interdisciplinary discourse on resilient through adaptive resource management and campus governance. It has been observed that the organic and sustainable farming has the potential to create new structures within the campus that actively work towards innovative solutions and protecting the use of indigenous knowledge.

According to the International Federation of Organic Agriculture Movements,

> Organic farming is a production system that sustains the health of soil, ecosystems and people. It relies on ecological processes, biodiversity and natural cycles adapted to local conditions, rather than the use of inputs with adverse effects. Organic agriculture combines tradition, innovation and science to benefit the shared environment and promote fair relationships and a good quality of life for all involved.

It is imbibed in the curriculum so that the campus community can actively engage with those who maintain the initiative and as a result learn how to grow food crops, traditional knowledge of native plants, their health benefits, etc. Apart from this, detailed plans are being developed for implementing seed banks for traditionally cultivated crops and exchange of seeds as an outreach at JNV.

4. Office of Sustainability

   There is a great controversial argument between the existing campus environment management system versus green initiative versus sustainable development which the office of sustainability at JNV tries to address. It serves as one of the most important components of this program that inspires and engages the faculty, students, staffs and also the community in a voluntary, self-guided initiative that sustains and promotes best environmental practices in the campus and the surrounding environment. The office coordinates and supports initiatives that helps reduce the campus's ecological footprint.

   Stakeholders within schools, colleges and universities exhibit allegiance to multiple professional and managerial cultures (Rothblatt 1995). Every objective for GCI has a specific research methodology with quantitative and qualitative analysis, recommendations based on the outcome, policy planning, action research and community outreach. Any green initiative is always meant to abate environmental issues or problems associated with it. Therefore the approach adopted for GCI plays a very important role for the sustenance and accountability of the initiative. The following are some of the approaches and theories adopted for the implementation of the ongoing GCI:

i. **Goal setting approach**: The major aim of the goal setting through education is to make campus community wiser, more knowledgeable, better informed, ethical, responsible, critical and capable of continuing to learn. Were all people to possess such abilities and qualities, the world's problems would not be automatically solved, but the means and willingness to address them would be at hand, by promoting greater consciousness and awareness, exploring new visions and concepts, and inventing new techniques and tools as reported by Shriberg (2002a); education, in short, is humanity's best hope and most effective means in the quest to achieve sustainable development (UNESCO 1997). While setting the goal the promoter one should consider the three components of the goal namely- conceivable, believable, achievable (Locke and Latham 2002) for practical solutions:

- Conceivable → the promoter should conceptualize the goal so that it is understandable clearly for the stakeholder/key personnel.
- Believable → the promoter and stakeholder/key personnel should believe that the goal can be reached and also other people believe in it.
- Achievable → One should ask themselves if given their strengths and weaknesses can they reach their desired goal

ii. **Motivational approach**: Corporate environmental management and social responsibility theory and practice are beginning to converge on the importance of "enlightened self-interest" as a motivator for management of environmental and social issues (Gladwin et al. 1995; Whitman 1999; Shriberg 2002a). There is a great controversial argument between the exiting campus environment management system versus green initiative versus sustainable development. One of the major aims of this action research oriented initiative is to identify and assess differences between environmental management systems based on campus stakeholder perceptions of "sustainability" as opposed to those between "environmental", "greening", "stewardship" or other related concepts and to provide schools, college and university, change agents with a set of guidelines to enact cultural change and to understand the difficulties in pursuing organizational change specific to environmental issues and sustainability as suggested by some pioneering workers in this area (Shriberg 2001, 2002a, b). In this way, this study tests the motivational potential and practical influence of orienting towards sustainability in a campus setting. The basics in the GCI approach of motivation is that promoters and stakeholders should come together "as partners - as one group" to produce enormous quantities of products at high production rates (Snell 1972). Government of Puducherry's vision for 2020 clearly describes the land, water, air related sustainable management problems such as ground water depletion, salt water intrusion, urban sprawl, urbanization, industrial pollution, sewage pollution, coastal disasters, etc. and it has called for the attention of educational and research institutions to supplement their effort on minimizing such environmental issues and to develop and adopt strategies to abate and mitigate climate change. "this initiative propose that every school, college, and university should stand up and be counted on the issue of climatic change by beginning now to develop plans to reduce and eventually eliminate or offset the emission of heat-trapping gases by the year 2020 as suggested by several earlier workers (Orr 2000; Shriberg 2002b). Even though the literature provides many excellent case studies of environmental initiatives that have been implemented throughout the world, most of the information available is in the form of examples of 'this is what we did on our campus" (Herremans and Allwright 2000), and hence in this GCI we also participated not as the mere promoters but as the implementers and co-participants.

iii. **Choice Based Approach**: Motivation is the set of combined forces that cause the community/stakeholders/key personnel to choose and adopt any of the initiatives. Normally in all green initiatives there will be either one or two objectives focused, and while implementing apart from the benefits associated this objective, there are possibilities to get benefitted by additional value added benefits (co benefits). Since sustainability, depends on integrated knowledge and attempts to understand complexity as interconnected systems (Henson et al. 2007), through GCI, we have developed a novel integration of multiple components with which one can include any green components at any walk of campus life either as intermediary components or connecting components, Because of this novel components with initial integration and future additional possibilities, many alternative options are open to them to choose. Hence this type of approach is called as "Choice Based Approach" (CBA).

iv. **Multilevel Human Resource Approach (MHRA)**: Natural resources will get depleted when improperly used and on the other hand the human resources will get depleted when properly not used. Each and every individual has instinct likes and dislikes. The GCI promoters responsibility is not to force the stakeholders/key personnel to work on the objective which he/she dislikes; but rather to provide multiple platforms to choose, fostering multilevel human resource participation Now a days the human resource approach guides most thinking about motivation, conceptualize more completely on need-based (WSU 2014). Greene and Caracelli (1997, p. 7) conclude: "There is wide consensus that mixing different types of methods at the technical level, or the level of method, is not problematic and can often strengthen a given study." Simply put, "Multiple viewpoints allow for greater accuracy" (Jick 1979). Clearly, a mixed methodological approach is the preferred option for many researchers in diverse fields and we have used this method.

v. **Top down and bottom up approach**: A "top-down" approach is one of the best and initiating approach, where the highest authority/authority in charge/an executive, decision maker, or other person or a body makes a decision and implements. This can also be clubbed with bottom-up strategies once the system is established and started delivering results. However for initial stages, the primary focus of sustainability initiatives should be restricted to top-down and bottom- up strategies with no intermediate levels (John 2003). Presumably fairer and more effective way to tackle climate change today is by bringing together the major greenhouse gas (GHG) emitters; to solve the climate related issue with bottom-up approach (Leal Filho 2012). If humans are going to have a global economy, a global media, a global technology, there must also be global ethics to which all nations and peoples of the most varied backgrounds and beliefs can commit themselves (Kung 1998). The findings in the campus environmentalism literature with the higher education and transformational leadership literature reveals that the strongest campus sustainability initiatives have support from individuals at lower levels in the organizational hierarchy (e.g., students and staff) as well as institutional leaders (e.g., the president), and at least one charismatic environmental "champion" (Shriberg 2002b) . Hence we have used an integrated top down and bottom up approach.

vi. **Need-Based Approach (NBA)**: Currently due to want of time there are many needs pending to be fulfilled in almost all institutional campuses. Identifying the need which can be fulfilled through need-based approach, motivates stakeholders/key personnel to choose certain behaviors. This approach is divided under two basic categories:

- **Basic needs**:
  - Initiatives for solving waste related issues (Ecological, Physiological need)
  - Changing unsustainable activities to sustainable initiatives (Water, Food, Energy, Ecosystems safety/security pertaining to social acceptance)
  - Change in campus environment/atmosphere (Esteem/Honour)
- **Growth needs**:

  One must satisfy lower level basic needs which are mandatory for the campus environment before progressing on to meet higher level growth needs. Every person is capable and has the desire to move up, unfortunately, progress is often disrupted by failure to meet lower level basic needs. Satisfaction in the basic needs will ultimately progress forward toward a level of self-actualization, resulting in the sustenance of the campus sustainability.

7 Discussion

As part of this initiative, a student centered motivational approach is institutionalized to foster the campus sustainability agenda. Shriberg (2002a, b) studied the perceptions of sustainability in order to provide the change agents from schools, colleges and universities, with a set of guidelines to enact cultural change and to understand the difficulties in pursuing organizational change specific to environmental issues and sustainability. In general, the initiatives oriented toward "greening", "stewardship", environmentalism" or a similar term or concept are more likely to focus solely on ecological issues (ignoring social and economic issues) thereby reducing the effectiveness to motivate the stakeholders. In agreement with Henson et al. (2007) this study focuses on the sustainability model that depends on integrated knowledge and attempts to be flexible enough to evolve continually and also to understand complexity as interconnected systems. Thus it needs to be addressed from various angles by involving all the stakeholders to work on as a group. It has been observed that constant motivation among the different stakeholders is needed to sustain the programme for a longer term. In reality, motivation and problems associated with it are neither easily addressed nor measured due to its intangible nature. Effective leadership is also a much needed criteria for campus sustainability initiatives to be successful. Moreover GCI are site-specific and must be understood on a case-by-case basis. The GCI programme is in place now for more than a year and it has now become very much part of the daily routine of the campus community. Since this green campus initiative, is a first of its kind among JNV's of India and for the Puduchery region as well, this can ideally be the model school, which can be scaled up for other campuses. Constant motivation of the stakeholders played a vital role from meticulous planning to successful implementation of all the components within 100 working days. It has also paved the way for further action research, development and dissemination for the rest of the years to come.

8 Outcome of GCI

Indian Environment Protection Act was enacted in 1986, for the protection and improvement of the environment, following the Bhopal tragedy. Specific standards are laid down from Schedule 1 to Schedule 4 of the Environment Protection Act, for the emission or discharge of environmental pollutants. These standards are not only applicable to the industrial sector but also to any 'operation' or 'process' of similar nature. Since this legislation was enacted in pursuance with the decisions taken at the International conference at Stockholm in June 1972, such legislation may override any existing state laws (Article 253 of the Indian constitution). Water being the vital resource and prime life support system of our planet, is misused as a medium for transporting pollutants degrading the natural resources crossing boundaries, etc.

In order to embrace campus sustainability through environmental stewardship, the Schools/HEI's/Departments/Buildings/Blocks should cooperate in a spirit of global partnership to conserve, protect and restore the health and integrity of the earth's ecosystem (campus ecosystem), with common but differentiated responsibilities. When such differentiated responsibilities are exercised effectively in good faith between each Schools/HEI's/Departments/Buildings/Blocks, three important outcomes are anticipated:

- **Outcome 1**. The common but differentiated responsibilities involve all stakeholders to participate in the local/national/International challenges adopted to mitigate local/national/Global Environmental issues.
- **Outcome 2**. The common but differentiated responsibilities lead all stakeholders to adopt stringent environmental standards which impose differing obligations.
- **Outcome 3**. The students will leave the School/HEI/University with a broad understanding about the importance of water, energy and other environmental issues/policies/rules/laws with enough information to pursue more specific detailed action research projects in the course of their academic or professional career.

9 Conclusion

Globally, the eminent scientists and professionals are constantly warning the governments about the possible crisis in 2030 "a whole series of events come together" the "**Perfect Storm**": rise in world's population by 33 % triggering the demand for food by 50 %, water by 30 %, energy by 50 % coupled with manifold multiple disasters. To anticipate this, initiatives like this must be embraced like a movement. This approach will not only addresses global environmental challenges but also prepares this generation to face the unpredictable and unprecedented disasters which are bound to come.

References

Ali MN, Ghatak S, Ragul T (2011) Biochemical analysis of Panchagavya and Sanjibani and their effect in crop yield and soil health. J Crop Weed 7(2):84–86

Altieri M (1995) Agroecology: the science of sustainable agriculture, 2nd edn. Westview Press, Boulder, CO

Branca G, McCarthy N, Lipper L, Jolejole MC (2011) Climate-smart agriculture: a synthesis of empirical evidence of food security and mitigation benefits from improved cropland management. In: Mitigation of climate change in agriculture series 3, FAO, Rome, Italy, pp 35

Branca G, Tennigkeit T, Mann W, Lipper L (2012) Identifying opportunities for climate-smart agriculture investments in Africa. FAO-EPIC, pp 129

Brix H (1993) Wastewater treatment in constructed wetlands: system design, removal processes, and treatment performance. In: Constructed wetlands for water quality improvement. Lewis Publishers, Boca Raton, Florida, USA, pp 9–22

Duan N, Lin C, Wang P, Meng J, Chen H, Li X (2014) Ecological analysis of a typical farm-scale biogas plant in China. Front Earth Sci 1–10

FAO (2011) Organic agriculture and climate change mitigation. A report of the Round Table on Organic Agriculture and Climate Change. Rome, Italy, Nov 2011

Garg VK, Kaushik P, Dilbaghi N (2006) Vermiconversion of wastewater sludge from textile mill mixed with anaerobically digested biogas plant slurry employing *Eisenia foetida*. Ecotoxicol Environ Saf 65(3):412–419

Gladwin TN, Kennelly JJ et al (1995) Shifting paradigms for sustainable development: implications for management theory and research. Acad Manag Rev 20(4):874–907

Golda AE, Poyyamoli G, Arun Prasath R, Nandhivarman M (2012) Water management and reuse strategies at Pondicherry University- sustainable alternatives. In: Leal Filho W (ed) Sustainable Development at Universities: New Horizons. Peter Lang

Greene JC, Caracelli VJ (1997) Defining and describing the paradigm issue in mixed-method evaluation. In: Greene JC, Caracelli VJ (eds) Advances in mixed-method evaluation: the challenges and benefits of integrating diverse paradigms. Josey-Bass Publishers, San Francisco, pp 5–17

Henderson K, Tilbury D (2004) Whole-school approaches to sustainability: an international review of sustainable school programs. Report Prepared by the Australian Research Institute in Education for Sustainability (ARIES) for the Department of the Environment and Heritage, Australian Government

Henson M, Missimer M, Muzzy S (2007) The campus sustainability movement: a strategic perspective. Master of Strategic Leadership towards Sustainability, Blekinge Institute of Technology, Karlskrona, Sweden

Herremans I, Allwright DE (2000) Environmental management systems at North American Universities: what drives good performance? Int J Sustain High Educ 1(2):168–181

Hilaire RS, Arnold MA, Wilkerson DC, Devitt DA, Hurd BH, Lesikar BJ, … , Zoldoske DF (2008) Efficient water use in residential urban landscapes. HortScience 43(7):2081–2092

IAASTD (2009) Synthesis report with executive summary: a synthesis of the global and sub-global IAASTD reports/Edited by McIntyre BD, Herren HR, Wakhungu J, Watson RT, pp 95

IFAD (2012) Climate-smart smallholder agriculture: What's different? IFAD Occasional paper 3, pp 19

Indian Programs (2014) Available at www.jatrophabiodiesel.org/ accessed on 05 Feb 2014

Jick TD (1979) Mixing qualitative and quantitative methods: triangulation in action. Adm Sci Q 24:602–611

John C Jr (2003) Integrating top-down/bottom-up sustainability strategies: an ethical challenge. Ethics Sci Environ Polit 6:1–6

Kung H (1998) A global ethic for global politics and economics. Oxford University Press, Oxford

Leal Filho W (2012) Future challenges for sustainable development. http://www.universityworldnews.com/article.php?story=20120613184239690 accessed on 20 Jan 2014

Locke EA, Latham GP (2002) Building a practically useful theory of goal setting and task motivation: a 35-year odyssey. Am Psychol 57(9):705–717

Moore SR (2010) Energy efficiency in small-scale biointensive organic onion production in Pennsylvania, USA. Renew Agric Food Syst 25(03):181–188

NTNU (2014) http://greenuni.ntnu.edu.tw/en_about01.html accessed on 5 Feb 2014

Nandhivarman M, Golda AE, Arun Prasath R, Poyyamoli G (2012) Integrated organic kitchen waste management for campus sustainability- a case study of Pondicherry University, India. In: Leal Filho W (ed) Sustainable development at universities: New Horizons. Peter Lang

O'Brien R, Heim SM (2009) Impact of garden-based youth nutrition intervention programs: a review. J Am Diet Assoc 109(2):273–280

Olsen S, Amundsen D, Varga B, Minch D, Anderson D (1999) The Utah botanical gardens: an educational resource for the university and the community. HortTechnology 9(4):562–565

Orr D (2000) 2020: a proposal. Conserv Educ 14(2):338–341

Padmavathy A, Poyyamoli G (2013) Biodiversity comparison between paired organic and conventional fields in Puducherry, India, Pakistan. J Biol Sci 16:1675–1686

Rajaona AM, Sutterer N, Asch F (2012) Potential of waste water use for jatropha cultivation in arid environments. Agriculture 2(4):376–392

Rothblatt S (1995) An historical perspective on the university's role in social development. In: Dill DD, Sporn B (eds) Emerging patterns of social demand and university reform: through a glass darkly. IAU Press, Tarrytown, NY, pp 1–19

Samples KM (2013) Environmental ethics and urban permaculture in central texas (Doctoral dissertation, Texas State University)

Sangeetha V, Thevanathan R (2010) Biofertilizer potential of traditional and Panchagavya amended with seaweed extract. Am Sci 6(2):61–67
Scherr SJ, Shames S, Friedman R (2012) From climate-smart agriculture to climate-smart landscapes. Agric Food Secur 1(12):1–15
Shanmugasundaram J, Arunachalam S, Gomathinayagam S, Lakshmanan N, Harikrishna P (2000) Cyclone damage to buildings and structures—a case study. J Wind Eng Ind Aerodyn 84 (3):369–380
Shriberg M (2001) Creating organizational change for sustainability: lessons from the student-led "Sustainable University of Michigan Initiative". In: Greening of the campus IV conference: moving to the mainstream, Muncie, IN, Ball State University
Shriberg M (2002a) Institutional assessment tools for sustainability in higher education: strengths, weaknesses, and implications for practice and theory. High Educ Policy 15(2):153–167
Shriberg M (2002b) Sustainability in United States higher education: organizational factors influencing campus environmental performance and leadership. (Doctoral dissertation, University of Michigan). Retrieved 9 June 2008, from Dissertations & Theses: A&I database. (Publication No. AAT 3058044)
Sinha RK, Agarwal S, Chauhan K, Valani D (2010) The wonders of earthworms & its vermicompost in farm production: Charles Darwin's' friends of farmers', with potential to replace destructive chemical fertilizers from agriculture. Agric Sci 1(2)
Snell E (1972) Japanese approach to motivation. SAE Technical Paper 720378, doi:10.4271/720378
Stabinsky D, Ching LL (2012) Ecological agriculture, climate resilience and the road to get there. TWN, Penang, Malaysia, p 44
Suthar S (2010) Potential of domestic biogas digester slurry in vermitechnology. Bioresour Technol 101(14):5419–5425
UNEP (2013) Greening universities toolkit. Transforming universities into green and sustainable campuses: a toolkit for implementers. Available at http://www.unep.org/training/docs/Greening_University_Toolkit.pdf
UNESCO (1997) Educating for a sustainable future: a transdisciplinary vision for concerted action. In: Report of the international conference on environment and society: education and public awareness for sustainability, Thessaloniki: Greece
WSU (2014) http://courses.washington.edu/inde495/lece.htm accessed on 05 Feb 2014
Wei X, Wang X, Dong B, Li X, Plappally AK, Mao Z et al (2013) Simplified residence time prediction models for constructed wetland water recycling systems. Desalin Water Treat 51 (7–9):1494–1502
Whitman MvN (1999) New world, new rules: the changing role of the American corporation. Harvard Business School Press, Boston
World Bank (2010) Climate-smart agriculture a call to action, pp 24

Authors Biography

Muthu Nandhivarman received his Masters degree in Zoology and has over 10 years of experience as an environmentalist, researcher and an activist for sustainable development. He is currently pursuing his doctorate from Pondicherry University, India. His research interest includes an extensive study to evolve policies and implement Green Campus Initiative in Pondicherry University. The main objective of this research is to assess and provide factual solutions to the educational institutions like schools and colleges to make them sustainable with special reference to water, energy and waste management.

Shanmugam Bhaskaran has worked in the field of sustainable development for over 10 years. He worked with Heritage Educational Trust, an NGO where he contributed to the planning and implementation of education to the underprivileged children in the coastal district of Cuddalore, India. Moving on to the Association for Promoting Sustainability in Campuses and Communities as a founder member, he actively promotes sustainability education in the schools and colleges of South India. He also works as a consultant for the implementation of Education for Sustainable Development.

Dr. G. Poyyamoli Associate Professor, Department of Ecology and Environmental Sciences Pondicherry University, Puducherry, India; PhD (Ecology—Madurai Kamaraj University, Madurai); Areas of specialization: Social Ecology, climate change and sustainability–sustainable water/waste management, green campus, Agro-Ecology and Industrial Ecology, sustainable Tourism/Ecotourism, Integrated Coastal Zone Management, Environmental Education for sustainable development; member Secretary, Pondicherry university renewable energy and sustainability cell; member of the State Expert Appraisal Committee; member of the Commission on Ecosystem Management, IUCN, Switzerland; a member of the Global Experts Directory on Ecosystem Services (IUCN, WRI, WBCSD and Earth watch Institute); a network member of the National Ecosystem Services Research Partnership, US EPA; Published 36 international journal articles and contributed 14 invited articles to SAGE Series on Green Society, SAGE Publications, USA; written 12 book chapters; Guided 12 PhD candidates; guiding 8 PhD students.

Golda A. Edwin is a researcher at Pondicherry University. She has been involved in a number of green campus projects in her region and had implemented meaningful sustainability projects. She decided to pursue her research and career in water science because of her passion for addressing the challenges of sustainable water management in developing countries. Her core area of research is abatement of water pollution using eco-technologies. She authored several articles and presented her findings in several national/International conferences and workshops. She was awarded Gold medal from Pondicherry University for her outstanding academic excellence. She is also one of the Founders and Executive Directors of APSCC (Association for promoting sustainability in Campuses and Communities).

Thinking Differently About Sustainability: Experiences from the UK Open University

Christine Blackmore, Ray Ison and Martin Reynolds

Abstract

Systems thinking is often invoked as a panacea for dealing with issues of sustainable development. Imperatives towards being more holistic—getting the bigger picture—are often coupled with a need for greater interdisciplinarity—joined-up-thinking—particularly amongst triple bottom line disciplines of economics, social studies and natural sciences. So why are systems thinking courses not more prevalent? And how might the teaching of systems thinking enhance the value of thinking differently about sustainable development? The Open University, UK, is a recognised international leader in the provision of Systems education for over 40 years. More recent experiences with the launch of a postgraduate Systems Thinking in Practice suite of qualifications at Certificate, Diploma, and Masters level, suggest an appetite for systems thinking amongst mature-age part-time students from a variety of professional backgrounds with an interest in learning for sustainability. This paper outlines three key features of the two core modules of the programme—epistemic understanding, active pedagogy, and design praxis. Significantly, these attributes have helped to complement rather than replace existing skill-sets amongst professionals from different sectors working in the field of sustainable development.

Keywords

Learning for sustainability · Systems thinking in practice · Systems thinking for sustainability · Epistemic understanding · Active pedagogy · Design praxis

This paper builds on two previous papers each dealing with one of the two Open University modules (OU codes TU811 and TU812) covered together here; one published by Reynolds (2011) relating to TU811, and another published by Ison and Blackmore (2014) relating to TU812.

C. Blackmore · R. Ison · M. Reynolds (✉)
Applied Systems Thinking in Practice Group, Department of Engineering and Innovation, The Open University, Walton Hall, Milton Keynes MK7 6AA, UK
e-mail: martin.reynolds@open.ac.uk

W. Leal Filho et al. (eds.), *Integrating Sustainability Thinking in Science and Engineering Curricula*, World Sustainability Series, DOI 10.1007/978-3-319-09474-8_43

1 Introduction: 'Wacky' Systems Thinking for Sustainability

Whilst 'systems thinking' skills may have gained currency in discourse on sustainable development, rarely is the term used in further education beyond invoking the need for more holistic 'joined-up-thinking' and 'whole systems' design (e.g., Cortese 2003; Holman 2009). Attempts made at Higher Education Institute (HEI) level to explore the implications of systems thinking beyond thinking holistically can instead be regarded as unconventional. For example, one journalist writing for a UK-based national newspaper, referred to the *Systems Thinking in Practice* (STiP) programme at The Open University in the UK (OU) as constituting one of the more constructive but 'wacky' sources of postgraduate subject matter (Finn 2013).[1]

This paper reports on the STiP programme referred to by Finn in her article. The set of postgraduate STiP qualifications at Certificate (60 credits), Diploma (120 credits) and Masters (180 credits) level which began in 2010 has two main foci manifest through two core modules each commanding 30 credits: *Thinking strategically: systems tools for managing change* (OU module code TU811) and *Managing systemic change: inquiry, action and interaction* (OU module code TU812).[2] As well as being core to the STiP programme, the two modules provide elective options for other postgraduate programmes including environmental management, technology management, international development, amongst others. The two modules have been developed to build praxis capability in relation to complex issues that confront mature-age students in their professional and personal lives using the internationally recognized OU supported-open learning approach. Each module is studied part-time over a 6-month period, requiring a commitment of approximately 10-hour study a week. STiP students have a wide array of professional interests associated with managing complex situations, ranging from commissioning nuclear plants to public health administration to international development and gender-sensitive interventions.

Table 1 and Table 2 provide data on STiP student registrations and student origins, retrieved through an internal review of the two modules (see Ison and Blackmore 2014 for further details)

Some evidence of STiP impact to date can be seen through sales figures for the set of four co-published books produced by the Open University with Springer (UK) for use in the STiP programme (Table 3).

[1] In a national UK newspaper article entitled *A World of Wacky Subjects* the journalist, Widget Finn, in a review of several UK University programmes included the postgraduate programme Systems Thinking in Practice as a particular exemplar of what she describes as "how to think with the Open University" (Finn 2013, p. 6).

[2] Until recently The Open University referred to the modules that form a degree qualification as courses. Nomenclature is now becoming more in line with other UK universities, and courses are often now referred to as 'modules'. Many OU students though still regard individual modules which they undertake, including TU811 and TU812 referred to in this paper, as stand-alone courses. They may not wish to necessarily fulfil a complete formalised qualification (such as, for example, a postgraduate certificate comprising of two 30 credit modules).

Table 1 Data on students registering on STiP core module presentations (TU811 and TU 812) 2010–2014

Year	TU811	TU812	Total
2010	91	107	198
2011	134	83	217
2012	111	78	189
2013	110	97	207
2014	**102**	not available	102
Total	548	365	913

TU812 starts in November so total registration numbers at time of going to Press were not available. Note also that since historically registration at the OU is module, not award based, data applying to each module do not necessarily apply to the same students

Table 2 Core STiP module student origins 2011–2012

Module	Presentation	Non-UK (%)	EU(%)	Ireland(%)	Outside EU(%)
TU811	2011	31	18	3	9
TU811	2012	28	11	5	12
TU812	2011	40	28	4	8
TU812	2012	18	15	n/a	1

Table 3 Book sales (includes print sales, MyCopy sales, bulk sales and individual eBook sales—as of April 2013) and chapter downloads Jun 06, 2010–March 2013 of the four co-published books

Title	2010	2011	2012	2013	Total	
	Chapters	Chapters	Chapters	Chapters	Books	Chapters
Systems thinkers (ST) (Ramage and Shipp 2009)	3,344	2,548	3,621	574	1,437	10,903
Systems approaches (SA) (Reynolds and Holwell 2010)	1,101	1,171	1,499	424	1,022	4,195
Systems practice (SP) (Ison 2010)	346	439	582	107	477	1,474
Social learning systems (SLS) (Blackmore 2010)	969	1,281	1,451	406	465	4,107
Total	5,760	5,429	6,753	1,511	3,401	20,679

This paper first outlines some of the key challenges associated with conventional teaching of systems thinking at HE level in supporting learning for sustainability. Three particular challenges to learning are identified—epistemic learning, social learning, and systemic design learning. They provide the backdrop for three peculiar features of the STiP core modules; innovative features that take the pedagogy of systems thinking beyond merely holistic thinking about inter-relationships. The paper goes on to describe briefly how these features are relevant to learning for sustainability and how they are expressed in each of the two core modules of the STiP programme. Some indicative further challenges ahead for mainstreaming systems thinking for sustainability are discussed in the final section **Taking stock: challenges of systems thinking for sustainability**.

Learning for sustainability has been characterised as comprising an understanding of complexity, an active engagement with conflicting beliefs and values, and the formulation of new values where "existing values are incompatible with the needs of a sustainability transition" (Henry 2009, p. 134). Such learning requires not only *understanding* disciplinary perspectives but the *practice* of interdisciplinarity (moving beyond 'understanding' content within disciplinary boundaries), and the *praxis* of transdisciplinarity (moving beyond the divide between theory and practice to the world of learning as action). A systems approach has long been identified at higher education level as helpful for supporting transdisciplinarity (for example, Jantsch 1970), a tradition that continues, (Sterling 2004; Blackmore and Ison 2012) albeit at the margins of HEIs. For mature-age students who may have considerable practical experience in different professions but lack formal academic training, three significant challenges in higher education appear to hinder systems thinking for sustainability: the entrenchment of existing disciplinary boundaries; pedagogic traditions that fail to engage learners' existing work experiences; and lastly, institutional assessment strategies based on summative as against more formative or developmental evaluation.

Despite espoused needs for interdisciplinarity, conventional higher education offerings on sustainability more often perpetuate disciplinary silos. The 2002 World Summit on Sustainable Development (WSSD) prompted a recommitment towards "a collective responsibility to advance and strengthen the interdependent and mutually reinforcing pillars of sustainable development—economic development, social development and environmental protection—at local, national and global level" (Hens and Nath 2005, p. 386). Such mainstream exhortations of the triple bottom line (cf. Elkington 1998) can be helpful in prompting attention beyond economic concerns and may also invite joined-up-thinking of the type conventionally understood in terms of systems thinking. But it can similarly also reify existing disciplinary boundaries and reify the economic as distinct from the social, rather than as a special form of the social.

Academic disciplines may themselves be regarded as bounded systems. Such an idea fits well with Thomas Kuhn's notion of 'paradigms' (cf. Kuhn 1962); areas of beliefs, assumptions and practices which are partially fixed in order to enable the ongoing pursuit of particular understandings and practices. Kuhn did not explicitly equate paradigms with disciplines, but as an historian he demonstrated how paradigms often have a survival or duration longer than perhaps required given the changes in context (including 'real world evidence' that counters the beliefs and assumptions underpinning a particular paradigm). In relating disciplines as bounded paradigms, the challenge is in moving from an understanding of disciplines as *fixed purposive* 'real-world bounded systems'—towards disciplines as more *transient purposeful* systems circumscribed by boundary judgements.

The (multi)disciplinary challenge mirrors that of a significant shift in systems thinking during the late twentieth century from what was termed 'hard systems thinking'—systems regarded as real world (reified) entities or objects as with common understandings of 'the' economic system, 'the' legal system, 'the' eco-system, etc.—towards 'soft systems thinking'—systems as conceptual constructs useful for learning about and changing the real world (cf. Churchman 1971; Checkland 1978; Jackson 1982; Ulrich 1983). The challenge is in using 'systems' as an epistemological rather than a purely ontological tool. Prigogine (1989) provides a lens on epistemic learning in his discussion of "dissipative structures"; an alternative model of the dynamics of learning where each learner goes through a period of chaos, confusion and being overwhelmed by complexity before new conceptual information brings about a spontaneous restructuring of mental models at a higher level of complexity thereby allowing a learner to understand concepts that were formally opaque.

Systems thinking for sustainability requires an *epistemic understanding* in exploring 'systems' as conceptual heuristic constructs rather than ontological real-world entities; constructs for learning in a radical constructivist tradition of epistemology (cf. Von Glasersfeld 1995). Thinking for sustainability requires moving beyond understanding of disciplines as fixed entities—fixed systems. Salner (1986) found in her study of systems teaching in the US, that many people are not able to fully grasp relatively simple systemic concepts (such as non-linear processes, or self-reflexive structures). They will therefore not be able to rethink organizational dynamics in terms of "managing" complexity or systemic change without substantial alteration in the worldviews (their "applied" epistemology).

A second challenge is in the conventional HEI *modus operandi* of a presumed divide between theory (thinking) and action (doing); between understanding and practice. Robertson (1998) identified 'complexity skills' as being significantly missing in a study of HEIs in the UK—particularly abilities to manage ambiguity and connectivity and to be comfortable with provisionality (uncertainty) and emergence (unforeseen consequences). Assumptions that human and societal behaviour can be changed just through knowledge transfer is a common predisposition often associated as John Seddon exemplifies with a belief in 'command and control management' (Seddon 2003). Blackmore (2012) points out that practitioners who are dealing with issues of sustainability in a particular context are likely to have different perspectives and different values and that the process of bringing diverse perspectives together is often challenging and requiring particular attention to practice as well as understanding. Engaging with multiple perspectives invites attention to praxis—reflection on practice—and issues of social learning—the dynamic between understanding and practice (cf. Röling 2002). Systems, cybernetic and complexity research are historically connected in their concern for understanding communication and control, emergence, self-organization, feedback and connectivity. However, to enable innovative systems thinking for sustainability, it is helpful to distinguish between learning concepts abstracted from their context of origin, and their adaptation for potential use as part of what might be called an *active pedagogy*—part of praxis in the learner's own context/lifeworld (cf. Nagda et al. 2003).

The third challenge relates to the institutionalized structures of HEIs that are inimical to the teaching of systems thinking for sustainability. The general industry of HEI certification and validation of education attainment tends to be wedded to assessment strategies based on principles of summative evaluation. Students learning about sustainability, for example, are usually required to evaluate their learning through a post hoc reflection process that feeds into an exam or project with set externally generated criteria of competencies. David Robertson—the reviewer of HEI provision in UK (1998)—claimed later that there had been "no serious attempt to capture complexity skills in competency statements" noting that in any case "competency statements tend to over formalize things when they are still emerging" (Pers. Comm. Ison 2004).

Jantsch (1970) in a paper with great resonance today entitled "Inter- and Transdisciplinary University: a Systems Approach to Education and Innovation" commented on the dysfunctional organization of Universities that retained a misguided belief of an internal logic or ecology which remains unaffected from social innovation. He went on to claim that:

> [a] systems approach…would consider science, education and innovation, above all, as general instances of purposeful human activity, whose dynamic interactions have come to exert a dominant influence on the development of society and its environments. A new policy as well as new structures for the university may be expected to emerge…designed explicitly with a view toward their innate capability for flexible change in accordance with the dynamically evolving situation" (ibid, p. 406).

Jantsch viewed systems thinking in terms of providing purposeful design support. Such a view of *design praxis* chimes well with contemporary ideas of the need for HEIs to teach 'design thinking' for embedded creativity across higher education curriculum outside of specific 'design' courses (Lloyd 2013). It articulates with what Ison (2010) refers to as systems thinking constituting a design-turn. The design-turn shifts attention away from an institutionalized endeavor for the pursuit of 'truth' towards an approach rooted in traditions of philosophical pragmatism (for example, Charles Peirce, William James and John Dewey) for purposeful learning and reflection (cf. James 1943).[3]

The three issues—epistemic understanding, active pedagogy, and design praxis—represent particular challenges for systems learning for sustainability. They are challenges that both core modules in the STiP programme address as briefly illustrated in the following three sections. The core STiP modules—TU811 *Thinking strategically* (Open University 2010a) and TU812 *Managing systemic change* (Open University 2010b) (henceforth referred to by their OU course codes—TU811 and TU812 respectively)—are designed as stand-alone modules. They provide alternative though companion routes towards an understanding of systems thinking in practice—TU811 being more 'methods' orientated (typical first

[3] The title of Al Gore's award winning documentary 'Inconvenient Truths' is, arguably, perhaps counter-productive in this context.

attraction amongst technical-based students) and TU812 being more 'philosophy' orientated (typical first attraction amongst humanities-based students).

2 STiP and Epistemic Understanding: From Understanding to Practice

Both TU811 and TU812 draw on a rich tradition of constructivist systems thinking exemplified by a four book-set series co-published by Springer used as Readers for the two courses (Table 3). *Systems approaches to managing change: a practical guide* (Reynolds and Holwell 2010) is the core reader for TU811. Five systems approaches are covered in the Reader representing the crux of TU811. Each of the five core chapters were commissioned to experts by the STiP course team on the basis of recognised expertise with either originating and/or adapting the approach towards constructivist thinking:

1 *System dynamics* (author: John Morecroft): an approach to understanding non-linear behaviour of complex systems over time, using ideas on feedback loops, stocks and flows, and time delays that affect the behaviour of the entire system in focus.
2 *Viable systems model* (author: Patrick Hoverstadt): an approach to understanding the necessary and sufficient conditions for the viability of systems in order to keep an independent existence, using principles of recursion (whereby a viable system itself can be seen as either part of a wider system or constitutive of many viable systems), and Ashby's law of requisite variety (capacity to exhibit diversity)
3 *Strategic options development and analysis* (authors: Fran Ackermann and Colin Eden): an approach for revealing and actively shaping the mental models, or belief systems (mind maps, cognitive models) that people use to perceive, contextualize, simplify, and make sense of otherwise complex issues, using cognitive mapping.
4 *Soft Systems Methodology* (author: Peter Checkland): an approach to process modelling in the analysis of complex situations where there are divergent views about the definition of the problem - 'soft problems' (e.g. how to improve health services delivery, or what to do about homelessness amongst young people?).
5 *Critical systems heuristics* (authors: Werner Ulrich and Martin Reynolds): an approach providing both a philosophical foundation and a practical framework for critical systems thinking, using ideas of reflective practice and boundary critique.

TU812 in contrast has two core co-published readers. *Systems Practice: how to act in a climate-change world* (Ison 2010) is structured into four parts: Part 1 introduces the societal need to move towards a more systemic and adaptive governance against the backdrop of human-induced climate change; Part 2 unpacks what is involved in systems practice by means of a juggler metaphor; Part 3 identifies the main factors that constrain the uptake of systems practice, introducing the notion of systemic inquiry; Part 4 critically examines how systems practice is, or might be, utilised at different levels. *Social Learning Systems and Communities of Practice* (Blackmore 2010) is a collection of classical and contemporary writings associated with learning and systemic change in different contexts, including key contributors to constructivist epistemology such as Sir Geoffrey Vickers, Richard Bawden and Etienne Wenger amongst others.

The fourth book—*Systems Thinkers* (Ramage and Shipp 2009)—presents a biographical history of the field of systems thinking by examining the life and work of thirty of its major thinkers. *Systems Thinkers* is used by both courses for secondary reference purposes.

Both modules introduce and use 'systems' as explicit conceptual models—epistemological devices rather than ontological realities. Both employ variations of an heuristic exemplifying the epistemic-turn required with employing systems. Figure 1 represents the TU811 heuristic in two formats; Fig. 1a is an introductory heuristic and Fig. 1b is a more elaborated version presented later in the module. The heuristic is presented to students as a metaphor of 'conversation' in two dimensions; first-order between systems ideas and situations, and a second-order conversation between practitioners about the situations of focus. Systems of interest are regarded as language tools for enabling good conversation about the improvement of situations.

The TU811 heuristic makes a clear distinction between three entities (situations, people, and ideas—including 'systems' as particular expressions of ideas) alongside three activities (*understanding* inter-relationships, *engaging* with multiple perspectives, and *reflecting* on boundary judgements). TU811 offers two parallel streams of learning—the primary Tools stream and a secondary People stream. The Tools stream is where practical knowledge and competence is gained in experimenting with systems tools from the five systems approaches illustrated in the primary reader. The People stream provides a complementary awareness of how such practice is shaped by the often idiosyncratic ways people think and interact. The People stream is wholly online because it makes use of multimedia and interactive possibilities appropriate for an experiential, multiperspective approach to understanding the complexity of human thinking and communication. The two streams run through all parts of the module. The People stream can be regarded as providing a second order epistemic reflection, helping to place systems ideas themselves in a wider intellectual context, enabling appreciation of how interpersonal differences are likely to affect how 'tools' are used.

Figure 2 represents the TU812 heuristic. Like the TU811 heuristic, there is a clear though dynamic 'conversation' between the domain of ideas and situations. A second-order epistemic reflection is signaled here by the explicit representation of (systems) thinking about (systems) thinking.

The key feature here is in the practitioner using an embodied framework of ideas and method as part of a 'systems thinking' process separate from the 'situation'. Systems are not explicitly referred to in the heuristic though 'frameworks' and

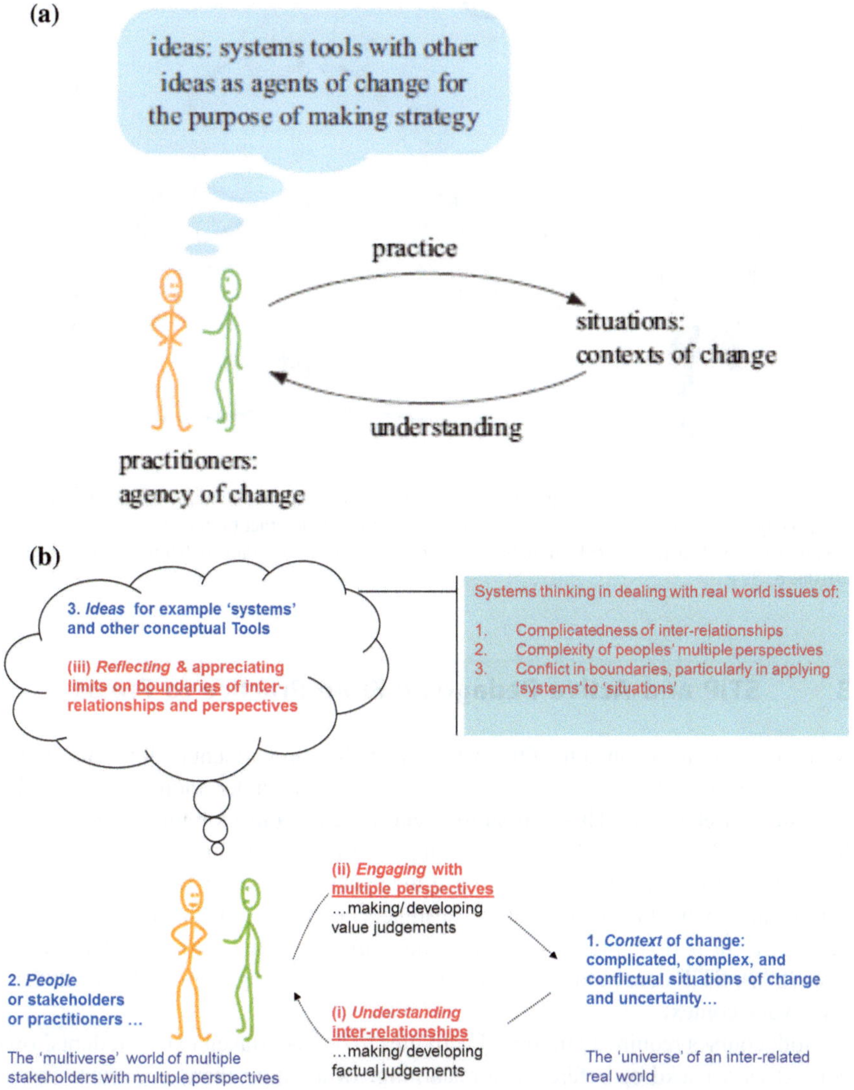

Fig. 1 TU811 systems thinking in practice heuristic for strategy making: two levels of resolution. **a** Low resolution heuristic (Open University 2010a, p. 20). **b** High level resolution heuristic (Reynolds 2013)

'methods' can be conceived as 'systems' in the first-order conversation, and the idea of a conceptual model itself can be regarded as a system in the second-order reflective conversation.

Fig. 2 TU812. A conceptual model or heuristic which can be applied to many forms of practice—comprising a person thinking about a real world situation a practitioner (*P*) engaging with a situation (*S*) with a framework of ideas (*F*) and a method (*M*) (adapted from Open University 2010b, p. 37)

3 STiP and Active Pedagogy: From Practice to Praxis

Systems teaching began at the OU in 1971 with the appointment of John Beishon as the first Systems Professor and the development of the undergraduate course T241—Systems Behaviour. This and subsequent systems courses built on the distance learning model at the OU known as supported-open learning. The aim from the beginning was to enable students to use their own context for learning in creative combination with tutors and course designers; part of what the OU calls active pedagogy—a part of praxis involving the learner's own lifeworld (Maiteny and Ison 2000). STiP core modules ensure that systems learning is grounded in students own work context.

Both courses require a number of activities to be undertaken using students' own context as a backdrop reference for learning. In addition there is a more formal assessment strategy which explicitly invokes the students' work experience; divided equally in weighting for each course between (i) continuous assessment consisting of three tutor-marked assignments (TMAs) and (ii) an end-of-module assessment (EMA) in the form of a project.

Figure 3 provides a model of the active pedagogy used to guide TU811 students in terms of a 'system for making strategy'.

Three distinct levels of student context are introduced at the outset, and central to the pedagogy of TU811: (i) *Area of practice* (AoP) - a high level domain of loosely defined professional or personal interest which individual students choose to use for

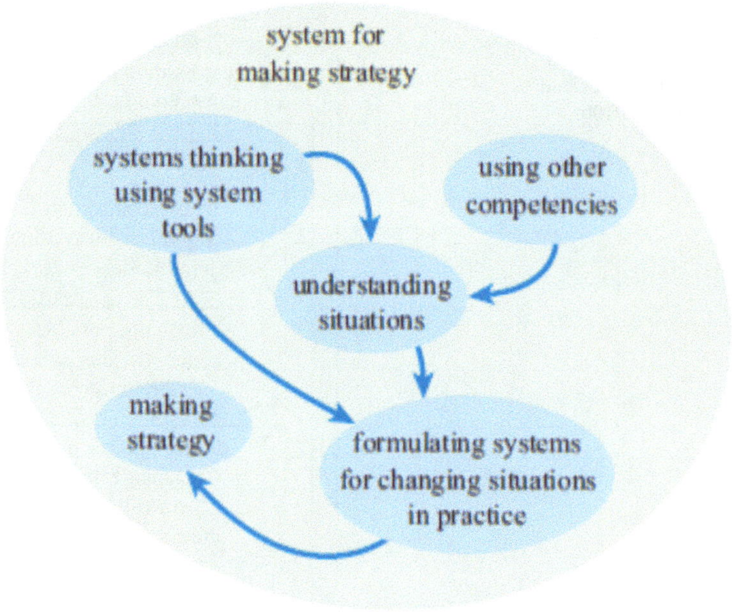

Fig. 3 TU811 simple conceptual model of strategy making (Open University 2010a, p. 20)

activities and assessments throughout the course; (ii) *Situations of interest*—a range of specific situations within a declared broad AoP that would benefit from 'strategic thinking', with different contemporary 'situations' chosen for each TMA; and (iii) *Systems of interest*—acting as proxies to perspectives regarding a 'system to do something' within any particular situation of interest.

Students are asked to choose situations of interest to explore within their AoPs as the experimental ground for practicing tools and techniques from each of the five systems approaches. They do this through Activities in the module Study Guide, Tutor Marked Assignments which constitute the continuous assessment (TMAs) and the end of module assessment (EMA). In the latter assessment work systems of interest are explored both as a proxy to perspectives (analytical mode) and as constructive devices (design mode)—strategic plans—for transforming the situations in the AoP in which they work.

TU812 draws on Wenger's social theory of learning (Wenger 1998), elaborated in his work on Communities of Practice (CoPs). For Wenger, social learning is about learning in a social context. Learning can be viewed as a journey through landscapes of practices (LoPs).

> "As learning gives rise to a multiplicity of interrelated practices, it shapes the human world as a complex landscape of practices. Each community is engaged in the production of its own practice—in relation to the whole system, of course, but also through its own local negotiation of meaning. This process is therefore inherently diverse." (Wenger in Blackmore 2010, p. 140)

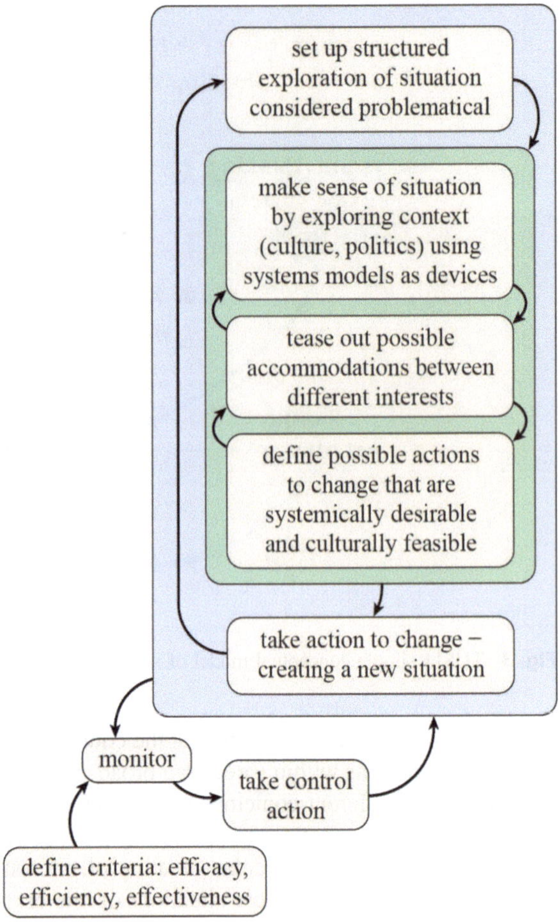

Fig. 4 TU812. An activity model of a system to conduct a systemic inquiry based on the work of Peter Checkland (Open University 2010b, p. 61)

In a similar mode to TU811, TU812 students use the ideas of CoPs and LoPs in relation to situations of their own choice. Figure 4 provides a model of the active pedagogy used to guide TU812 students in terms of a systemic inquiry for 'managing systemic change'.

The three core activities in the model and the TU812 pathway starts with the practitioner and their situation (Part 1), expands to include the dynamics of practitioner, situation, frameworks and methods (Part 2) and then expands to include material that develops skills and understanding and interaction through social learning and communities of practice to make transformations (Part 3). The model recognises that as more stakeholders become involved the complexity expands as do the demands for practice involving interaction of some form with others (stakeholders, clients, employees, employers etc.).

The LoPs concept enables students to review their own future learning trajectories by helping them review their multi-membership of communities, and recognise the multiple levels of scale with which they identified. LoPs generally provide students with a potential way of considering what they perceive beyond the communities and practices with which they most identified from their own experience. Students found particularly inspiring Wenger's suggestion that

> …we each have a unique trajectory through the landscape of practices. This trajectory has created a unique point of view, a location with specific possibilities for enhancing the learning capability of our sphere of participation. From this perspective, our identity, and the unique perspective it carries is our gift to the world (Wenger in Blackmore 2012, p. 197).

Iterative use of a 'learning contract' in successive TMAs for TU812 was designed to test if students can make the shift from a systematic to a systemic design of their own 'learning system'. This is more than a shift in representation, though this is also needed. The evolving learning contract forms the foundation for their engagement with the course concepts so that the students own learning needs and desires for situational transformation can be accommodated within the module context.

4 STiP and Design Praxis: Towards Pragmatic Reflexivity

Both modules are about reflexive practice or 'praxis'—a second-order practice involving reflection on reflection. Figure 5 illustrates first-order reflective practice as illustrated in TU811.

TU811 students are required to make several loops throughout their assessments using contemporary news items associated with their chosen area of practice. Part of the EMA, aside from the core project report, requires a further second-order reflection on their practice.

Figure 6 illustrates second-order, reflexive practice as depicted in TU812.

By using systems thinking in practice it is possible to appreciate potential changes in a situation of concern that are systemically desirable and, if managed appropriately, become culturally feasible. The strategic opportunity offered by STiP is that through this combination of processes, it is possible to alter the trajectory of change.

TU811 and TU812 require students to develop a project for their EMA using the systems concepts and other ideas introduced in the course in a constructive, reflexive, design-mode manner. The driver is one of formative rather than summative evaluation. For TU811, a significant point of departure from conventional teaching of systems approaches is the encouragement to play with different ideas and tools from different systems approaches as a way of escaping the reification and fetishisation of methods and methodologies. Rather than falling into the trap of a contingency approach—strict rules for designating particular methods for particular situations, TU811 students are allowed to choose whatever tools they feel

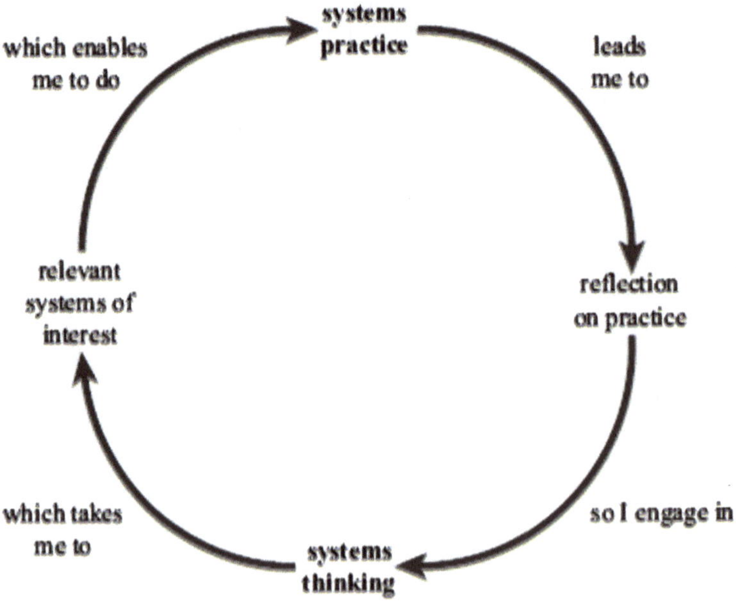

Fig. 5 TU811 the learning process as a virtuous cycle associated with reflective practice (*source* Open University 2010a, p. 21)

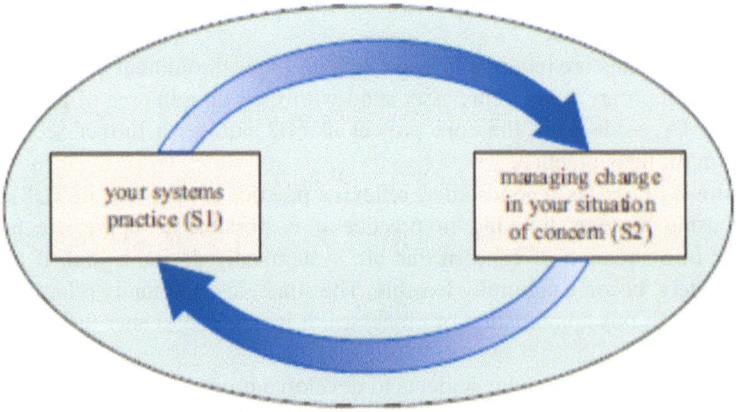

Fig. 6 A virtuous cycle of inquiry in which an appreciation of systems practice (*S1*) when enacted can contribute to managing change in a situation or situations of concern (*S2*) that is systemic, at the same time as deepening understanding and practice of systems practice (*S1*) which can be applied in new situations (*Sn*) (Open University 2010b, p. 58)

appropriate for their task in designing a strategy for their chosen final situation of interest, based on experimentation undertaken in earlier assignments. For TU812 all three TMAs again feed into the EMA which comprises the design of a 'briefing paper' relating to systemic inquiry for managing and effecting purposeful systemic change. As with TU811, TU812 similarly has a reflexive component as well as core component (project report or briefing paper) to the final submission. The reflexive component enables students to demonstrate their second-order reflexivity.

5 Discussion: Moving Beyond 'Empty Management Talk'

A testimony from one of our alumni on the *Wacky subjects* newspaper article stated: "I was keen to learn about complex systems and how to use them, but didn't want empty management talk…." (Finn 2013, p. 6). The empty management talk referred to might have several manifestations in relation to conventional learning for sustainability. Firstly, the recurring call for multidisciplinarity coupled often with a promotion of multi-methods. The prevalence of multidisciplinary teams associated with sustainable development interventions signals a concern for inter-relationships between different substantive areas of knowledge. But it does not necessarily address the epistemic challenge of knowledge generation through interdisciplinary constructivism. Instead it may often serve to reify disciplinary boundaries with ontological assumptions of there being some objective 'economy', 'society' and 'natural environment' existing as 'systems' outside of our knowledge construction. From a STiP viewpoint, an epistemic-turn is required: systems are not real world ontologies, but rather epistemic constructs used to learn about and improve situations of uncertainty, complexity and conflict. Secondly, practitioners exploring and engaging with issues of sustainability as learners cannot be divorced from the reality of other stakeholder perspectives. From a STiP viewpoint, learning for sustainability requires a pragmatic-turn: active engagement and experimentation through continual interaction and feedback with other practitioners as co-learners including where possible colleagues in the workplace. Thirdly, and relating to the other two manifestations of conventional learning for sustainability, learners cannot be divorced from the situations that they are co-creating in the learning process. Conventions of what Ison (2010) calls a 'projectified world'—a world of project planning with orthodox practices of command and control mindsets (cf. Seddon, 2003)—often assume methods and tools to have some innate value for transformation (technical fix) outside of the users experiences as part of a learning process. Incidences of systemic failure or success are not other-worldly events but events to which acts of learning contribute. Ralph Stacey put this well in observing how managers adapt to uncertainty and complexity: "The changes occurred, not because [of the] planning, but because [of the] learning in a manner provoked by the very ambiguity and conflict [they] were trying to remove" (Stacey 1993). From a STiP viewpoint, a systemic design-turn is required. That is, systems used as heuristic devices not just for understanding systemic change and engaging systemically with other stakeholder perspectives, but for shaping the world through innovative, experimental modes of creative learning; systems learning for sustainability.

Several significant challenges remain in the teaching of STiP as a means of enhancing learning for sustainability. Despite considerable favourable feedback from a majority of STiP students undertaking the courses, some students feel alienated by the demands on epistemic reflection in the use of systems. An epistemic-turn is often discomforting particularly amongst students from positivist traditions of engineering and technology where 'systems' are taken as given realities.

A second challenge and source of alienation is in the demands on praxis and interaction. Many traditional part-time students are attracted to distance learning because of the 'independent learning' features offered by the OU. Whilst many take advantage of the vibrant online discussion forums and video-conferencing tutorial events offered in both courses, some choose not to engage actively or at best engage, albeit quite legitimately, as passive observers. (All online tutorials whilst offered at differing times are recorded.).

Perhaps the most pervasive challenge though is institutional—both in terms of HEI norms and practices and wider institutional norms and practices experienced by our students in their workplace. Conventional demands of summative assessment procedures at University level with associated embedded measures of performance associated with, say, the number of passes, can sometimes be slow to keep up with innovative designs around more formative developmental assessments in learning systems designed at the module/course level. But over the years we have also become aware that developing new ways of teaching may be insufficient to develop STiP competencies. Such competencies might only be sustained if the institutional structures and relationships where a learner is employed are not inimical to the further development and testing of systems ideas. In short, communities of practice established during the courses can be difficult to sustain after the course finishes unless there is purposeful innovation beyond the university. In our case an innovative response to this challenge is the emergence of a vibrant self-organising LinkedIn on-line community of nearly 600 STiP alumni. This has arisen through the ongoing desire of our students to 'walk the talk!'.

6 Conclusion

TU811 and TU812 are heuristic devices (Figs. 1 and 2), each designed by a team of Systems academics with a conscious and concerted need to break out of conventional 'management talk'. Both heuristics deal with all three misgivings of conventional learning for sustainability described above, but with different emphases. TU811 attends more to the context of change and the experimental adaptability of ideas and tools (epistemic understandings) for systems praxis. TU812 focuses more on the agents of change (active pedagogy) and the practitioners involved with social learning and communities of practice in systemic inquiry. Both modules share the same pivotal attention to issues of reflexive responsibility in creating improved situations of sustainability (design learning).

Alongside the success of the STiP programme, there remains a need to consider what characteristics are most likely to be needed for the design of learning systems with a high degree of connection between learner, tutor, course, work context, and academic management of the curriculum. A system capable of sustaining STiP competencies in learning for sustainability may require different structures and organization than is currently found in most formal and non-formal education and training settings. A systemic research inquiry to surface these concerns—concerns which often militate against emergence and self-organization—is currently under development at the OU involving the STiP authors, tutors, students, and employers (Open University 2014).

References

Blackmore C (ed) (2010) Social learning systems and communities of practice. Springer and The Open University, London and Milton Keynes
Blackmore C (2012) Sustainability. In: Chadwick R et al (eds) Encyclopedia of applied ethics, 2nd edn, vol 4. Elsevier, Oxford
Blackmore CP, Ison RL (2012) Designing and developing learning systems for managing systemic change in a climate change world. In Wals A, Corcoran PB (eds) Learning for sustainability in times of accelerating change. Wageningen Academic Publishers, Education and Sustainable Development Series, Wageningen, pp 347–364
Churchman CW (1971) The design of inquiring systems: basic concepts of systems and organizations. Basic Books, New York
Checkland P (1978) The origins and nature of 'hard' systems thinking. J Appl Syst Anal 5:99–110
Cortese AD (2003) The critical role of higher education in creating a sustainable future. Plan High Educ 31(3):15–22
Elkington J (1998) Cannibals with forks: the triple bottom line of 21st century business (conscientious commerce). New Society Publishers, Gabriola Island
Finn W (2013) A world of wacky subjects. The Independent (UK Newspaper: Post Graduate Supplement), 20th Nov 2013, pp 5–6
Holman P (2009) The change handbook: the definitive resource on today's best methods for engaging whole systems: Easyread Edition, ReadHowYouWant. com
Henry AD (2009) The challenge of learning tor sustainability: a prolegomenon to theory. Human Ecol Rev 16(2):131–140
Hens L, Nath B (2005) The world summit on sustainable development. Springer, New York
Ison R (2010) Systems practice: how to act in a climate-change world. Springer and The Open University, London and Milton Keynes
Ison R, Blackmore C (2014) Designing and developing a reflexive learning system for managing systemic change. Systems 2(2):119–136
Jackson M (1982) The nature of soft systems thinking: the work of Churchman, Ackoff and Checkland. J Appl Syst Anal 9:17–28
James W (1943) Pragmatism, a New name for some old ways of thinking: together with 4 related essays selected from: the meaning of truth. Longmans, Green & Company
Jantsch E (1970) Inter- and transdisciplinary university: a systems approach to education and innovation. Policy Sci 1(1):403–428
Kuhn T (1962) The structure of scientific revolutions. University of Chicago, Chicago
Lloyd P (2013) Embedded creativity: teaching design thinking via distance education. Int J Technol Des Educ 23(3):749–765
Maiteny P, Ison R (2000) Appreciating systems: critical reflections on the changing nature of systems as a discipline in a systems-learning society. Syst Practice Action Res 13(4):559–586

Nagda BA et al (2003) Transformative pedagogy for democracy and social justice. Race Ethn Educ 6(2):165–191
Open University (2010a) TU811 thinking strategically: systems tools for managing change (study guide). The Open University, Milton Keynes
Open University (2010b) TU812 managing systemic change: inquiry, action and interaction (study guide). The Open University, Milton Keynes
Open University (2014) Post-graduate student recruitment and retention: design criteria for a learning system based on community of practice building and employer engagement and eSTEeM project. The Open University. http://www.open.ac.uk/about/teaching-and-learning/esteem/projects/themes/other/post-graduate-student-recruitment-and-retention-design-criteria-learning-syste (accessed, 10th April 2014)
Prigogine I (1989) The philosophy of instability. Futures 21(4):396–400
Ramage M, Shipp K (2009) Systems thinkers. Springer and The Open University, London and Milton Keynes
Reynolds M (2011) Bells that still can ring: systems thinking in practice. moving forward with complexity. In: Tait A, Richardson KA (eds) Proceedings of the 1st international workshop on complex systems thinking and real world applications. Emergent Publications, Litchfield Park, pp 327–349
Reynolds M (2013) Managing systemic risk using systems thinking in practice. AWERProcedia Adv Appl Sci 1:217–224
Reynolds M, Holwell S (2010) Systems approaches to managing change: a practical guide. Springer and The Open University, London and Milton Keynes
Robertson D (1998) The emerging political economy of higher education. Stud High Educ 23 (2):221–228
Röling N (2002) Beyond the aggregation of individual preferences: moving from multiple to distributed cognition in resource dilemmas. In: Leeuwis C , Pyburn R (eds) Wheelbarrows full of frogs: social learning in rural resource management. Koninklijke Van Gorcum, Assen
Salner M (1986) Adult cognitive and epistemological development in systems education. Syst Res 3(4):225–232
Seddon J (2003) Freedom from command and control. Vanguard Education, Buckingham
Stacey R (1993) Strategy as order emerging from chaos. Long Range Plan 26(1):10–17
Sterling, S. (2004). Higher education, sustainability, and the role of systemic learning. Higher education and the challenge of sustainability, Springer: 49-70
Ulrich W (1983) Critical heuristics of social planning: a new approach to practical philosophy. Wiley, Stuttgart (paperback version)
Von Glasersfeld E (1995) A constructivist approach to teaching. Construct Educ 3:15
Wenger E (1998) Communities of practice. Cambridge University Press, Cambridge

Authors Biography

Christine Blackmore is a Senior Lecturer in Environmental Systems and Qualifications co-Leader on Systems Thinking in Practice at The Open University, UK

Ray Ison is Professor, Systems for Sustainability and Program Leader for Systemic Governance Research Program, Monash University, and Professor of Systems at The Open University, UK

Martin Reynolds is Senior Lecturer in Systems Thinking and Qualifications co-Leader on Systems Thinking in Practice at The Open University, UK

MIX
Papier aus verantwortungsvollen Quellen
Paper from responsible sources
FSC® C105338
FSC
www.fsc.org

If you have any concerns about our products,
you can contact us on
ProductSafety@springernature.com

In case Publisher is established outside the EU,
the EU authorized representative is:
Springer Nature Customer Service Center GmbH
Europaplatz 3, 69115 Heidelberg, Germany

Printed by Libri Plureos GmbH
in Hamburg, Germany